Relaxing and Contracting Factors

The Endothelium

Relaxing and Contracting Factors
Edited by ***Paul M. Vanhoutte,*** 1988

Relaxing and Contracting Factors

Biological and Clinical Research

Edited by

Paul M. Vanhoutte

Library of Congress Cataloging in Publication Data

Main entry under title:

Relaxing and Contracting Factors.

(Contemporary biomedicine) (The Endothelium)
Includes bibliographies and index.
1. Endothelium. 2. Blood—Circulation—Deregulation.
3. Vasodilators. 4. Blood vessels—Dilatation. 4. Blood
vessels—Contraction. I. Vanhoutte, Paul M. II. Series.
III. Series: Endothelium. [DNLM: 1. Endothelium—physiol-
ogy. 2. Vasoconstrictor Agents—secretion. 3. Vasodi-
lator Agents—secretion. QV 150 R382]
QP88.45.R45 1988 612'.18 87-22541
ISBN 0-89603-128-4

Crescent Manor
PO Box 2148
Clifton, NJ 07015

Printed in the United States of America

Preface

It is an exciting task to be the editor of the first monograph covering a new area of the biomedical sciences. Since the first report in 1980 by Robert Furchgott and colleagues (*see* Chapter 1) of the evidence of endothelium-dependent relaxation in isolated arteries, there are ever-increasing numbers of vascular physiologists and pharmacologists who are scraping away the endothelium to look into its role in cardiovascular control. And the more one looks, the more one discovers. Not only is the list of substances that can induce endothelium-dependent relaxations impressively long, but these intriguing cells can also secrete vasoconstrictor substances. The ability of the endothelium to modulate the degree of contraction of the underlying smooth muscle is an ancestral property of the blood vessel wall, illustrating the logic of nature, since the endothelial cells are located in the best possible strategic location to continuously monitor the properties (chemical or physical) of the blood. And more and more data emerge suggesting that in several cardiovascular diseases perturbations in endothelium-dependent responses are one of the early signs of the abnormal process. Thus, the importance of endothelium-dependent responses, triggered by the intellectual curiosity of one of the pioneers of vascular physiology and pharmacology, is now recognized not only by basic scientists, but also by all concerned with the cardiovascular diseases. The purpose of this monograph is to provide them with a reference work, so that they know where to start.

As editor, I recognized the overlapping of information among the authors. Because it is crucial, however, at such an early stage of this particular quest for knowledge to confirm, reconfirm, and extend the original observations, and with each author having a different approach, the outcome is a richer knowledge of this growing field.

I would like to thank the authors for their exciting chapters. My secretaries, Mrs. J. Beckman and Mrs. C. Camrud, deserve my gratitude for helping me to coach the effort. And my special thanks goes to Mr. Lanigan and his staff, at Humana Press, who masterly converted the edited chapters into a publication of the highest technical quality.

Paul M. Vanhoutte

Contents

CHAPTER 3
The Nature of Endothelium-Derived Relaxing Factor
T. M. Griffith and A. H. Henderson

CHAPTER 4
Metabolism of Arachidonic Acid and Release of Endothelium-Derived Relaxing Factors
Roger A. Johns and Michael J. Peach

CHAPTER 5
Modulation of the Release and Biological Activity of Endothelium-Derived Relaxing Factor by Oxygen-Derived Free Radicals

Gabor M. Rubanyi and Paul M. Vanhoutte

CHAPTER 6
Endothelial Cells in Culture and Production of Endothelium-Derived Relaxing Factor

T. M. Cocks, J. A. Angus, J. H. Campbell, and G. R. Campbell

CHAPTER 7
Endothelial Cells in Culture and Production of Endothelium-Derived Constricting Factors

Robert F. Highsmith, David Aichholz, Oliver FitzGerald, Richard Paul, Gabor M. Rubanyi, and Kristine Hickey

CHAPTER 8
Basal Release of Endothelium-Derived Relaxing Factor
William Martin

CHAPTER 9
Calcium Transport Mechanisms in Endothelial Cells Regulating the Synthesis and Release of Endothelium-Derived Relaxing Factor
Gabor M. Rubanyi, Arnold Schwartz, and Paul M. Vanhoutte

CHAPTER 10
Release of Endothelium-Derived Relaxing Factor(s) by Physicochemical Stimuli
Eberhard Bassenge, Rudi Busse, and Ulrich Pohl

CHAPTER 11
Role of Cyclic GMP in Endothelium-Dependent Relaxation of Vascular Smooth Muscle
Robert M. Rapoport and Ferid Murad

CHAPTER 13
Endothelium-Derived Relaxing Factor Relaxes Vascular Smooth Muscle by Cyclic GMP-Mediated Effects on Calcium Movements
P. Collins, M. J. Lewis, and A. H. Henderson

CHAPTER 14
Heterogeneity in Endothelium-Dependent Relaxations: *Acute, Chronic, and Evolutionary Modulations*
Virginia M. Miller and Paul M. Vanhoutte

CHAPTER 15
Endothelium-Dependent Regulation of Resting Levels of Cyclic GMP and Cyclic AMP and Tension in Pulmonary Arteries and Veins

Louis J. Ignarro, Russell E. Byrns, and Keith S. Wood

CHAPTER 16
Endothelium-Dependent Responses of Cerebral Arteries
Z. S. Katusic and J. T. Shepherd

CHAPTER 17
Endothelium, Blood Flow, and Vascular Responses in Large Coronary and Iliac Arteries of the Conscious Dog
Mark A. Young and Stephen F. Vatner

CHAPTER 18
Endothelium-Dependent Responses in Large Arteries and in the Microcirculation
J. A. Angus, T. M. Cocks, C. E. Wright, K. Satoh, and G. R. Campbell

CHAPTER 19
Endothelium-Dependent Responses in the Peripheral Microcirculation

T. M. Griffith, D. H. Edwards, R. L. Davies, T. J. Harrison, and K. T. Evans

CHAPTER 20
Endothelium-Dependent Vasodilatation in the Cerebral Microcirculation
Hermes A. Kontos

CHAPTER 21
Platelets and Endothelium-Dependent Responses
Donald S. Houston and Paul M. Vanhoutte

CHAPTER 22
Endothelium-Dependent Responses and the Release of Endothelium-Derived Relaxing Factor in Atherosclerotic Blood Vessels

Tony J. Verbeuren and Arnold G. Herman

CHAPTER 23
Endothelium-Dependent Relaxations in Hypertensive Blood Vessels
Raymond J. Winquist

CHAPTER 24
Mechanisms of Altered Endothelium-Dependent Responses in Hypertensive Blood Vessels
Thomas F. Luscher and Paul M. Vanhoutte

CHAPTER 25
Endothelium-Dependent Responses in Human Arteries
S. Thom, A. Hughes, and P. S. Sever

Contributors

DAVID AICHHOLZ · *Department of Physiology and Biophysics, University of Cincinnati College of Medicine, Cincinnati, Ohio*

J. A. ANGUS · *Barker Medical Research Institute, Victoria, Australia*

EBERHARD BASSENGE · *Department of Applied Physiology, University of Freiburg, Freiburg, Federal Republic of Germany*

RUDI BUSSE · *Department of Applied Physiology, University of Freiburg, Freiburg, Federal Republic of Germany*

RUSSELL E. BYRNS · *Department of Pharmacology, University of California Los Angeles School of Medicine, Los Angeles, California*

G. R. CAMPBELL · *Department of Anatomy, University of Melbourne, Victoria, Australia*

J. H. CAMPBELL · *Barker Medical Research Institute, Victoria, Australia*

T. M. COCKS · *Barker Medical Research Institute, Victoria, Australia*

P. COLLINS · *Department of Cardiology, University of Wales College of Medicine, Heath Park, Cardiff, Wales, UK*

R. L. DAVIES · *Department of Diagnostic Radiology, University Hospital of Wales, Heath Park, Cardiff, Wales, UK*

D. H. EDWARDS · *Department of Cardiology, University Hospital of Wales, Heath Park, Cardiff, Wales, UK*

K. T. EVANS · *Department of Diagnostic Radiology, University Hospital of Wales, Heath Park, Cardiff, Wales, UK*

OLIVER FITZGERALD · *Department of Physiology and Biophysics, University of Cincinnati College of Medicine, Cincinnati, Ohio*

ROBERT F. FURCHGOTT · *Department of Pharmacology, State University of New York, Health Science Center at Brooklyn, Brooklyn, New York*

T. M. GRIFFITH · *Department of Diagnostic Radiology, University Hospital of Wales, Heath Park, Cardiff, Wales, UK*

T. J. HARRISON · *Department of Diagnostic Radiology, University of Wales, Heath Park, Cardiff, Wales, UK*

A. H. HENDERSON · *Department of Cardiology, University Hospital of Wales, Heath Park, Cardiff, Wales, UK*

ARNOLD G. HERMAN · *Division of Pharmacology, Faculty of Medicine, University of Antwerp, Wilrijk, Belgium*

KRISTINE HICKEY • *Department of Physiology and Biophysics, University of Cincinnati College of Medicine, Cincinnati, Ohio*
ROBERT F. HIGHSMITH • *Department of Physiology and Biophysics, University of Cincinnati College of Medicine, Cincinnati, Ohio*
DONALD S. HOUSTON • *University of Winnipeg, Winnipeg, Manitoba, Canada*
A. HUGHES • *Department of Clinical Pharmacology, St. Mary's Hospital Medical School, London, UK*
LOUIS J. IGNARRO • *Department of Pharmacology, University of California Los Angeles School of Medicine, Los Angeles, California*
ROGER A. JOHNS • *Departments of Pharmacology and Anesthesiology, University of Virginia School of Medicine, Charlottesville, Virginia*
Z. S. KATUSIC • *Department of Pharmacology, Medical Faculty, Belgrade, Yugoslavia*
HERMES A. KONTOS • *Department of Medicine, Medical College of Virginia, Virginia Commonwealth University, Richmond, Virginia*
M. J. LEWIS • *Department of Pharmacology and Therapeutics, University of Wales College of Medicine, Heath Park, Cardiff, Wales, UK*
THOMAS F. LUSCHER • *Departments of Internal Medicine and Research, University Hospital, Basel, Switzerland*
WILLIAM MARTIN • *Department of Cardiology, University of Wales College of Medicine, Cardiff, Wales, UK*
R. C. MILLER • *Universite Louis Pasteur, Laboratoire de Pharmacodynamie Cellulaire et Moleculaire, Strasbourg, Cedex, France*
VIRGINIA M. MILLER • *Department of Physiology and Biophysics, Mayo Clinic and Mayo Foundation, Rochester, Minnesota*
FERID MURAD • *Department of Medicine and Pharmacology, Stanford University and Veterans Medical Center, Palo Alto, California*
RICHARD PAUL • *Department of Physiology and Biophysics, University of Cincinnati College of Medicine, Cincinnati, Ohio*
MICHAEL J. PEACH • *Department of Pharmacology, University of Virginia School of Medicine, Charlottesville, Virginia*
ULRICH POHL • *Department of Applied Physiology, University of Freiburg, Freiburg, Federal Republic of Germany*
ROBERT M. RAPOPORT • *Departments of Pharmacology and Cell Biophysics, Veterans Administration Medical Center and University of Cincinnati College of Medicine, Cincinnati, Ohio*
GABOR M. RUBANYI • *Department of Physiology and Biophysics, Mayo Clinic and Mayo Foundation, Rochester, Minnesota*
K. SATOH • *Department of Pharmacology, School of Medicine, Tohoku University, Sendai, Japan*

V. SCHINI • *Universite Louis Pasteur, Laboratoire de Pharmacodynamie, Cellulaire et Moleculaire, Strasbourg, Cedex, France*
P. SCHOEFFTER • *Universite Louis Pasteur, Laboratoire de Pharmacodynamie Cellulaire et Moleculaire, Strasbourg, Cedex, France*
ARNOLD SCHWARTZ • *Department of Physiology and Biophysics, Mayo Clinic and Mayo Foundation, Rochester, Minnesota*
P. S. SEVER • *Department of Clinical Pharmacology, St. Mary's Hospital Medical School, London, UK*
J. T. SHEPHERD • *Department of Physiology and Biophysics, Mayo Clinic, Rochester, Minnesota*
S. THOM • *Department of Clinical Pharmacology, St. Mary's Medical School, London, UK*
PAUL M. VANHOUTTE • *Department of Physiology and Biophysics, Mayo Clinic and Mayo Foundation, Rochester, Minnesota*
STEPHEN F. VATNER • *New England Regional Primate Research Center, Southboro, Massachusetts*
TONY J. VERBEUREN • *Division of Pharmacology, Faculty of Medicine, University of Antwerp, Wilrijk, Belgium*
RAYMOND J. WINQUIST • *Department of Pharmacology, Merck Institute for Therapeutic Research, Merck, Sharp and Dohme Research Laboratories, West Point, Pennsylvania*
KEITH S. WOOD • *Department of Pharmacology, University of California Los Angeles School of Medicine, Los Angeles, California*
C. E. WRIGHT • *The Baker Medical Research Institute, Prahran, Victoria, Australia*
MARK A. YOUNG • *New England Regional Primate Research Center, Southboro, Massachusetts*

Chapter 1

Endothelium-Dependent Relaxation in Systemic Arteries

Robert F. Furchgott

1. Historical Perspective

1.1. The Discovery of Acetylcholine-Induced Endothelium-Dependent Relaxation

The ability of acetylcholine and other muscarinic agonists to produce marked vasodilatation in various vascular beds in vivo was well established many years ago. In view of the potent vasodilating activity of muscarinic agonists in vivo, it was to be expected that these agents would produce relaxation of isolated blood vessels in vitro. More than 13 years ago it was demonstrated in a number of laboratories that acetylcholine and other muscarinic agonists could indeed produce relaxation (dilatation) of isolated perfused or superfused blood vessels contracted by stimulation of perivascular adrenergic nerves (e.g., Rand and Varma, 1970; Hume et al., 1972; Steinsland et al., 1973; Vanhoutte, 1974, 1977). This relaxation was attributed to an action of muscarinic agonists on the muscarinic "inhibitory receptor" of adrenergic nerve endings (Loffelholz and Muscholl, 1969), resulting in inhibition of release of norepinephrine from these terminals. This prejunctional inhibitory action of muscarinic agonists could explain in part the vasodilatation that they produced in innervated vascular beds in vivo; but it could not explain the ability of these agents to still cause marked dilatation in denervated vascular beds. In view of such vasodilatation, it was to be expected that muscarinic agonists would produce relaxation of spontaneous or drug-induced contractions of isolated preparations of blood vessels. Until about eight years ago, however, such relaxation had been observed in only a minority of the many vascular preparations tested in vitro (for relevant references, *see* Furchgott, 1982). Among the many preparations that had been reported as failing to relax in response to muscarinic agonists was the helical strip of thoracic aorta of the rabbit (Furchgott and Bhadrakom, 1953). In the author's laboratory, that preparation, which had been used for many years for quantitative studies of drug-receptor interactions, had never exhibited relaxation in response to muscarinic agonists. The only observed response to acetylcholine and carbachol had been a concentration-dependent contraction, regardless of whether the strip was initially at resting tone or in drug-induced contraction (Furchgott, 1955). The maximum acetylcholine-induced contraction varied considerably, being usually about 10–50% of the maximum with epinephrine (Furchgott and Bhadrakom, 1953).

It should be noted that the helical strip was originally developed to be used with the recording device of the time, namely the isotonic lever writing on a kymograph. After replacement of the lever and kymograph by the isometric strain gage and recording oscillograph in the 1960s, the helical strip was still used for many years. Then around 1976 the author's laboratory began to use rings of aorta rather than helical strips for studies on receptors, since the ring preparation is easy to prepare, conserves tissue, and is easily mounted on hooks for recording tension. In 1978, during an experiment designed to compare potencies of different catecholamines in activating beta-adrenoceptors of the aorta of the rabbit, the author's technician, misreading the protocol for the experiment, accidentally added carbachol to a ring that was already contracted by norepinephrine—and the ring underwent partial relaxation. Since this was a completely unexpected response to carbachol acting on the rabbit aorta, all rings in the same experiment were then tested, and all exhibited partial relaxation of norepinephrine-induced tone when treated with either acetylcholine or carbachol. Additional experiments were soon carried out to compare responses of helical strips and rings from the same aorta to these agents. The strips showed only the expected contraction, whereas the rings gave relaxation.

After the initial discovery of carbachol- and acetylcholine-induced relaxation of rings of rabbit aorta, several kinds of experiments soon indicated that the capacity of aortic preparations to relax in response to these agents was lost if the intimal surface of the preparation had been rubbed excessively against a foreign surface or against itself during the making of the preparation. (For more information about these early experiments, *see* Furchgott and Zawadzki, 1980a; Furchgott, 1982). Once this relationship was understood, it became apparent that the reason helical strips prepared by the usual procedure in the author's laboratory did not exhibit relaxation in response to muscarinic agonists was that during this procedure the intimal surface of the strips always underwent some rubbing against the fingertips of the operator, as well as against other foreign surfaces. When the usual procedure was modified so that this rubbing was avoided, the resulting helical strips exhibited good relaxation in response to low concentrations of acetylcholine. Once the relationship between loss of relaxing capacity and rubbing of the intimal surface was appreciated and additional precautions were taken to minimize such rubbing during the entire procedure, it became common to obtain both rings and strips (helical or transverse) that would undergo 90–100% relaxation from a level of moderate tone (usually produced by norepinephrine) when exposed to an optimal concentration (usually 1 μM) of acetylcholine (Fig. 1).

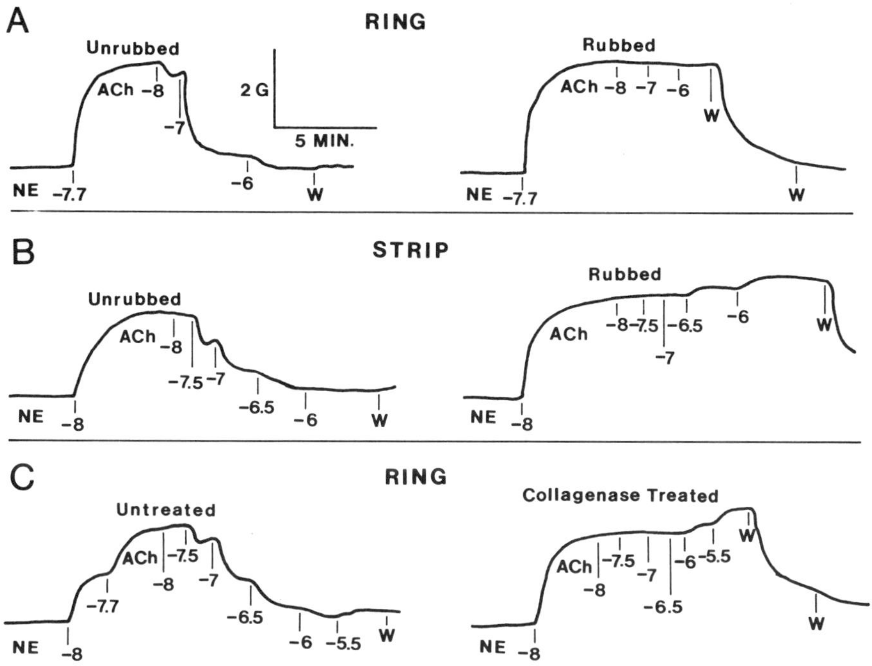

Fig. 1. Loss of relaxing response of preparations of rabbit aorta after rubbing of intimal surface or exposure to collagenase. (A) Recordings from same ring before and 30 min after rubbing the intimal surface with a small wooden stick. (B) Recordings from an unrubbed and a rubbed transverse strip from the same aorta (rubbing achieved by dragging intimal surface over filter paper). (C) Recording from rings from same aorta before and after exposure of the intimal surface to 0.2% collagenase for 15 min. In this and all other figures, concentrations are expressed as logarithms of molar concentrations. W, wash; NE, (−)-norepinephrine; ACh, acetylcholine (from Furchgott et al., 1981, with permission).

The mechanism by which rubbing the intimal surface of an aortic preparation abolished the ability of the preparation to relax in response to acetylcholine was not immediately clear. Among the several possible mechanisms considered was that rubbing removes endothelial cells, and that such cells have an obligatory role in mediating the relaxation. Early histological studies, however, of transverse sections of unrubbed and rubbed rings stained with hematoxylin and eosin showed neither unrubbed nor rubbed rings to have significant numbers of endothelial cells. Unfortunately, these early studies were apparently subject to technical errors because later it was clearly shown by an *en face* silver staining technique and light microscopy (Furchgott and Zawadzki, 1980a), and also by scanning electron microscopy (Furchgott et al., 1981), that there is a direct relationship between the extent of relaxation of aortic preparations by acetylcholine and the fraction of endothelial cells retained. The *en face* microscopic examination of preparations immediately after pharmacological experiments showed that those preparations giving 90–100% relaxation of moderate tone (as in Fig. 1) usually had about 60–80% of their endothelial cells present, that preparations with only a few percent of these cells present still gave moderate relaxation, but that preparations completely devoid of endothelial cells gave no relaxation. Another procedure besides rubbing that effectively removed endothelial cells, namely exposure of the intimal surface of a segment of aorta to the enzyme collagenase, also abolished relaxation by acetylcholine (Furchgott and Zawadzki, 1980a; Fig. 1C).

Removal of endothelial cells did not interfere with the sensitivity of preparations of rabbit aorta to contracting agents (e.g., norepinephrine in Fig. 1) if care was taken not to overstretch the smooth muscle during the rubbing procedure or not to expose the preparation too long to collagenase. Indeed, the contractile effect of acetylcholine at higher concentrations in the absence of any other contracting agent was actually greater after removal of the endothelial cells—as would be expected, since there would now be no opposing relaxing activity mediated through these cells.

Removal of endothelial cells did not alter the sensitivity of thoracic aortic preparations of the rabbit to a variety of nonmuscarinic relaxing agents, including glyceryl trinitrate, sodium azide, adenosine, adenylic acid (AMP), and isoproterenol, or to photorelaxation (Furchgott and Zawadzki, 1980a).

The obligatory role of endothelial cells in acetylcholine-induced relaxation was soon found to be applicable not only in the case of rabbit aorta, but also in the case of all other arteries tested, including a variety of arteries from the rabbit, dog, cat, rat, guinea pig, and humans (Furchgott and Zawadzki, 1980a; Furchgott et al. 1981; Cherry et al., 1982). From these results it became clear that the lack of relaxation of contracted preparations of arteries in many early studies was the result of an inadvertent removal of endothelial cells from these preparations prior to pharmacological testing.

1.2. Other Early Findings, Conclusions, and Speculation

1.2.1. Characteristics of the Relaxation by Acetylcholine

That the endothelial cell receptor on which acetylcholine acts is of the muscarinic type was indicated in early experiments on rabbit aorta (Furchgott and Zawadzki, 1980a; Furchgott et al., 1981). The apparent K_B obtained for atropine as a competitive antagonist against acetylcholine was 0.35 n*M*. The relative potencies of the muscarinic agonists acetylcholine, methacholine, and carbachol for producing relaxation were approximately 1, 0.5, and 0.1, respectively. Early quantitative studies with antagonists and agonists did not differentiate into separate subtypes the muscarinic receptor on the endothelium mediating relaxation and that on smooth muscle mediating the direct activation of contraction. [More recent results with a larger number of agonists and antagonists, including pirenzepine, suggest that the two are actually of the same muscarinic subtype (Furchgott and Cherry, 1984; Furchgott, 1986), but that the receptor-effector system in the case of the endothelial cells has a "receptor reserve" for acetylcholine, whereas that in the case of smooth muscle cells does not.]

The extent of relaxation by acetylcholine of norepinephrine-induced contractions was shown to depend on the degree of tone, with the percent of relaxation by acetylcholine being greater at low to moderate levels than at high levels of tone. For example, against maximal tone produced by norepinephrine (30 μM), acetylcholine (1 μM) produced about 35% relaxation, whereas against half-maximal tone, it usually produced over 80% relaxation (Furchgott et al., 1981).

Acetylcholine was about equally effective in relaxing contractions of the same level of tone initially produced in aortic rings of the rabbit by any of a number of vasoconstrictor agents, including norepinephrine, phenylephrine, histamine, serotonin, angiotensin II, and prostaglandin $F_{2\alpha}$. Against contractions produced by elevating K^+ in the bathing Krebs solution, however, or by replacing the ordinary Krebs solution with a "K_2SO_4-Krebs solution," the maximal relaxation by acetylcholine was always considerably less (10–50%) than it was against equivalent contractions produced by norepinephrine (Furchgott et al., 1981).

1.2.2. Other Agents Evoking Endothelium-Dependent Relaxation

1.2.2.1. A23187. Within two years of the discovery that the acetylcholine-induced relaxation of isolated arteries was endothelium-dependent, work in the author's laboratory had revealed several additional agents that depended on endothelium to produce relaxation of aorta of the rabbit or of other arteries. The second agent studied was the Ca^{2+}-ionophore A23187 (Zawadzki et al., 1980). The agent was at least 10 times more potent than acetylcholine in producing endothelium-dependent relaxation of rabbit aorta (EC_{50} < 10 n*M*). It was also more powerful, giving a greater maximal relaxation than acetylcholine against high levels of contraction to norepinephrine. Also, if a relatively high concentration (1 μM) of A23187 was added for more than several minutes to an aortic preparation, its stimulation of endothelium-dependent relaxation persisted for long periods after washout of the agent from the muscle chamber (Furchgott, 1983). In part because of findings with A23187, it was proposed that an increase of calcium ions in the region of some key Ca^{2+}-activated enzyme might be an early step in the reactions mediating release of the endothelium-derived relaxing factor (EDRF) (Furchgott et al., 1981). Such an increase might be brought about by A23187 acting as an ionophore or by acetylcholine opening a membrane Ca^{2+} channel coupled to the muscarinic receptor. Later work in several other laboratories (e.g., Singer and Peach, 1982; Griffith et al., 1986) clearly showed that extracellular calcium is required for stimulation of release of EDRF by acetylcholine and A23187.

1.2.2.2. ATP and ADP. About the time that endothelium-dependent relaxation by A23187 was found, adenosine triphosphate (ATP) and adenosine diphosphate (ADP) were also found to exert most of their relaxing effect on isolated rings of rabbit aorta through an endothelium-dependent mechanism (Furchgott and Zawadzki, 1980b; Furchgott, 1981). Independently, De Mey and Vanhoutte (1980, 1981) showed endothelium-dependent relaxation of canine arteries by ATP and ADP. In the aorta of the rabbit and the canine arteries, adenosine and AMP produced relaxation only in an endothelium-independent manner. The remaining relaxing capacity of ADP and ATP on rabbit and canine arteries denuded of endothelial cells may actually be a result of the direct action of their metabolic products, AMP and adenosine, on the smooth muscle. Interestingly, in the porcine aorta, Gordon and Martin (1983) showed that a significant part of the relaxation evoked by AMP and adenosine, as well as that by ATP and ADP, is endothelium-dependent.

1.2.2.3. Substance P. Substance P was the next agent shown to strictly depend on the endothelium for producing relaxation of arteries from the rabbit and other species (Zawadzki et al., 1981). This polypeptide is the most potent endothelium-dependent relaxing agent found in the author's laboratory, having a threshold concentration around 30 p*M* in rabbit aorta and around 1 p*M* in some canine arteries (Fig. 2). Later it was found that other polypeptide tachykinins related structurally to substance P, namely

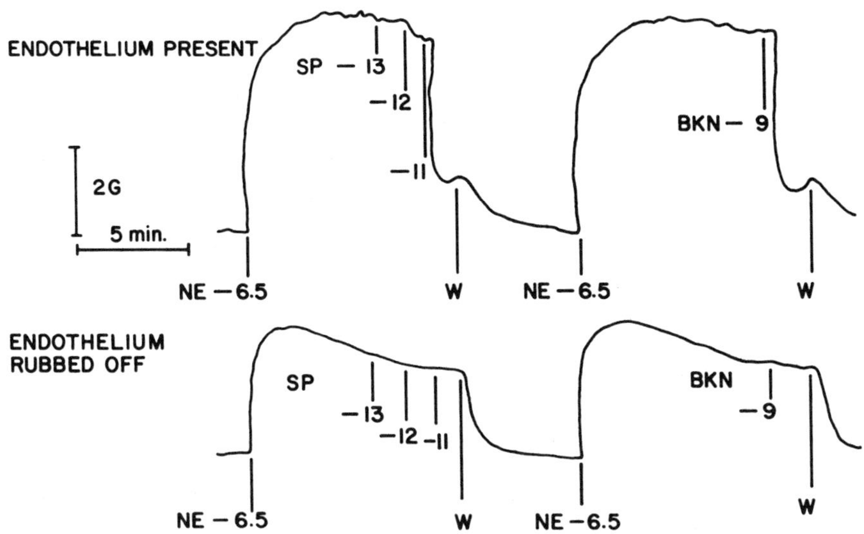

Fig. 2. Endothelium-dependent relaxation of canine superior mesenteric artery by substance P (SP) and bradykinin (BKN). Tests were made on paired rings with and without endothelium (from Furchgott et al., 1983, with permission).

kassanin, physalemin, and eledoisin, are also very potent endothelium-dependent relaxing agents (Zawadzki et al., 1983; Furchgott, 1984). A characteristic of the relaxation caused by substance P in all arteries was the occurrence of desensitization—that is, within a short time (usually 3–10 min) after addition of a maximally effective relaxing dose, the treated artery recovered its initial tone and failed to relax on further additions. Sensitivity to the substance returned readily after washout of a desensitizing dose. The desensitization was selective, for it did not alter the sensitivity to acetylcholine or A23187. There was cross-desensitization between the other tachykinins and substance P. Still another peptide that was found to give endothelium-dependent relaxation of the rabbit-aorta was octa-cholescystokinin (Zawadzki et al., 1983). On the other hand, relaxation of rabbit aorta by vasoactive intestinal polypeptide (VIP) was endothelium-independent.

1.2.2.4. Bradykinin. Bradykinin was found to give no relaxation of rings of rabbit aorta, whether or not endothelium was present and whether or not cyclooxygenase was inhibied (to prevent synthesis of prostaglandins). However, bradykinin caused excellent endothelium-dependent relaxations of a variety of canine arteries and of human mesenteric arteries, whether or not cyclooxygenase was inhibited (to prevent synthesis of prostaglandins). Experimental evidence was obtained supporting the postulate that bradykinin releases the same nonprostanoid endothelium-derived relaxing factor

(EDRF) in canine arteries as do acetylcholine and A23187 (Cherry et al., 1982; Furchgott, 1983). Later, Gordon and Martin (1983) showed that relaxation of porcine aorta by bradykinin was also endothelium-mediated and did not depend on stimulation of release of prostaglandins. On the other hand, the relaxation of certain arteries of rabbit and cat (celiac and superior mesenteric) by bradykinin appeared to depend on stimulation of release of prostaglandins from both endothelial and nonendothelial cells, rather than on release of an EDRF (Cherry et al., 1982).

1.2.3. Demonstration of Release of an Endothelium-Derived Relaxing Factor (EDRF)

One hypothesis to explain the obligatory role of endothelial cells in the relaxation of the rabbit aorta by acetylcholine was that this agent acting on the muscarinic receptor in these cells stimulates them to release a substance that in turn acts on the smooth muscle cells in the media to activate relaxation. The first experiments to test this hypothesis, employing cascade superfusion procedures, only occasionally gave positive results. A procedure was developed, however, that consistently gave direct evidence for the hypothesis (Furchgott and Zawadzki, 1980a). In this procedure, a transverse strip freed of endothelial cells by rubbing was tested when mounted separately and also when mounted, intimal surface against intimal surface, with a longitudinal strip of the same width and length with endothelial cells present ("sandwich" mount). Because of the orientation of its muscle cells, the longitudinal strip could contribute only trivial changes in tension on the strain gage during contraction. Acetylcholine, which failed to inhibit contractions of the transverse strip mounted separately, always gave good relaxation of the same strip when it was mounted as a "sandwich" with the longitudinal strip. Thus, release of a relaxing substance (later termed EDRF) was demonstrated. This "sandwich" procedure, or a modification of it, was also used to demonstrate release of EDRF by A23187 in the rabbit aorta and by acetylcholine and bradykinin in several canine arteries (Furchgott, 1983).

Although cascade perfusion and superfusion procedures to demonstrate release of EDRF from endothelial cells were not initially successful in the author's laboratory, such procedures were later successfully developed (Griffith et al., 1984a; Cocks and Angus, 1985; Forstermann et al., 1984; Rubanyi et al., 1985; Gryglewski et al., 1986; Furchgott and Khan, 1986).

1.2.4. Speculation on the Nature of EDRF and Its Mechanism of Action

Within a short time after demonstration of endothelium-dependent relaxation of arteries by acetylcholine, it was possible to rule out the following substances as being EDRF: cyclic AMP, cyclic GMP, adenosine, AMP, prostacyclin, and other prostanoid products of cyclooxygenase. It was shown that anoxia completely inhibited relaxation of rabbit aorta by acetylcholine in a reversible manner and that both 5,8,11,14-eicosatetraynoic acid (ETYA) and quinacrine were rapid-acting selective inhibitors of acetylcholine-induced relaxation (Fig. 3; Furchgott and Zawadzki, 1980a; Furchgott et al., 1981; Furchgott, 1981). In addition, exposure of aortic

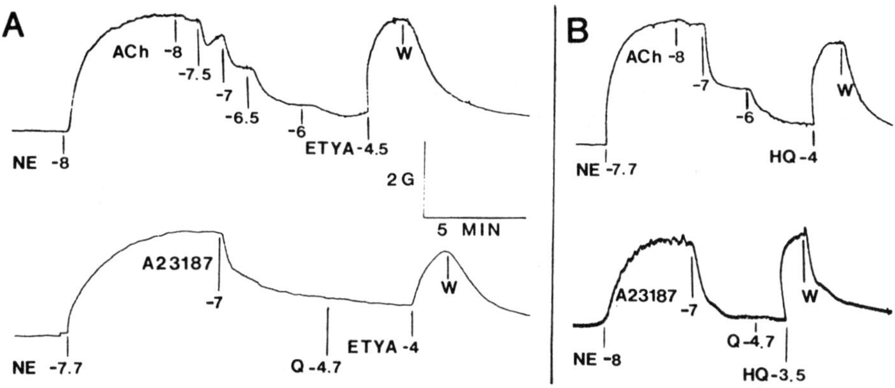

Fig. 3. Relaxation of rings of rabbit thoracic aorta by acetylcholine (ACh) and by A23187 and the acute inhibition of each relaxation on addition of 5,8,11,14-eicosatetraynoic acid (ETYA) (A) or of hydroquinone (HQ) (B). Note that quinacrine (Q) does not inhibit the A23187-induced relaxation as it does the ACh-induced relaxation (from Furchgott, 1981, with permission).

preparations to a high concentration of ETYA (100 μM) for 30–60 min, followed by washout, completely and irreversibly inhibited the relaxation by acetylcholine. The inhibitory effects of anoxia, ETYA (an inhibitor of lipoxygenase as well as cyclooxygenase), and quinacrine (an inhibitor of the release of arachidonic acid from certain phosphatides) suggested that the stimulation of the muscarinic receptor of endothelial cells by acetylcholine somehow activates a reaction sequence in which arachidonic acid (or some other unsaturated acid) is liberated from phosphatides and then oxidized by lipoxygenase to a labile intermediate product that acts on the smooth muscle to cause relaxation. The possibility that this product might be a free radical was proposed, partly because another agent found to very rapidly inhibit relaxation by acetylcholine was hydroquinone, which is a free radical scavenger (Fig. 3; Furchgott, 1981).

Some additional early findings were consistent with this proposal about the precursor and general nature of EDRF (Furchgott, 1983; Furchgott et al., 1983). For example, nordihydroguiaretic acid (NDGA), which is both a lipoxygenase inhibitor and an antioxidant, also rapidly inhibited endothelium-dependent relaxations produced by acetylcholine, A23187, ATP, substance P, and bradykinin. The inhibition was largely reversible on short exposure, but irreversible on prolonged exposure (e.g., 30 min at 100 μM). Also, exposure of arteries to *p*-bromophenacyl bromide (BPB), a potent irreversible inhibitor of phospholipase A_2, for short exposure periods (e.g., 3 μM for 20 min) produced complete and irreversible loss of endothelium-dependent relaxation by all the relaxants noted above.

Prior to the findings on inhibition of endothelium-dependent relaxation by anoxia and ETYA, there had been several reports by others that in certain smooth muscles relaxation by certain agents was accompanied by increases in cyclic GMP (for references, *see* Murad et al., 1979). Also, there were reports that guanylate cyclase was markedly stimulated by hydroperoxides of arachidonic acid (Hidaka and Asano, 1977; Goldberg et al., 1978) and by free radicals, particularly nitric oxide (Murad et al., 1979). It had been proposed by Murad et al. (1979) that all of the so-called nitrovasodilators, such as sodium nitroprusside, organic nitrates, azide, and inorganic nitrite, activate guanylate cyclase indirectly via nitric oxide, which they release as a reaction product. Considerations of these various findings led to the speculative scheme in which EDRF, like the nitrovasodilators, stimulates guanylate cyclase of the vascular smooth muscle, causing an increase in cyclic GMP, which then somehow activates relaxation (Furchgott et al., 1981).

2. Recent Developments

2.1. The Role of Cyclic GMP in Endothelium-Dependent Relaxation

The speculation that EDRF released by acetylcholine and other endothelium-dependent relaxants would stimulate an increase in cyclic GMP has been supported by independent studies in several laboratories—e.g., Rapoport and Murad (1983) using the aorta of the rat, Diamond and Chu (1983) and Furchgott et al. (1984a) using the aorta of the rabbit, Holzmann (1982) using the bovine coronary artery, and Ignarro et al. (1984) using the bovine pulmonary artery. In the study in the author's laboratory, rings or strips of rabbit aortas were mounted in specially designed muscle chambers that allowed for both continuous recording of tension and rapid freezing of the preparation at any desired time in an experiment. Relaxation during norepinephrine-induced contraction by acetylcholine (1 μM) and by A23187 (0.1 μM) added to endothelium-containing rings for 3 min was accompanied by fivefold and sevenfold increases in cyclic GMP, respectively. In endothelium-free rings, acetylcholine and A23187 produced neither relaxation nor any change in cyclic GMP. Glyceryl trinitrate relaxed endothelium-free rings and intact rings equally well and produced marked increases in cyclic GMP in both types of preparations. It should be noted that the level of cyclic GMP of aortic rings denuded of endothelial cells fell to less than one-third of the basal level of about 165 pmol/g found in intact rings. The increase in cyclic GMP after addition of acetylcholine to intact rings peaked within 15–30 s after the addition, and then remained essentially at the peak level up to 3 min (Fig. 4). The fact that the rise in cyclic GMP preceded the onset of relaxation was consistent with the proposal that cyclic GMP has a causal role in the relaxation. That the rise in cyclic GMP precedes endothelium-dependent relaxation produced by acetylcholine was also shown independently by others (Rapoport and Murad, 1983; Diamond and Chu, 1983; Ignarro et al., 1984). In all of the studies on endothelium-dependent relaxation, no significant change in cyclic AMP occurred. The experimental evidence available demonstrates that the increase in cyclic GMP occurring during endothelium-dependent relaxation by acetylcholine is in the smooth muscle rather than in the endothelial cells (Fig. 5).

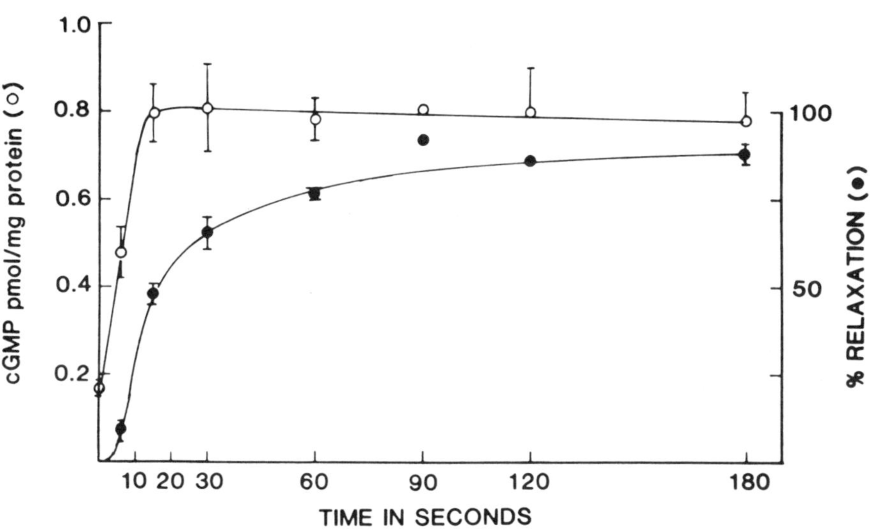

Fig. 4. Time course of change of cyclic GMP (cGMP) and of relaxation in endothelium-containing rings of rabbit aorta after addition of acetylcholine (ACh). Norepinephrine (0.1 μM) was first added to produce a steady contraction. ACh (1 μM) was then added, and the tissue was frozen after desired period of exposure to ACh (from Furchgott et al., 1984a, with permission).

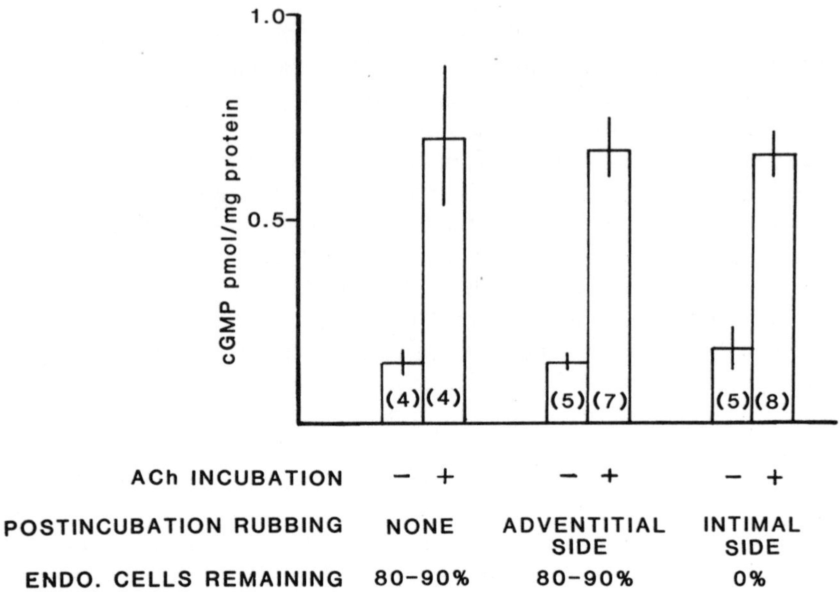

Fig. 5. Results of three experiments showing that increase of cyclic GMP in rabbit aorta produced by acetylcholine is in the smooth muscle cells and not in the endothelial cells. In each experiment, 16–20 matched transverse strips (rings cut open) with endothelium present were suspended from individual clips in Krebs solution at 37 °C in muscle chambers. After equilibration of more than 2 h, each of 10–14 strips was removed from its muscle chamber after either a 1-min incubation with 1 μM acetylcholine or no incubation. On removal, each strip was either frozen immediately in liquid N_2, dragged for 3 s on its adventitial side on moistened filter paper and then frozen, or dragged for 3 s on its intimal side and then frozen. Other strips were put through each of the removal procedures except for the freezing and were then subjected to *en face* silver staining to determine the percent of endothelial cells remaining after each removal procedure. Numbers in parentheses show numbers of strips.

The mechanism by which an increase in cyclic GMP might produce relaxation of vascular smooth muscle is not yet definitely known; however, Rapoport et al. (1983) have proposed that relaxation is mediated through cyclic GMP-dependent protein phosphorylation and dephosphorylation or myosin light chains.

2.2. Additional Agents Producing Endothelium-Dependent Relaxation

2.2.1. A General Compilation

Over the past few years a large number of agents that produce endothelium-dependent relaxation of arteries has been found. Table 1 is a com-

Table 1
Agents That Have Been Shown to Produce Endothelium-Dependent Relaxation of Isolated Arteries

Agent	First reports	Kinds of arteries[a]
Acetylcholine	Furchgott and Zawadzki, 1980a	RA and many other arteries
A23187 (calcium ionophore)	Zawadzki et al., 1980	RA and many other arteries
ATP and ADP	Furchgott and Zawadzki, 1980b De Mey and Vanhoutte, 1980, 1981	RA and many other arteries
Substance P and cholescystokinin	Zawadzki et al., 1981	RA and many other arteries
Bradykinin	Cherry et al., 1981, 1982 Chand and Altura, 1981	Dog, human, and pig arteries (not rabbit, cat)
Thrombin	De Mey et al., 1982	Dog arteries
Histamine	Van de Voorde and Leusen, 1983 Satoh and Inui, 1984	Rat arteries (H_1-receptor), guinea pig pulmonary artery
Serotonin	Cohen et al., 1983 Cocks and Angus, 1983	Dog coronary artery (possibly $5\text{-}HT_1$ receptor)
Norepinephrine	Cocks and Angus, 1983	Dog coronary artery (α_2-receptor)
Arachidonic acid	Singer and Peach, 1983 Cherry et al., 1983	RA and dog arteries
Other unsaturated fatty acids	Cherry et al., 1983	RA and dog arteries
Saturated fatty acids	Furchgott et al., 1985	RA (not tested on others)
Ergometrine	Griffith et al., 1984b	RA
VIP	Davies and Williams, 1983	Rat aorta (not RA)
Vasopressin	Katusic et al., 1984	Dog basilar artery (not dog femoral or RA)
Calcitonin gene-related peptide	Brain et al., 1985	Rat aorta
Mellitin	Forstermann and Neufeng, 1985	RA

[a]RA, rabbit aorta.

pilation of the first reports for each agent through 1985. [For tables giving more details about specific arteries and species, *see* reviews by Peach et al. (1985) and Busse et al. (1985).] In most of the studies referred to in Table 1, the relaxing substance released by the agent was shown not to be a prostanoid on the basis of the finding that a cyclooxygenase inhibitor did not inhibit the relaxation.

With respect to the agents listed in Table 1, it should be emphasized that, depending on the agent, the nonprostanoid-mediated, endothelium-dependent mechanism of relaxation may be limited to certain species or even to certain arteries from certain species. For example, bradykinin acts by this indirect mechanism in relaxing all arteries tested from dogs and humans, but acts by another indirect mechanism involving release of vasodilator prostaglandins from both endothelial and other types of cells in arteries from cats and rabbits (Cherry et al., 1982). Histamine fails to produce endothelium-dependent relaxation of rabbit aorta and other arteries, but it produces such relaxation in arteries of the rat (Van de Voorde and Leusen, 1983) and the guinea pig (Satoh and Inui, 1984). Serotonin acts through the nonprostanoid, endothelium-dependent mechanism to relax canine and porcine coronary arteries (Cocks and Angus, 1983), but produces no such relaxation in the case of a number of different arteries of the rabbit. Vasoactive intestinal polypeptide (VIP), which has been reported to elicit endothelium-dependent relaxation in the aorta of the rat (Davies and Williams, 1983), only elicits endothelium-independent relaxation in that of the rabbit. Recently, vasopressin has been found to be an endothelium-dependent relaxing agent of the canine basilar artery, but not of the femoral artery of the same species (Katusic et al., 1984). Finally, it has been reported that acetylcholine, which heretofore has been found to produce only endothelium-dependent relaxations in a variety of arteries from a variety of species, can produce an endothelium-independent relaxation of certain cephalic arteries from the rabbit (Brayden and Bevan, 1985). These various examples should caution us against concluding that the mechanism for producing relaxation by any one agent is always the same in all arteries.

2.2.2. Fatty Acids

Since arachidonic acid had been proposed as a possible precursor of EDRF (Furchgott and Zawadzki, 1980a; Furchgott et al., 1981), the effect of this unsaturated fatty acid on isolated arteries was investigated. In the presence of cyclooxygenase inhibitors (to prevent formation of vasoactive prostaglandins), arachidonic acid was shown to produce an endothelium-dependent relaxation in canine arteries (Cherry et al., 1983; Furchgott et al., 1984b) and in the aorta of the rabbit (Singer and Peach,

1983; Cherry et al., 1983; Furchgott et al., 1984a, 1985). The endothelium-dependent relaxation by arachidonic acid was generally small compared to that evoked by acetylcholine or A23187, and, especially in the case of the rabbit aorta, was transient in nature, tending to fade within 5–10 min (Fig. 6, middle tracing). Other unsaturated fatty acids, including *cis*-4,7, 10,13,16,19-docosohexaenoic, oleic, elaidic, palmitoleic, and *cis*-vaccenic acid, were also shown to produce transient endothelium-dependent relaxations (Cherry et al., 1983). Of the unsaturated fatty acids tested, arachidonic acid was the most potent, being on average about 0.004 as potent as acetylcholine for producing moderate degrees of relaxation.

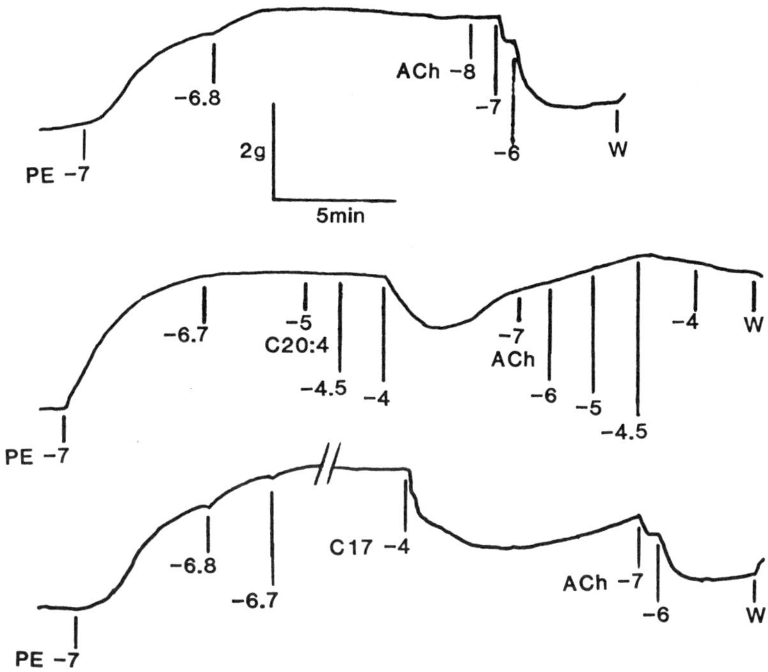

Fig. 6. Comparison of relaxations of rings of rabbit aorta by acetylcholine (ACh), arachidonic acid (C20:4), and heptadecanoic acid (C17); and demonstration of inhibition of ACh-induced relaxation by treatment with arachidonic acid. Endothelium-containing rings were all from the same aorta. PE, phenylephrine (from Furchgott et al., 1985, with permission).

Saturated fatty acids containing from 10 to 22 carbon atoms also produced endothelium-dependent relaxation of the rabbit aorta (Furchgott et al., 1985). The acids having 16 and 17 carbons (palmitic and heptadecanoic) were the most potent in the series, being about equal in potency to arachidonic acid. The relaxation produced by addition of saturated fatty acids was better sustained than that of unsaturated fatty acids (compare heptadecanoic acid and arachidonic in Fig. 7). In the case of heptadecanoic acid acting on the rabbit aorta, it was possible to show a good correlation between the degree of endothelium-dependent relaxation and the increase in cyclic GMP (Furchgott et al., 1985), indicating that the fatty acid was indeed causing the release of an EDRF similar to that released by acetylcholine and A23187.

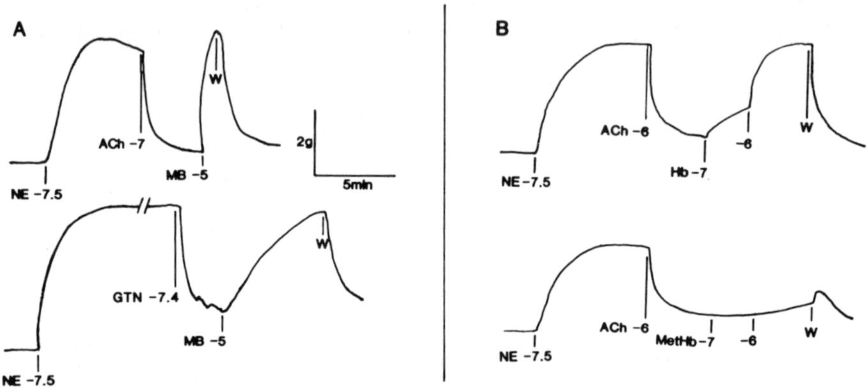

Fig. 7. Acute inhibition of acetylcholine (ACh)-induced relaxation of rings of rabbit aorta by methylene blue (MB) and hemoglobin (Hb). (A) MB at 10 μM inhibited rapidly and completely ACh-induced relaxation, but inhibited glyceryl trinitrate (GTN)-induced relaxation much more slowly. (B) Hb at 1 μM produced rapid inhibition of ACh-induced relaxation, but methemoglobin (MetHb) failed to produce significant inhibition (from Furchgott et al., 1984a, with permission).

In view of the ability of many unsaturated and saturated fatty acids to elicit endothelium-dependent relaxations, the finding that arachidonic acid elicited such relaxations was no longer tenable evidence in favor of this acid as the precursor of EDRF. In addition, it was found that treatment of rabbit aortic rings with arachidonic acid actually inhibited the response to acetylcholine (Fig. 6, middle tracing). Indeed, arachidonic acid, as an inhibitor or acetylcholine-induced relaxation, behaved very much like eicosatetraynoic acid, ETYA, the triple-bond analog of arachidonic acid (Furchgott et al., 1984c, 1985). Other unsaturated fatty acids capable of eliciting endothelium-dependent relaxations shared with arachidonic acid this capacity to inhibit the response to acetylcholine. In contrast, pretreatment with saturated fatty acids (palmitic or heptadecanoic) did not inhibit the relaxation evoked by acetylcholine (Fig. 6, lower tracing).

2.3. Inhibitors of Endothelium-Dependent Relaxation

2.3.1. Methylene Blue

Methylene blue was shown by Gruetter et al. (1981) to inhibit stimulation of guanylate cyclase by nitrovasodilators in cell-free systems and to inhibit the relaxation as well as the increase in cyclic GMP elicited by these agents in vascular smooth muscle. Subsequently, methylene blue was found to inhibit both the endothelium-dependent relaxation and the accompanying increase in cyclic GMP produced by acetylcholine in bovine arteries (Holzmann, 1982; Ignarro et al., 1984), and by acetylcholine and A23187 in the rabbit aorta (Furchgott et al., 1984c; Martin et al., 1985a). Holzmann (1982) and Ignarro et al. (1984) concluded that the inhibition of the relaxation was the result of the inhibition of guanylate cyclase in the vascular smooth muscle; and Griffith et al. (1985), using a perfusion cascade to determine the site of action of methylene blue, came to the same conclusion. On the other hand, although agreeing that such enzyme inhibition accounts for the sustained inhibition by methylene blue developed during prolonged exposure of the rabbit aorta, Furchgott et al. (1984c) and Martin et al. (1985a) proposed that the extremely fast reversal (inhibition) by methylene blue of acetylcholine- or A23187-induced relaxation (*see* Fig. 7A) was more likely the result of either a rapid inhibition of endothelial synthesis and release of EDRF or a direct reaction of the oxidant with EDRF.

2.3.2. Hemoglobin

Hemoglobin had been found by Murad et al. (1978) to inhibit guanylate cyclase in cellular extracts, and by Bowman and Gillespie (1982) to inhibit the relaxation of the bovine retractor penis muscle and rabbit aorta elicited by both nitroprusside and a partially purified, unidentified inhibitory factor extracted from the former. In the author's laboratory, hemoglobin was found to be a potent, rapidly acting inhibitor of acetylcholine- and A23187-induced relaxations of the rabbit aorta (Fig. 7B; Furchgott et al., 1984a,b,c; Martin, 1985a). A short treatment with 10 μM hemoglobin almost completely prevented the relaxation by any concentration of acetylcholine or A23187. The inhibition could be reversed by washout of the hemoprotein. Hemoglobin also inhibited relaxations induced by glyceryl trinitrate, but to a lesser extent than that by acetylcholine, and did not inhibit relaxations induced by isoproterenol. The inhibition by hemoglobin of endothelium-dependent relaxations to acetylcholine, A23187, or heptadecanoic acid was accompanied by a corresponding inhibition of the increase in cyclic GMP that is ordinarily associated with such relaxation

(Martin et al., 1985a; Furchgott et al., 1985). It was proposed that hemoglobin inhibits relaxation and the rise in cyclic GMP by binding EDRF before the latter diffuses to and activates the guanylate cyclase of the arterial smooth muscle. Direct evidence for strong, rapid binding of EDRF by hemoglobin has recently been obtained in cascade experiments in which relaxation of an endothelium-free bioassay ring of aorta by EDRF released from endothelial cells was rapidly abolished by introduction of hemoglobin between the endothelial cells and the bioassay ring (Furchgott and Khan, 1986).

The ferrous hemoprotein myoglobin proved to be as effective as hemoglobin in inhibiting acetylcholine-induced relaxations and the associated rise in cyclic GMP in the rabbit aorta. In contrast, both methemoglobin and metmyoglobin gave only relatively weak inhibition (Fig. 7B; Martin et al., 1985b), and neither the ferrous nor ferric form of cytochrome C exerted any inhibition. The ability of hemoglobin to inhibit almost completely endothelium-dependent relaxations was also demonstrated in rat aorta (Martin et al., 1986a; Khan et al., 1986) and in canine blood vessels (Furchgott, unpublished results). In endothelium-containing rat aorta, hemoglobin not only inhibited acetylcholine-induced increases in cyclic GMP, but also significantly reduced the basal level of cyclic GMP—as would be expected because of its binding of basally (spontaneously) released EDRF.

Because of its high potency, selectivity, and ready reversibility as an inhibitor of EDRF, hemoglobin has become a very useful agent for: (1) determining whether relaxation of an artery induced by a vasodilator is mediated by EDRF; (2) ascertaining whether there is a basal (spontaneous) release of EDRF; and (3) determining the extent to which such basal release depresses contractile responses to added vasoconstrictor agents and potentiates relaxation by agents that inhibit phosphodiesterases that inactivate cyclic GMP (Martin et al., 1985a, 1986a,b). Hemoglobin was found to inhibit in large part vasodilatation produced by acetylcholine in the perfused mesenteric vascular bed of the rat, thus providing evidence that the dilatation of resistance vessels (small arteries and arterioles) by acetylcholine is also largely the result of local release of EDRF (Khan et al., 1986).

2.4. Further Characterization of EDRF

The development of cascade perfusion and superfusion procedures, in which EDRF released from endothelial cells upstream is bioassayed for relaxing activity on an endothelium-denuded arterial preparation downstream, has provided new information about the chemical reactivity of EDRF. Griffith et al. (1984a), using such a procedure in which the transit time of the perfusate between endothelial cells and bioassay artery could be varied, estimated the half-life of EDRF released by acetylcholine from the rabbit aorta to be 6.3 ± 0.6 s. Half-life values of about 6 s were also reported by Cocks and Angus (1985) for EDRF released from cultured bovine endothelial cells on microcarrier beads by bradykinin and by Rubanyi et al. (1985) for EDRF released by acetylcholine from the endothelium of the canine femoral artery. In the author's laboratory, the half-life of EDRF released by acetylcholine from the endothelium of the rabbit aorta averaged 10.16 ± 1.6 s (Furchgott and Khan, 1986). On the other hand, Forstermann et al. (1984) estimated longer half-lives for EDRF released by acetylcholine from endothelium of the rabbit aorta (24 ± 3 s) and the canine femoral artery (49 ± 5 s). They attributed the longer half-lives in their experiments to the fact that they used a lower O_2 concentration than others in their perfusion solution.

Two recent independent studies using cascade perfusion procedures may help explain differences in half-lives for EDRF obtained in different laboratories. In these studies it was found that superoxide dismutase markedly increased the half-life of EDRF released by acetylcholine from the endothelium of dog femoral artery (Rubanyi and Vanhoutte, 1986) and of EDRF released by bradykinin from cultured porcine endothelial cells (Gryglewski et al., 1986), indicating that superoxide anions can rapidly inactivate EDRF. Thus, in different laboratories with different endothelial preparations for releasing EDRF and different experimental designs for cascade perfusion or superfusion, there could well be different levels of superoxide anions in the fluid leaving the endothelial cells, and this could lead to different half-lives. The probability of a fast inactivation of EDRF by superoxide anion does not preclude slower inactivation by other mechanisms (e.g., oxidation by O_2) in cascade perfusion systems.

Griffith et al. (1984a), using their cascade perfusion system, were able to demonstrate that KBH_4, dithiothereitol, phenidone, and nordihydroguiaretic acid (reducing agents), and also phenylhydrazine (carbonyl reagent), blocked EDRF action by actually reacting with EDRF. These results and others suggested that EDRF might be a labile carbonyl-containing compound (e.g., an aldehylde, ketone, or lactone), and that it was unlikely to be a free radical arising from lipoxygenase oxidation of arachidonic or some other unsaturated fatty acid (Griffith et al., 1984a). By the time of the report by Griffith and colleagues, the present author had also moved away from his initial hypothesis that EDRF was a lipoxygenase product, partly because of results obtained with fatty acids (*see* section 2.2.2) and partly because of new evidence indicating that the rapid acute inhibition of endothelium-dependent relaxation by eicosatetraynoic acid (ETYA) and nordihydroguiaretic acid (NDGA) was not caused by inhibition of lipoxygenase, and that inhibition by bromophenacyl bromide (BPB) was likely caused by nonspecific damage to endothelial cells (Furchgott, 1984a; Furchgott et al., 1985). The possibility of EDRF being a free radical was still considered likely, however.

Griffith et al. (1986) in another interesting study found that production of EDRF by endothelial cells of the aorta of the rabbit was rapidly inhibited by agents that inhibit mitochondrial electron transport or uncouple mitochondrial oxidative phosphorylation. These results suggested that mitochondrial ATP synthesis may be needed in these cells for the production and/or release of EDRF.

On the basis of the inhibition of EDRF by hemoglobin (*see* section 2.3.2) and the fact that soluble guanylate cyclase is linked to a hemoprotein with which nitric oxide binds when it activates this enzyme (Craven and De Rubertis, 1978), the proposal was made that EDRF, like nitric oxide, might react directly with this hemoprotein to activate the enzyme (Martin et al., 1985a). Very recently two groups have independently shown that EDRF released from endothelial cells of blood vessels can indeed react directly with soluble cytoplasmic guanylate cyclase and activate the synthesis of cyclic GMP (Forstermann et al., 1986; Ignarro et al., 1986).

As of this writing, the chemical identity of EDRF has not yet been established. Some investigators have felt that their results indicate more than one EDRF (Forstermann et al., 1984; Rubanyi et al. 1985), but there may be other explanations for these results.

At the fourth symposium on "Mechanisms of Vasodilatation" held in Rochester, Minnesota, July 10–12, 1986, the author proposed that EDRF is nitric oxide (Furchgott, 1987). This proposal was based on a study of the characteristics of relaxation of rabbit aorta produced by an endothelium-independent relaxing factor, which is quickly generated as an intermediate on acidification of sodium nitrite solutions. This factor is presumably nitric oxide, which forms as an intermediate during the spontaneous decomposition of nitrous acid. The factor is labile—as indicated by the transiency of the relaxation induced by it when it is added to a ring of aorta bathed in Krebs-bicarbonate solution; its relaxing effect is completely blocked by hemoglobin and only partially by methemoglobin; its relaxing effect is shortened by generators of superoxide anions and prolonged by superoxide dismutase. All of these characteristics are also characteristics of EDRF. Further research is now underway to test critically the proposal that EDRF is nitric oxide.

Acknowledgments

The recent research in the author's laboratory reported here has been supported by USPHS grant HL-21860.

References

Bowman, A. and Gillespie, J. S.: Block of some non-adrenergic inhibitory responses of smooth muscle by a substance from haemolysed erythrocytes. *J. Physiol.* (Lond.) **328:**11–25, 1982.

Brain, S. D., Williams, T. J., Tippins, J. R., Morris, H. R., and MacIntyre, I.: Calcitonin gene-related peptide is a potent vasoliator. *Nature* **313:**54–56, 1985.

Brayden, J. E. and Bevan, J. A.: Neurogenic muscarinic vasodilation in the cat. An example of endothelial cell independent cholinergic relaxation. *Circ. Res.* **56:**205–211, 1985.

Busse, R., Trogisch, G., and Bassenge, E.: The role of endothelium in control of vascular tone. *Basic Res. Cardiol.* **80:**475–490, 1985.

Chand, N. and Altura, B. M.: Acetylcholine and bradykinin relax intrapulmonary arteries by acting on endothelial cells: Role in lung vascular diseases. *Science* **213:**1376–1379, 1981.

Cherry, P. D., Furchgott, R. F., and Zawadzki, J. V.: The indirect nature of bradykinin relaxation of isolated arteries: Endothelial dependent and independent components. *Fed. Proc.* **40:**589, 1981.

Cherry, P. D., Furchgott, R. F., Zawadzki, J. V., and Jothianandan, D.: The role of endothelial cells in the relaxation of isolated arteries by bradykinin. *Proc. Natl. Acad. Sci. USA* **79:**2105–2110, 1982.

Cherry, P. D., Furchgott, R. F., and Zawadzki, J. V.: The endothelium-dependent relaxation of vascular smooth muscle by unsaturated fatty acids. *Fed. Proc.* **42:**619, 1983.

Cocks, T. M. and Angus, J. A.: Endothelium-dependent relaxation of coronary arteries by noradrenaline and serotonin. *Nature* **305:**627–630, 1983.

Cocks, T. M. and Angus, J. A.: Bioassay of the Release of Endothelium-Derived Relaxing Factor (EDRF) from Isolated Endothelial Cells *In Vitro,* in *Vascular Neuroeffector Mechanisms* (Bevan, J., Godfraind, T., Maxwell, R., Stoclet, J. C., and Worcel, M., eds.) Elsevier, Amsterdam, 1985.

Cohen, R. A., Shepherd, J. T., and Vanhoutte, P. M.: 5-Hydroxytryptamine can mediate endothelium-dependent relaxation of coronary arteries. *Am. J. Physiol.***245:**H1077–H1080, 1983.

Craven, P. A. and De Rubertis, F. R.: Restoration of the responsiveness of purified guanylate cyclase to nitrosoguanidine, nitric oxide, and related activators by heme and hemeproteins. *J. Biol. Chem.* **253:**8433–8443, 1978.

Davies, J. M. and Williams, K. I.: Relaxation of the rat aorta by vasoactive intestinal polypeptide is endothelial cell dependent. *J. Physiol.* (Lond.) **343:**65P, 1983.

De May, J. and Vanhoutte, P. M.: Endothelium and relaxation of isolated canine arteries. *Pharmacologist* **22:**282, 1980.

De Mey, J. and Vanhoutte, P. M.: Role of the intima in cholinergic and purinergic relaxation of isolated canine femoral arteries. *J. Physiol.* **316:**437–455, 1981.

De Mey, J., Claeys, M., and Vanhoutte, P. M.: Endothelium-dependent inhibitory effects of acetylcholine, adenosine triphosphate, thrombin and arachidonic acid in the canine femoral artery. *J. Pharmacol. Exp. Ther.* **222:**166–173, 1982.

Diamond, J. and Chu, E. B.: Possible role for cyclic GMP in endothelium-dependent relaxation of rabbit aorta by acetylcholine. Comparison with nitroglycerin. *Res. Comm. Chem. Pathol. Pharmacol.* **41:**369–381, 1983.

Forstermann, U. and Neufang, B.: Endothelium-dependent vasodilation by melittin: Are lipoxygenase products involved? *Am. J. Physiol.* **249:**1114–1119, 1985.

Forstermann, U., Trogisch, G., and Busse, R.: Species-dependent differences in the value of endothelium-derived vascular relaxing factor. *Eur. J. Pharmacol.* **106:**639–643, 1984.

Forstermann, U., Mulsch, A., Bohme, E., and Busse, R.: Stimulation of soluble guanylate cyclase by an acetylcholine-induced endothelium-derived factor from rabbit and canine arteries. *Circ. Res.* **58:**531–538, 1986.

Furchgott, R. F.: Pharmacology of vascular smooth muscle. *Pharmacol. Rev.* **7:**183–265, 1955.

Furchgott, R. F.: The requirement for endothelial cells in the relaxation of arteries by acetylcholine and some other vasodilators. *Trends Pharmacol. Sci.* **2:**173–176, 1981:

Furchgott, R. F.: Acetylcholine and Blood Vessel Relaxation: Complications and Clarifications, in *Autonomic Pharmacology* vol. 2 (Kalsner, S., ed.) Urban and Schwartzenberg, Baltimore, 1982.

Furchgott, R. F.: Role of endothelium in responses of vascular smooth muscle. *Circ. Res.* **53:**557–573, 1983.

Furchgott, R. F.: Role of endothelium in responses of vascular smooth muscle to drugs. *Ann. Rev. Pharmacol. Toxicol.* **24:**175–197, 1984.

Furchgott, R. F.: Will Pharmacological Procedures for Receptor Classification Continue To Be Useful in the Future? Recent Research Supporting an Affirmative Answer, in *Perspectives on Receptor Classification* (Black, J., Gerskowitch, P., and Jenkinson, D., eds.) Alan R. Liss, New York, 1986.

Furchgott, R. F.: Studies on Relaxation of Rabbit Aorta by Sodium Nitrite: The Basis for the Proposal that the Acid-Activatable Inhibitory Factor from Bovine Retractor Penis Is Inorganic Nitrite and the Endothelium-Derived Relaxing Factor Is Nitric Oxide, in *Mechanisms of Vasodilation* (Vanhoutte, P. M., ed.) Raven, New York, 1987, in press.

Furchgott, R. F. and Bhadrakom, S.: Reactions of strips of rabbit aorta to epinephrine, isopropylarterenol, sodium nitrite and other drugs. *J. Pharmacol. Exp. Ther.* **108:**129–143, 1953.
Furchgott, R. F. and Cherry, P. D.: The muscarinic receptor of vascular endothelium that subserves vasodilation. *Trends Pharmacol. Sci.* (suppl.): 42–45, 1984.
Furchgott, R. F. and Khan, M. T.: Some properties of endothelium-derived relaxing factor (EDRF) released from endothelial cells of rabbit aorta. *Fed. Proc.* **45:**425, 1986.
Furchgott, R. F. and Zawadzki, J. V.: The obligatory role of endothelial cells in the relaxation of arterial smooth muscle by acetylcholine. *Nature* **288:**373–376, 1980a.
Furchgott, R. F. and Zawadzki, J. V.: ATP relaxes rabbit aortic smooth muscle by both an indirect action via endothelial cells and a direct action. *Pharmacologist* **22:**271, 1980b.
Furchgott, R. F., Zawadzki, J. V., and Cherry, P. D.: Role of Endothelium in the Vasodilator Response to Acetylcholine, in *Vasodilation* (Vanhoutte, P. and Leusen, I., eds.) Raven, New York, 1981.
Furchgott, R. F., Cherry, P. D., and Zawadzki, J. V.: Endothelium-Dependent Relaxation of Arteries by Acetylcholine, Bradykinin and Other Agents, in *Vascular Neuroeffector Mechanisms* (Bevan, J., ed.) Raven, New York, 1983.
Furchgott, R. F., Cherry, P. D., and Zawadzki, J. V.: Endothelial cells as mediators of vasodilation of arteries. *J. Cardiovasc. Pharmacol.* **6** (suppl. 2)**:**S336–S344, 1984a.
Furchgott, R. F., Jothianandan, D., and Cherry, P. D.: Endothelium-Dependent Responses: The Last Three Years, in *Vasodilator Mechanisms* (Vanhoutte, P. M. and Vatner, S. F., eds.) Karger, Basel, 1984b.
Furchgott, R. F., Martin, W., Jothianandan, D., and Villani, G. M.: Comparison of Endothelium-Dependent Relaxation by Acetylcholine and Endothelium-Independent Relaxation by Light in the Rabbit Aorta, in *Proceedings of the International Congress of Pharmacology* vol. 1 (Paton, W., Mitchell, J., and Turner, P., eds.) Macmillan, London, 1984c.
Furchgott, R. F., Martin, W., Cherry, P. D., Jothianandan, D., and Villani, G. M.: Endothelium-Dependent Relaxation, Photorelaxation and Cyclic GMP, in *Vascular Neuroeffector Mechanisms* (Bevan, T., Godfraind, R., Maxwell, R., Stoclet, J. C., and Worcel, M., eds.) Elsevier, Amsterdam, 1985.
Goldberg, N. D., Graff, C., Haddox, M. K., Stephenson, J. H., Glass, D. B., and Moser, N. E.: Redox modulation of splenic cell soluble guanylate cyclase activity: activation by hydrophillix and hydrophobic antioxidants represented by ascorbic and dehydroascorbic acids, fatty acid hydroperoxides, and prostaglandin endoperoxides. *Adv. Cyclic Nucleotide Res.* **9:**101–130, 1978.

Gordon, J. L. and Martin, W.: Endothelium-dependent relaxation of the pig aorta: Relationship to stimulation of Rb efflux from isolated endothelial cells. *Br. J. Pharmacol.* **79:**531–541, 1983.

Griffith, T. M., Edwards, D. H., Lewis, M. J., Newby, A. C., and Henderson, A. H.: The nature of the endothelium-derived vascular relaxant factor. *Nature* **308:**645–647, 1984a.

Griffith, T. M., Edwards, D. H., Lewis, M. J., and Henderson, A. H.: Ergometrine-induced arterial dilatation: An endothelium-mediated effect. *J. Mol. Cell Cardiol.* **16:**479–482, 1984b.

Griffith, T. M., Edwards, D. H., Lewis, M. J., and Henderson, A. H.: Evidence that cyclic guanosine monophosphate (cGMP) mediates endothelium-dependent relaxation. *Eur. J. Pharmacol.* **112:**195–202, 1985.

Griffith, T. M., Edwards, D. H., Newby, A. C., Lewis, M. J., and Henderson, A. H.: Production of endothelium derived relaxant factor is dependent on oxidative phosphorylation and extracellular calcium. *Cardiovasc. Res.* **0:**7–12, 1986.

Gruetter, C. A., Gruetter, D. Y., Lyon, J. E., Kadowitz, P. J., and Ignarro, L. F.: Relationship between cyclic guanosine 3′:5′-monophosphate formation and relaxation of coronary arterial smooth muscle by glyceryl trinitrate, nitroprusside, nitrite and nitric oxide: Effects of methylene blue and methemoglobin. *J. Pharmacol. Exp. Ther.* **219:**181–186, 1981.

Gryglewski, P. J., Palmer, R. M., and Moncado, S. A.: Superoxide anion is involved in the breakdown of endothelium-derived relaxing factor. *Nature* **320:**454–456, 1986.

Hidaka, H. and Asano, T.: Stimulation of platelet guanylate cyclase by unsaturated fatty acid peroxides. *Proc. Natl. Acad. Sci. USA* **74:**3657–3661, 1977.

Holzmann, S.: Endothelium-induced relaxation by acetylcholine associated with larger rises in cyclic GMP in coronary arterial strips. *J. Cyclic Nucleotide Res.* **8:**409–419, 1982.

Hume, W. R., de la Lande, I. S., and Waterson, J. G.: Effect of acetylcholine on response of the isolated rabbit ear artery to stimulation of the perivascular sympathetic nerves. *Eur. J. Pharmacol.* **17:**227–233, 1972.

Ignarro, L. J., Burke, T. M., Wood, K. S., Wolin, M. S., and Kadowitz, P. J.: Association between cyclic GMP accumulation and acetylcholine-elicited relaxation of bovine intrapulmonary artery. *J. Pharmacol. Exp. Ther.* **228:**682–690, 1984.

Ignarro, L. J., Harbison, R. G., Wood, K. S., and Kadowitz, P. J.: Activation of purified soluble guanylate cyclase by endothelium-derived relaxing factor from intrapulmonary artery and vein: Stimulation by acetylcholine, bradykinin and arachidonic acid. *J. Pharmacol. Exp. Ther.* **237:**893–900, 1986.

Katusic, Z. S., Shepherd, J. T., and Vanhoutte, P. M.: Vasopressin causes endothelium-dependent relaxation of the canine basilar artery. *Circ. Res.* **55:**575–579, 1984.

Khan, M. T., Furchgott, R. F., Hung, I-L., and Jothianandan, D.: Endothelium-dependent and independent vasodilation in rat aorta and mesenteric vasculature. *Fed. Proc.* **45:**425, 1986.

Loffelholz, K. and Muscholl, E.: A muscarinic inhibition of noradrenaline release evoked by postganglionic sympathetic nerve stimulation. *Naunyn Schmiedebergs Arch. Pharmacol.* **265:**1–15, 1969.

Martin, W., Villani, G. M., Jothianandan, D., and Furchgott, R. F.: Selective blockade of endothelium-dependent and glyceryl trinitrate-induced relaxation by hemoglobin and by methylene blue in the rabbit aorta. *J. Pharmacol. Exp. Ther.* **232:**708–716, 1985a.

Martin, W., Villani, G. M., Jothianandan, D., and Furchgott, R. F.: Blockade of endothelium-dependent and glyceryl trinitrate-induced relaxation of rabbit aorta by certain ferrous hemoproteins. *J. Pharmacol. Exp. Ther.* **233:**679–685, 1985b.

Martin, W., Furchgott, R. F., Villani, G. M., and Jothianandan, D.: Depression of contractile responses in rat aorta by spontaneously released endothelium-derived relaxing factor (EDRF). *J. Pharmacol. Exp. Ther.* **237:**529–538, 1986a.

Martin, W., Furchgott, R. F., Villani, G. M., and Jothianandan, D.: Phosphodiesterase inhibitors induce endothelium-dependent relaxation of rat and rabbit aorta by potentiating the effects of spontaneously released endothelium-derived relaxing factor (EDRF). *J. Pharmacol. Exp. Ther.* **237:**539–547, 1986b.

Murad, F., Mittal, C. K., Arnold, W. P., Katsuki, S., and Kimura, H.: Guanylate cyclase: Activation by azide, nitro compounds, nitric oxide, and hydroxyl radical and inhibition by hemoglobin and myoglobin. *Adv. Cyclic Nucleotide Res.* **9:**145–158, 1978.

Murad, F., Arnold, W. P., Mittal, C. K., and Braughier, J. M.: Properties and regulation of guanylate cyclase and some proposed functions of cyclic GMP. *Adv. Cyclic Nucleotide Res.* **11:**175–204, 1979.

Peach, M. J., Loeb, A. L., Singer, H. A., and Saye, J.: Endothelium-derived vascular relaxing factor. *Hypertension* **7** (suppl. l):I94–I100, 1985.

Rand, M. J., and Varma, B.: The effects of cholinominetic drugs on responses to sympathetic nerve stimulation and noradrenaline in the rabbit ear artery. *Br. J. Pharmacol.* **38:**758–770, 1970.

Rapoport, R. M. and Murad, F.: Agonist-induced endothelium-dependent relaxations in rat thoracic aorta may be mediated through cGMP. *Circ. Res.* **52:**352–357, 1983.

Rapoport, R. M., Draznin, M. B., and Murad, F.: Endothelium-dependent relaxation in rat aorta may be mediated through cyclic GMP-dependent protein phosphorylation. *Nature* **306:**174–176, 1983.

Rubanyi, G. M. and Vanhoutte, P. M.: Superoxide anions and hyperoxia inactivate endothelium-derived relaxing factor. *Am. J. Physiol.* **250:**H822–H827, 1986.

Rubanyi, G. M., Lorenz, R. R., and Vanhoutte, P. M.: Bioassay of endothelium-derived relaxing factor(s): Inactivation by catecholamines. *Am. J. Physiol.* **215:**H1077–H1080, 1985.

Satoh, H. and Inui, J.: Endothelial cell-dependent relaxation and contraction induced by histamine in the isolated guinea-pig pulmonary artery. *Eur. J. Pharmacol.* **97:**321–324, 1984.

Singer, H. A. and Peach, M. J.: Calcium and endothelial-mediated vascular smooth muscle relaxation in rabbit aorta. *Hypertension* **4** (suppl. 2): II19–II25, 1982.

Singer, H., A. and Peach, M. J.: Endothelium-dependent relaxation of rabbit aorta. I. Relaxation stimulated by arachidonic acid (AA). *J. Pharmacol. Exp. Ther.* **226:**790–795, 1983.

Steinsland, O. S., Furchgott, R. F., and Kirpekar, S. M.: Inhibition of adrenergic neurotransmission by parasympathomimetics in the rabbit ear artery. *J. Pharmacol. Exp. Ther.* **184:**346–356, 1973.

Van de Voorde, J. and Leusen, I.: Role of endothelium in the vasodilator responses of rat thoracic aorta to histamine. *Eur. J. Pharmacol.* **87:**113–120, 1983.

Vanhoutte, P. M.: Inhibition by acetylcholine of adrenergic neurotransmission in vascular smooth muscle. *Circ. Res.* **34:**317–326, 1974.

Vanhoutte, P. M.: Cholinergic inhibition of adrenergic transmission. *Fed. Proc.* **36:**2444–2449, 1977.

Zawadzki, J. V., Cherry, P. D., and Furchgott, R. F.: Comparison of endothelium-dependent relaxation of rabbit aorta by A23187 and by acetylcholine. *Pharmacologist* **22:**271, 1980.

Zawadzki, J. V., Furchgott, R. F., and Cherry, P.: The obligatory role of endothelial cells in the relaxation of arterial smooth muscle by substance P. *Fed. Proc.* **40:**689, 1981.

Zawadzki, J. V., Furchgott, R. F., and Cherry, P. D.: Endothelium-dependent relaxation of arteries by acta-substance P, kassinin and octa-cholecystokinin.; *Fed. Proc.* **42:**619, 1983.

Chapter 2

Endothelium-Dependent Contractions in Veins and Arteries

Paul M. Vanhoutte

1. Introduction

While searching for possible heterogeneity of endothelium-dependent relaxations to acetylchloride, adenosine diphosphate, arachidonic acid, and thrombin, the author's laboratory discovered that the endothelial cells not only play an obligatory role in a number of dilator responses (Furchgott and Zawadzki, 1980; Furchgott, 1983, 1984; Vanhoutte et al., 1986), but also can induce contractions of isolated blood vessels when exposed to certain stimuli (De Mey and Vanhoutte, 1982, 1983). This chapter summarizes the data demonstrating that endothelial cells can indeed produce at least two different types of constricting substances (endothelium-derived contracting factors) (*see* Vanhoutte, 1987a).

2. Metabolite(s) of Arachidonic Acid ($EDCF_1$)

2.1. Veins

In canine systemic and pulmonary veins, arachidonic acid augmented the contractions evoked by norepinephrine; the augmentation was not observed in preparations in which the endothelium had been removed mechanically (Fig. 1; De Mey and Vanhoutte, 1982). This observation was the first demonstration that the presence of endothelial cells can augment, rather than depress, contractile responses of isolated blood vessels.

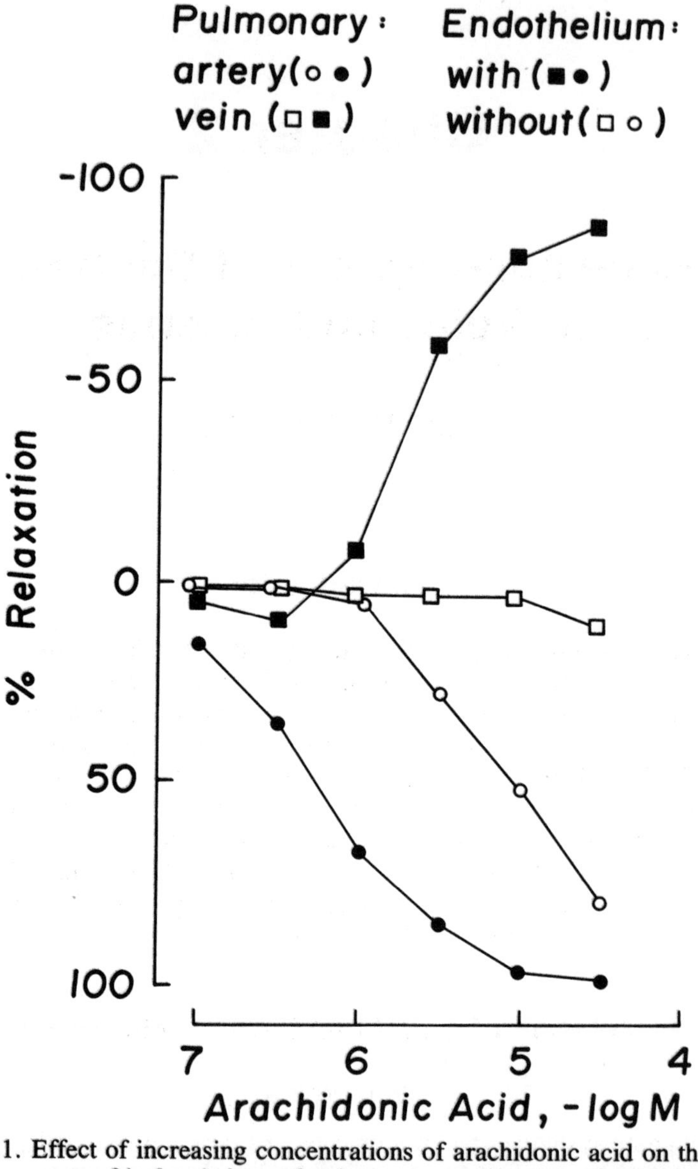

Fig. 1. Effect of increasing concentrations of arachidonic acid on the contractile responses of isolated rings of pulmonary arteries and veins of the same dogs to norepinephrine (ED_{30}). Rings with and without endothelium were studied in parallel. Data are expressed as percentage depression of the response to norepinephrine, and shown as means. SEM omitted for the sake of clarity (from De Mey and Vanhoutte, 1982; from Vanhoutte and McGoon, 1986, with permission).

The endothelium-dependent facilitation by arachidonic acid of the responses of systemic and pulmonary veins was prevented by inhibitors of the enzyme cyclooxygenase, but not by those of prostacyclin-synthetase, thromboxane-synthetase, or lipoxygenase (Fig. 2; Miller and Vanhoutte, 1985). These findings indicated that venous endothelial cells can metabolize arachidonic acid into a vasoconstrictor prostanoid, other than prostacyclin and thromboxane A_2. Since the liberation of arachidonic acid occurs under many physiological and pathological conditions, they suggested the possibility that the endothelium could become a source of vasoconstrictor signals. This appears to be the case in cerebral arteries under normal conditions (*see* below and chapter by Katusic and Shepherd, in this volume), and in systemic arteries, in certain types of essential hypertension (*see* chapter by Luscher and Vanhoutte).

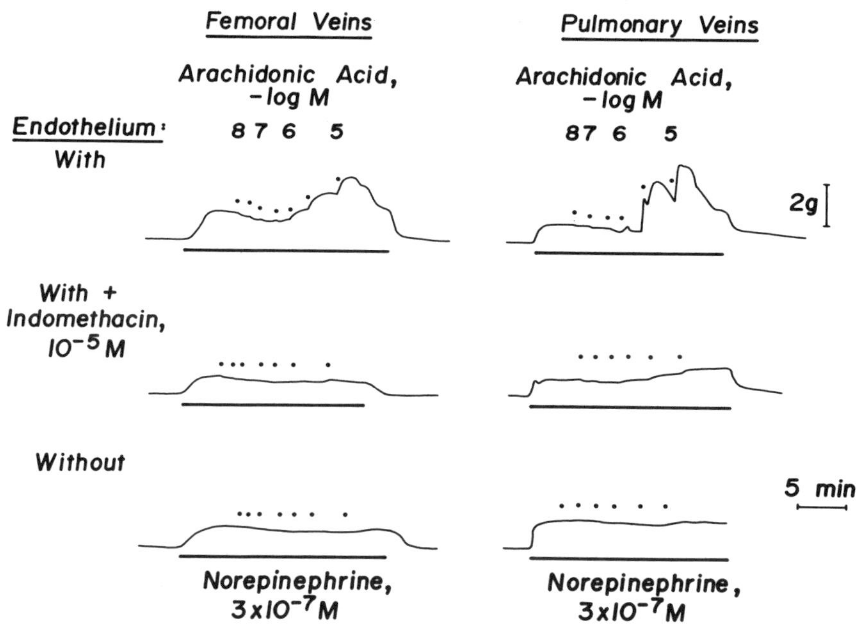

Fig. 2. Isometric tension recordings in canine femoral (left) and pulmonary (right) veins. Arachidonic acid evoked contractions in both veins contracted with the ED_{30} of norepinephrine (upper). In pulmonary vein, relaxations and/or transient contractions were often observed at low concentrations of arachidonic acid. Contractions to arachidonic acid were eliminated in both vessels by indomethacin (10^{-5} *M*; middle) or removal of endothelium (lower) (from Miller and Vanhoutte, 1985, with permission).

2.2. *Cerebral Arteries*

2.2.1. Arachidonic Acid

In the basilar artery of the dog, arachidonic acid caused endothelium-dependent contractions, which are prevented by inhibitors of cyclooxygenase (Shirahase et al., 1986; Katusic et al., 1987). The endothelium-dependent contraction to the fatty acid can be antagonized by inhibitors of thromboxane-synthetase and thromboxane receptors, suggesting that thromboxane A_2 mediates the response.

2.2.2. Ca^{2+} Ionophores

In the canine basilar artery, the Ca^{2+} ionophore A23187 causes endothelium-dependent contractions that are prevented by inhibitors of cyclooxygenase and thromboxane-synthetase (Katusic et al., 1986). This suggests that the increased cytosolic Ca^{2+} concentration caused by the ionophore activates the production of endogenous arachidonic acid; this is then metabolized further by the endothelial cells to the thromboxane A_2, which in turn activates the contractile process in the cerebrovascular smooth muscle. Alternatively, as is the case for endothelium-derived relaxing factor, the release of the prostanoid endothelium-derived contracting factors may be a Ca^{2+}-dependent process (*see* chapter by Rubanyi et al.).

2.2.3. Acetylcholine

In the canine basilar artery, increasing concentrations of acetylcholine cause weak, endothelium-dependent contractions (Katusic et al., 1987); repeated exposures of a single dose of the cholinergic transmitter also cause contractions, which disappear after removal of the intima (Usui et al., 1983; Shirahase et al., 1986). In both cases, the contractions are prevented by inhibition of cyclooxygenase; the prostaglandin involved may (Usui et al., 1983; Shirahase et al., 1986) or may not (Katusic et al., 1987) be thromboxane A_2.

2.2.4. Potassium

Moderate increases (3–20 m*M*) in the extracellular potassium-concentration induce rhythmic contractions of isolated basilar arteries of the dog that are prevented by the removal of the endothelium, inhibitors of cyclooxygenase and Ca^{2+} antagonists (Katusic et al., 1985); the rhythmic activity was augmented by prostaglandin E_2 and $F_{2\alpha}$, but abolished by prostacyclin (Katusic et al., 1985; *see* chapter by Katusic and Shepherd, in this volume). These results suggested that increasing the extracellular concentration of potassium activates endothelial cyclooxygenase and augments the release of vasoconstrictor prostaglandins, and of prostacyclin; the latter exerts an inhibitory feedback on the activated cyclooxygenase, which ensures the rhythmic pattern of contraction.

2.2.5. Stretch

In canine basilar arteries with endothelium, rapid stretch evokes a contraction, which is sustained for several minutes. The response is prevented by removal of the endothelium, the inhibitor of cyclooxygenase indomethacin, and by the Ca^{2+} antagonist diltiazem (Katusic et al., 1987; *see* chapter by Katusic and Shepherd, in this volume). These observations suggest that, in cerebral arteries, endothelial cells can act as sensors of changes in distension, and that the release of a prostanoid endothelium-derived contracting factor may help to explain physical autoregulation in cerebral blood vessels (Fig. 3; Vanhoutte, 1987a,b).

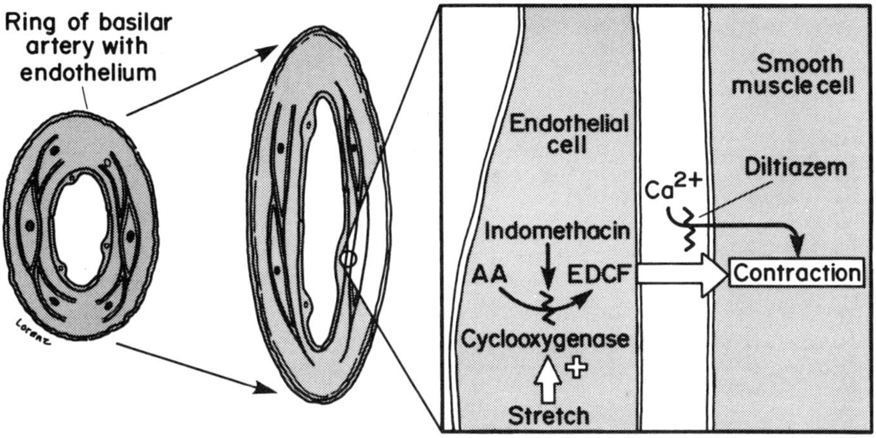

Fig. 3. In isolated basilar arteries of the dog sudden stretch causes the release from the endothelium of vasoconstrictor prostanoids that activate the entry of Ca^{2+} into the vascular smooth muscle. If this were to occur in the intact organism, it could help to explain autoregulation (from Vanhoutte, 1987, with permission).

3. Hypoxia

In a variety of arteries of the dog, including peripheral, coronary, and cerebral arteries, hypoxia causes endothelium-dependent increases in tension (Fig. 4; De Mey and Vanhoutte, 1983). Hypoxia prevents endothelium-dependent relaxations to acetylcholine, adenosine diphosphate, thrombin, and vasopressin (De Mey and Vanhoutte, 1980, 1983; Katusic and Vanhoutte, 1986). However, inhibition by hypoxia of the basal release of endothelium-derived relaxing factors (*see* chapter by Martin, in this volume) does not explain the endothelium dependency of the hypoxic facilitation, since inhibitors of the relaxing factors do not prevent it (Fig. 5; Rubanyi and Vanhoutte, 1985). Experiments with strips of canine cor-

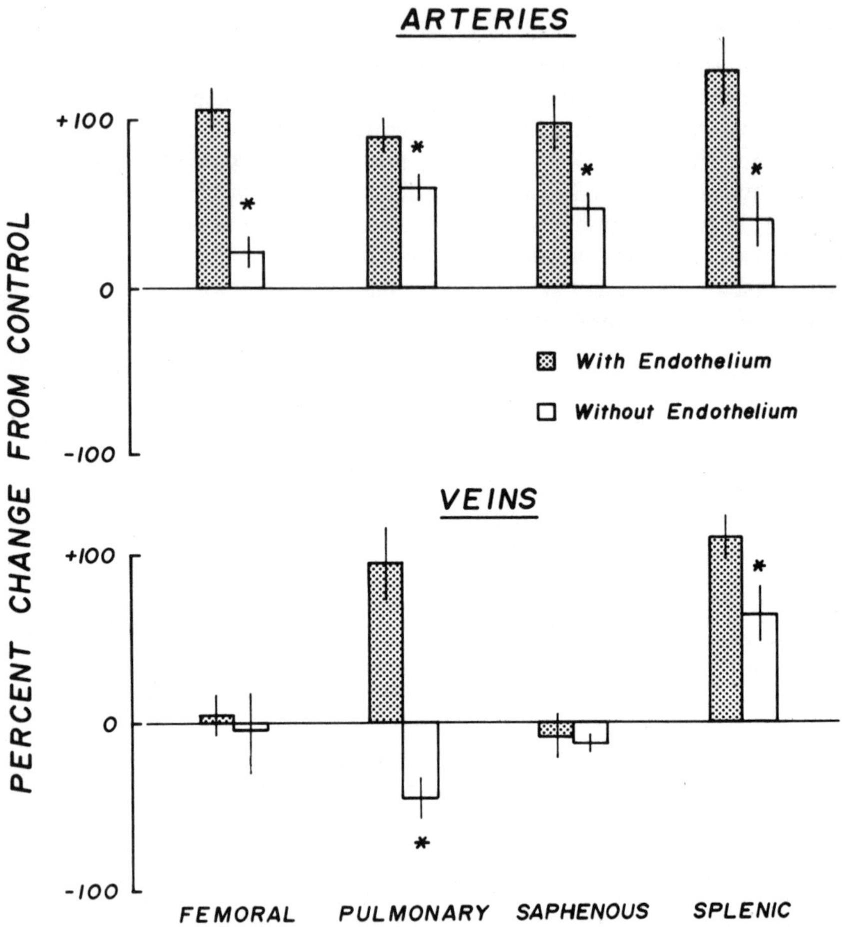

Fig. 4. Effect of anoxia (10 min) on the contractile responses to norepinephrine (ED_{30}) in canine arteries (upper) and veins (lower). Rings with (stippled bars) and without (clear bars) endothelium were studied in parallel. Changes in tension are expressed as percent of the contractile response to norepinephrine, and shown as means ± SEM. *The difference between rings with and without endothelium is statistically significant (from De Mey and Vanhoutte, 1982, with permission of the American Heart Association).

onary arteries without endothelium (in which isometric force is measured) layered (''sandwiched''; see Furchgott and Zawadzki, 1980 and the first chapter in this volume) with strips containing endothelial cells demonstrated that the hypoxic facilitation is caused by the diffusion of a vasoconstrictor substance from the endothelium to the smooth muscle (Fig. 6; Rubanyi and Vanhoutte, 1985).

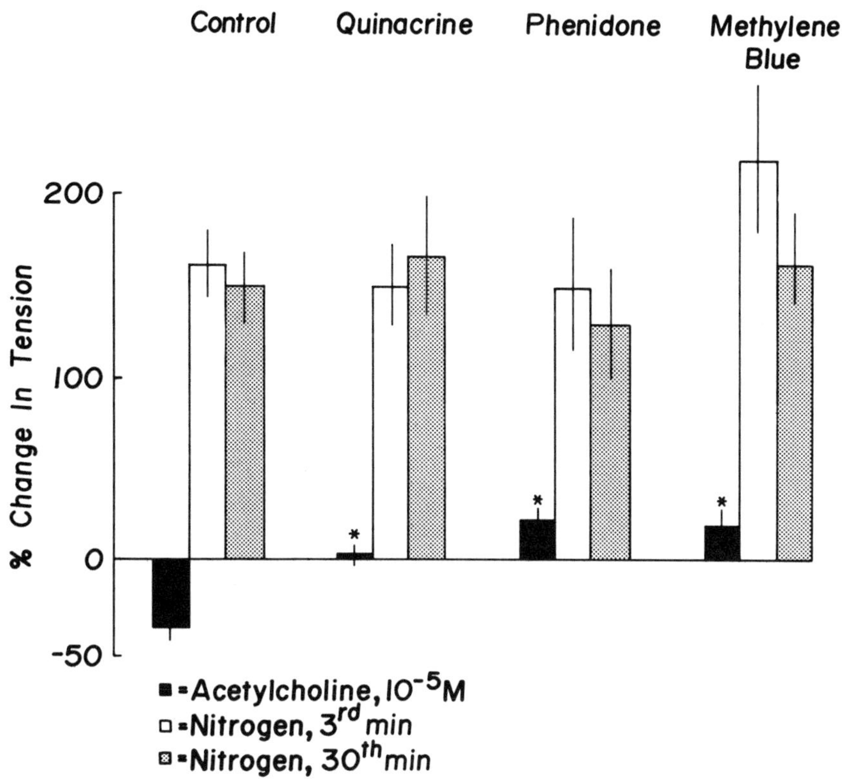

Fig. 5. Effects of quinacrine ($2 \times 10^{-5}M$), phenidone ($5 \times 10^{-5}M$), and methylene blue ($10^{-5}M$) on the relaxation induced by acetylcholine ($10^{-5}M$: filled bars) and the increase in tension caused by anoxia (N_2) of layered preparations (with endothelium) of canine coronary arteries. The preparations were contracted with prostaglandin $F_{2\alpha}$ and shown as means ± SEM. *Differences from control responses are statistically significant ($p < 0.05$) (from Rubanyi and Vanhoutte, 1985, with permission).

The nature of the endothelium-derived contracting factor(s) released from the hypoxic endothelium is unknown. The metabolism of arachidonic acid or other autacoids are not involved (De Mey and Vanhoutte, 1983; Rubanyi and Vanhoutte, 1985; Katusic and Vanhoutte, 1986). The onset of the endothelium-dependent hypoxic contraction is very rapid; upon return to oxygen, a very rapid relaxation ensues. This time course makes it unlikely, at least for the author, that the endothelium-derived contracting factors involved are the same as the peptide released from cultured endothelial cells (Hickey et al., 1985; Gillespie et al., 1986; *see* chapter by Highsmith et al., in this volume). The hypoxic contractions are inhibited by Ca^{2+} antagonists (e.g., Fig. 7, Katusic and Vanhoutte, 1986). To judge

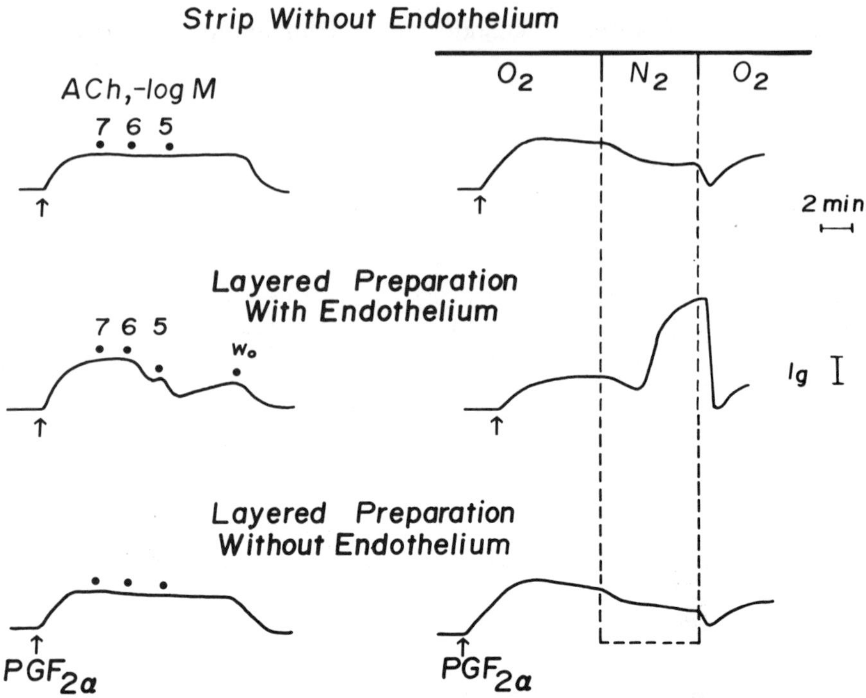

Fig. 6. Effects of acetylcholine (ACh) and anoxia (N_2) on the contractile response to prostaglandin $F_{2\alpha}$; $2 \times 10^{-6} M$) in a circumferential strip of coronary artery (without endothelium) mounted alone (upper) or layered with a longitudinal strip of coronary artery with endothelium (lower) (from Rubanyi and Vanhoutte, 1985, with permission).

from preliminary experiments with layered preparations, these drugs prevent the endothelium-dependent response by their action on vascular smooth muscle, rather than on endothelial cells (Iqbal and Vanhoutte, 1987).

Endothelium-dependent hypoxic contractions (or facilitation of contraction) were not observed in limb veins (De Mey and Vanhoutte, 1982). When venous endothelium was layered with coronary arterial smooth muscle, however, the layered preparation contracted when made hypoxic (Fig. 8; Rubanyi and Vanhoutte, 1985). This, then, demonstrates that the absence of endothelium-dependent hypoxic facilitation in systemic veins is not because of lack of production of endothelium-derived contracting factors, but rather because of a lack of sensitivity of venous smooth muscle to the factor. The pulmonary artery and vein are among the blood vessels in which endothelium-dependent facilitation is prominent (De Mey and Vanhoutte, 1982). Thus, in the pulmonary vasculature, the phenomenon may play a physiological role and contribute to hypoxic pulmonary vasoconstriction (Vanhoutte and McGoon, 1986). In other blood vessels, in particular in cerebral and coronary arteries, the release of endothelium-derived contracting factors during hypoxia may aggregate vasospasm.

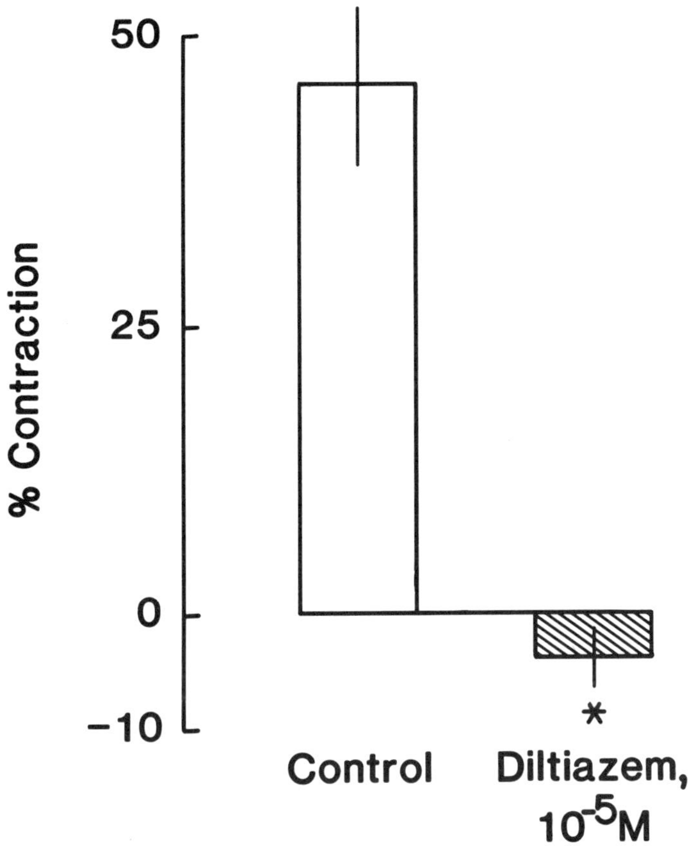

Fig. 7. Effect of diltiazem ($10^{-5}M$) on anoxic contractions of canine basilar arteries with endothelium. Data shown as means ± SEM ($n = 6$) and expressed as percent. The asterisk indicates that the effect of diltiazem was statistically significant ($p < 0.05$) (from Katusic and Vanhoutte, 1986, with permission).

4. Conclusion

A number of physiological or pathological stimuli can cause endothelium-dependent contractions of isolated blood vessels. These contractions can be attributed to the release of vasoconstrictor substances from

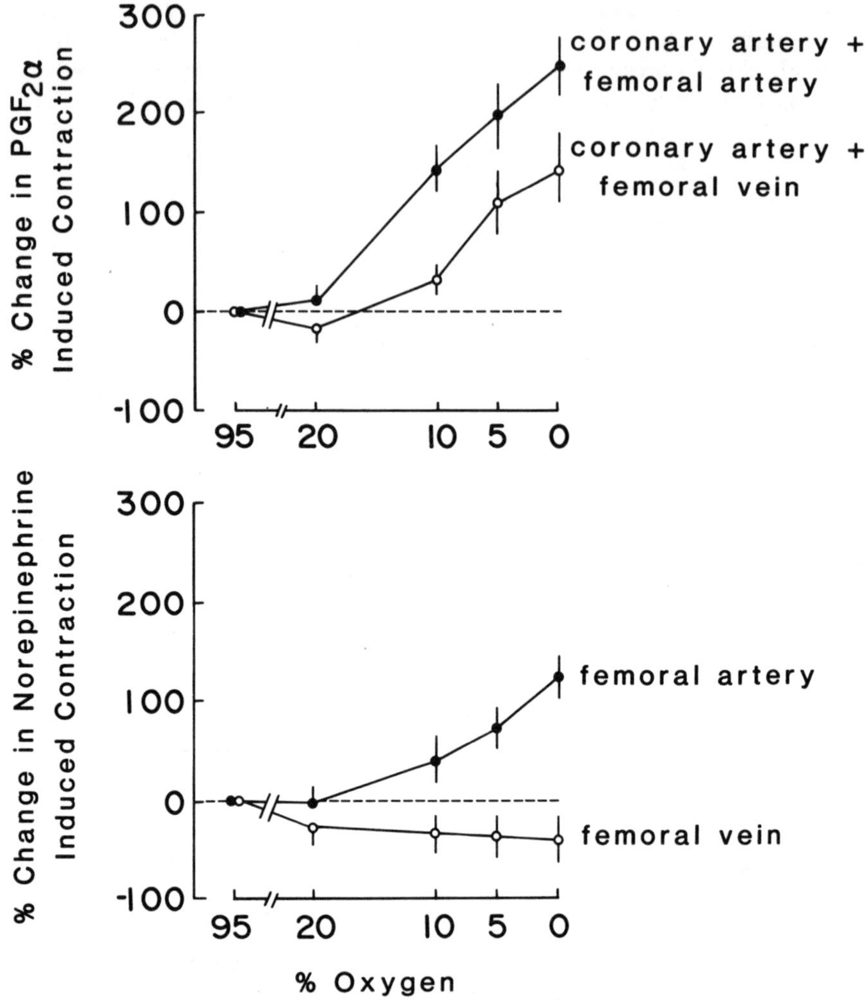

Fig. 8. Upper panel: Effect of stepwise decreases in oxygen content on contractions of circumferential strips of canine coronary arteries (without endothelium) layered with longitudinal strips (with endothelium) of either femoral arteries (●; $n = 4$) or femoral veins (○; $n = 5$). The preparations were contracted with prostaglandin $F_{2\alpha}$ ($2 \times 10^{-6}M$) prior to the exposure to low oxygen. The experiments were performed in the presence of indomethacin ($10^{-5}M$), propranolol ($5 \times 10^{-6}M$), and phentolamine ($10^{-5}M$). The data are expressed as percentage of the initial response to prostaglandin $F_{2\alpha}$ (100% = 2.3 ± 0.4 and 2.2 ± 0.3 g for femoral arteries and veins, respectively), and shown as means ± SEM. Lower panel: Effect of stepwise decreases in oxygen content on contractions of rings with endothelium of femoral arteries (●; $n = 4$) and veins (○; $n = 5$) of the

the endothelial cells. There are at least two different factors involved: one is a product of cyclooxygenase and is prominent in cerebral arteries and in certain hypertensive blood vessels, whereas the other is not a metabolite of arachidonic acid and is released by hypoxic endothelial cells (Fig. 9).

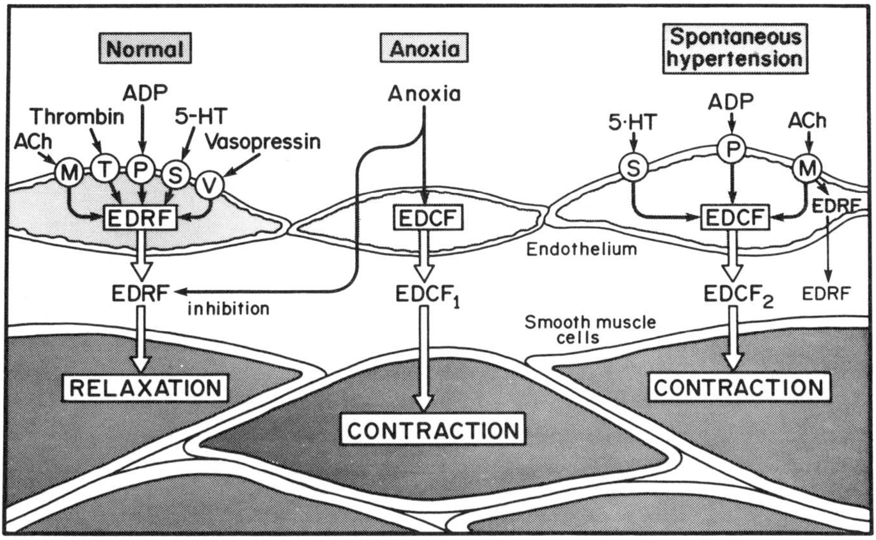

Fig. 9. Under normal conditions (left), a number of vasoactive agonists can act on the endothelial cells to cause the release of endothelium-derived relaxing factor(s), which causes relaxation of underlying smooth muscle cells. Under anoxic conditions (middle), endothelium-dependent relaxations are inhibited, and the anoxic endothelium causes the release of endothelium-derived contracting factor(s), which activates the underlying smooth muscle, which could greatly favor the occurrence of, or prolong, vasospasm. In spontaneous hypertension (right), the endothelium still releases the relaxing factor(s), but it cannot prevent the activation of the smooth muscle by the simultaneously liberated endothelium-derived contracting factor(s). Abbreviations: EDRF, endothelium-derived relaxing factor; $EDCF_1$ and $EDCF_2$, two distinct endothelium-derived contracting factors; 5-HT, serotonin; P, purinergic receptor; S, serotonergic receptor; M, muscarinic receptor; V, vasopressinergic receptor; T, thrombin receptor (from Vanhoutte, 1986, with permission).

← Fig. 8 (*Continued*)

same dog. The rings were contracted with norepinephrine ($3 \times 10^{-7} M$). The experiments were performed in the presence of indomethacin ($10^{-6} M$). The data are expressed as percentage of the initial contractions to norepinephrine (7.2 ± 0.6 at 4.2 ± 0.4 g, for the arteries and the veins, respectively), and shown as means ± SEM (from Rubanyi and Vanhoutte, 1985, with permission).

References

De Mey, J. G. and Vanhoutte, P. M.: Anoxia and endothelium-dependent reactivity of the canine femoral artery. *J. Physiol.* **335:**65–74, 1983.

De Mey, J. G. and Vanhoutte, P. M.: Heterogeneous behavior of the canine arterial and venous wall: Importance of the endothelium. *Circ. Res.* **51:**439–447, 1982.

De Mey, J. and Vanhoutte, P. M.: Interaction between Na^+, K^+ exchanges and the direct inhibitory effect of acetylcholine on canine femoral arteries. *Circ. Res.* **46:**826–836, 1980.

Furchgott, R. F.: Role of endothelium in responses of vascular smooth muscle to drugs. *Circ. Res.* **53:**557–573, 1983.

Furchgott, R. F.: The role of endothelium in the response of vascular smooth muscle to drugs. *Ann. Rev. Pharmacol. Toxicol.* **24:**175–197, 1984.

Furchgott, R. F. and Zawadzki, J. V.: The obligatory role of endothelial cells in the relaxation of arterial smooth muscle by acetylcholine. *Nature* **228:**373–376, 1980.

Gillespie, M. N., Owasoyo, J. O., McMurtry, I. F., and O'Brien, R. F.: Sustained coronary vasoconstriction provoked by a peptidergic substance released from endothelial cells in culture. *J. Pharmacol. Exp. Ther.* **236:**339–343, 1986.

Hickey, K. A., Rubanyi, G., Paul, R. J., and Highsmith, R. F.: Characterization of a coronary vasoconstrictor produced by cultured endothelial cells. *Am. J. Physiol.* **248:**C550–C556, 1985.

Iqbal, A. and Vanhoutte, P. M.: The Ca^{2+}-antagonist flunarizine inhibits endothelium-dependent hypoxic facilitation through an action on vascular smooth muscle. *Fed. Proc.*, in press, 1987.

Katusic, Z. S. and Vanhoutte, P. M.: Anoxic contractions in isolated canine cerebral arteries. Contribution of endothelium-derived factors, metabolites of arachidonic acid and calcium entry. *J. Cardiovasc. Pharmacol.* **8:**S97–S101, 1986.

Katusic, Z. S., Shepherd, J. T., and Vanhoutte, P. M.: Endothelium-dependent contraction to stretch in canine basilar arteries. *Am. J. Physiol.* **252:**H1–H3, 1986.

Katusic, Z. S., Shepherd, J. T., and Vanhoutte, P. M.: Potassium causes endothelium-dependent rhythmic activity in canine basilar artery. *Pharmacologist* **27:**223, 1985.

Katusic, Z. S., Shepherd, J. T., and Vanhoutte, P. M.: Calcium ionophore A23187, arachidonic acid and acetylcholine cause endothelium-dependent contractions in the canine basilar arteries. *Fed. Proc.*, in press, 1986.

Miller, V. M. and Vanhoutte, P. M.: Endothelium-dependent contractions to arachidonic acid are mediated by products of cyclooxygenase in canine veins. *Am. J. Physiol.* **248:**H432–H437, 1985.

Rubanyi, G. M. and Vanhoutte, P. M.: Hypoxia releases a vasoconstrictor substance from the canine vascular endothelium. *J. Physiol.* **364:**45–56, 1985.

Shirahase, H., Usui, H., Kurahashi, K., and Fujiwara, M.: Possible role of endothelial thromboxane A_2 in the resting tone and contractile responses to acetylcholine and arachidonic acid in canine cerebral arteries; submitted for publication, 1986.

Usui, H., Kurahashi, K., Ashida, K., and Fujiwara, M.: Acetylcholine-induced contractile response in canine basilar artery with activation of thromboxane A_2 synthesis sequence. *IRCS Med. Sci. Physiol.* **11**:418–419, 1983.

Vanhoutte, P. M.: Endothelium-Dependent Contractions in Arteries and Veins, in *Blood Vessels*, in press, 1987a.

Vanhoutte, P. M.: Endothelium and the control of vascular tone. *News Physiol. Sci.*, in press, 1987b.

Vanhoutte, P. M. and McGoon, M. D.: The Endothelial Cell, in *Abnormal Pulmonary Circulation* (E. H. Bergofsky, ed.) Churchhill-Livingstone, New York, 1986.

Vanhoutte, P. M., Rubanyi, G. M., Miller, V. M., and Houston, D. S.: Modulation of vascular smooth muscle contraction by the endothelium. *Ann. Rev. Physiol.* **48**:307–320, 1986.

Chapter 3

The Nature of Endothelium-Derived Relaxing Factor

T. M. Griffith and A. H. Henderson

1. Introduction

The phenomenon of endothelium-dependent relaxation, first described in detail by Furchgott and colleagues (Furchgott and Zawadzki, 1980), has now been demonstrated in a wide range of blood vessels from a variety of species studied, including human (for reviews, *see* Furchgott, 1983; Griffith et al., 1985a; Busse et al., 1985). Endothelium-dependent relaxation can be elicited in vivo (Angus et al., 1983) as well as in standard in vitro pharmacological preparations by a number of agents at pharmacological concentrations. The inhibitory potential of the endothelium on vascular smooth muscle can be potent enough to virtually abolish responses in some arteries (Fig. 1) (Griffith et al., 1984a). The important role of endothelium is further emphasized when considering that the total mass of endothelial cells in the human body is approximately equivalent to that of the liver (Gerlach et al., 1985). Initial investigations into the mechanisms by which the phenomenon occurred suggested that it was mediated by a humoral agent, on the basis of "sandwich" experiments in which an aortic strip with intact endothelium could be shown to relax a closely apposed strip in which the endothelium had been removed (Furchgott and Zawadzki, 1980). An alternative suggestion was that the relaxation was brought about by a hyperpolarizing current originating from the endothelium (Chand and Altura, 1981), and endothelium-dependent relaxation is indeed accompanied by hyperpolarization of underlying smooth muscle cells (Bolton et al., 1984). It is now established, however, that endothelium-dependent

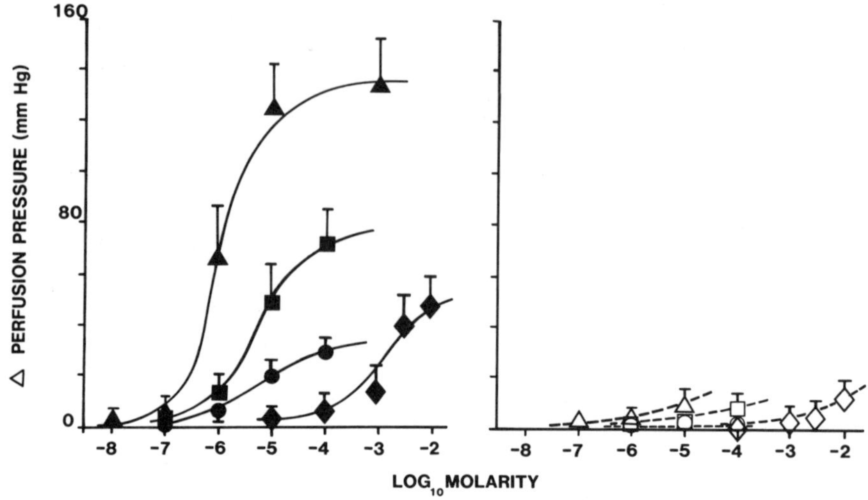

Fig. 1. Cumulative concentration–response curves to histamine (▲), phenylephrine (♦), serotonin (●), and acetylcholine (■) in isolated perfused rabbit coronary arteries denuded of endothelium (left-hand panel, solid lines, and symbols) and with intact endothelium (right-hand panel, dotted lines, and open symbols) (data are mean ± SEM, $n > 12$).

dilatation is mediated by the release from endothelial cells of a humoral agent, endothelium-derived relaxing factor (EDRF) (Griffith et al., 1984b). Much is known about the production and mode of action of EDRF, but its chemical identity has yet to be elucidated. Likewise we stand only on the threshold of understanding its physiological role, which seems likely to prove a central one in vascular control.

2. Physical Properties

The labile nature of EDRF in physiological buffer has been demonstrated by means of bioassay experiments in which a source of EDRF is perfused in series with a vascular preparation from which the endothelium has been removed (Fig. 2). Arteries acting as the source of EDRF have included intact segments of rabbit aorta (Griffith et al., 1984b; Forstermann et al., 1985), canine femoral artery (Forstermann et al., 1985; Rubanyi et al., 1985a), and rat aorta (Long and Stone, 1985); endothelial cell cultures (grown on microcarrier beads) have been used from bovine (Cocks et al., 1985; Busse et al., 1985) or porcine (Gryglewski et al., 1986) aortas. With bioassay techniques, it is possible to demonstrate not only stimulated release of EDRF in response to agents such as acetylcholine, but also a continuous basal release of EDRF that can be maintained for hours (Griffith et al., 1984b; Rubanyi et al., 1985a). From bioassay experiments in which the source of EDRF is rapidly introduced into the perfusion circuit, it is apparent that its dilator action on vascular smooth muscle is of rapid onset, reaching a maximum at about 60 s; the effect is equally rapidly lost when the source of EDRF is removed from the circuit (Fig. 2B). The biological half-life of EDRF (in physiological buffer) has been estimated variously in different laboratories using a number of different bioassay systems and techniques. Since the concentration of EDRF is unknown, estimates of its half-life necessarily depend on using the response to EDRF instead of its actual concentration. They are therefore critically dependent on the relationship between the concentration of EDRF and the response over the range being used. Decay of EDRF is assumed to be exponential (which it may not be). The concentration–response curve is very unlikely to be strictly linear, so that the response cannot be related simply and linearly to the concentration and substituted validly for it. Where $t_{½}$ is estimated from (unidentified) portions of the concentration–response curve, the shortest value will be derived from measurements made in the midportion of such a standard sigmoid-type curve where the slope is at its steepest and approximately linear; this portion of the curve, however, represents the fastest change of response with respect to concentration, and theoretically could lead to an underestimate of the true value for $t_{½}$. Estimations of the $t_{½}$ of EDRF by different workers have ranged from about 6 to 50 s (Griffith et al., 1984b; Forstermann et al., 1985; Cocks et al., 1985; Busse et al., 1985; Rubanyi et al., 1985a; Cocks et al., 1986). Furthermore, the biological half-life may also be affected by a number of other factors, such as agents present in the buffer or released together with EDRF from endothelial cells either in the basal state or when stimulated; and these may either shorten or prolong the half-life, either by direct interaction with EDRF or by modulating other factors that affect the half-life of EDRF. There is, for example, evidence that indicates that superoxide anions interact with the EDRF molecule, in that the enzyme superoxide dismutase, which destroys this free radical by dismutation, prolongs EDRF half-life (Rubanyi and Vanhoutte, 1985a; Gryglewski et al., 1986; Rubanyi and Vanhoutte, 1986). Consistent with this is the observation that the $t_{½}$ of EDRF is longer when experiments are performed at physiological PO_2 levels of about 140 mm Hg than at 500 mm Hg (Forstermann et al., 1985; Rubanyi and Vanhoutte, 1986), although an alternative explanation for these data might be that the sensitivity of the smooth muscle is influenced by PO_2 with consequent changes in the response curve to EDRF and thus in the derivation of a value for $t_{½}$. The

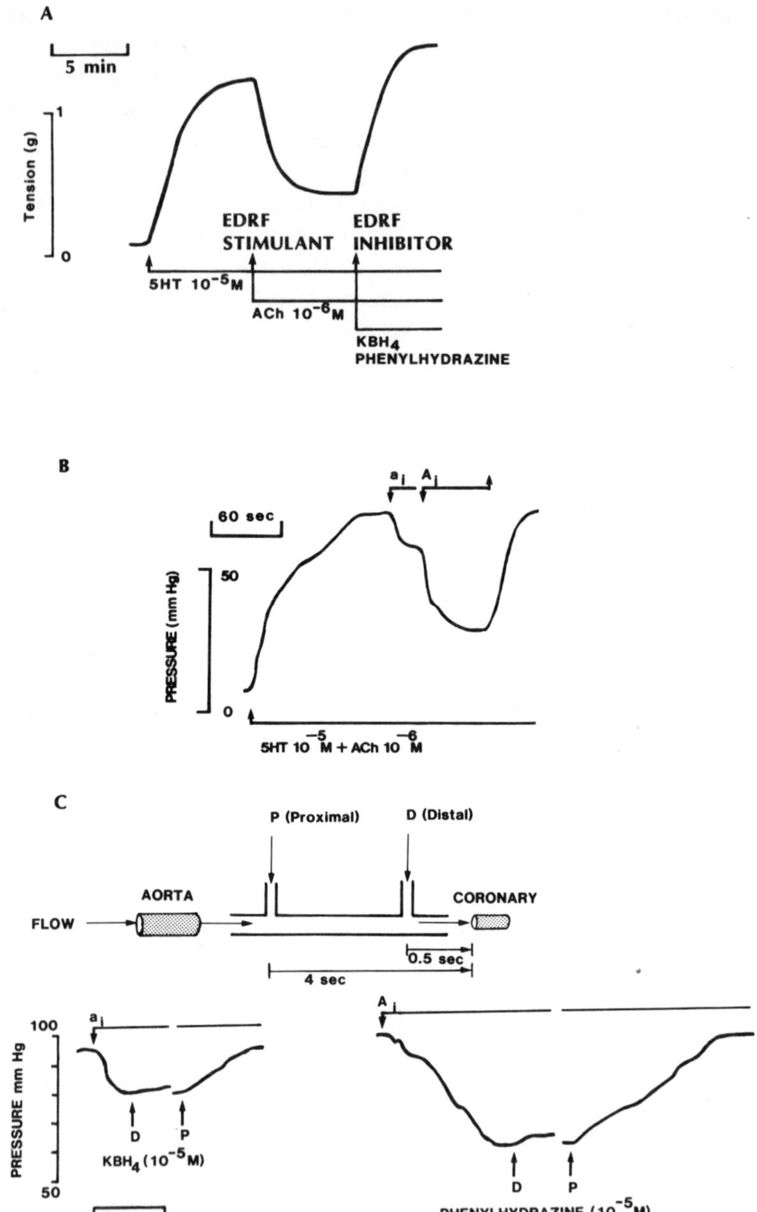

Fig. 2. (A) Diagram of isometric tension responses of rabbit aortic strip with intact endothelium constricted by serotonin (5-HT), showing endothelium-dependent dilatation by the EDRF stimulant, acetylcholine (ACh), which can then

fact that a range of half-lives has been obtained in different laboratories may therefore not be incompatible with the view that EDRF is a unique substance, with a molecular structure that is common to all mammalian species and vessels. This would certainly be the simplest explanation of the current data. Until definitive chemical information about the structure of EDRF is known, however, it remains theoretically possible that EDRF from different species or vessel types may have different chemical structures, or indeed that EDRF is a mixture of several dilator substances.

There is evidence that EDRF possesses a negative charge since its dilator action is abolished by passage through columns of anion, but not cation, exchange resin (Cocks et al., 1985). Passage through lipid absorption columns similarly indicates that it is hydrophilic (Cocks et al., 1985). Preliminary evidence indicates that the estimated $t_{1/2}$ is not significantly affected by (1) temperature change between 30 and 37 °C (Cocks et al., 1985) and (2) visible light at ambient room level (Griffith, 1985).

3. Chemical Properties

3.1. Experimental Approach

Pending the definitive identification of EDRF, some possible clues to (1) its mode of production, (2) its chemical nature, and (3) its mechanism of action have been obtained through the use of pharmacological and chemical probes that inhibit the activity of EDRF (Table 1). The use of isolated intact vascular segments (Fig. 2A) does not allow for discrimination between different sites of action, since such arterial preparations are

← Fig. 2 (*Continued*)

be blocked by EDRF inhibitors, e.g., potassium borohydride (KBH_4) or phenylhydrazine. (B) Representative pressure trace of isolated denuded rabbit coronary artery constricted by serotonin and acetylcholine, used as bioassay for effluent from perfused aorta with endothelium, introduced into the circuit at "a_i" and stimulated to release more EDRF by acetylcholine at "A_i". (C) Diagram of bioassay system, which allows infusion of agents before the aorta (not shown), into intervening tubing past the aorta, proximally (P) (4 s interaction time), or distally (D) (0.5 s interaction time). Representative traces below (left-hand trace) illustrate introduction of effluent from unstimulated aorta resulting in dilatation caused by basal EDRF release and (right-hand trace) effluent from acetylcholine-stimulated aorta (A_i) resulting in greater dilatation caused by stimulated EDRF release; in each case the dilatation can be inhibited by infusion of potassium borohydride (KBH_4) or phenylhydrazine into intervening tubing to a much greater extent when infused proximally than distally.

a composite of effector cells (endothelium), mediator (EDRF), and target cells (smooth muscle). Bioassay techniques (Fig. 2B) overcome this difficulty by providing for temporal and spatial separation of the release of EDRF and the relaxation of the smooth muscle. Indeed, the very fact that such experiments can be performed virtually excludes the possibility that EDRF is a free radical, as originally proposed (Furchgott and Zawadzki, 1980), since biological free radicals are generally extremely short-lived. Its lability also rules out many known endogenous vasodilators such as adenosine that are intrinsically stable. With cascade bioassay systems, putative inhibitors may be infused at different sites in the circuit (Fig. 2C), so distinguishing between (1) compounds that inhibit the release of EDRF (Edwards et al., 1985a; Griffith et al., 1986), (2) compounds that interact with EDRF (or with agents that influence its half-life) in transit (Griffith et al., 1984b; Griffith, 1985; Rubanyi et al., 1985a; Rubanyi and Vanhoutte, 1986), and (3) compounds that inhibit the response of the smooth muscle to EDRF (Griffith et al., 1985a,b). The last two possibilities can be distinguished from one another by infusing inhibitors at two different infusion sites in the intervening tubing and comparing their effects at steady state: the two infusion sites allow for different interaction times in transit, but provide for the same exposure of vascular smooth muscle to the infused agent (Griffith et al., 1984b; Griffith, 1985; Rubanyi et al., 1985a; Rubanyi and Vanhoutte, 1986). Appropriate control experiments exclude direct effects of these agents (at the concentrations and exposure times used) on vascular smooth muscle (Griffith et al., 1984b; Rapoport et al., 1985).

3.2. Inhibition of Antioxidants

Knowledge of which agents inhibit EDRF by interaction in transit (Table 1), and which do not, offers some insight into its chemical nature. Effective agents include borohydride and phenylhydrazine (Fig. 2), phenidone, hydroquinone, nordihydroguaiaretic acid, thiol reagents, ascorbate, and norepinephrine (Griffith et al., 1984b; Griffith, 1985; Rubanyi et al., 1985a). Consistent with such data from bioassay experiments, the endothelium-dependent relaxations, whether caused by basal or stimulated release of EDRF, also are inhibited by the same compounds in intact vascular segments (Furchgott and Zawadzki, 1980; Furchgott, 1983; Griffith et al., 1984a,b; Griffith, 1985). An important common feature of these inhibitors is the possession of antioxidant properties, and the simplest interpretation of the data is that EDRF itself is inactivated by antioxidants. A further common feature of most, if not all, of the agents shown to be effective as inhibitors in this way is that they react with carbonyl groups, leading to the suggestion that EDRF may possess a carbonyl group at or near its active site (Griffith et al., 1984b).

An alternative explanation for the data is that these antioxidants act on some other agent arising during the experiments that influence the half-life of EDRF, although the only evidence for such modulation of half-life points to an inactivating effect of oxygen-derived free radicals, in that superoxide dismutase prolongs half-life (Rubanyi and Vanhoutte, 1985a; Gryglewski et al., 1986; Rubanyi and Vanhoutte, 1986). At first sight, oxygen-derived free radicals might have been expected to have the opposite effects to antioxidants. The opposing effects of superoxide dismutase and of the antioxidants that interact with EDRF, which are also known to scavenge superoxide anions, are thus difficult to reconcile on this basis unless the superoxide anion is able to reduce EDRF under the experimental conditions.

Many of the antioxidant inhibitors of EDRF are also able to undergo autooxidation (Table 1), a process that results in production of both superoxide anion and other organic free radicals. This has been demonstrated in the case of phenylhydrazine (Misra and Fridovich, 1976), thiols (Misra, 1974), hemoglobin (Misra and Fridovich, 1972a), catecholamines (Cohen and Heikkila, 1974), and hydroquinone (Misra and Fridovich, 1972b). It is of considerable interest that these compounds are either ineffective or show greatly diminished potency as inhibitors of endothelium-dependent relaxations in intact rabbit aortic strips when the experiments are conducted

Table 1
EDRF Inactivators and Their Chemical Properties

	Antioxidant	Scavengers of free radicals	Autooxidation producing free radicals	Inhibition of lipoxygenase	Interaction with carbonyl groups
Phenylhydrazine	+	+	+		+
KBH_4	+				+
Thiols	+	+	+		+
Hydroquinone	+	+	+		
Phenidone	+	+	+	+	+
Nordihydroguaiaretic acid	+	+	+	+	
Hemoglobin	+	+	+		
Ascorbate	+	+	+		
Catecholamines	+	+	+		
Superoxide anion	+				

at a PO_2 of 40–50 mm Hg (unpublished observations). This suggests that initiation of autooxidation by oxygen may be necessary for their chemical interaction with EDRF. Interestingly, their inhibitory potency in intact aortic preparations is not affected by superoxide dismutase at PO_2 levels of approximately 450 mm Hg (unpublished observations), implying that their inhibitory action is not mediated by the generation of superoxide anion: it may be that it is caused by the generation of other free radicals. Attack by free radicals may constitute a generalized mechanism by which EDRF is destroyed.

EDRF is believed to directly stimulate the enzyme guanylate cyclase. Many other agents that result in activation of this enzyme, such as nitric oxide and other nitro-compounds, are potent oxidizing agents, and their ability to activate the enzyme is abolished by antioxidants. This also suggests that EDRF is likely to be a highly oxidized species.

3.3. Metabolism of Arachidonate and EDRF

There has been much discussion as to whether EDRF is a product of the metabolism of arachidonic acid — through its cyclooxygenase or lipoxygenase pathways or through a third pathway involving cytochrome P-450 monoxygenase. The possibility that EDRF might be a lipoxygenase product was originally proposed by Furchgott and colleagues on the basis of data obtained with a variety of inhibitors and pharmacological probes in arterial strips (Furchgott and Zawadzki, 1980). Thus endothelium-dependent relaxation elicited by acetylcholine was found to be inhibited by 5,8,11,14-eicosatetraenoic acid (ETYA), which is known to inhibit both cyclooxygenase and lipoxygenase, and also by quinacrine, which is known to inhibit phospholipase A_2 (Furchgott and Zawadzki, 1980; Furchgott, 1983). Furthermore, endothelium-dependent relaxation, not inhibited by blockade of cyclooxygenase, can be induced by arachidonic acid (Furchgott, 1983; Singer and Peach, 1983a) and also by mellitin, which activates Ca^{2+}-sensitive phospholipase A_2 and thus stimulates the metabolism of arachidonate (Forstermann and Neufang, 1985). Other workers, however, have suggested that it is production of prostacyclin through the cyclooxygenase pathway of arachidonic acid metabolism that is involved in arachidonic acid-induced endothelium-dependent relaxations (De Mey et al., 1982). Metabolism of arachidonic acid by a third pathway, the endothelial cytochrome P-450 monoxygenase enzyme, can result in endothelium-dependent relaxations (Singer et al., 1984; Pinto et al., 1986). Inhibitors of this oxygenase enzyme are, however, not uniformly successful in blocking such relaxations (Pinto et al., 1986), and it remains to be shown that the dilator substance produced is the same as that released from the endothelium by agents such as acetylcholine and the Ca^{2+} ionophore A23187.

Arguments against the possibility that EDRF is a lipoxygenase derivative may be summarized as follows: (1) A number of saturated and unsaturated fatty acids that are not substrates for lipoxygenase can induce endothelium-dependent relaxation in arteries of the rabbit and the dog, suggesting that their action and that of arachidonic acid is nonspecific, possibly by altering membrane fluidity (Furchgott et al., 1984); (2) Not all stimulants of phospholipase A_2 are effective; exogenous phospholipase A_2 obtained from the venom of the cobra (Griffith et al., 1985c) or *Vipera russelli* (Huang and Lee, 1985) does not influence endothelium-dependent relaxation; (3) Prolonged exposure to high concentrations of glucocorticoids, which decrease phospholipase A_2 activity, does not attenuate endothelium-dependent relaxations (Griffith et al., 1984b); (4) Quinacrine has no effect on the relaxation elicited by A23187, whereas it inhibits relaxation induced by acetylcholine or adenosine triphosphate (ATP) (Furchgott, 1983; Singer and Peach, 1983b; Griffith et al., 1984b). In bioassay systems it can be shown to act at the level of the endothelial cells (Griffith et al., 1984b), implying that its mechanism of action is either at the receptor level or on processes that couple receptor occupation to release of EDRF and not on the release of the factor itself; (5) Quinacrine does not inhibit the mechanical effects (Griffith et al., 1984a) nor the elevation of cyclic GMP (Rapoport and Murad, 1983) associated with basal release of EDRF, which is consistent with the previous conclusion; (6) Although two inhibitors of acetylcholine-induced relaxations in the aorta of the rabbit, phenidone and nordihydroguaiaretic acid (Forstermann and Neufang, 1984), inhibit lipoxygenase, bioassay experiments indicate that they interfere with the activity of EDRF by interacting with it in transit (Griffith et al., 1984b; Rubanyi et al., 1985a; Griffith, 1985), in accord with their antioxidant properties (Table 1); (7) Not all lipoxygenase inhibitors are inhibitory in arterial strips (Griffith et al., 1984b); others have to be used at such high concentrations that their effects are likely to be nonspecific (Forstermann and Neufang, 1984); (8) EDRF is not lipophilic (Cocks et al., 1985), which indicates it is not a fatty acid derivative.

There is uniform agreement that EDRF is unlikely to be a product of cycloxygenase since there is no inhibition of endothelium-dependent relaxations by indomethacin or flurbiprofen (Furchgott and Zawadzki, 1980; De Mey et al., 1982; Singer and Peach, 1983b; Furchgott, 1983; Gordon and Martin, 1983; Griffith et al., 1984b).

At the present time, the biochemical pathways involved in the synthesis of EDRF and the cellular location of these pathways remain unknown.

4. Mechanisms of Production

The release of EDRF can occur not only in response to exogenous pharmacological stimulation of the endothelium, but also at a basal level in the absence of such stimulation (Fig. 2). There is, moreover, evidence for polarity of release: the liberation of EDRF in response to acetylcholine is considerably greater toward the smooth muscle side of the arterial wall than into the lumen of the blood vessel (Busse et al., 1985).

4.1. Stimulated Release

4.1.1. Dependency upon Calcium

The stimulated release of EDRF by acetylcholine and other agents depends on the presence of extracellular calcium (Singer and Peach, 1982). Indeed the fact that the calcium ionophore A23187 is a potent stimulator of endothelium-dependent relaxation strongly suggests that Ca^{2+} influx into the endothelial cell is of primary importance (Furchgott, 1983). Bioassay experiments show that the time course of this requirement for calcium is very rapid, since the intraluminal release of EDRF is abruptly turned off by the removal of Ca^{2+} and equally rapidly restored by the return of the ions into the buffer perfusing the endothelial cells (Fig. 3) (Griffith et al., 1985a; Long and Stone, 1985; Busse et al., 1985; Griffith et al., 1986). Consistent with these findings, the calcium ''agonists'' Bay K 8644 and (+)-202,791 have been shown to stimulate EDRF release, seemingly by stimulating Ca^{2+} influx through endothelial voltage-dependent channels, since their action was blocked by the calcium antagonist nitrendipine (Rubanyi et al., 1985b). Reported data are conflicting about whether Ca^{2+}-antagonists such as verapamil and nifedipine inhibit acetylcholine- and A23187-stimulated endothelium-dependent relaxations (Singer and Peach, 1982). Other workers have failed to confirm a major role for Ca^{2+} influx mediated through voltage-dependent channels in the endothelium-dependent action of acetylcholine and have suggested that a Na^{+}-Ca^{2+} exchange may be involved in the Ca^{2+} dependency of the release of EDRF so that pharmacological agents may act by interfering with this process (Winquist et al., 1985). Thus, though it is clear that extracellular calcium is directly involved in the production of EDRF, it remains to be shown whether this necessarily involves the influx of Ca^{2+} through voltage-dependent channels.

4.1.2. Dependency upon ATP

Stimulated release of EDRF critically depends also on mitochondrial ATP synthesis. The activity of EDRF can be rapidly and in most cases

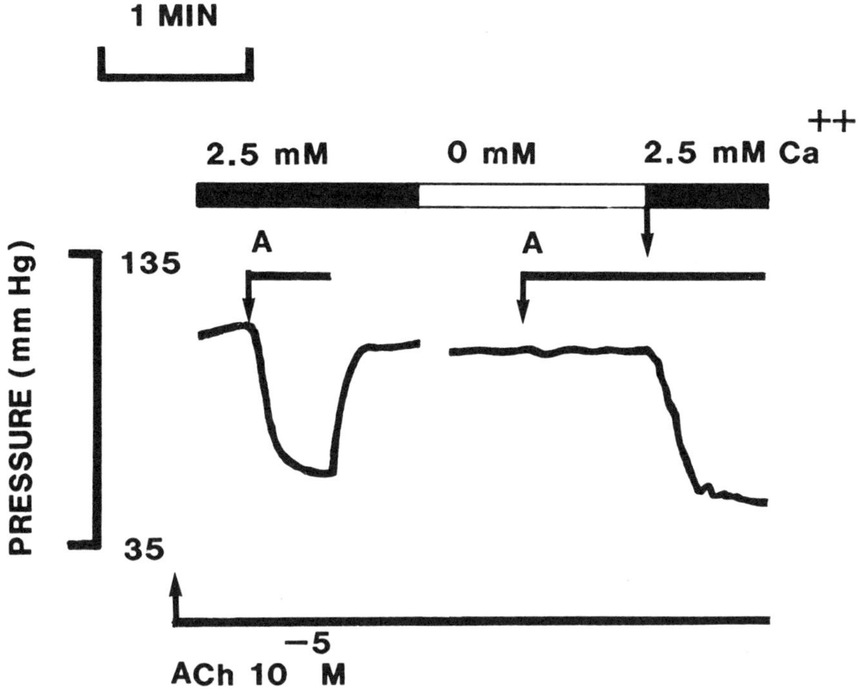

Fig. 3. Bioassay experiments showing the temporal dependence of intraluminal EDRF release on extracellular calcium. Perfusion pressure of a de-endothelialised coronary artery constricted by continuous perfusion with buffer containing 2.5 m*M* Ca^{2+} and 10^{-5}*M* acetylcholine (ACh) is shown. Introduction of an acetylcholine-stimulated intact aorta into the circuit (A) causes EDRF-induced dilatation when the aortic perfusate contains 2.5 m*M* Ca^{2+} but not when an aorta, preincubated in buffer containing 2.5 m*M* Ca^{2+}, is introduced (within 5 s) into a bioassay circuit, where it is perfused without added Ca^{2+} (0 m*M*). Reintroduction of Ca^{2+} into the aortic perfusate results in the rapid return of EDRF production (published with permission from Griffith et al., 1986).

irreversibly abolished by agents that inhibit electron transport, uncouple oxidative phosphorylation, or inhibit mitochondrial F_1-ATPase (Fig. 4); bioassay experiments confirm that these agents act at the level of the endothelium (Edwards et al., 1985a; Griffith et al., 1986). This dependency of stimulated EDRF release on ATP synthesis may be related to observations that endothelium-dependent relaxation is inhibited by severe anoxia (PO_2 < 10 mm Hg) below the level needed to support oxidative phosphorylation (Furchgott and Zawadzki, 1980; De Mey and Vanhoutte, 1983). Competitive inhibition of glycolysis by 2-deoxyglucose only slowly and incompletely inhibits the stimulated release of EDRF and only if glucose

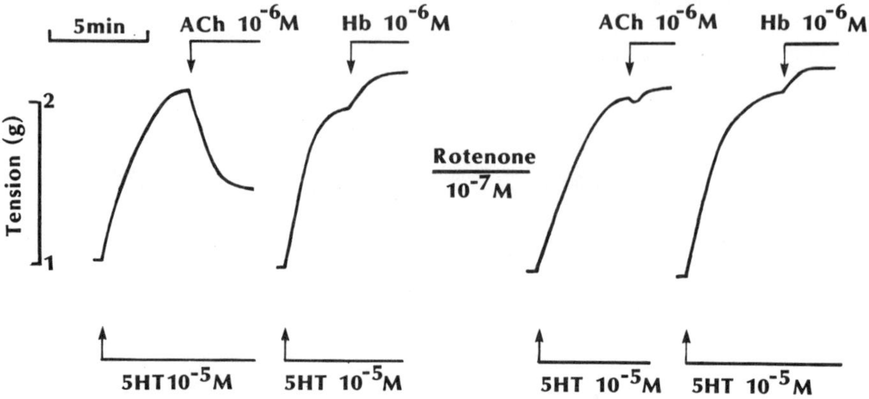

Fig. 4. Representative isometric tension traces of aortic strips of the rabbit with intact endothelium constricted with serotonin (5-HT). The endothelium-dependent dilator response to acetylcholine (ACh) and the endothelium-dependent constrictor response to hemoglobin are compared before (left traces) and after (right traces) incubation with rotenone.

is omitted from the buffer (Edwards et al., 1985a; Griffith et al., 1986); its inhibitory effects may therefore be caused indirectly by the consequences of glycolytic inhibition. The dependency of the stimulated release of EDRF on both ATP and Ca^{2+} is consistent with the involvement of an exocytotic release mechanism. It is equally possible, however, that ATP and Ca^{2+} are required for alternative cellular processes. Indeed, the available experimental techniques do not distinguish between mechanisms of synthesis and release of EDRF.

4.2. Basal Release

In contrast with stimulated release, the basal release of EDRF is not attenuated by blockade of mitochondrial ATP synthesis or inhibition of glycolysis (Fig. 4), although it still has a requirement for the presence of Ca^{2+} (Rapoport et al., 1984). There is circumstantial evidence that basal release of EDRF occurs in vivo since localized removal of the endothelium in the canine coronary artery predisposes to both spontaneous and enhanced vasoconstrictor responses (Brum et al., 1984; Lamping et al., 1985). Furthermore, the phenomenon of flow-dependent dilatation may be a manifestation of basal release of EDRF. Thus it seems likely that basal release is not an in vitro artifact. The differences in the mechanism of stimulated and basal release may simply reflect a quantitative difference in the requirements for ATP: cellular ATP stores may be adequate for basal, but not for stimulated, release of EDRF. Indeed, endothelial cells contain extremely high levels of ATP relative to other cell types and have a high turnover of nucleotides, as is evident from a continuous release of dephosphorylated purine metabolites (Gerlach et al., 1985). The rapid and complete reversal of stimulated, but not basal, release of EDRF by mitochondrial inhibitors suggests, however, not a gradual depletion of the energy supply, but a more direct and specific effect on stimulated, as opposed to basal, EDRF activity. Basal EDRF activity could result from an energy-independent release from endothelial cells caused by continuous synthesis, whereas stimulated release of EDRF may be an energy-requiring process, such as exocytosis from intracellular stores.

5. Endogenous Inactivation

5.1. Hemoglobin

Hemoglobin is a very potent inhibitor of EDRF (Martin et al., 1985a). An experimental equivalent of the two-site infusion technique has shown that hemoglobin bound to an agarose column abolishes the dilator effect of EDRF released from cultured cells (Cocks and Angus, 1985). Hemoglobin therefore must inhibit EDRF by direct interaction with the factor. The mechanism by which it does so is unknown, but appears to involve oxygen, since at low PO_2 levels it is ineffective (unpublished observations). Studies in arterial strips have shown that only ferrous hemoproteins possessing ligand binding properties behave in this way (Martin et al., 1985b).

5.2. Haptoglobin–Hemoglobin Complex

Plasma (from humans or rabbits) is also a potent inhibitor of endothelium-dependent relaxations. The component responsible has been identified using gel chromatography, ion exchange, and immunoelectrophoretic techniques, as the haptoglobin–hemoglobin complex. Haptoglobin itself is ineffective, but it becomes active after incubation with hemoglobin (Edwards et al., 1985b, 1986). Previous reports that albumin inhibited EDRF-activity may have been a result of the presence of impurities (Edwards et al., 1986). Free hemoglobin is not normally present in plasma in vivo, but is combined with circulating haptoglobin to form the haptoglobin–hemoglobin complex. The concentration of hemoglobin present as haptoglobin–hemoglobin complex in normal human plasma is high enough to result in inhibition of EDRF-activity, measured in vitro in aortic strips of the rabbit. The apparent paradox that plasma inhibits endothelium-dependent relaxation in vitro, whereas the phenomenon of endothelium-dependent dilatation is demonstrable in vivo, is probably explained by an in vitro artifact, namely loss of the intimal permeability barrier from incubation or perfusion in the absence of albumin and immunoglobulins that normally bind to the endothelial cell glycocalyx and maintain this barrier (Schneeberger and Hamelin, 1984).

The implication of this finding is that EDRF has no downstream effect in the intact circulation and that it functions as a local autocoid, exerting an influence only on immediately subjacent smooth muscle (Edwards et al., 1986). The very localized nature of endothelium-dependent relaxations has been demonstrated in vivo by Angus et al. (1983).

6. Mode of Action and Inferences for the Identity of EDRF

The mechanism by which EDRF and nitrovasodilators relax arterial smooth muscle is related to an elevation of the levels of cyclic GMP (Holzmann, 1982; Rapoport and Murad, 1983; Diamond and Chu, 1983; Ignarro et al., 1984; Miller et al., 1985). Cyclic GMP activates endogenous cyclic GMP-dependent protein kinase (Fiscus et al., 1983), which is associated with dephosphorylation of a number of proteins, including myosin light chains (Rapoport et al., 1983). An elevation in cyclic GMP mediates relaxation probably by reducing intracellular free Ca^{2+}, particularly via influx through "receptor operated channels" (Collins et al., 1985a,b; 1986a,b). Manipulation of the cyclic GMP levels in arteries by methylene blue (which inhibits guanylate cyclase and thus depresses cyclic GMP levels) and MB 29948 (which inhibits cyclic GMP phosphodiesterase and thus augments cyclic GMP levels) has shown that changes in the level of the cyclic nucleotide correlate with the observed relaxations (Holzmann, 1982). Bioassay experiments confirm that these two opposing pharmacological tools exert their effect by an action on smooth muscle and not by chemical interaction with EDRF or by a direct inhibitory or stimulatory effect on the endothelium (Griffith et al., 1985b).

When arterial segments with intact endothelium are incubated with purified soluble guanylate cyclase and its substrate GTP, both the basal and the acetylcholine-stimulated guanylate cyclase activity is increased (Busse et al., 1985). No such rise occurs if the arterial segments are denuded of endothelium. These data probably indicate that EDRF or one of its presumed decay products directly stimulates soluble guanylate cyclase (Busse et al., 1985). In this context, the fact that hemoglobin and other ferrous hemoproteins with ligand-binding properties inactivate EDRF (Cocks and Angus, 1985; Martin et al., 1985a,b) is of considerable interest because hemoglobin also inhibits the stimulation of soluble guanylate cyclase in cell-free systems (Miki et al., 1977; Mittal et al., 1978; Murad et al., 1978). Since soluble guanylate cyclase contains a heme prosthetic group that appears to be necessary for activation of the enzyme (Craven and De Rubertis, 1978), the inference is that hemoglobin and guanylate cyclase compete for interaction with the active site of EDRF. EDRF released from endothelial cells may cross the smooth muscle cell membrane and activate soluble guanylate cyclase by interacting with its heme prosthetic group.

As noted above, chemical agents that appear to inactivate EDRF also interact with carbonyl groups, leading to the hypothesis that a carbonyl group is necessary for the vasodilator activity of EDRF (Griffith et al., 1984b; Griffith, 1985). This hypothesis receives some implicit support from the observation that carbonyl compounds derived from lipid peroxidation also stimulate soluble guanylate cyclase at low concentrations (Sobolev et al., 1983). Nitric oxide (which may be the final mediator of relaxation by nitro-vasodilators including sodium nitroprusside, glyceryl trinitrate, and azide) also directly stimulates soluble guanylate cyclase. Interestingly, some of the antioxidants that directly inactivate EDRF (borohydride and thiols) also rapidly (within 1–2 min) abolish the dilator properties of sodium nitroprusside by direct chemical interaction (unpublished observations). The possibility that EDRF depends on a nitro group for its activity thus also warrants consideration.

7. Modulation of Vasomotor Responses by Basal Release of EDRF

7.1. Pharmacological Implications

If intact aortic strips of the rabbit are constricted by an agent that does not stimulate the release of EDRF, a further increase in tone can be elicited by acute addition of inactivators of EDRF (Griffith et al., 1984a; Martin et al., 1985a) or of guanylate cyclase (Griffith et al., 1985b; Martin et al., 1985a). Addition of these inhibitors can similarly increase the resting tone of unconstricted preparations (Griffith et al., 1984a, Griffith et al., 1985b). The increases in tone that result from the inhibition of basal release of EDRF vary with the artery studied, reflecting wide differences in basal activity of the factor (Griffith et al., 1984a; Martin et al., 1986; Collins et al., 1986a). The presence of basally released EDRF is manifest more in agonist-constricted than in potassium-constricted or resting preparations (Collins et al., 1986b).

The presence of the endothelium can have a major modulating influence on the response to vasoactive drugs (Fig. 1). The presence of a basal release of EDRF thus introduces a further complexity into the interpretation of concentration–responses to vasoconstrictors (*see* also section 8). The net response will represent the sum of the direct action of the vasoconstrictor agent being studied and the dilator action of any released EDRF. The amount of EDRF may be constant at the basal level or be increased by the vasoconstrictor agent tested. It may be high or low depending on the preparation studied (Collins et al., 1986a). The released EDRF may also covertly lower the baseline level of tone and thus increase the difference between the resting tone and the common ceiling (maximal response) of contraction, with the result that the data may be interpreted as indicating a paradoxical augmentation of constrictor responses in the presence of endothelium (Fig. 5) (Griffith et al., 1984a; White et al., 1986). Concentration–relaxation curves will be similarly influenced (Fig. 5) and subject to misinterpretation (Shirasaki and Su, 1985; Rubanyi and Vanhoutte, 1985b).

7.2. Physiological Implications

Peripheral dilatation of a vascular bed results in secondary dilatation of the more proximal arteries. The phenomenon was first described in the canine femoral bed (Schretzenmayr, 1933), and more recently in the coronary circulation (Gerova et al., 1981). Holtz and colleagues (1983, 1984) have demonstrated that this flow-related effect is mediated via the endothelium of the proximal vessels. The likely mechanism is that increased shear stress on the endothelium stimulates the production of EDRF. This is a potentially important positive feedback control system that may help to integrate vasomotor tone throughout a vascular bed.

It also complicates the interpretation of pharmacological experiments in vascular beds. In addition, other endogenous vasomotor influences will of course affect the diameter of the blood vessel and thus, in turn, alter EDRF-mediated responses (*see* Fig. 3, with respect to endothelium-dependent responses in the presence of differing basal tone) including flow-dependent dilatation.

Holtz and colleagues (1983, 1984) have reported an interesting distinction in the coronary circulation of conscious dogs between the responses to: (1) adenosine, dipyridamole, and reactive hyperemia, which act directly on resistance vessels and only indirectly (via the flow-dependent, endothelium-mediated mechanism) on conduit vessels; (2) Ca^{2+} antagonists, which dilate conduit vessels both directly and through a flow-dependent effect secondary to dilatation of resistance vessels; and (3) nitrovasodilators, whose only effect is a direct one, with no secondary flow-dependent dilatation of conduit arteries, presumably because cyclic GMP-mediated dilatation is already activated by the nitrovasodilators.

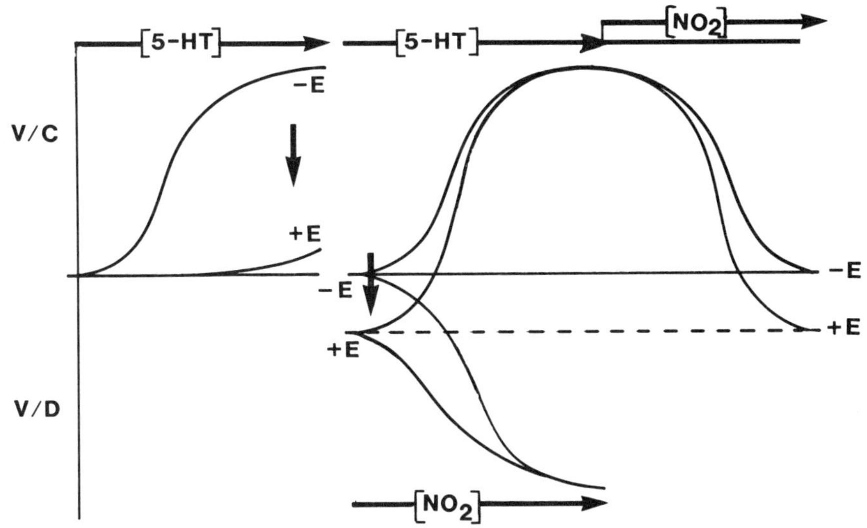

Fig. 5. Composite diagram illustrating possible ways in which basal release of EDRF may affect cumulative concentration–responses to vasoconstrictor (V/C) and/or vasodilator (V/D) agents, e.g., serotonin (5-HT) and/or nitrovasodilators (NO_2), respectively. Arrowed horizontal lines represent increasing concentrations of 5-HT and of NO_2 (i.e., cumulative concentration–response curves); continuing horizontal line for 5-HT in upper right panel represents continuing presence of 5-HT at constant (maximal) concentration. Responses in presence (+E) or absence (−E) of endothelium are shown.

Left trace illustrates major effect of basally released EDRF in preparations with no resting tone (e.g., perfused coronary artery of the rabbit). Right trace illustrates minor effect of basal EDRF release in artery with resting tone (e.g., aortic strip of the rabbit). Note that basally released EDRF can almost abolish net constrictor responses (left trace). Alternatively, the basal production of EDRF leads to lower baseline, but unaltered, ceiling of V/C or V/D responses (right traces), so that the *responses* are influenced in opposite directions by the level of the baseline, whereas the difference between maximal constriction and maximal dilatation is independent of the presence of endothelium.

7.3. Pathophysiological Implications

Local loss of the production of EDRF can result in unopposed constrictor influences in vivo. Endogenous inhibition of EDRF by the haptoglobin–hemoglobin complex in the vascular lumen, together with the polarity of the release, will direct EDRF exclusively to the subjacent smooth muscle. A potent mechanism thus exists whereby local loss of endothelial function can lead to localized arterial spasm, as demonstrated in coronary arteries in vitro (Griffith et al., 1984a; Bing et al., 1985) and in vivo (Brum et al., 1984; Lamping et al., 1985). Blood vessels with high basal release of EDRF, such as coronary arteries (Griffith et al., 1984a; Collins et al.,

1986a), may be particularly susceptible to this phenomenon. Causes may include mechanical damage (Griffith et al., 1984a; Brum et al., 1984; Lamping et al., 1985), the imposition of a diffusion barrier between endothelium and vascular smooth muscle (e.g., by lipid deposition; Chappell et al., 1985, 1986; Verbeuren et al., 1985, 1986), or access of an inhibitor such as plasma into the subendothelial layer (e.g., intimal or extravascular hemorrhage).

8. EDRF in Vivo

In vivo data on the role of the endothelium in vasodilator responses are scant. Endothelium-dependent dilatation of conduit arteries in response to acetylcholine and substance P has, however, been demonstrated in the canine femoral artery (Angus et al., 1983). Removal of the endothelium can lead to spontaneous local vasoconstriction and enhanced vasoconstrictor responses (Brum et al., 1984; Lamping et al., 1985). It follows that the potential exists for both basal and stimulated release of EDRF to play an important role in modulating vascular tone in the intact organism.

9. Conclusions

EDRF is a labile, reactive agent whose identity has yet to be elucidated. Available evidence suggests that it is not a product of the metabolism of arachidonate, but a novel and probably highly oxidized compound. Its potentially potent relaxant effects are probably limited in vivo to immediately subjacent vascular smooth muscle. Its biological action is rapid. It acts, like nitrovasodilators, by stimulating soluble guanylate cyclase and increasing cyclic GMP levels in smooth muscle, which results in relaxation probably by reducing intracellular free calcium. The stimulated release of EDRF from the endothelium is Ca^{2+}- and energy-dependent, whereas the basal release of the factor is Ca^{2+}-dependent, but not energy-dependent. These are the characteristics of a substance capable of modulating vascular tone in response to rapidly changing physiological and metabolic demands.

EDRF appears to be basally produced, but a wide variety of endogenous agents can stimulate its release. This suggests three possible roles for EDRF: (1) Basal release, probably modulated by flow, provides a positive feedback mechanism to counterbalance intrinsic myogenic tone and a dynamic means of integrating the calibre of blood vessels throughout the vascular tree. (2) Release in response to substances in the blood, such as products released during platelet activation and thrombosis, provides a transendothelial messenger system capable of conveying a vasomotor signal across what is normally an impermeable barrier. (3) Release in response to neurogenic messengers, such as acetylcholine or peptides (some of which are cotransmitters in autonomic nerves), suggests a neurogenic response in small arteries where the diffusion distance between the adventitia where the neurons are located and the endothelium is short.

Acknowledgments

The work was supported by the British Heart Foundation (BHF). Wendy Simons and Andrea Hill assisted with the preparation of the manuscript.

References

Angus, J. A., Campbell, G. R., Cocks, T. M., and Manderson, J. A.: Vasodilatation by acetylcholine is endothelium-dependent: A study by sonomicrometry in canine femoral artery in vivo. *J. Physiol.*(Lond.) **344:**209–222, 1983.

Bing, R. J., Burger, W., Chemnitius, J. M., Saeed, M., and Metz, M. Z.: Effect of endothelium against platelet-induced coronary artery spasm in isolated rabbit hearts. *Am. J. Cardiol.* **55:**1596–1600, 1985.

Bolton, T. B., Lang, R. J., and Takewaki, T.: Mechanisms of action of noradrenaline and carbachol on smooth muscle of guinea-pig anterior mesenteric artery. *J. Physiol.*(Lond.) **351:**549–572, 1984.

Brum, J. M., Sufan, Q., Lane, G., and Bove, A.: Increased vasoconstrictor activity of proximal coronary arteries with endothelial damage in intact dogs. *Circulation* **70:**1066–1073, 1984.

Busse, R., Trogish, G., and Bassenge, E.: The role of endothelium in the control of vascular tone. Basic *Res. Cardiol.* **80:**475–490, 1985.

Chand, N., and Altura, B. M.: Acetycholine and bradykinin relax intrapulmonary arteries by acting on endothelial cells: Role in lung vascular diseases. *Science* **213:**1376–1379, 1981

Chappell, S. P., Griffith, T. M., Henderson, A. H., and Lewis, M. J.: Influence of cholesterol feeding on endothelium-dependent vasomotor response in rabbit aortic strips. *Br. J. Pharmacol.* **85:**266P, 1985.

Chappell, S. P., Lewis, M. J., and Henderson, A. H.: Effect of lipid feeding on endothelium-dependent relaxation in rabbit aortic preparation. *Cardiovasc. Res.* 1986, in press.

Cocks, T. M. and Angus, J. A.: Bioassay of the Release of Endothelium-Derived Relaxing Factor (EDRF) from Isolated Endothelial cells In Vitro, in *Vascular Neuroeffector Mechanisms* (Bevan, J. A., Godfraind, T., Maxwell, R. A., Stodet, J. C., and Worcel, M., eds.) Elsevier, Amsterdam, New York, Oxford, 1985.

Cocks, T. M., Angus, J. A., Campbell, J. H., and Campbell, G. R.: Release and properties of endothelium-derived relaxing factor (EDRF) from endothelial cells in culture. *J. Cell. Physiol.* **123:**310–320, 1985.

Cocks, T. M., Angus, J. A., Campbell, J., and Campbell, G.: Nature and role of endothelium derived relaxing factor (EDRF). *J. Mol. Cell. Cardiol.* **18** (suppl. 1):17, 1986.

Cohen, G. and Heikkla, R. E.: The generation of hydrogen peroxide, superoxide radical and hydroxyl radical by 6-hydroxydopamine, dialuric acid and related cytotoxic agents. *J. Biol. Chem.* **249:**2447–2452, 1974.

Collins, P., Griffith, T. M., Henderson, A. H., and Lewis, M. J.: 8-Bromo-cGMP and calcium flux in arterial smooth muscle. *Br. J. Pharmacol.* **85:**279P, 1985a.

Collins, P., Griffith, T. M., Henderson, A. H., and Lewis, M. J.: Endothelium and calcium flux in rabbit aortic preparations. *Br. J. Pharmacol.* **85:**342P, 1985b.

Collins, P., Chappell, S. P., Griffith, T. M., Lewis, M. J., and Henderson, A. H.: Differences in basal endothelium-derived relaxing factor activity in different artery types. *J. Cardiovasc. Pharmacol.* **8:**1158–1162, 1986a.

Collins, P., Griffith, T. M., Henderson, A. H., and Lewis, M. J.: Endothelium-derived relaxing factor alters calcium fluxes in rabbit aorta: A cyclic guanosine monophosphate-mediated phenomenon. *J. Physiol.* (Lond.) **381:** 427–437, 1986b.

Craven, P. A. and De Rubertis, F. R.: Restoration of the responsiveness of purified guanylate cyclase to nitrosoguanidine, nitric oxide, and related activators by heme and hemoproteins. *J. Biol. Chem.* **253:**8433–8443, 1978.

De Mey, J. G., Claeys, M., and Vanhoutte, P. M.: Endothelium-dependent inhibitory effects of acetylcholine, adenosine triphosphate, thrombin and arachidonic acid in the canine femoral artery. *J. Pharmacol. Exp. Ther.* **222:**166–173, 1982.

De Mey, J. G. and Vanhoutte, P. M.: Anoxia and endothelium-dependent reactivity of the canine femoral artery. *J. Physiol.* (Lond.) **335:**65–74, 1983.

Diamond, J. and Chu, E. U.: Possible role for cyclic GMP in endothelium-dependent relaxation of rabbit aorta by acetylcholine. Comparison with nitroglycerin. *Res. Comm. Chem. Path. Pharmac.* **41:** 369–381, 1983.

Edwards, D. H., Griffith, T. M., Henderson, A. H., Lewis, M. J., and Newby, A. C.: Production of endothelium derived relaxant factor is both ATP and calcium dependent. *Br. J. Pharmacol.* **85:**344P, 1985a.

Edwards, D. H., Griffith, T. M., Henderson, A. H., Lewis, M. J., and Ryley, H. C.: Endothelium dependent relaxation is inhibited by a high molecular weight protein fraction of whole human plasma. *Br. J. Pharmacol.* **85:**341P, 1985b.

Edwards, D. H., Griffith, T.M., Ryley, H. C., and Henderson, A. H.: Haptoglobin–haemoglobin complex in human plasma inhibits endothelium dependent relaxation: Evidence that endothelium derived relaxing factor acts as a local autocoid. *Cardiovasc. Res.* **20:**549–556, 1986.

Fiscus, R. R., Rapoport, R. M., and Murad, F.: Endothelium-dependent and nitrovasodilator-induced activation of cyclic GMP-dependent protein kinase in rat aorta. *J. Cyclic Nucleotide Protein Phosphor. Res.* **9:**415–425, 1983.

Forstermann, M. and Neufang, B.: The endothelium-dependent vasodilator effect of acetylcholine: Characterization of the endothelial relaxing factor with inhibitors of arachidonic acid metabolism. *Eur. J. Pharmacol.* **103:**65–70, 1984.

Forstermann, J. and Neufang, B.: Endothelium-dependent vasodilatation by mellitin: Are lipoxygenase products involved? *Am. J. Physiol.* **249:**H14–H19, 1985.

Forstermann, U., Trogisch, G., and Busse, R.: Species-dependent differences in the nature of endothelium-derived vascular relaxing factor. *Eur. J. Pharmacol.* **106:**639–43, 1985.

Furchgott, R. F.: Role of endothelium in responses of vascular smooth muscle. *Circ. Res.* **53:**557–573, 1983.

Furchgott, R. F. and Zawadzki, J. V.: The obligatory role of endothelial cells in the relaxation of arterial smooth muscle by acetylcholine. *Nature* **288:**373–376, 1980.

Furchgott, R. F., Cherry, P. D., Zawadzki, J. V., and Jothianandan, D.: Endothelial cells as mediators of vasodilatation of arteries. *J. Cardiovasc. Pharmacol.* **6:**S336–343, 1984.

Gerlach, E., Nees S., and Becker, B. F.: The vascular endothelium: A survey of some newly evolving biochemical and physiological features. *Basic Res. Cardiol.* **80:**459–474, 1985.

Gerova, M., Gero, J., Barta, E., Dolezel, S., Smiesko, V., and Levicky, V: Neurogenic and myogenic control of conduit coronary artery: A possible interference. *Bas. Res. Cardiol.* **76:**503–507, 1981.

Gordon, J. L. and Martin, W.: Stimulation of endothelial prostacyclin production plays no role in endothelium-dependent relaxation of the pig aorta. *Br. J. Pharacol.* **80:**179–186, 1983.

Griffith, T. M.: Studies of endothelium-derived relaxant factor (EDRF), its nature and mode of action. *Eur. Heart J.* **6:**37–49, 1985.

Griffith, T. M., Edwards, D. H., Collins, P., Lewis, M. J., and Henderson, A. H.: Endothelium derived relaxant factor. *J. Roy. Coll. Physicians Lond.* **19:**74–79, 1985a.

Griffith, T. M., Edwards, D. H., Lewis, M J., and Henderson, A. H.: Evidence that cyclic guanosine monophosphate (cGMP) mediates endothelium-dependent relaxation. *Eur. J. Pharmacol.* **112:**195–202, 1985b.

Griffith, T. M., Edwards, D. H., Lewis, M. J., and Henderson, A. H.: Endothelium Influences Coronary and Aortic Vasomotion by Release of an Unstable Humoral Factor, in *Advances in Myocardiology,* vol. 6 (Dhalla, N. S. and Hearse, D. J.: eds.) Plenum, New York and London, 1985c.

Griffith, T. M., Henderson, A. H., Hughes Edwards, D., and Lewis, M. J.: Isolated perfused rabbit coronary artery and aortic strip preparations: The role of endothelium-derived relaxant factor. *J. Physiol.* (Lond.) **351:**13–24, 1984a.

Griffith, T. M., Edwards, D. H., Lewis, M. J., Newby, A. C., and Henderson, A. H.: The nature of endothelium-derived relaxant factor. *Nature* (Lond.) **308:**645–647, 1984b.

Griffith, T. M., Edwards, D. H., Lewis, M. J., Newby, A. C., and Henderson, A. H: Production of endothelium-derived relaxant factor is dependent on oxidative phosphorylation and extracellular calcium. *Cardiovasc. Res.* **20:**7–12, 1986.

Gryglewski, R. J., Palmer, R. M. J., and Moncada, S.: Superoxide anion is involved in the breakdown of endothelium-derived vascular relaxing factor. *Nature* **320:**454–456, 1986.

Holtz, J., Giesler, M., and Bassenge, E.: Two dilatory mechanisms of antianginal drugs on epicardial coronary arteries in vivo: Indirect, flow-dependent, endothelium-mediated dilation and direct smooth muscle relaxation. *Z. Kardiol.* **3** (suppl.):98–106, 1983.

Holtz, J., Forstermann, U., Pohl, U., Giesler, M., and Bassenge, E.: Flow-dependent, endothelium-mediated dilation of epicardial coronary arteries in conscious dogs: Effects of cyclooxygenase inhibition. *J. Cardiovasc. Pharmacol.* **6:**1161–1169, 1984.

Holzmann, S.: Endothelium-induced relaxation by acetylcholine associated with larger rises in cyclic GMP in coronary arterial strips. *J. Cyclic Nucl. Res.* **8:**409–419, 1982.

Huang, H-C. and Lee, C. Y.: Relaxant effect of phospholipase A_2 from *Vipera russelli* snake venom on rat aorta. *Eur. J. Pharmacol.* **118:**139–146, 1985.

Ignarro, L. J., Burke, T. M., Wood, K. S., Wolin, M. S., and Kadowitz, P. J.: Association between cyclic GMP accumulation and acetylcholine-elicited relaxation of bovine intrapulmonary artery. *J. Pharmacol. Exp. Ther.* **228:**682–690, 1984.

Lamping, K. G., Marcus, M. L., and Dole, W. P.: Removal of the endothelium potentiates canine large coronary artery constrictor responses to 5-hydroxtryptamine in vivo. *Circ. Res.* **57:**46–54, 1985.

Long, C. J. and Stone, T. W.: The release of endothelium-derived relaxant factor is calcium dependent. *Blood Ves.* **22:**205–208, 1985.

Martin, W., Villani, G. M., Jothianandan, D., and Furchgott, R. F.: Selective blockade of endothelium-dependent and glyceryl trinitrate-induced relaxation by haemoglobin and by methylene blue in the rabbit aorta. *J. Pharmacol. Exp. Ther.* **232:**708–716, 1985a.

Martin, W., Villani, G. M., Jothianandan, D., and Furchgott, R. F.: Blockade of endothelium-dependent and glyceryl trinitrate-induced relaxation of rabbit aorta by certain ferrous haemoproteins. *J. Pharmacol. Exp. Ther.* **233:**679–685, 1985b.

Martin, W., Furchgott, R. F., Villani, G. M., and Jothianandan, D.: Depression of contractile responses in rat aorta by spontaneously released endothelium-derived relaxing factor. *J. Pharmacol. Exp. Ther.* **237:**529–538, 1986.

Miki, N., Kawabe, Y., Kuriyama, K.: Activation of cerebral guanylate cyclase by nitric oxide. *Biochem. Biophys. Res. Commun.* **75:**851–856, 1977.

Miller, R. C., Schoeffter, P., and Stoclet, J. C.: Insensitivity of calcium-dependent endothelial stimulation in rat isolated aorta to the calcium entry blocker flunarizine. *Br. J. Pharmacol.* **85:**481–487, 1985.

Misra, H. P.: Generation of superoxide free radical during the autooxidation of thiols. *J. Biol. Chem.* **249:**2151–2155, 1974.

Misra, H. P. and Fridovich, I.: The generation of superoxide radical during the autooxidation of hemoglobin. *J. Biol. Chem.* **247:**6960–6962, 1972a.

Misra, H. P. and Fridovich, I.: The univalent reduction of oxygen by reduced flavins and quinones. *J. Biol. Chem.* **247:**188–192, 1972b.

Misra, H. P. and Fridovich, I.: The oxidation of phenylhydrazine: Superoxide and mechanism. *Biochemistry* **15:**681–687, 1976.

Mittal, C. K., Arnold, W. P., and Murad, F.: Characterization of protein inhibitors of guanylate cyclase activation from rat heart and bovine lung. *J. Biol. Chem.* **253:**1266–1271, 1978.

Murad, F., Mittal, C. K., Arnold, W. P., Katsuki, S., and Kimura, H.: Guanylate cyclase: Activation by azide, nitro compounds, nitric oxide and hydroxyl radical and inhibition by hemoglobin and myoglobin. *Adv. Cyclic Nucleotide Res.* **9:**145–158, 1978.

Pinto, A., Abraham, N. G., and Mullane, K. M.: Cytochrome P-450-dependent monoxygenase activity and endothelial-dependent relaxations induced by arachidonic acid. *J. Pharmacol. Exp. Ther.* **236:**445–451, 1986.

Rapoport, R. M. and Murad, F.: Agonist-induced endothelium-dependent relaxation in rat thoracic aorta may be mediated through cyclic GMP. *Circ. Res.* **52:**352–357, 1983.

Rapoport, R. M., Draznin, M. B., and Murad, F. : Endothelium-dependent relaxation in rat aorta may be mediated through cyclic GMP-dependent protein phosphorylation. *Nature* **306:**174–176, 1983.

Rapoport, R. M., Draznin, M. B., and Murad, F.: Mechanisms of adenosine triphosphate-, thrombin- and trypsin-induced relaxation of rat thoracic aorta. *Circ. Res.* **55:**468–479, 1984.

Rapoport, R. M., Waldman, S. A., Schwartz, K., Winquist, R. J., and Murad, F.: Effects of atrial natriuretic factor, sodium nitroprusside and acetylcholine on cyclic GMP levels and relaxation in rat aorta. *Eur. J. Pharmacol.* *115:*219–229, 1985.

Rubanyi, G. M. and Vanhoutte, P. M.: Superoxide dismutase prolongs the half-life of endothelium-derived relaxing factor(s). *Clin. Res.* **33:**522A, 1985a.

Rubanyi. G. M. and Vanhoutte, P. M.: Endothelium-removal decreases relaxations of canine coronary arteries caused by beta-adrenergic agonists and adenosine. *J. Cardiovasc. Pharmacol.* **7:**139–144, 1985b.

Rubanyi, G. M. and Vanhoutte, P. M.: Superoxide anions and hyperoxia inactivate endothelium-derived relaxing factor. *Am. J. Physiol.* **250:**H822–H827, 1986.

Rubanyi, G. M., Lorenz, R. R., and Vanhoutte, P. M.: Bioassay of endothelium-derived relaxing factor(s): inactivation by catecholamines. *Am. J. Physiol.* **249:**H95–H101, 1985a.

Rubanyi, G. M., Schwartz, A., and Vanhoutte, P. M.: The calcium antagonists Bay K 8644 and (+)202, 791 stimulate the release of endothelial relaxing factor from canine femoral arteries. *Eur. J. Pharmacol.* **117:**143–144, 1985b.

Schneeberger, E. E. and Hamelin, M.: Interaction of serum proteins with lung endothelial glycocalyx: Its effect on endothelial permeability. *Am. J. Physiol.* **247:**H206–H217, 1984.

Schretzenmayr, A.: Uber kreislaufregulatorische Vorgange an den grossen Arterien bei der Muskelarbeit. *Pfluegers Arch.* **232:**743–748, 1933.

Shirasaki, Y. and Su, C.: Endothelium removal augments vasodilation by sodium nitroprusside and sodium nitrite. *Eur. J. Pharmac.* **114:**93–96, 1985.

Singer, H. A. and Peach, M. J.: Calcium- and endothelial-mediated vascular smooth muscle relaxation in rabbit aorta. *Hypertension* **4** (suppl. II):II19–II25, 1982.

Singer, H. A. and Peach, M. J.: Endothelium-dependent relaxation of rabbit aorta. 1. Relaxation stimulated by arachidonic acid. *J. Pharmacol. Exp. Ther.* **226:**790–795, 1983a.

Singer, H. A. and Peach, M. J.: Endothelium-dependent relaxation of rabbit aorta. II. Inhibition of relaxation stimulated by methacholine and A23187 with antagonists of arachidonic acid metabolism. *J. Pharmacol. Exp. Ther.* **226:**796–801, 1983b.

Singer, H. A., Saye, J. A. , and Peach, M. J.: Effects of cytochrome P450 inhibitors on endothelium-dependent relaxation in rabbit aorta. *Blood Vess.* **21:**223–230, 1984.

Sobolev, A. S., Tertov, V. V., and Rybalkin, S. D.: A study of rat liver guanylate cyclase activation by peroxides of fatty acids, carbonyl compounds and biogenic amines. *Biochim. Biophys. Acta* **756:**92–105, 1983.

Verbeuren, T. J., Coene, M-C., Jordaens, F., Van Hove, C., Zonnekeyn, L., and Herman, A. G.: Endothelium-dependent relaxations in isolated arteries of control and hypercholesterolaemic rabbits. *Br. J. Pharmacol.* **85:**267P, 1985.

Verbeuren, T. J., Jordaens, F., Zonnekeyn, L., Van Hove, C. E., Coene, M-C., and Herman, A. G.: Endothelium-dependent and endothelium-independent concentrations and relaxations in isolated arteries of control and hypercholesterolemic rabbits. *Circ. Res.* **58:**552–564, 1986.

White, D. G., Lewis, M. J., Griffith, T. M., Edwards, D. H., and Henderson, A. H.: Influence of endothelium on drug-induced relaxation of the rabbit aorta. *Eur. J. Pharmacol.* **121:**19–23, 1986.

Winquist, R. J., Bunting, P. B., and Schofield, T. L.: Blockade of endothelium-dependent relaxation by the amiloride analog dichlorobenzamil: Possible role of Na^+/Ca^{++} exchange in the release of endothelium-derived relaxant factor. *J. Pharmacol. Exp. Ther.* **235:**644–650, 1985.

Chapter 4

Metabolism of Arachidonic Acid and Release of Endothelium-Derived Relaxing Factors

Roger A. Johns and Michael J. Peach

1. Introduction

In the past five to ten years, studies that have focused on the endothelium have taken on a very significant role in the expansion of our understanding of vascular pharmacology. Three major areas of interest have been endothelium-dependent production of vasoactive substances (conversion of angiotensin I); the ability of endothelial cells to metabolize arachidonic acid to various eicosanoid (epoxide and endoperoxide) products; and inactivation of vasoactive compounds (bradykinin, adenosine, norepinephrine, and so on). An interaction or relationship among these areas was established when it was shown that the relaxation or modulation of vascular tone by angiotensin II and bradykinin was mediated by the endothelial production and release of prostacyclin (Dusting et al., 1981; Mullane and Moncada, 1980). In 1980, Furchgott and Zawadzki reported the obligatory role of endothelial cells in the in vitro relaxation of vascular smooth muscle by acetylcholine. Subsequently, numerous other vasodilating substances have been found to have complete or partial endothelium dependence, and the term "endothelium-derived relaxing factor" (EDRF) was coined for the putative mediator of relaxation (Furchgott, 1983; *see* chapter 1 in this volume).

At the present time, the chemical nature of this factor is not known, and it is not clear whether a single or multiple factor(s) exist. Numerous studies have provided indirect evidence of a role for arachidonic acid metabolites in the mechanism of endothelium-dependent relaxation, but strong opposing arguments have also been proposed.

This chapter will begin with an overview of the current knowledge of arachidonic acid metabolism by endothelial cells. It will then review the evidence for and against the role of arachidonic acid metabolites in endothelium-dependent relaxation. Experimental problems with the use of exogenous arachidonic acid will be discussed, and areas for further research on the nature of EDRF(s) will be considered.

2. Endothelial Cell Metabolism of Arachidonic Acid

Eicosanoids and a wide variety of endoperoxides and epoxides are produced by the enzymatic oxygenation of certain polyunsaturated fatty acids released from the lipids of intracellular and plasma membranes. The most common fatty acids are 8,11,14-eicosatrienoic acid (dihomo-γ-linolenic acid), 5,8,11,14-eicosatetraenoic acid (arachidonic acid), and 5,8,11,14,17-eicosapentaenoic acid (Needleman and Isakson, 1980). Oxygenation must first be preceded by the activation of lipases (e.g., phospholipase A_2 or diacylglyceride lipase) that release the esterified fatty acid from membrane lipids. The free fatty acids are then subject to oxidative metabolism by one of three enzymatic pathways: cyclooxygenase, lipoxygenases, and cytochrome P-450-dependent mixed function oxidases (Schwartzman et al., 1985; Fig. 1). The presence of all three of these pathways has been documented for endothelial cells (see below). The membrane fatty acid metabolism discussed in this review will focus specifically on arachidonic acid.

2.1. Cyclooxygenase Pathways

Metabolism of arachidonic acid via cyclooxygenase results in the formation of cyclic endoperoxides PGG_2 and PGH_2. These endoperoxides are unstable in aqueous solution and are further metabolized to the stable prostaglandins (PG)D_2, PGE_2, and $PGF_{2\alpha}$, as well as thromboxane A_2 (TXA_2) and prostacyclin (PGI_2) (Marcus et al., 1982), as shown in Fig. 2. Although endothelial cells produce mainly prostacyclin, all of these products have been demonstrated in endothelial cell culture preparations using chromatographic methods and/or prostaglandin bioassays (Marcus et al., 1982; Gerritsen and Chelli, 1983; Ingerman-Wojenski et al., 1981; Bunting et al., 1976; Dusting et al., 1977).

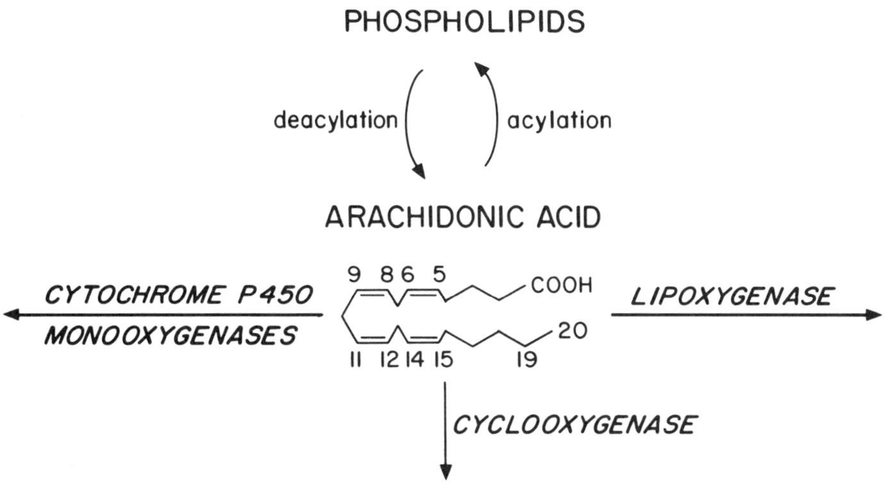

Fig. 1. Arachidonic acid released from phospholipids undergoes oxidative metabolism via three enzymatic pathways.

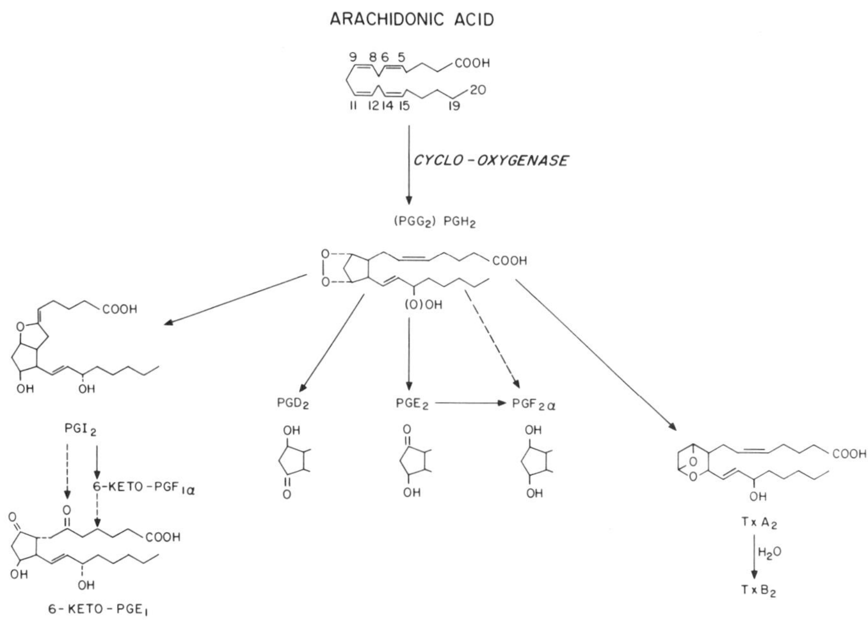

Fig. 2. The cyclooxygenase pathway of arachidonate metabolism.

The relative capacity to produce PGI_2 varies across the layers of the arterial wall. In aorta of the rabbit there is a progressive decrease in PGI_2 synthetase activity from the internal elastic lamina to the adventitia (Moncada et al., 1977), with at least a tenfold difference in conversion of PGH_2 to PGI_2 in the intimal layer compared to the media or adventitia (MacIntyre et al., 1978).

2.2. Lipoxygenase Pathways

The second major pathway for arachidonic acid metabolism involves the enzymatic placement of molecular oxygen at positions C-7, C-10, and C-13 by a 5-, 12-, or 15-lipoxygenase, respectively, to produce 5-, 12-, or 15-hydroperoxyeicosatetraenoic acid (HPETE) (Ford-Hutchinson, 1985; Fig. 3). These HPETEs can then be further metabolized to corresponding hydroxyeicosatetraenoic acids (HETEs), di-hydroxyeicosatrienoic acids (di-HETEs), and leukotrienes (LT; Fig. 3).

The activity of all three lipoxygenase enzymes has been demonstrated in a variety of endothelial cell and vascular preparations. Gorman et al. (1985) and Buchanan et al. (1985), using high-pressure liquid chromatography and gas chromatography–mass spectrometry analysis, reported the production of 5-, 12-, and 15-HETEs in cultured endothelial cells of the human umbilical vein. Similar techniques have provided evidence for the production of HETEs in aorta of the rabbit (Greenwald et al., 1979).

The biological activity of many of these products remains unknown, although several of the leukotrienes have been shown to modulate vascular tone. LTC_4, LTD_4, and LTE_4 are potent pulmonary and systemic vasoconstrictors (Samuelson, 1983). LTC_4 and LTD_4 together make up the components of slow-reacting substance of anaphylaxis (Jakschik and Lee, 1980; Johnson et al., 1985), whereas only LTF_4 and LTD_4 have been shown to have vasodilating properties (Letts et al., 1985; Secrest et al., 1985). The potential exists for other vasodilators among the many lipoxygenase products, and additional study is required.

2.3. Cytochrome P-450 Pathways

The metabolism of arachidonic acid via cytochrome P-450 monoxygenases was first reported in the kidney (Morrison and Pascoe, 1981) and liver (Capdevila et al., 1981; Oliw et al., 1982). Three types of reactions have been described: (1) allylic oxidation producing HETEs, (2) olefin epoxidation resulting in four different epoxyeicosatrienoic acids that can undergo hydrolysis by epoxide hydrolase to form corresponding dihydroxyeicosatrienoic acids, and (3) oxidation at w and w-1 positions to form the 20-HETEs and 19-HETEs, respectively (Schwartzman et al., 1985; Fig.

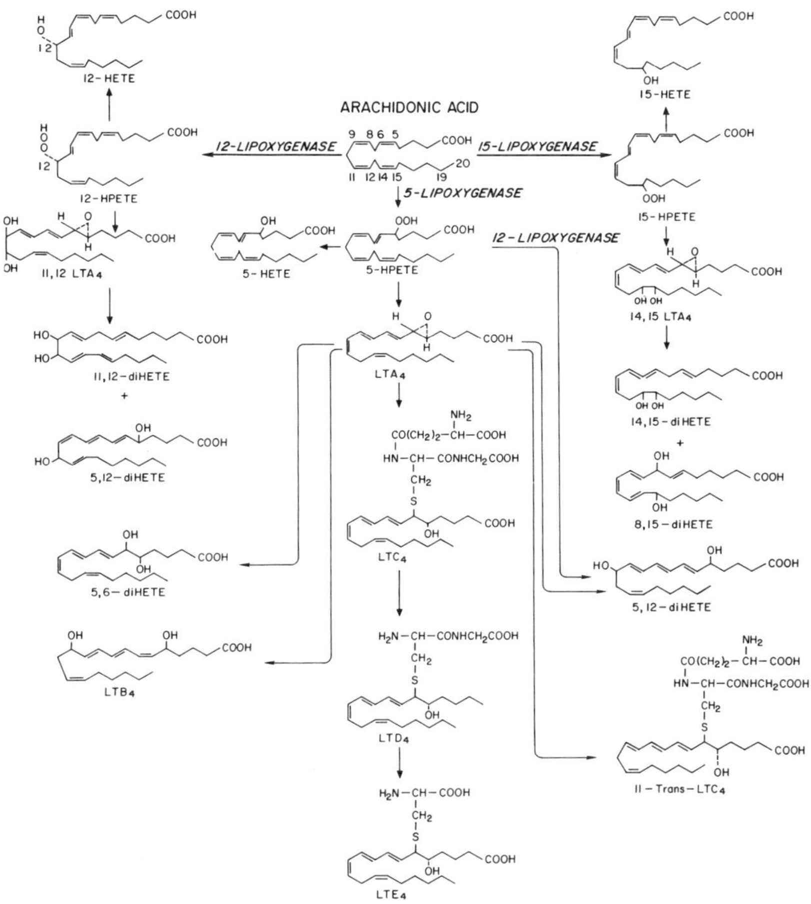

Fig. 3. The lipoxygenase pathways of arachidonate metabolism.

4). As was the case for the lipoxygenase pathways, the vascular actions of most of these products are unknown.

Cytochrome P-450 spectra and enzymatic activity have been detected in aortic microsomes from rabbits, monkeys, and humans (Juchau et al., 1976); and its presence in bovine aortic endothelial cell cultures has been inferred by the metabolism of benzo(a)pyrene (Baird et al., 1980). The presence of inducible isoenzymes of cytochrome P-450 in rabbit renal, hepatic, and pulmonary arteries and veins has been demonstrated, and the enzyme was localized to the intima using immunohistochemical techniques (Dees et al., 1982). A recent study assaying aryl-hydrocarbon hydroxylase activity in hog aorta has shown that cytochrome P-450 is localized primarily in the endothelial cells of the intima, as opposed to the muscle cells of the media (Abraham et al., 1985).

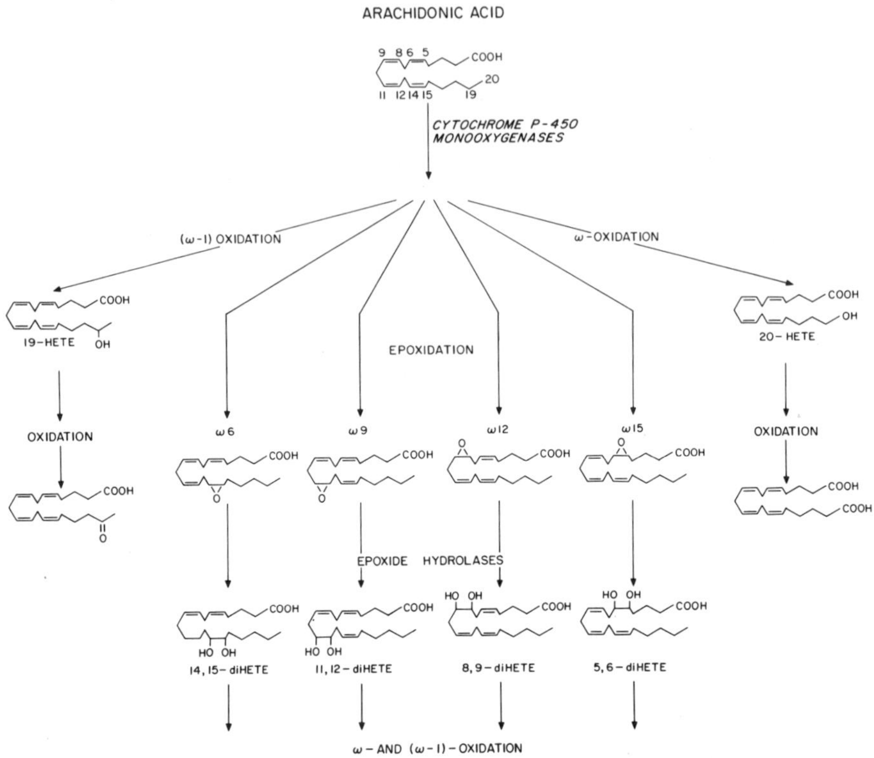

Fig. 4. The cytochrome P-450 pathway of arachidonate metabolism.

3. Evidence Relating Arachidonate Metabolites to EDRF

Numerous studies have provided indirect evidence of a role for arachidonic acid and its metabolites in endothelium-dependent relaxation. These include blockade of the endothelium-dependent response with compounds known to inhibit arachidonic acid metabolism and phospholipase activation, as well as the ability to mimic endothelium-dependent relaxation with agents that are known to activate phospholipases or by the administration of exogenous phospholipase or exogenous arachidonic acid.

3.1. Inhibition of Arachidonic Acid Metabolism

3.1.1. Cyclooxygenase Blockade

EDRF is not prostacyclin or any other prostaglandin, since the use of inhibitors of cyclooxygenase such as indomethacin and aspirin does not prevent endothelium-dependent relaxation in response to multiple agents in a variety of species (Furchgott and Zawadzki, 1980; Singer and Peach, 1983b; Gordon and Martin, 1983a). In addition, prostacyclin produces relaxation in vascular smooth muscle by activation of adenylate cyclase to increase cyclic AMP levels (Ignarro and Kadowitz, 1985), whereas EDRF appears to relax arterial smooth muscle by activation of soluble guanylate cyclase and subsequent accumulation of cyclic GMP (Rapoport and Murad, 1983a,b; Loeb et al., 1985; Rapoport et al., 1985).

3.1.2. Lipoxygenase Blockade

5,8,11,14-Eicosatetraenoic acid (ETYA), an inhibitor of cyclooxygenase and lipoxygenase was one of the first agents shown to antagonize the endothelium-dependent relaxation to acetylcholine (Furchgott and Zawadzki, 1980). Subsequently, several other inhibitors of lipoxygenase, including BW755C (Forstermann and Neufang, 1984b), nordihydroguaiaretic acid (NDGA) (Singer and Peach, 1983b; Forstermann and Neufang, 1984b; Furchgott, 1981), nafazatrom (Forstermann and Neufang, 1984b), and phenidone (Forstermann and Neufang, 1984b), have been demonstrated to inhibit endothelium-dependent relaxation in the aorta of the rabbit, as well as other muscular arteries of several species (Peach et al., 1985a). Lipoxygenase products of arachidonic acid, therefore, have received strong consideration as the potential chemical mediator(s) of endothelium-dependent relaxation.

3.1.3. Cytochrome P-450 Blockade

Although cytochrome P-450 had been implicated in hypoxic pulmonary vasoconstriction (Sylvester and McGowan, 1978; Miller and Hales, 1979), the possible role of the cytochrome P-450 pathway in the mechanism of endothelium-dependent relaxation was first suggested by Singer et al. (1984). They demonstrated that two inhibitors of cytochrome P-450, SKF-525A or metyrapone, were each capable of inhibiting endothelium-dependent relaxation in response to methacholine, A23187, or exogenous arachidonic acid in the aorta of the rabbit. The blockade with SKF-525A also has been successful against substance P and A23187 in the porcine aorta (Dusting and MacDonald, 1986).

Recently, Pinto et al. (1986) reported a series of experiments that strongly support the involvement of the P-450 pathway. Arachidonic acid-induced relaxations in intact rings of the pulmonary artery of the rabbit were antagonized by SKF-525A. Vascular cytochrome P-450 activity was then increased by enzyme induction by treating rabbits for 3 d with 3-methylcholanthrene and β-napthoflavone (confirmed by aryl hydrocarbon hydroxylase assay). Cytochrome P-450 induction resulted in a tenfold increase in endothelium-dependent relaxations to arachidonic acid. This relaxation was blocked by SKF-525A. In addition, cobalt chloride, an agent that is known to deplete cytochrome P-450 enzymes was shown to attenuate the arachidonic acid-induced relaxations.

The ability of several potential inhibitors of lipoxygenase to block endothelium-dependent responses does not rule out a role for cytochrome P-450 enzymes. In fact, several of these compounds also block cytochrome P-450 metabolism of arachidonate. ETYA blocks the metabolism of arachidonic acid in renal cortex (Morrison and Pascoe, 1981), and NDGA may inhibit cytochrome P-450 metabolism in anterior pituitary cells (Snyder et al., 1983). It has been shown that treatment of rat liver microsomes with NDGA ($1 \times 10^{-4}M$) caused a shift in the cytochrome P-450 spectra to that of the denatured enzyme (Izzo et al., 1983).

3.2. Involvement of Phospholipase and Calcium Dependence

Inhibitors of phospholipase activation, including quinacrine (mepacrine), trifluoperazine, and *p*-bromophenacylbromide, block endothelium-dependent relaxations in canine, rat, and rabbit aorta (Furchgott, 1983; Singer and Peach, 1983a,b; Forstermann and Neufang, 1985). Also, when exogenous melittin (*see* section 3.4) or phospholipase C are applied to rabbit aortic rings, endothelium-dependent relaxation is observed (Forstermann and Neufang, 1985).

PGI_2 and other cyclooxygenase products of arachidonic acid are released in response to several endothelium-dependent agonists, as demonstrated by radioimmunoassay (Gordon and Martin, 1983a; Forstermann and Neufang, 1985; Loeb et al., 1986). Such findings imply that endothelium-dependent agents induce activation of phospholipase(s) that hydrolyze phospholipids to release arachidonic acid.

There is strong evidence that the synthesis and/or release of EDRF is calcium (Ca^{2+}) dependent (Peach et al., 1987). The removal of external Ca^{2+} or pretreatment with Ca^{2+} channel blockers (verapamil, nifedipine, SKF-525A) in aortic rings of rabbits or rats attenuates endothelium-dependent relaxation to acetylcholine, methacholine, and A23187 (Singer and Peach, 1982; Singer et al., 1984; Long and Stone, 1985; Loeb et al., 1986). Although removal of Ca^{2+} consistently causes marked impairment of responses to EDRF (Peach et al., 1985a; Loeb et al., 1986), not all vessels or endothelium-dependent relaxations are inhibited by Ca^{2+} entry blockers (Miller et al., 1985; Winquist et al., 1985). Several agents known to cause endothelium-dependent relaxation, including norepinephrine (Saida and Van Breemen, 1983), histamine (Hudgins and Weiss, 1968; Gordon and Martin, 1983b), bradykinin (Gordon and Martin, 1983a), and thrombin (DeGroot et al., 1985) have been associated with increases in intracellular Ca^{2+} concentrations. It therefore appears likely that receptor-mediated translocation of intracellular Ca^{2+} (and/or influx) may be the initial step in the release or synthesis of EDRF. It was reported recently that an increase in intracellular Ca^{2+} accompanied the release of EDRF by cultured endothelial cells in response to bradykinin and melittin (Izzo et al., 1986; Peach et al., 1987). Treatment of cultured endothelium with ATP and ADP also results in an increase in free Ca^{2+} concentration in the cells (Luckhoff and Busse, 1986).

This Ca^{2+} dependence further supports the theory that endothelium-dependent relaxation is mediated by metabolites of arachidonic acid. The activation of phospholipase in endothelium that is required for arachidonate release clearly has been shown to be initiated by an increase in cytoplasmic free Ca^{2+} concentration (Whorton et al., 1984).

3.3. Exogenous Arachidonic Acid Studies

Exogenous arachidonate can cause endothelium-dependent vasodilatation. Several laboratories have investigated the effects of exogenous arachidonic acid applied to a variety of vessels from several different species (Table 1). Although many of these studies are consistent with the arachidonate metabolite hypotheses for endothelium-dependent relaxation, there have been conflicting results that make the studies difficult to interpret. The presence or absence of vascular tone prior to treatment with arachidonate, the vessel studied, the dose of arachidonic acid employed, autooxidation, and other problems inherent to the exogenous arachidonic acid are among the factors contributing to variability among studies.

Arachidonate is unstable, and commercially available preparations can vary widely in purity and deteriorate with storage. Many of the spontaneous degradation products are vasoactive. The actions of arachidonate also vary widely with concentration, and this could account for some of the disparate results noted. When applied exogenously, the relatively large concentrations of arachidonate frequently employed can produce excess free radicals that might damage membranes, inactivate enzymes such as cyclo- or lipoxygenases, or even inactivate EDRF in solution. As discussed below, exogenously applied arachidonate may result in the production of several factors in addition to EDRF, many of which are contractile, further confusing the interpretation of experimental results in vascular segments.

Singer and Peach (1983a) applied arachidonate (1–100 μM) to rings of thoracic aorta of the rabbit contracted with phenylephrine. A transient, dose-dependent relaxation was observed in rings with intact endothelium, but not in those preparations in which the endothelium had been removed. This relaxation response induced by arachidonate was potentiated by in-

Table 1
Summary of In Vitro Exogenous Arachidonate Studies[a]

Species	Vessel	Precontracted	[AA] μM	Vascular response	+/−ENDO	Effective inhibitors	Ineffective inhibitors	Reference
Rabbit	Aorta	Yes	10–100	r	+	E, N	I	Singer and Peach (1983a)
				e	−	I		
		No	3–100	c (rapid) c (slow, small)	+	I, E	N	Singer and Peach (1983a)
Dog	Coronary art.	Yes	0.1–1.0	r	+	I, F		Furchgott (1983)
	Sup. mes. art.	Yes	10–100	r	+		I, F	
	Femoral art.	Yes	0.1–30	r	+	I, E	T, 15-HPETE	De Mey et al. (1982)
		Yes	0.1–30	r (small)	−			
	Femoral vein.	Yes	0.01–10	c	+	I	T, IM, N	Miller and Vanhoutte (1984, 1985)
		Yes	0.01–10	0	−			
	Coronary art.	Yes	1–100	r	+		I, N	Rubanyi et al. (1985)
		Yes	1–100	r (small)	−		N	
		Yes	0.1, 0.3	r	+	I		
		Yes	0.1, 0.3	0	−	I		
	Pulmonary vein	Yes	0.1–10	c	+	I	T, IM, N	Miller and Vanhoutte (1985)
		Yes	0.01–10	0	−			
Rat	Aorta	Yes	0.67	r	+		I	Davies and Williams (1984)
		No	6.7–17.4	c	+	I		
Cow	Pulmonary art.	Yes	0.01–1.0	r	+	I, S		Ignarro et al. (1985)
		Yes	10	e	+		I, S (potentiated relaxation)	
		Yes	0.1–10	c	−	I, S		
	Pulmonary vein	Yes	0.1–10	c	+	I		
		Yes	0.1–10	c	−	I		

[a]Includes only studies where the status of endothelial cell integrity is known. Vascular response: r, relaxation; c, contraction; e, equivocal; 0, no response. Inhibitors: I, indomethacin; E, ETYA; N, NDGA; T, tranylcypromine; F, flubiprofen; IM, imidazole; S, SKF-525A.

domethacin (inhibitor of cyclooxygenase) and blocked by ETYA (inhibitor of cyclooxygenase and lipoxygenase) and NDGA (inhibitor of lipoxygenase), implying the involvement of noncyclooxygenase metabolites. In intact aortic rings without active tone, arachidonate (3–100 *μM*) caused a rapidly developing contraction. A slowly developing, small contraction was observed, however, when arachidonate was applied to rings without endothelium. In this situation, indomethacin and ETYA pretreatment inhibited the contraction, indicating that the contractile response to arachidonate was caused by cyclooxygenase metabolites. Arachidonate-induced contractions that were dependent on the endothelium were potentiated by NDGA, suggesting that arachidonate was being diverted to the cyclooxygenase pathway. Extracts of buffer from organ baths with rabbit aorta treated with arachidonate followed by elution from a C18 column and radioimmunoassay indicate that the eicosanoid formed is PGE_2 (Peach and Saye, unpublished observation). Similar results were found in aortic strips of the rabbit (Furchgott, 1984) and thoracic aorta of the rat (Davies and Williams, 1984).

De Mey et al. (1982), using lower concentrations of arachidonate (0.3–30 *μM*) in canine femoral arteries contracted with norepinephrine, produced partial endothelium-dependent vasodilatation that was blocked completely by inhibitors of cyclooxygenase. The responses were obtained with 15-HPETE (not a substrate for cyclooxygenase) that were dependent on the presence of endothelium. In dog coronary and superior mesenteric arteries, the endothelium-dependent relaxation to submicromolar concentrations of archidonate (0.1–1 *μM*) was prevented by inhibition of cyclooxygenase, but that to higher concentrations (10–100 *μM*) was not (Furchgott, 1984).

It is probable that exogenous arachidonic acid causes the elaboration and release of multiple EDRFs. One is clearly cyclooxygenase-dependent, most likely PGI_2 (De Mey et al., 1982; Ignarro et al., 1985); other(s) could be lipoxygenase (Forstermann and Neufang, 1985; Salzman et al., 1980; De Mey et al., 1982), cytochrome P-450 products (Singer et al., 1984; Schwartzman et al., 1985; Pinto et al., 1986), or another agent altogether (Rubanyi et al., 1985; De Mey et al., 1982; Cherry et al., 1983).

Ignarro et al. (1985) studied the effects of arachidonate (0.1–10 *μM*) on bovine intrapulmonary arteries and veins. In arteries, they observed an endothelium-dependent relaxation accompanied by a time- and concentration-dependent increase of both cyclic GMP and cyclic AMP. Indomethacin inhibited relaxations at low concentrations of arachidonate (0.1–1 *μM*) and enhanced relaxations at concentrations of greater than 10 *μM*. Treatment with indomethacin also inhibited the rise in intracellular levels of cyclic AMP, but not cyclic GMP. Arteries without endothelium and veins with or without endothelium, as well as intact arteries, that had no tone only contracted when exposed to arachidonate. These contractions of arteries and veins were inhibited by indomethacin, implying that the responses are mediated by the production of contractile prostaglandins or other eicosanoids from arachidonate. Agents that inhibited endothelium-dependent relaxation induced by acetylcholine (including ETYA, NDGA, quinacrine, and SKF-525A) were relatively ineffective in reversing the relaxation response to arachidonic acid, suggesting that acetylcholine and arachidonic acid act via different EDRFs (Furchgott, 1983; Singer and Peach, 1983b; Ignarro et al., 1985).

Miller and Vanhoutte (1985) studied canine femoral and intrapulmonary veins and found endothelium-dependent contractions in response to arachidonic acid (0.01–10 μM). These contractions were blocked by indomethacin (cyclooxygenase inhibitor), but not by inhibitors of thromboxane A_2 synthetase (imidazole and BW149H), prostacyclin synthetase (tranylcypromine), or lipoxygenase (NDGA).

3.4. Melittin and Endogenous Arachidonate Studies

Recently, the role of endogenous arachidonic acid in endothelium-dependent relaxation has been examined using the putative activator of phospholipase melittin (Forstermann and Neufang, 1985; Loeb et al., 1986), a peptide derived from bee venom (Hassid and Levine, 1977). This agent potentially avoids many of the problems associated with exogenous arachidonate discussed above. Melittin stimulated a dose-related (0.5–5 μg/mL), endothelium-dependent relaxation in contracted rings and strips of thoracic aorta of the rabbit. This response was not blocked by indomethacin, but was inhibited by ETYA, NDGA, mepacrine, *p*-bromophenacyl bromide, BW755C, and hydroquinone. These results are consistent with the pattern of blockade for EDRF released by acetylcholine, A23187, and other agents. The release of arachidonic acid by melittin was confirmed by the concurrent formation of prostacyclin, as measured by its stable hydrolysis metabolite, 6-keto-prostaglandin $F_{1\alpha}$.

The release of EDRF by melittin was confirmed by the rise in cyclic GMP levels of smooth muscle that preceded or was coincident with relaxation (Loeb et al., 1986). The release and transfer of a relaxing factor from the endothelium to vascular smooth muscle was demonstrated also using transfer/superfusion experiments involving cultured endothelial cells and the aorta of the rabbit. Bovine aortic endothelial cells were grown in roller cultures on microcarrier beads that could be packed into a column and superfused with a buffer solution. The superfusate from the column was then bioassayed for EDRF on rings of aorta from which the endothelium had been removed. Melittin relaxed the rings when it was added to the superfusate and allowed to pass over the column of endothelial cells, but not when it was applied directly to the rings denuded of endothelium.

3.5. Antioxidants and Nonspecific Radical Scavengers

Consistent with the hypothesis that EDRF is a noncyclooxygenase derivative of arachidonic acid is the large body of evidence that implies that the factor is a free radical or oxygen scavenger. This includes the fact that anoxia inhibits the endothelium-dependent relaxation of rabbit aorta to acetylcholine or A23187, and the relaxation of dog intrapulmonary and renal arteries to acetylcholine or bradykinin (Furchgott, 1984). Also, EDRF has been shown to be extremely labile (Griffith et al., 1984; Rubanyi et al., 1985; Peach et al., 1985b; Cocks et al., 1985), as would be expected for a mechanism involving a free radical. Early studies of endothelium-dependent relaxation determined that hydroquinone, a potent free radical scavenger, rapidly antagonized the relaxation of aortic rings of the rabbit to acetylcholine or A23187 (Furchgott, 1981). It has long been known that other free radicals (e.g., nitric oxide and the hydroxyl radical) stimulate guanylate cyclase to mediate the relaxation of smooth muscle by glyceryl trinitrate and other organic or inorganic nitrates (Murad et al., 1979). The finding that rises in intracellular cyclic GMP levels accompany endothelium-dependent relaxation (Rapoport and Murad, 1983a,b; Ignarro et al., 1985) further support the hypothesis that a free radical is involved. The acute inhibition of acetylcholine and A23187 by hemoglobin (Holzman, 1982; Furchgott, 1984; Martin et al., 1985) and methylene blue (Furchgott, 1984; Martin et al., 1985) (each capable of binding free radicals) is consistent with the idea that endothelium-dependent relaxation involves a free radical. It is also possible that the inhibition of the EDRF response is a result of the blockade of guanylate cyclase, since the rise in cyclic GMP that is coincident with relaxations to acetylcholine, A23187, and sodium nitroprusside or glyceryl trinitrate also is prevented (Martin et al., 1985; Rapoport et al., 1985).

A variety of antioxidants have been applied to arterial preparations in an attempt to prevent acetylcholine-induced endothelium-dependent relaxation with mixed results (Table 2). Butylated hydroxytoluene, dithiothreitol, and α-tocopherol all reversed endothelium-dependent relaxations in a dose-dependent manner (Forstermann and Neufang, 1984b). A different group of oxygen metabolite scavengers (superoxide dismutase, catalase, dimethyl sulfoxide, and mannitol) had no effect on acetylcholine vasodilatation (Silin et al., 1985). Griffith et al. (1984) inhibited the acetylcholine response with several antioxidants (KBH_4, dithiothreitol, cysteine, hydroquinone, phenidone, and NDGA), but failed to inhibit with several others (spin trap reagents phenyl-*t*-butylnitrone and 2-nitroso-2-methylpropane, 1,4-diazo-

Table 2
Effects of Antioxidants and Nonspecific Radical Scavengers on Endothelium-Dependent Relaxation

Reagent or compound	Block or Reverse EDRF	References
Superoxide dismutase	No	Silin et al. (1985); Griffith et al. (1984)
Catalase	No	Silin et al. (1985); Griffith et al. (1984)
Mannitol	No	Silin et al. (1985); Griffith et al. (1984)
Dimethyl sulfoxide	No	Singer and Peach (1983a); Silin et al. (1985); Griffith et al. (1984)
Sodium metabisulfite	No	Peach et al. (1985a)
Ascorbate	Yes/No	Rubanyi et al. (1985); Furchgott et al. (1983); Peach et al. (1985a)
KBH_4	Yes	Furchgott (1983); Griffith et al. (1984)
Butylated hydroxytoluene	Yes	Forstermann and Neufang (1984b)
Methylene blue	Yes	Furchgott (1984); Martin et al. (1985); Rapoport et al. (1985)
Phenidone	Yes	Griffith et al. (1984)
Methionine	No	Griffith et al. (1984)
Cysteine	Yes	Griffith et al. (1984)
Dithiothreitol	Yes	Griffith et al. (1984); Forstermann and Neufang (1984b)
Nitroblue tetrazolium	Yes	Johns and Peach (unpublished observation)
Sodium benzoate	No	Griffith et al. (1984)
Hemoglobin	Yes	Holzman (1982); Furchgott (1984); Martin et al. (1985); Rapoport et al. (1985)
Methemoglobin	No	Martin et al. (1985)
Myoglobin	No	Martin et al. (1985)
α-Tocopherol	Yes	Forstermann and Neufang (1984b)
Phenyl-*t*-butylnitrone	Yes/No	Griffith et al. (1984); Sata et al. (1986)
2-Nitroso-2-methyl Propane	No	Griffith et al. (1984)
Catecholamine	yes	Rubanyi et al. (1985); Bacq and Peach (unpublished)

Table 2 (*continued*)
Effects of Antioxidants and Nonspecific Radical Scavengers on Endothelium-Dependent Relaxation

Reagent or compound	Block or Reverse EDRF	References
Catechol	Yes	Bacq and Peach (unpublished observation)
Resorcinol	Yes	Bacq and Peach (unpublished observation)
Hydroquinone	Yes	Furchgott (1981); Griffith et al. (1984); Loeb et al. (1986)
Pyrogallol	Yes	Bacq and Peach (unpublished observation)
Phloroglucinol	Yes	Bacq and Peach (unpublished observation)
Phenylalanine	No	Griffith et al. (1984)
Histidine	No	Griffith et al. (1984)
NDGA	Yes	Singer and Peach (1983b); Furchgott (1983); Forstermann and Neufang (1984b)
1,4-Diazobicyclo (2,2,2) octane	No	Griffith et al. (1984)
Arachidonate	Yes	Peach et al. (1985b)
ETYA	Yes	Furchgott (1983); Singer and Peach (1983b); Griffith et al. (1984); Loeb et al. (1985); Forstermann and Neufang (1984b)

bicyclo(2,2,2)-octane, phenylalanine, histidine, methionine, dimethyl sulfoxide, sodium benzoate, mannitol, superoxide dismutase, and catalase). DMSO, in concentrations known to inactivate the hydroxyl radical, does not block EDRF (Singer and Peach, 1983a). It also has been shown that nitroblue tetrazolium, a yellow dye that precipitates and turns blue in the presence of free radicals (Beauchamp and Fridovich, 1971), is reduced and precipitates on the intima when an intact artery is exposed to acetylcholine, A23187, or melittin, and the relaxation response to these agents is blocked (Peach and Johns, unpublished observations).

3.6. Electron Spin Resonance Spectroscopy

Recently, Sata et al. (1986) have applied electron spin resonance spectroscopy and the spin trapping agent phenyl-*t*-butylnitrone to trap and characterize free radicals involved in endothelium-dependent relaxation.

They studied the relaxations of the thoracic aorta of the rat and the pulmonary artery of the guinea pig to acetylcholine and adenosine triphosphate (ATP) with and without the addition of phenyl-*t*-butylnitrone. Phenyl-*t*-butylnitrone reduced by 50% the maximal relaxation of both arteries to acetylcholine, but had no effect on ATP-induced inhibition. Electron spin resonance spectroscopy detected a carbon-centered free radical in lipid extracts of guinea pig pulmonary artery incubated with phenyl-*t*-butylnitrone and acetylcholine, but not ATP. Phenyl-*t*-butylnitrone also was shown to abolish the acetylcholine-induced, but not the ATP-induced, accumulation of cyclic GMP in the rat tail artery. These investigators concluded that endothelium-dependent relaxation caused by acetylcholine is mediated via a free radical mechanism, and that ATP induces relaxation either by a different radical or a completely different mechanism that does not involve free radicals.

4. Evidence Against Arachidonate Metabolite Hypothesis

The evidence presented above forms the basis for the hypothesis that Ca^{2+}-dependent activation of phospholipase causes the release from endothelium or smooth muscle of arachidonate or other unsaturated fatty acids. The fatty acid(s) subsequently undergo metabolism via *non*cyclooxygenase pathways localized in the endothelium to produce a labile substance that is released from the endothelium and acts on smooth muscle to mediate endothelium-dependent vasodilatation. A number of observations, however, draw question to this hypothesis.

4.1. Nonspecific Action of Many Blockers

Many of the agents that have been used to block arachidonic acid metabolism and phospholipase activation are nonspecific in their activity. Also, several studies with the same reagents have produced conflicting results. Seldom have the drugs been shown to have exerted the specific effect for which they were given.

For instance, the lipoxygenase inhibitor BW755C did not have any effect on the endothelium-dependent relaxation by acetylcholine in the aorta of the rabbit (Furchgott and Zawadzki, 1980), whereas others reported that this compound blocked EDRF (Griffith et al., 1984; Forstermann and Neufang, 1985). Quinacrine, a proposed phospholipase inhibitor, while successfully blocking acetylcholine-induced inhibition, did not alter the response caused by A23187 (Singer and Peach, 1983b). Because contractile responses to acetylcholine following removal of the endothelium also were inhibited by quinacrine, it was clear that the effects of quinacrine are at least partially related to inhibition of muscarinic receptors (Peach et al., 1985a). Other inhibitors of phospholipase (or calmodulin), dexamethasone and *p*-bromophenacyl bromide or trifluoperazine also have yielded conflicting results regarding EDRF (Furchgott, 1983; Griffith et al., 1984; Peach et al., 1985a,b).

Doses of ETYA (>100 μM) that have been used and are required to block endothelium-dependent relaxations are in vast excess of reported K_i values of these agents for cyclooxygenase and/or lipoxygenase, but are appropriate K_i values for cytochrome P-450 (Peach et al., 1985a,b). At these high concentrations, both ETYA and NDGA are strong antioxidants that could be acting by blocking guanylate cyclase or by increasing the lability of EDRF in solution (Griffith et al., 1984). ETYA (300 μM) and NDGA (50–100 μM) attenuate the rise in cyclic GMP induced by sodium nitroprusside (Peach, unpublished observation; Rapoport and Murad, 1983b). Griffith et al. (1984), using perfusion transfer experiments, have demonstrated that these agents and other putative lipoxygenase inhibitors with antioxidant activity inactivated EDRF in solution without ever coming into contact with the endothelium.

4.2. Other Fatty Acids Cause Endothelium-Dependent Relaxation

Several unsaturated and saturated fatty acids other than arachidonate (*cis*-4,7,10,13,16,19-docosohexaenoic, oleic, elaidic, and *cis*-vaccenic acids) produce endothelium-dependent relaxations in arteries of rabbits and dogs (Cherry et al., 1983). It is proposed that the nonspecific action of unsaturated fatty acids in causing endothelium-dependent relaxation implies that they are not precursors of EDRF, but rather act indirectly to effect its synthesis or release. They could potentially do this by increasing membrane fluidity (Furchgott, 1984) or via nonspecific membrane effects, including detergent-like actions or endothelium damage (Pinto et al., 1986). The effects of these "other fatty acids" to induce relaxation requires Ca^{2+} in the external medium, which is consistent with the hypothesis that they cause EDRF release. Arachidonate-induced relaxation does not require extracellular Ca^{2+} (Michener et al., 1983), however, suggesting a different mechanism of action.

Pinto et al. (1986) have also questioned whether all fatty acids work in the same manner, pointing out that the concentrations of arachidonic acid that were required for relaxation are much lower (10^{-8} to $10^{-6}M$) in comparison to the other fatty acids that were tested. The increased sensitivity to arachidonate suggests that its action is not a nonspecific effect, as is the case with other fatty acids. They also point out that docosohexanoic acid, the fatty acid second most potent to arachidonate in producing endothelium-dependent relaxation, is a substrate for cytochrome P-450 enzymes and could potentially produce an endothelium-dependent metabolite.

4.3. Possible Hydrophilic Nature of EDRF

Cocks et al. (1985) recently investigated the chemical nature of EDRF using cultured bovine endothelial cells grown on microcarrier beads packed into columns and superfused. These cells were shown to produce EDRF during treatment with bradykinin and A23187 by monitoring an endothelium-dependent response when the perfusate was dripped over bioassay rings of canine coronary artery from which the endothelium had been removed. Hydrophobic (BondElut C-18) exchange columns placed between the endothelial cell source and the bioassay ring failed to prevent endothelium-dependent relaxation. When an anion exchange column was used in place of the C-18 column, however, endothelium-dependent relaxation was markedly inhibited. Thus EDRF was hydrophilic and negatively charged at neutral pH, leading Cocks and coworkers to propose that the factor could not be an oxidative product of arachidonate. It should be pointed out, however, that several metabolites of arachidonic acid (e.g., LTD_4 and LTF_4) would not be retained on a C-18 column under the conditions of study. Their results are quite consistent with the chemical characteristics of some epoxide derivative eicosanoids.

4.4. Exogenous Application of Arachidonate Metabolites

Several noncyclooxygenase products of arachidonate have been tested for endothelium-dependent relaxation without success. The hydroxylated eicosatetraenoic acids 5-HETE, 12-HETE, 15-HETE, and 5,12-diHETE had no vasoactive effect on intact rabbit aorta strips (Forstermann and Neufang, 1984c). Likewise, the exogenous leukotrienes LTA_4, LTB_4, LTC_4, and LTD_4 failed to produce relaxation (Forstermann and Neufang, 1984a). 15-HPETE, LTD_4, and LTF_4, however, are vasodilating metabolites of arachidonic acid (De Mey et al., 1982; Secrest et al., 1985; Letts et al., 1985) in a limited number of vessels and species.

5. Future Studies of Endothelium-Derived Relaxing Factor(s)

Although a great deal is now known about endothelium-dependent relaxation, the chemical nature of EDRF and more precise details regarding its synthesis, release, and mechanism of action remain to be established.

Indeed, it is quite possible that more than one "factor" may be elaborated by endothelium. It is clear, however, that most if not all indomethacin *in*sensitive relaxation responses that require the endothelium, correlate with activation of the guanylate cyclase of the smooth muscle. Therefore, if there are multiple EDRFs, they have a common mechanism of action.

Several recent findings may prove to be pivotal in future investigations into these issues. The demonstration that melittin releases endogenous arachidonate and stimulates endothelium-dependent relaxation in rings and in cultured endothelial cells will help to avoid some of the problems associated with the use of exogenous arachidonic acid. The ability to produce EDRF from cultured endothelial cells may allow for its production in sufficient quantities to permit its isolation and characterization, particularly in light of its labile nature. The use of melittin in combination with endothelial cell cultures will avoid many of the problems with variability of agonist-receptor expression and rapid refractoriness to several agents (ATP, bradykinin) that have been encountered in cultured cells (Ager and Martin, 1983; Loeb et al., 1985).

The recent work by Pinto et al. (1986) (discussed above) on the role of cytochrome P-450-dependent enzymes in endothelium-dependent relaxation provides new impetus for investigation of this pathway of arachidonic acid metabolism. Continuing advances in eicosanoid and endoperoxide chemistry will be helpful. Indeed, Ullrich and Graf (1984) have reported that prostacyclin synthetase and thromboxane synthetase belong to the family of cytochrome P-450 enzymes.

A great deal of evidence has been presented supporting a role for non-cyclooxygenase metabolites of arachidonic acid in endothelium-dependent vasodilatation. Unfortunately, all of it is indirect, and other studies are inconsistent with such a conclusion. A final decision regarding the nature of EDRF will require further investigation.

References

Abraham, N. G., Pinto, A., Mullane, K. M., Revere, R. A., and Spokes, E.: Presence of cytochrome P450-dependent monoxygenase in intimal cells of the hog aorta. *Hypertension* **7**:899–904, 1985.

Ager, A. and Martin, W.: Loss of receptor-mediated ^{86}Rb efflux from pig aortic endothelial cells in culture. *Br. J. Pharmacol.* **80**:5–6, 1983.

Baird, W. N., Chemerys, R., Grinspan, J. B., Mueller, S. N., and Levine, E. M.: Benzo(a)pyrene metabolism in bovine aortic endothelial and bovine lung fibroblast-like cultures. *Cancer Res.* **40**:1781–1786, 1980.

Beauchamp, C. and Fridovich, I.: Superoxide dismutase: Improved assays and an assay applicable to acrylamide gels. *Anal. Biochem.* **44:**276–287, 1971.

Buchanan, M. R., Batt, R. W., Magas, Z., Van Ryan,, J., Hirsh, J., and Nazir, D. J.: Endothelial cells produce a lipoxygenase derived chemo-repellant which influences platelet–endothelial cell interactions—effect of aspirin and salycilate. *Thromb. Haemost,* **53:**306–311, 1985.

Bunting, S., Gryglewski, R., Moncada, S., and Vane, J. R.: Arterial walls generate from prostaglandin endoperoxides a substance (prostaglandin X) which relaxes strips of mesenteric and cardiac arteries and inhibits platelet aggregation. *Prostaglandins* **12:**897–913, 1976.

Capdevilla, J., Chacos, N., Werringloer, J.,Prough, R. A., and Estabrook, R. W.: Liver microsomal cytochrome P450 and the oxidative metabolism of arachidonic acid. *Proc. Natl. Acad. Sci. USA* **78:**5362–5366, 1981.

Cherry, P. D., Furchgott, R. F., and Zawadzki, J. V.: The endothelium-dependent relaxation of vascular smooth muscle by unsaturated fatty acids. *Fed. Proc.* **42:**619, 1983.

Cocks, T. M., Angus, J. A., Campbell, J. H., and Campbell, G. R.: Release and properties of endothelium-derived relaxing factor (EDRF) from endothelial cells in culture. *J. Cell. Physiology* **123:**310–320, 1985.

Davies, J. M. and Williams, K. I.: Endothelial-dependent relaxant effects of vasoactive intestinal polypeptide and arachidonic acid in rat aortic strips. *Prostaglandins* **27:**195–202, 1984.

Dees, J. H., Masters, B. S. S., Muller-Eberhard, U., and Johnson, E. F.: Effect of 2,3,7,8-tetrachlorodibenzo-*p*-dioxin and phenobarbital on the occurrence and distribution of four cytochrome P-450 isozymes in rabbit kidney, lung, and liver. *Cancer Res.* **42:**1423–1432, 1982.

DeGroot, P. G., Brinkman, H. J. M., Gonsalves, M. D., and Van Mourik, J. A.: The role of thrombin in the regulation of the endothelial protaglandin production. *Biochem. Biophys. Acta* **846:**342–349, 1985.

De Mey, J. G., Claeys, M., and Vanhoutte, P. M.: Endothelium-dependent inhibitory effects of acetycholine, adenosine tri-phosphate, thrombin and arachidonic acid in the canine femoral artery. *J. Pharmacol. Exp. Ther.* **222:**166–173, 1982.

Dusting, G. J. and MacDonald, P. S.: Endothelium-dependant vasodilation: Role of beta-adrenoceptors, calcium and cytochrome P-450. *Blood Vess.* **23:**66, 1986.

Dusting, G. J., Moncada, S., and Vane, J. R.: Prostacyclin (PGX) is the endogenous metabolite responsible for relaxation of coronary arteries induced by arachidonic acid. *Prostaglandins* **13:**3–15, 1977.

Dusting, G. J., Mullins, E. M., and Nolan, R. D.: Prostacyclin (PGI_2) release accompanying angiotensin conversion in rat mesenteric vasculature. *Eur. J. Pharmacol.* **70:**129–137, 1981.

Ford-Hutchinson, A. W.: Leukotrienes: Their formation and role as inflammatory mediators. *Fed. Proc.* **44:**25–29, 1985.

Forstermann, U. and Neufang, B: Inhibition by diethylcarbamazine of acetylcholine induced endothelium-dependent relaxation of rabbit aorta: Are leukotrienes involved? *J. Pharm. Pharmacol.* **36:**696–697, 1984a.

Forstermann, U. and Neufang, B: The endothelium-dependent vasodilator effect of acetylcholine: A characterization of the endothelial relaxing factor with inhibitors of arachidonic acid metabolism. *Eur. J. Pharmacol.* **103:**65–70, 1984b.

Forstermann, U. and Neufang, B.: The endothelium-dependent relaxation of rabbit aorta: Effects of antioxidants and hydroxylated eicosatetrenoic acids. *Br. J. Pharmacol.* **82:**765–767, 1984c.

Forstermann, U. and Neufang, B: endothelium-dependent vasodilation by melittin: Are lipoxygenase products involved? *Am. J. Physiol.* **249** (*Heart Circ. Physiol.* **18**):H14–H19, 1985.

Furchgott, R. F.: The requirement for endothelial cells in the relaxation of arteries by acetylcholine and some other vasodilators. *Trends Pharmacol. Sci.* **2:**173–176, 1981.

Furchgott, R. F.: Role of endothelium in response of vascular smooth muscle. *Circ. Res.* **53:**557–573, 1983.

Furchgott, R. F.: The role of endothelium in the responses of vascular smooth muscle to drugs. *Ann. Rev. Pharmacol. Toxicol.* **24:**175–197, 1984.

Furchgott, R. and Zawadzki, D.: The obligatory role of endothelial cells in the relaxation of arterial smooth muscle by acetylcholine. *Nature* **288:**373–376, 1980.

Gerritsen, M. C. and Chelli, C. D.: Arachidonic acid and prostaglandin endoperoxide metabolism in isolated rabbit and coronary microvessels and isolated and cultivated coronary microvessel endothelial cells. *J. Clin. Invest.* **72:**1658–1671, 1983.

Gordon, J. L. and Martin, W.: Stimulation of endothelial prostacyclin production plays no role in endothelium-dependent relaxation of the pig aorta. *Br. J. Pharmacol.* **80:**179–186, 1983a.

Gordon, J. L. and Martin, W.: Endothelium-dependent relaxation of the pig aorta: Relationship to stimulation of ^{86}Rb efflux from isolated endothelial cells. *Br. J. Pharmacol.* **79:**531–541, 1983b.

Gorman, R. R., Oglesby, T. D., Bundy, G. L., and Hopkins, A. K.: Evidence for 15-HETE synthesis by human umbilical vein endothelial cells. *Circulation* **72:**708–712, 1985.

Greenwald, J. E., Bianchine, J. R. and Wong, L. K.: The production of the arachidonate metabolite HETE in vascular tissue. *Nature* **281:**588–589, 1979.

Griffith, T. M., Edwards, D. H., Lewis, M. J., Newby, A. L., and Henderson, A. H.: The nature of endothelium-derived relaxing factor. *Nature* **308:**645–647, 1984.

Hasaid, A. and Levine, L.: Stimulation of phospholipase activity and prostaglandin biosynthesis by melittin in cell culture and *in vivo*. *Res. Commun. Chem. Pathol. Pharmacol.* **18:**507–517, 1977.

Holzman, S.: Endothelium induced relaxation by acetylcholine associated with larger rises in cGMP in coronary arterial strips. *J. Cyclic Nucleotide Res.* **8:**409–419, 1982.

Hudgins, P. M. and Weiss, G. B.: Differential effects of calcium removal upon vascular smooth muscle contraction induced by norepinephrine, histamine, and potassium. *J. Pharmacol. Exp. Ther.* **159:**91–97, 1968.

Ignarro, L. J. and Kadowitz, P. J.: The pharmacological and physiological role of cyclic GMP in vascular smooth muscle relaxation. *Ann. Rev. Pharmacol. Toxicol.* **25:**171–191, 1985.

Ignarro, L. J., Harbison, R. G., Wood, K. S., Wolin, M. S., McNamara, D. B., Nyman, A. L., and Kadowitz, P. J.: Differences in responsiveness of intrapulmonary artery and vein to guanosine 3′,5′-monophosphate and cyclic adenosine 3′,5′-monophosphate. *J. Pharmacol. Exp. Ther.* **233:**560–569, 1985.

Ingerman-Wojenski, C., Silver, M. J., Smith, J. B., and Macarak, E.: Bovine endothelial cells in culture produce thromboxane as well as prostacyclin. *J. Clin. Invest.* **67:**1292–1296, 1981.

Izzo, N. J., Loeb, A. L., Johns, R. A., and Peach, M. J.: Intracellular calcium flux accompanies the release of endothelium derived relaxing factor (EDRF) and prostacyclin (PGI_2) from cultured endothelial cells. *Fed. Proc.* **45:**198, 1986.

Izzo, N. J., Singer, H. A., Saye, J. A., and Peach, M. J.: Cytochrome P450 inhibitors block endothelium-dependent aortic relaxation responses. *Fed. Proc.* **42:**651, 1983.

Jakschik, B. A. and Lee, L. H.: Enzymic assembly of slow reacting substance. *Nature* **287:**51–52, 1980.

Johnson, A. R., Revtyak, G., and Campbell, W. B.: Arachidonic acid metabolites and endothelial injury: Studies with cultures of human endothelial cells. *Fed. Proc.* **44:**19–24, 1985.

Juchau, M. R., Bond, J. A., and Benditt, E. P.: Aryl-4-monooxygenase and cytochrome P-450 in the aorta: Possible role in atherosclerosis. *Proc. Natl. Acad. Sci. USA* **73:**3723–3725, 1976.

Letts, L. G., Cirino, M., Yusko, P., Fitzsimmons, B., Ford-Hutchinson, A. W., and Rokach, J.: Actions of synthetic leukotrienes on platelets and blood vessels in the anesthetized pig: The release of a platelet derived vasodilator. *Prostaglandins* **29:**1049–1062, 1985.

Loeb, A. L., Owens, G. K., and Peach, M. J.: Evidence for endothelium-derived relaxing factor in cultured cells. *Hypertension* **7:**804–807, 1985.

Loeb, A. L., Johns, R. A., and Peach, M. J.: Extracellular Calcium is Not Required for Melittin-Induced Release of Endothelium Derived Relaxing Factor from Intact Tissues or Cultured Endothelial Cells in *Proceedings of Mechanisms of Vasodilation Symposium* (Vanhoutte, P. M., ed.), in preparation.

Long, C. J. and Stone, T. W.: The release of endothelium-derived relaxing factor is calcium dependent. *Blood Vess.* **22:**205–208, 1985.

Luckhoff, A. and Busse, R.: Increased free calcium in endothelial cells under stimulation with adenine nucleotides. *J. Cell. Physiol.* **126:**414–420, 1986.

MacIntyre, D. E., Pearson, J. D., and Gordon, J. C.: Localisation and stimulation of prostacyclin production in vascular cells. *Nature* **271:**549–551, 1978.

Marcus, A. J.,Broekman, M. J., Weksler, B. B., Jaffe, E. A., Safier, L. B., Ullman, H. L., Islam, N., and Tack-Goldman, K.: Arachidonic acid metabolism in endothelial cells and platelets. *Ann. NY Acad. Sci.* **401:**195-202, 1982.

Martin, W., Villani, G. M., Jothianandan, D., and Furchgott, R. F.: Blockade of endothelium-dependent and glyceryl trinitrate-induced relaxation of rabbit aorta by certain ferrous hemoproteins. *J. Pharm. Exp. Ther.* **233:**675–679, 1985.

Michener, M. L., Singer, H. A., and Peach, M. J.: Endothelium-dependent relaxation by arachidonic acid in rabbit thoracic aorta. *Fed. Proc.* **42:**651, 1983.

Miller, M. A. and Hales, C. A.: Role of cytochrome P450 in alveolar hypoxic pulmonary vasoconstriction in dogs. *J. Clin. Invest.* **64:**666–673, 1979.

Miller, R. C., Schoeffter, P., and Stoclet, J. C.: Insensitivity of calcium-dependent endothelial stimulation in rat isolated aorta to the calcium entry blocker, flunarazine. *Br. J. Pharmacol.* **85:**481–487, 1985.

Miller, V. M. and Vanhoutte, P. M.: Endothelium dependent contractions to arachidonic acid are mediated by products of cyclooxygenase. *Am. J. Physiol.* **248:**H432–H437, 1985.

Moncada, S., Herman, A. G., Higgs, E. A., and Vane, J. R.: Differential formation of prostacyclin (PGX or PGI_2) by layers of arterial wall. An explanation for the anti-thrombotic properties of vascular endothelium. *Thrombosis Res.* **11:**323–344, 1977.

Morrison, A. R. and Pascoe, N.: Metabolism of arachidonate through NADPH-dependent oxygenase of renal cortex. *Proc. Natl. Acad. Sci. USA* **78:**7375–7378, 1981.

Mullane, K. M. and Moncada, S: Prostacyclin release and modulation of some vasoactive hormones. *Prostaglandins* **20:**25–49, 1980.

Murad, F., Arnold, W. P., Mittal, C. K., and Braughler, J. M.: Properties and regulation of guanylate cyclase and some proposed functions for cyclic GMP. *Adv. Cyclic Nucleotide Res.* **11:**175–204, 1979.

Needleman, P. and Isakson, P. C.: Intrinsic Prostaglandin Biosynthesis in Blood Vessels, in *Handbook of Physiology* vol. II *Vascular Smooth Muscle,* American Physiological Society, Bethesda, Maryland, 1980.

Oliw, E. H., Guengerich, F. P., and Oates, J. A.: Oxygenation of arachidonic acid by hepatic monooxygenases. *J. Biol. Chem.* **257:**3771–3787, 1982.

Peach, M. J., Loeb, A. L., Singer, H. A., and Saye, J. A.: Endothelium-derived relaxing factor. *Hypertension* **7**(suppl. I)**:**I94–I100, 1985a.

Peach, M. J., Singer, H. A., and Loeb, A. L.: Mechanisms of endothelium dependent vascular smooth muscle relaxation. *Biochem. Pharmacol.* **34:**1867–1874, 1985b

Peach, M. J., Singer, H. A., Izzo, N. J., and Loeb, A. L.: The role of calcium in endothelium-dependent relaxation of arterial smooth muscle. *Am. J. Cardiol.* **59:**35A–43A, 1987.

Pinto, A., Abraham, N. G., and Mullane, K. M.: Cytochrome P-450-dependent relaxations induced by arachidonic acid. *J. Pharmacol. Exp. Ther.* **236:**445–451, 1986.

Rapoport, R. M. and Murad, F.: Endothelium-dependent and nitrovasodilator-induced relaxation of vascular smooth muscle: Role of cyclic GMP. *J. Cyclic Nucleotide Protein Phosphor. Res.* **9:**281–296, 1983a.

Rapoport, R. M. and Murad, F: Agonist-induced relaxation in rat thoracic aorta may be mediated through cGMP *Circ. Res.* **521:**352–357, 1983b.

Rapoport, R. M., Waldman, S. A., Schwartz, K., Winquist, R. J., and Murad, F.: Effects of atrial natriuretic factor, sodium nitroprusside, and acetylcholine on cyclic GMP levels and relaxation in rat aorta. *Eur. J. Pharmacol.* **115:**219–229, 1985.

Rubanyi, G. M., Lorenz, R. R., and Vanhoutte, P. M.: Bioassay of endothelium-derived relaxing factor(s): inactivation by catecholamines. *Am. J. Physiol.* **249** (*Heart. Circ. Physiol.* 18) H95–H101, 1985.

Saida, K. and vanBreeman, C.: Mechanisms of Ca^{++} antagonist-induced vasodilation. Intracellular actions. *Circ. Res.* **52:**137–142, 1983.

Salzman, P. M., Salmon, J. A., and Moncada, S.: Prostacyclin and thromboxane A_2 synthesis by rabbit pulmonary artery. *J. Pharmacol. Exp. Ther.* **215:**240–247, 1980.

Samuelson, B.: Leukotrienes: Mediators of immediate hypersensitivity reactions and inflammation. *Science* **220:**568–575, 1983.

Sata, T., Misra, H. P., Linden, J. M., Liu, L. W., Kubota, E., and Said, S. I.: A carbon-centered free radical mediates acetylcholine-induced relaxation of rat aorta and guinea pig pulmonary artery. *Clin. Res.* **34:**710A, 1986.

Schwartzman, M., Carroll, M. A., Abraham, N. G., Ferreri, N. R., Songu-Mize, E., and McGiff, J. C.: Renal arachidonic acid metabolism: The third pathway. *Hypertension* **7**(suppl. 1)**:**I136–I144, 1985.

Secrest, R. J., Olsen, E. J., and Chapnick, B. M.: Leukotriene D_4 relaxes canine renal and superior mesenteric arteries. *Circ. Res.* **57:**323–329, 1985.

Silin, P. J., Strulowitz, J. A., Wolin, M. S., and Belloni, F. L.: Absence of a role for superoxide anion, hydrogen peroxide and hydroxyl radical in endothelium-mediated relaxation of rabbit aorta. *Blood Vess.* **22:**65–73, 1985.

Singer, A. A. and Peach, M. J.: Calcium and endothelial-mediated vascular smooth muscle relaxation in rabbit aorta. *Hypertension* **4**(Suppl. II): 19–25, 1982.

Singer, H. A. and Peach, M. J.: Endothelium-dependent relaxation of rabbit aorta. I. Relaxation stimulated by arachidonic acid. *J. Pharmacol. Exp. Ther.* **226:**790–795,, 1983a.

Singer, H. A. and Peach, M. J.: Endothelium-dependent relaxation of rabbit aorta. II. Inhibition of relaxation stimulated by methacholine and A23187 with antagonists of arachidonic acid metabolism. *J. Pharmacol. Exp. Ther.* **226:**796–801, 1983b.

Singer, H. A., Saye, J. A., and Peach, M. J.: Effects of cytochrome P-450 inhibitors on endothelium-dependent relaxation in rabbit aorta. *Blood Vess.* **21**:223–230, 1984.

Snyder, G. D., Capdevila, J., Chacos, N., Mauna, S., and Falck, J. R.: Action of luteinizing hormone-releasing hormone: Involvement of novel arachidonate metabolites. *Proc. Natl. Acad. Sci. USA* **80**:3504–3507, 1983.

Sylvester, J. T. and McGowan, C.: The effects of agents that bind to cytochrome P450 on hypoxic pulmonary vasoconstriction. *Circ. Res.* **43**:429–437, 1978.

Ullrich, V. and Graf, H.: Prostacyclin and thromboxane synthase as P-450 enzymes. *Trends Pharmacol. Sci.* **5**:352–355, 1984.

Whorton, A. R., Willis, C. E., Kent, R. S., and Young, S. L.: The role of calcium in the regulation of prostacyclin synthesis by porcine aortic endothelial cells. *Lipids* **19**:17–24, 1984.

Winquist, R. J., Bunting, P. B., and Schofield, T. L.: Blockade of endothelium-dependent relaxation by the amiloride analog of dichlorobenzamil: Possible role of Na^+/Ca^{++} exchange in the release of endothelium-derived relaxing factor. *J. Pharmacol. Exp. Ther.* **235**:644–650, 1985.

Chapter 5

Modulation of the Release and Biological Activity of Endothelium-Derived Relaxing Factor by Oxygen-Derived Free Radicals

Gabor M. Rubanyi and Paul M. Vanhoutte

1. Introduction

The complete reduction of molecular oxygen to water involves the addition of four electrons that can occur univalently, resulting in a series of intermediates (oxygen-derived free radicals, Fig. 1), or quadrivalently by the mitochondrial cytochrome oxidase system, which avoids these reactive intermediates (DelMaestro, 1980; Freeman and Crapo, 1982). The presence and uniquity of enzymes (superoxide dismutase, catalase, and peroxidases) that scavenge the byproducts of the univalent pathway in aerobic cells suggest that superoxide anions and hydrogen peroxide are important byproducts of the oxidative metabolism. Actually there are several pathways in aerobic cells leading to the production of oxygen-derived free radicals (DelMaestro, 1980; Freeman and Crapo, 1982). Important sources are enzymes associated with the metabolism of arachidonic acid, such as cyclooxygenase, lipoxygenase, and cytochrome P-450 (DelMaestro, 1980; Freeman and Crapo, 1982; Pinto et al., 1986). Superoxide anion radicals generated intracellularly or extracellularly by photolysis of water (Freeman and Crapo, 1982) or by xanthine oxidase in the presence of xanthine (Beauchamp and Fridovich, 1970; Kellog and Fridovich, 1975), produce hydrogen peroxide by the dismutation reaction. This reaction can proceed spontaneously, or can be catalyzed by superoxide dismutase (McCord and

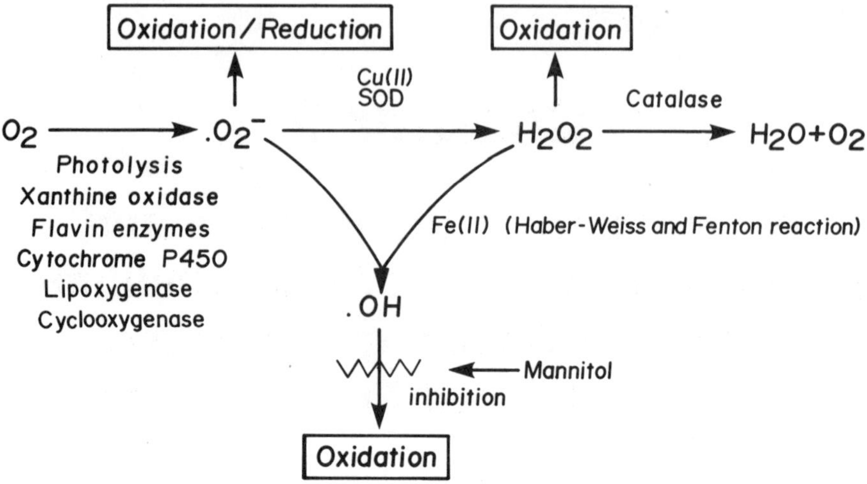

Fig. 1. Sources and scavengers of reactive intermediates [oxygen-derived free radicals: superoxide anion ($\cdot O_2-$); hydrogen peroxide (H_2O_2), and hydroxyl radical ($\cdot OH$)] produced in the univalent pathway of reduction of molecular oxygen (O_2) to water (H_2O) (modified from Rubanyi and Vanhoutte, 1986a, with permission).

Fridovich, 1969). The hydrogen peroxide formed can be scavenged with catalase (Beauchamp and Fridovich, 1970; Kellog and Fridovich, 1975). The superoxide anion radical and hydrogen peroxide can interact to generate the hydroxyl radical via the Haber-Weiss reaction (Beauchamp and Fridovich, 1970), which can be facilitated by ferrous ions (Fenton reaction; Freeman and Crapo, 1982) and scavenged by mannitol (McCord and Fridovich, 1973) (Fig. 1).

A regulatory role played by oxygen-derived free radicals in the local control of blood flow is a distinct possibility (Kontos et al., 1984; Proctor and Duling, 1982; Rosenblum, 1983; Wei et al., 1981). Certain situations are of particular interest, for example, those in which both the formation of oxygen-derived free radicals and alterations in local blood flow occur simultaneously, i.e., during inflammation (Baehner et al., 1982), reperfusion of ischemic tissues (Hess and Manson, 1984), increased pulmonary oxygen tension (Freeman and Crapo, 1981), and acute hypertension (Kontos et al., 1983). Although the exact mechanism by which oxygen-derived radicals induce changes in blood flow is uncertain, they may act directly on vascular smooth muscle or modify vascular tone by interacting with either adrenergic neurotransmission or endothelium-dependent responses. This chapter summarizes the actions of oxygen-derived free radicals on the different components of the vascular wall.

2. Direct Actions on Vascular Smooth Muscle

Generation of free oxygen radicals by the xanthine oxidase reaction (Rosenblum, 1983; Rubanyi and Vanhoutte, 1986a), electrical field stimulation (Lamb and Webb, 1984), humoral agents (Kontos et al., 1984), acute hypertension (Kontos et al., 1983), or trauma (Wei et al., 1981) dilate pial arterioles of the mouse (Rosenblum, 1983; Wei et al., 1981), rat (Kontos et al., 1983), and cat (Kontos et al., 1984) and relax isolated coronary arteries of the dog (Lamb and Webb, 1984; Rubanyi and Vanhoutte, 1986a). Arterioles of the hamster pouch do not respond to radicals formed by xanthine oxidase (Proctor and Duling, 1982). Reduced derivatives of oxygen metabolites stimulate the production of thromboxane and cause vasoconstriction is isolated perfused lungs (Tate et al., 1984). Thus, both the studies in intact vascular beds and those on isolated blood vessels demonstrate a wide range of vascular responses to oxygen-derived free radicals.

The direct actions of these radicals on vascular smooth muscle were analyzed in detail in isolated rings of canine coronary arteries (without endothelium) (Rubanyi and Vanhoutte, 1986a). The combination of xanthine oxidase and xanthine (which generates superoxide anions and other oxygen-derived free radicals; Fridovich, 1970; Kellog and Fridovich, 1975) inhibited the contractions evoked by prostaglandin $F_{2\alpha}$. The relaxations could not be attributed to the augmented synthesis of vasodilator prostanoids since the experiments were carried out in the presence of indomethacin, nor to beta-adrenergic stimulation by norepinephrine released from adrenergic nerve endings since propranolol did not affect the responses. The effect seemed to be the result of the direct action of the derivatives of oxygen on the vascular smooth muscle, as also observed in the same tissue during generation of oxygen-derived free radicals by electrical field stimulation (Lamb and Webb, 1984). The relaxations were prevented by catalase, but not by superoxide dismutase or mannitol; exogenous hydrogen peroxide evoked similar catalase-sensitive relaxations. These studies in the canine coronary artery indicate that hydrogen peroxide (but not superoxide anion or hydroxyl radical) is responsible for the direct inhibitory effect induced by the combination of xanthine oxidase and xanthine. Hydrogen peroxide causes relaxation of the aorta of the rabbit (Needleman et al., 1973) and of bovine intrapulmonary vessels (Wolin et al., 1985), but produces vasoconstriction in isolated perfused lungs (Tate et al., 1984) and the carotid artery of the rabbit (Heinle, 1984). Electrical stimulation induces relaxations of the tail artery of the rat and the coronary artery of the dog, which can be inhibited not only by catalase, but also by superoxide dismutase, glutathione, and dimethyl sulfoxide, suggesting that in addition to hydrogen peroxide, hydroxyl radicals are also generated by electrical stimulation and contribute to the inhibitory response (Lamb and Webb, 1984). Studies on pial arterioles of the mouse (Kontos et al., 1983; Rosenblum, 1983) suggest that the hydroxyl radical produces most of the relaxation in these blood vessels. This conclusion is supported by experiments in which the radicals were generated either by application of arachidonic acid or bradykinin, by acute hypertension, or by the combination of xanthine oxidase plus xanthine.

The exact mechanism by which hydrogen peroxide and hydroxyl radicals cause relaxation of vascular smooth muscle is not known. Oxygen-derived free radicals can both stimulate and inhibit production of prostaglandins (Deby and Deby-Dupont, 1981), and stimulate guanylate cyclase (Mittal and Murad, 1977; White et al., 1976; Hidaka and Asono, 1977). A causal relationship between these biochemical events and relaxation has not been established, however.

3. Adrenergic Neurotransmission

Reduced derivatives of oxygen produced by xanthine oxidase inhibit norepinephrine-induced contraction of rings of the thoracic aorta of the rabbit (Wolin and Belloni, 1985). Since contractions evoked by phenylephrine or KCl are not affected, a nonselective direct action of the radicals on the vascular smooth muscle can be excluded (Wolin and Belloni, 1985), and the inhibition must be caused by oxidation of catecholamines by superoxide anions (McCord and Fridovich, 1969). The oxidative inactivation of norepinephrine by superoxide anion would most likely occur in the extracellular space (Wolin and Belloni, 1985). Exposure of a tissue under adrenergic influence to a large amount of superoxide (e.g., inflammation, endotoxin shock, hyperoxia, or reperfusion after an ischemic episode) may inactivate the adrenergic neurotransmitter and, consequently, at least in part, be responsible for the observed vasodilatation. However, no inhibitory action of oxygen-derived free radicals on the response to endogenously released norepinephrine could be demonstrated in the tail artery of the rat (Lamb and Webb, 1984), suggesting that catecholamines released from adrenergic nerve endings may be protected from the oxidizing effect of free radicals. The protection may be affected by ascorbate, which is probably released together with norepinephrine (Viveros et al., 1983).

4. Endothelium-Dependent Relaxations

4.1. Endothelium-Derived Relaxing Factor(s) Is Not Likely To Be an Oxygen-Derived Free Radical

Soon after the discovery that endothelial cells can release a potent inhibitory substance, it was speculated that this endothelium-derived relaxing factor may be a free radical, perhaps derived from oxygen (Furchgott, 1983). In support of this hypothesis were the findings that endothelium-dependent relaxations to acetylcholine could be inhibited by anoxia (De Mey and Vanhoutte, 1980; Furchgott and Zawadzki, 1980; De Mey and Vanhoutte, 1983), and by hydroquinone, a free radical scavenger (Furchgott, 1983). The demonstration that the production of cyclic GMP accompanies endothelium-dependent relaxations (Rapoport and Murad, 1983; Furchgott et al., 1984; Ignarro et al., 1984) and that free oxygen radicals can stimulate guanylate cyclase (White et al., 1976; Hidaka and Asano, 1977; Mittal and Murad, 1977) also were consistent with the hypothesis that endothelium-derived relaxing factor may be an oxygen-derived free radical. Further indirect support for the hypothesis was provided by studies in which the production of free radicals was measured or in which activators or inhibitors of the generation of oxygen-derived free radicals were used. Cultured vascular smooth muscle (Heinecke et al., 1986) and endothelial cells (Rosen and Freeman, 1984) produce oxygen-derived free radicals, and the superoxide anions that they generate are released into the extracellular space (Heinecke et al., 1986). Endothelial cells contain the enzymes cyclooxygenase, lipoxygenase (De Mey et al., 1982; Johnson et al., 1985; Kuhn et al., 1983), and cytochrome P-450 monooxygenases (Pinto et al., 1986), and metabolize arachidonic acid to prostaglandins or intermediates produced by lipoxygenase and monooxygenase (De Mey et al., 1982; Kuhn et al., 1983; Johnson et al., 1985). Inhibitors of lipoxygenase or cytochrome P-450 and inhibitors of the release of arachidonic acid from certain phosphatides reduce the endothelium-dependent relaxations to acetylcholine (Singer et al., 1984; Furchgott et al., 1981; Rubanyi and Vanhoutte, 1986c; De Mey et al., 1982; Pinto et al., 1986). This suggests that the relaxing factor is either a labile oxidized metabolite of arachidonic acid (or other fatty acids) produced by lipoxygenase or monooxygenase or an oxygen-derived free radical generated as a byproduct of these enzymatic processes.

Thus, several studies analyzed the possibility that endothelium-dependent relaxations evoked by acetylcholine are mediated by the release of oxygen-derived free radicals from endothelial cells. Treatment of isolated rings of the rabbit thoracic aorta with a combination of scavengers (superoxide dismutase, plus catalase, plus mannitol) had no effect on acetylcholine-induced, endothelium-dependent relaxations (Silin et al., 1985). Similarly, the same scavengers (alone or in combination) did not prevent the endothelium-dependent relaxations to acetylcholine in canine coronary arteries (Rubanyi and Vanhoutte, 1986a). The blockade by these scavengers of the effects of xanthine oxidase plus hypoxanthine in the aorta of the rabbit (Silin et al., 1985) and the coronary artery of the dog (Rubanyi and Vanhoutte, 1986a) documented that the doses of scavengers used were effective. These studies did not, however, rule out the possibility that the relatively large scavenger molecules might not penetrate the preparations to reach the site(s) where the relaxing factor diffuses from the endothelial cells to the vascular smooth muscle. Bioassay studies with perfused thoracic aorta of the rabbit (Griffith et al., 1984) or the femoral artery of the dog (Rubanyi and Vanhoutte, 1986b) demonstrated that infusion of the scavengers beyond the site of production of endothelium-derived relaxing factor did not inhibit the relaxations of the bioassay tissues caused by acetylcholine. Thus, the relaxing mediator released from endothelial cells by acetylcholine is not an oxygen-derived free radical, such as superoxide anion, hydrogen peroxide, or hydroxyl radical. Other radicals or reduced derivatives of oxygen, however, that cannot be scavenged by the agents used so far remain possible candidates for the endothelium-derived relaxing factor(s).

4.2. Hydrogen Peroxide Triggers the Release of Endothelium-Derived Relaxing Factor(s)

Generation of free radicals by xanthine oxidase plus xanthine caused larger relaxations in canine coronary arterial rings with endothelium than in those without endothelium (Rubanyi and Vanhoutte, 1986a). A similar endothelium-dependent component was observed during relaxations induced by exogenous hydrogen peroxide. Catalase prevented these endothelium-dependent responses, suggesting that hydrogen peroxide, in addition to its direct depressing effect on vascular smooth muscle, also initiates the release of endothelium-derived relaxing factor(s). This conclusion is strengthened by the demonstration that in the isolated coronary artery of the dog, superoxide dismutase evoked catalase-sensitive, endothelium-dependent relaxations (Rubanyi and Vanhoutte, 1986a). These observations can be explained best by assuming that in the organ chamber, where these studies were performed, small amounts of superoxide anions were generated that did not trigger the release of endothelium-derived relaxing factor except after their accelerated transformation to hydrogen peroxide by superoxide dismutase (McCord and Fridovich, 1969). Superoxide dismutase or mannitol had no effect on the relaxations induced by xanthine oxidase plus xanthine in the rings with endothelium (Rubanyi and Vanhoutte, 1986a), suggesting that either free hydroxyl radicals are not generated in sufficient amounts under these experimental conditions or that they do not trigger the release of endothelium-derived relaxing factor(s). The analysis of the effect of generation and scavenging of oxygen-derived free radicals on the endothelium-dependent relaxations evoked by acetylcholine (*see* section 4.3) suggests that the latter explanation is the most likely.

The effects of hydrogen peroxide were analyzed also in a bioassay system, which allows for the separation of release, transit, and action of endothelium-derived relaxing factor(s) on the smooth muscle cells (Rubanyi et al., 1985). Rings of canine coronary arteries without endothelium were used as bioassay tissue and superfused with solution passing through the lumen of a femoral artery with endothelium taken from the same animals. Superoxide dismutase caused maximal relaxation of the bioassay ring when infused above, but not below, the segment of artery with endothelium. This indicated that the scavenger stimulated the release of endothelium-derived relaxing factor (Rubanyi and Vanhoutte, 1986b). Catalase prevented the relaxations of the bioassay rings when infused upstream (but not downstream) of the segment with endothelium, which suggested that superoxide dismutase triggered the release of endothelium-derived relaxing factor because it accelerates the formation of hydrogen peroxide from superoxide anions (Fig. 1; McCord and Fridovich, 1969). This observation is in agreement with those made on isolated rings of canine coronary arteries (Rubanyi and Vanhoutte, 1986a) and further emphasizes the ability of hydrogen peroxide to release endothelium-derived relaxing factor(s). Hydrogen peroxide also facilitates the release of endothelium-derived relaxing factor(s) evoked by acetylcholine, since catalase (when infused upstream of the artery segment with endothelium) depressed the relaxations of the bioassay rings induced by the cholinergic transmitter. The scavenger of hydrogen peroxide affected the production of relaxing factor(s) by the endothelial cells, since when infused downstream of the segment with endothelium it moderately augmented, rather than inhibited, relaxations of the bioassay rings evoked by acetylcholine (Rubanyi and Vanhoutte, 1986b).

4.3. Hydroxyl Radical Facilitates and Superoxide Anion Inhibits Endothelium-Dependent Relaxations to Acetylcholine

Superoxide dismutase or catalase given alone cause only a moderate rightward shift of the concentration–relaxation curve to acetylcholine in canine coronary arteries with endothelium (Rubanyi and Vanhoutte, 1986a). The more pronounced inhibitory affect of either the combination of the two enzymes or of mannitol suggests that hydroxyl radicals are generated under the experimental conditions used and facilitate the acetylcholine-induced relaxations (Rubanyi and Vanhoutte, 1986a). Neither of the scavengers affected relaxations induced by sodium nitroprusside in rings of the canine coronary artery without endothelium (Rubanyi and Vanhoutte, 1986a). Thus, the facilitation by the hydroxyl radicals must reflect an action at the level of synthesis or the release of the relaxing mediators, rather than the potentiation of the response of the smooth muscle to the factor(s).

Indeed, endothelium-derived relaxing factor(s) and sodium nitroprusside share the property to stimulate the production of cyclic GMP, which probably explains why they cause relaxation of the vascular smooth muscle (Rapoport and Murad, 1983; Furchgott et al., 1984; Ignarro et al., 1984).

Superoxide anions depress endothelium-mediated relaxations to acetylcholine; this effect is obvious only when they are produced in excess, for example, during the xanthine oxidase reaction. Indeed, xanthine oxidase plus xanthine markedly inhibits the relaxations evoked by acetylcholine in canine coronary arteries with endothelium (Rubanyi and Vanhoutte, 1986a). That superoxide anions may be responsible for the inhibition is suggested by the observation that superoxide dismutase (but not catalase or mannitol) prevents the inhibitory effect of the xanthine oxidase plus xanthine. The fact, that generation of free oxygen radicals by the xanthine oxidase reaction facilitates (rather than inhibits) the relaxations caused by sodium nitroprusside suggests that superoxide anions exert their inhibitory action on the endothelium and not on smooth muscle (Rubanyi and Vanhoutte, 1986a).

4.4. Superoxide Anions Inactivate Endothelium-Derived Relaxing Factors

The depression of the acetylcholine-induced relaxations of the canine coronary artery by superoxide anions (Rubanyi and Vanhoutte, 1986a) could be explained either by inhibition of the endothelial synthesis or release of the relaxing mediator(s) by the endothelial cells or by inactivation of the relaxing substances during its diffusion toward the vascular smooth muscle. To distinguish between these two possible mechanisms, experiments were performed under bioassay conditions.

Infusion of superoxide dismutase below the perfused segment with endothelium significantly prolonged the half-life of the relaxing mediator(s) from 6.3 to 15.7 s (Rubanyi and Vanhoutte, 1986b). The enzyme did not facilitate the action of the relaxing factor(s) on the vascular smooth muscle, since it had no effect when given directly to the bioassay ring. Superoxide dismutase did not alter the known properties of the endothelium-derived relaxing factor, since the relaxations caused by acetylcholine in its presence could be antagonized by various inhibitors (e.g., antioxidants and methylene blue) of endothelium-dependent relaxations. Thus the potentiating effect of superoxide dismutase must be caused by prevention or delay of the inactivation of the released factor(s) in the salt solution used in these experiments. Since catalase did not affect the protective action of superoxide dismutase, it must result from the removal of superoxide anions rather than the production of hydrogen peroxide. The lack of effect of catalase *per se* on the biological half-life of the relaxing factor(s) also implies that superoxide anions, but not hydrogen peroxide or hydroxyl radicals, contribute to the inactivation of endothelium-derived relaxing factor(s). The protective effect of superoxide dismutase also depends on the concentration of the enzyme, which determines the rate of elimination of the superoxide anions. The scavenger caused a further sixfold prolongation of the half-life of the relaxing factor(s) when the concentration of oxygen in the gas mixture used to aerate the solution was reduced from 95 to 10%. This indicates that the protective effect also depends upon the extent of generation of the oxygen-derived free radicals, which itself is a function of the partial pressure of oxygen in the salt solution used as perfusate. A reduction in oxygen concentration *per se* did not affect the half-life of endothelium-derived relaxing factor(s), and at maximally effective concentration (150 U/mL) superoxide dismutase restored only half of the acetylcholine-induced relaxation in the presence of 95% oxygen. Thus, the inactivation of the relaxing factor may not be caused solely by superoxide anions, but other radicals (not scavenged by the agents used) may also contribute to it (Rubanyi and Vanhoutte, 1986b).

Further evidence in support of the hypothesis that superoxide anions destroy endothelium-derived relaxing factor comes from observations in which the effluent from chromatographic columns, packed with microcarrier beads covered with cultured endothelial cells from the porcine aorta, was bioassayed with strips (without endothelium) of the thoracic aorta of the rabbit (Gryglewski et al., 1986). This study demonstrated that superoxide dismutase stabilizes the endothelium-derived relaxing factor(s) released in response to bradykinin. Superoxide dismutase was more effective when infused upstream than when infused downstream of the column. Catalase did not affect the stabilizing effect of superoxide dismutase, confirming that indeed the removal of superoxide anions rather than the production of hydrogen peroxide contributes to the protective effect. The endothelium-derived relaxing factor(s) could also be stabilized by $CuCl_2$ (although to a lesser extent than by superoxide dismutase); in contrast with the effect of superoxide dismutase, the protection could be observed up to 15 min after cessation of the infusion of the divalent metal ion (Gryglewski et al., 1986). These results suggest that Cu (II), which is the active species in the superoxide dismutase of the mammalian cytosol (a copper/zinc enzyme; Keele et al., 1971), acts by dismuting superoxide anions directly. It may also act, however, by reactivating exhausted superoxide dismutase of the endothelial cells (Heikkila and Cohen, 1977; Block et al., 1985). $ZnSO_2$ or $MnCl_2$ (the metal ions in superoxide dismutase from other sources, which can also scavenge superoxide anions; Epel and Neumann, 1973; Lumsden and Hall, 1975) did not protect endothelium-derived relaxing factor(s). $FeSO_4$ (which can generate the formation of superoxide anions in oxygenated buffer solutions; Michelson, 1977) selectively inhibited the endothelium-dependent relaxations of the bioassay tissues (Gryglewski et al., 1986). Since superoxide dismutase reduced the inhibitory action of $FeSO_4$, and catalase had no effect on the stability of the relaxing factor, Fe (II) must inactivate the mediator(s) by the generation of superoxide anions rather than by the production of hydroxyl radicals from hydrogen peroxide via the Fenton-reaction (Freeman and Crapo, 1982).

The source (or sources) of the superoxide anions generated in these in vitro studies is uncertain. They could be generated in the endothelial cells themselves (Rosen and Freeman, 1984), but also in the oxygenated salt solution by photolysis (McCord and Fridovich, 1973). Likewise, the chemical interaction underlying the exact nature of inactivation of endothelium-derived relaxing factor(s) by superoxide anions is still unknown. The superoxide anion may be the only oxygen-derived free radical that can reduce oxidized substances (McCord and Fridovich, 1973). Thus it may act as other reducing (antioxidant) agents (e.g., catecholamines, cysteine, dithiothreitol, hemoglobin, nordihydroguaiaretic acid, phenidone, and potassium borohydrid), which also inactivate endothelium-derived relaxing factor in transit (Griffith et al., 1984; Forstermann and Neufang, 1984; Rubanyi et al., 1985; Rubanyi and Vanhoutte, 1986b). Alternatively, generation of superoxide anions may be a common mechanism for all agents that inactivate the relaxing mediator(s).

The dependency of biological half-life of endothelium-derived relaxing factor upon the generation of superoxide anions may help to explain the differences in the half-life of the relaxing mediator(s) reported by various investigators (*see* Forstermann and Neufang, 1984; Griffith et al., 1984; Rubanyi et al., 1985; Cocks et al., 1985), since the production of superoxide anions may vary in the different experimental systems used. The stabilizing property of superoxide dismutase may be of practical importance for the further characterization of the chemical identity of the relaxing mediator(s). Since the synthesis and biological activity of endothelium-derived relaxing factor(s) can be influenced in opposite directions by trace metals [i.e., protection by Cu (II) and activation by Fe (II)] and by oxygen-derived free radicals (i.e., stimulation by hydrogen peroxide and hydroxyl radicals and inactivation by superoxide anions), these chemical substances may play an important modulatory role in the control of vascular function exerted by the endothelium.

5. Conclusions

Experiments using isolated blood vessels of the rabbit and the dog, and bioassay studies using femoral arteries of the dog and cultured aortic endothelial cells of the pig as source of endothelium-derived relaxing factor(s) have ruled out the possibility that the endothelial mediator(s) is an oxygen-derived free radical. Free oxygen radicals can affect endothelium-dependent responses in many ways, however (Fig. 2). Besides their direct

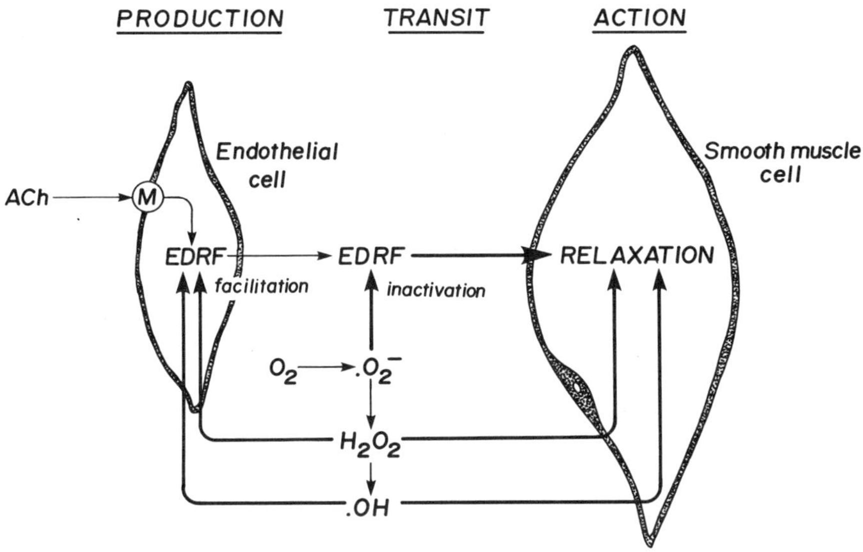

Fig. 2. Effect of oxygen-derived free radicals on the production and action on vascular smooth muscle and biological activity of endothelium-derived relaxing factor(s) (EDRF) released from canine femoral arteries during stimulation of muscarinic receptors (M) by acetylcholine (ACh).

inhibitory action on vascular smooth muscle, hydrogen peroxide and hydroxyl radicals can trigger and facilitate the evoked release of endothelium-derived relaxing factor(s). Thus oxidative mechanisms must play an important role in the endothelial synthesis and release of the factor(s) by the endothelial cells, which then would explain why anoxia can prevent endothelium-dependent relaxations induced by acetylcholine. Under bioassay conditions, when the relaxing factor is in prolonged contact with oxygenated artificial salt solutions, superoxide anions accelerate its destruction. As a consequence, removal of these radicals by superoxide dismutase or by Cu (II) stabilizes the factor and prolongs its biological half-life considerably, particularly if the oxygen concentration is lowered. These results and the fact that the factor(s) is inactivated by various antioxidants and by conditions that stimulate the generation of superoxide anions [i.e., Fe (II) ions or hyperoxia] suggest that the endothelium-derived relaxing factor(s) is an oxidized substance and that the oxidized state is essential for its biological activity (e.g., oxidative activation of guanylate cyclase). If endothelium-derived relaxing factor (or factors) have a homeostatic function in the vascular integrity and the regulation of blood flow, it can be assumed that free oxygen radicals and trace metals may play an important modulatory role in the control of the vascular system under physiological conditions and during pathological events such as hyperoxia, hypertension, ischemia, inflammation, and vasospasm.

References

Baehner, R. L., Boxer, L. A., and Ingraham, L. M.: Reduced oxygen byproducts and white blood cells. *Free Radicals Biol.* **5:**91N113, 1982.

Beauchamp, C. and Fridovich, I.: A mechanism for the production of ethylene from methional. The generation of the hydroxyl radical by xanthine oxidase. *J. Biol. Chem.* **245:**4641–4646, 1970.

Block, E. R., Patel, J. M., and Sheridan, N. P.: Effect of O_2 and endotoxin on lactate dehydrogenase release, 5-hydroxytryptamine, uptake and antioxidant enzyme activities in endothelial cells. *J. Cell. Physiol.* **122:**240–248, 1985.

Cocks, T. M., Angus, J. A., Campbell, J. H., and Campbell, G. R.: Release and properties of endothelium-derived relaxing factor (EDRF) from endothelial cells in culture. *J. Cell Physiol.* **123:**310–320, 1985.

Deby, C. and Deby-Dumont, G.: Natural factors modulating the intervention of an activated oxygen in the biosynthesis of prostanoids. *Bull. Eur. Physiol. Pathol. Respir.* (suppl.)**:**129–139, 1981.

De Mey, J. G. and Vanhoutte, P. M.: Interaction between Na^+, K^+ exchanges and the direct inhibitory effect of acetylcholine on canine femoral arteries. *Circ. Res.* **46:**826–836, 1980.

De Mey, J. G. and Vanhoutte, P. M.: Anoxia and endothelium dependent reactivity of the canine femoral artery. *J. Physiol.* **335:**65–74, 1983.

De Mey, J. G., Claeys, M., and Vanhoutte, P. M.: Endothelium-dependent inhibitory effect of acetylcholine, adenosine triphosphate, thrombin and arachidonic acid in the canine femoral artery. *J. Pharmacol. Exp. Ther.* **222:**166–173, 1982.

DelMaestro, R. F.: An approach to free radicals in medicine and biology. *Acta Physiol. Scand.* **492** (suppl.)**:**153–168, 1980.

Epel, B. L. and Neumann, J.: The mechanism of the oxidation of ascorbate and $Mn2^+$ by chloroplasts. The role of the radical superoxide. *Biochem. Biophys. Acta* **325:**520–529, 1973.

Forstermann, U. and Neufang, B.: The endothelium-dependent relaxation of rabbit aorta: Effects of antioxidants and hydroxylated eicosatetraenoic acids. *Br. J. Pharmacol.* **82:**765–767, 1984.

Freeman, B. A. and Crapo, J. D.: Hyperoxia increases oxygen radical production in rat lungs and lung mitochondria. *J. Biol. Chem.* **256:**10986–10992, 1981.

Freeman, B. A. and Crapo, J. D.: Biology of disease. Free radicals and tissue injury. *Lab. Invest.* **47:**412–426, 1982.

Fridovich, I.: Quantitative aspects of the production of superoxide anion radical by milkxanthine oxidase. *J. Biol. Chem.* **245:**4053–4057, 1970.

Furchgott, R. F.: Role of the endothelium in responses of vascular smooth muscle. *Circ. Res.* **53:**557–573, 1983.

Furchgott, R. F., Jothianandan, D., and Cherry, P. D.: Endothelium-Dependent Responses: The Last Three Years, in *Vasodilator Mechanisms* (Vanhoutte, P. M. and Vatner, S. F., eds.) Karger, Basel, 1984.

Furchgott, R. F. and Zawadzki, J. V.: The obligatory role of endothelial cells in the relaxation of arterial smooth muscle by acetylcholine. *Nature* **288:**373–76, 1980.

Furchgott, R. F., Zawadzki, J. V., and Cherry, P. D.: Role of Endothelium in the Vasodilator Response to Acetylcholine, in *Vasodilatation* (Vanhoutte, P. M. and Leusen, I., eds.) Raven, New York, 1981.

Griffith, T. M., Edwards, D. H., Lewis, M. J., Newby, A. C., and Henderson, A. H.: The nature of endothelium-derived vascular relaxant factor. *Nature* **308:**645–647, 1984.

Gryglewski, R. J., Palmer, R. M. J., and Moncada, S.: Superoxide anion plays a role in the breakdown of endothelium-derived relaxing factor. *Nature* **320:**454–456, 1986.

Heikkila, R. E. and Cohen, G.: The Inactivation of Copper Zinc Superoxide Dismutase by Diethyldithiocarbamate, in *Superoxide and Superoxide Dismutases* (Michelson, A. M., McCord, J. M., and Fridovich, I., eds.) Academic, London, 1977.

Heinecke, J. W., Baker, L., Rosen, H., and Chait, A.: Superoxide-mediated modification of low density lipoprotein by arterial smooth muscle cells. *J. Clin. Invest.* **77:**757–761, 1986.

Heinle, H.: Vasoconstriction of carotid artery induced by hydroperoxides. *Arch. Intl. Physiol. Biochim.* **92:**267–271, 1984.

Hess, M. L. and Manson, N. H.: Molecular oxygen: Friend and foe. The role of oxygen free radical system in the calcium paradox, the oxygen paradox and ischemia/reperfusion injury. *J. Mol. Cell. Cardiol.* **16:**969–985, 1984.

Hidaka, H. and Asono, T.: Stimulation of human platelet guanylate cyclase by unsaturated fatty acid peroxides. *Proc. Natl. Acad. Sci. USA* **74:**3657–3661, 1977.

Ignarro, L. J., Burke, T. M., Wood, K. S., Wolin, M. S., and Kadowitz, P. J.: Association between cyclic GMP accumulation and acetylcholine-elicited relaxation of bovine intrapulmonary artery. *J. Pharmacol. Exp. Ther.* **28:**682–690, 1984.

Johnson, A. R., Revtyak, G., and Campbell, W. B.: Arachidonic acid metabolites and endothelial injury: Studies with cultures of human endothelial cells. *Fed. Proc.* **44:**19–24, 1985.

Keele, B. B., McCord, J. M., and Fridovich, I.: Further characterization of bovine superoxide dismutase and its isolation from bovine heart. *J. Biol. Chem.* **246:**2875–2880, 1971.

Kellog, E. W., III and Fridovich, I.: Superoxide, hydrogen peroxide and singlet oxygen in lipid eroxidation by xanthine oxydase system. *J. Biol. Chem.* **250:**8812–8817, 1975.

Kontos, H. A., Wei, E. P., Christman, C. W., Levasseur, E., Povlishock, J. T., and Ellis, E. F.: Free oxygen radicals in cerebral vascular responses. *Physiologist* **26:**165–172, 1983.

Kontos, H. A., Wei, E. P., Povlishock, J. T., and Christman, C. W.: Oxygen radicals mediate the cerebral arteriolar dilation from arachidonate and bradykinin in cats. *Circ. Res.* **55:**292–303, 1984.

Kuhn, H., Ponicke, K., Halle, W., Scheme, T., and Forster, W.: Evidence for the presence of lipoxygenase pathway in cultured endothelial cells. *Biomed. Biochim. Acta* **42:**K1–K4, 1983.

Lamb, F. S. and Webb, R. C.: Vascular effects of free radicals generated by electrical stimulation. *Am. J. Physiol.* **217:**H709–H714, 1984.

Lumsden, J. and Hall, D. C.: Chloroplast manganese and superoxide. *Biochem. Biophys. Res. Commun.* **64:**595–602, 1975.

McCord, J. M. and Fridovich, I.: Superoxide dismutase. An enzymatic function for erythrocuprein (hemocurprein). *J. Biol. Chem.* **214:**6049–6055, 1969.

McCord, J. M. and Fridovich, I.: Production of O_2 in photolyzed water demonstrated through the use of superoxide dismutase. *Photochem. Photobiol.* **17:**115–121, 1973.

Michelson, A. M.: Production of Superoxide by Metal Ions, in *Superoxide and Superoxide Dismutases* (Michelson, A. M., McCord, J. M., and Fridovich, I., eds.) Academic, London, 1977.

Mittal, C. K. and Murad, F.: Activation of guanylate cyclase by superoxide dismutase and hydroxyl radical: A physiological regulator of guanosine 3′, 5′-monophosphate formation. *Proc. Natl. Acad. Sci. USA* **74:**4360–4364, 1977.

Needleman, P., Jakschik, B., and Johnson, E. M.: Sulfhydryl requirement for relaxation of vascular smooth muscle. *J. Pharmacol. Exp. Ther.* **187:** 324–331, 1973.

Pinto, A., Abraham, N. G., and Mullane, K. M.: Cytochrome P-450-dependent monooxygenase activity and endothelial-dependent relaxations induced by arachidonic acid. *J. Pharmacol. Exp. Ther.* **236:**445–451, 1986.

Proctor, K. G. and Duling, B. R.: Oxygen-derived free radicals and local control of striated muscle blood flow. *Microvasc. Res.* **24:**77–86, 1982.

Rapoport, R. M. and Murad, F.: Agonist-induced endothelium-dependent relaxation in rat thoracic aorta may be mediated through cGMP. *Circ. Res.* **52:**352–357, 1983.

Rosen, G. M. and Freeman, B. A.: Detection of superoxide generated by endothelial cells. *Proc. Natl. Acad. Sci. USA* **81:**7269–7273, 1984.

Rosenblum, W. I.: Effects of free radical generation on mouse pial arterioles; probable role of hydroxyl radicals. *Am. J. Physiol. (Heart Circ. Physiol.* **14***)*:H139–H142, 1983.

Rubanyi, G. M. and Vanhoutte, P. M.: Oxygen-derived free radicals, endothelium, and responsiveness of vascular smooth muscle. *Am. J. Physiol.* **250:**H815–H821, 1986a.

Rubanyi, G. M. and Vanhoutte, P. M.: Superoxide anions and hyperoxia inactivate endothelium-derived relaxing factor. *Am. J. Physiol.* **250:**H822–H827, 1986b.

Rubanyi, G. M. and Vanhoutte, P. M.: Acetylcholine stimulates the release of two chemically different endothelium-derived relaxing factors in canine femoral arteries, submitted, 1986c.

Rubanyi, G. M., Lorenz, R. R., and Vanhoutte, P. M.: Bioassay of endothelium-derived relaxing factor(s). Inactivation by catecholamines. *Am. J. Physiol.* **249:**H95–H101, 1985.

Silin, P. J., Strulowitz, J. A., Wolin, M. S., and Belloni, F. L.: Absence of a role for superoxide anion, hydrogen peroxide and hydroxyl radical in endothelium-mediated relaxation of rabbit aorta. *Blood Vessels* **22:**65–73, 1985.

Singer, H. A., Saye, J. A., and Peach, M. J.: Effects of cytochrome P-450 inhibitors on endothelium-dependent relaxation in rabbit aorta. *Blood Vessels* **21:**223–230, 1984.

Tate, R. M., Morris, H. G., Schroeder, W. R., and Repine, J. E.: Oxygen metabolites stimulate thromboxane production and vasoconstriction in isolated saline perfused lungs. *J. Clin. Invest.* **74:**608–613, 1984.

Viveros, O., Diliberto, E. J., Jr., and Daniels, A. J.: Biochemical and functional evidence for the cosecretion of multiple messengers from single and multiple compartments. *Fed. Proc.* **42:**2923–2928, 1983.

Wei, E. P., Kontos, H. A., Dietrich, W. D., Povlishock, J. T., and Ellis, E. F.: Inhibition by free radical scavengers and cyclooxygenase inhibitors of pial arteriolar abnormalities from concussive brain injury in cats. *Circ. Res.* **48:**95–103, 1981.

White, A. A., Crawford, K. M., Patt, C. S., and Lad, P. J.: Activation of soluble guanylate cyclase from rat lung by incubation or by hydrogen peroxide. *J. Biol. Chem.* **251:**7304–7312, 1976.

Wolin, M. S. and Belloni, F. L.: Superoxide anion selectively attenuates catecholamine-induced contractile tension in isolated rabbit aorta. *Am. J. Physiol.* **249:**H1127–H1133, 1985.

Wolin, M. S., Rodrigues, A. M., and Yu, J. M.: Peroxides cause dose-dependent relaxant and constrictor responses in isolated bovine intrapulmonary arterial and venous rings. *Fed. Proc.* **44:**821 (Abstract), 1985.

Chapter 6

Endothelial Cells in Culture and Production of Endothelium-Derived Relaxing Factor

T. M. Cocks, J. A. Angus, J. H. Campbell, and G. R. Campbell

1. Introduction

The chemical identity of endothelium-derived relaxing factor (EDRF) is unknown. Current hypotheses as to its structure are mostly based on pharmacological studies in which chemicals with diverse properties have been used to modify the relaxation caused by agonists releasing EDRF. Therefore any conclusions based on these approaches should necessarily be viewed with caution.

A more direct approach is to isolate EDRF from endothelial cells, purify it, and then analyze it chemically. The basic requirements here are the supply of large amounts of EDRF and a suitable assay technique. This chapter outlines the development of a sensitive transferral bioassay system for EDRF released from endothelial cells grown in suspension culture.

2. Isolation of EDRF

The most important requirement in the isolation of any biologically active substance is the availability of a sensitive and selective means of detection. At present the only way to detect EDRF is by bioassay in which the inherent ability of arterial smooth muscle to relax in response to the factor is used as the detector system. A ring or strip of isolated artery with intact endothelium can thus be regarded as a sensitive bioassay system for EDRF and has formed the basis for most of the pharmacological and physiological studies on this substance to date. Because the two components of the bioassay—the source of EDRF (endothelium) and the detector (smooth muscle)—in an isolated artery preparation cannot be anatomically separated, however, its usefulness is limited mainly to the study of pharmacological aspects of EDRF.

The more recently developed bioassays for EDRF (*see* section 4) still utilize the same components as in an intact ring or strip of artery, but their major advantage is that the source of EDRF can be separated from the detector. The isolated cell source is perfused with a physiological salt solution and the concentration of EDRF in the eluent assayed by passing it directly over a de-endothelialized arterial preparation. Release of EDRF can then be detected by a relaxation response of the detector artery. Endothelial cells grown in suspension culture are a potential source of large amounts of EDRF.

2.1. Endothelial Cells in Culture

Before any large-scale production of a particular line of endothelial cells is attempted, a useful preliminary experiment is to test for the capacity of the cells to release EDRF. One way to achieve this is to "regrow" the cultured cells back onto the endothelium-denuded surface of a segment of artery. The reconstituted arteries can then be examined in an organ bath to see if they are able to relax in response to endothelium-dependent agonists. A useful technique is as follows. Endothelium-denuded arterial rings are placed on their side in a thick suspension of multipassaged endothelial cells in culture medium and rotated about 90° every hour for 4–8 h at 37 °C to ensure a uniform covering of cells on all surfaces of the artery. Either the endothelial cells are omitted during incubation in the culture medium or endothelium-intact rings can be taken through the same procedure to test for any adverse effects of the culture medium (plus or minus cells) on responses to EDRF. A cyclooxygenase inhibitor such as indomethacin may also be added during both the incubation with cells and the assay to assess any role for prostanoid synthesis under these conditions. After the incubation period, the rings can be resuspended in the organ bath, washed thoroughly to remove traces of culture medium, and again tested for their responsiveness to EDRF-releasing agonists. The results of a typical experiment with bovine thoracic aortic endothelial cells and canine coronary arterial rings are shown in Figs. 1 and 2.

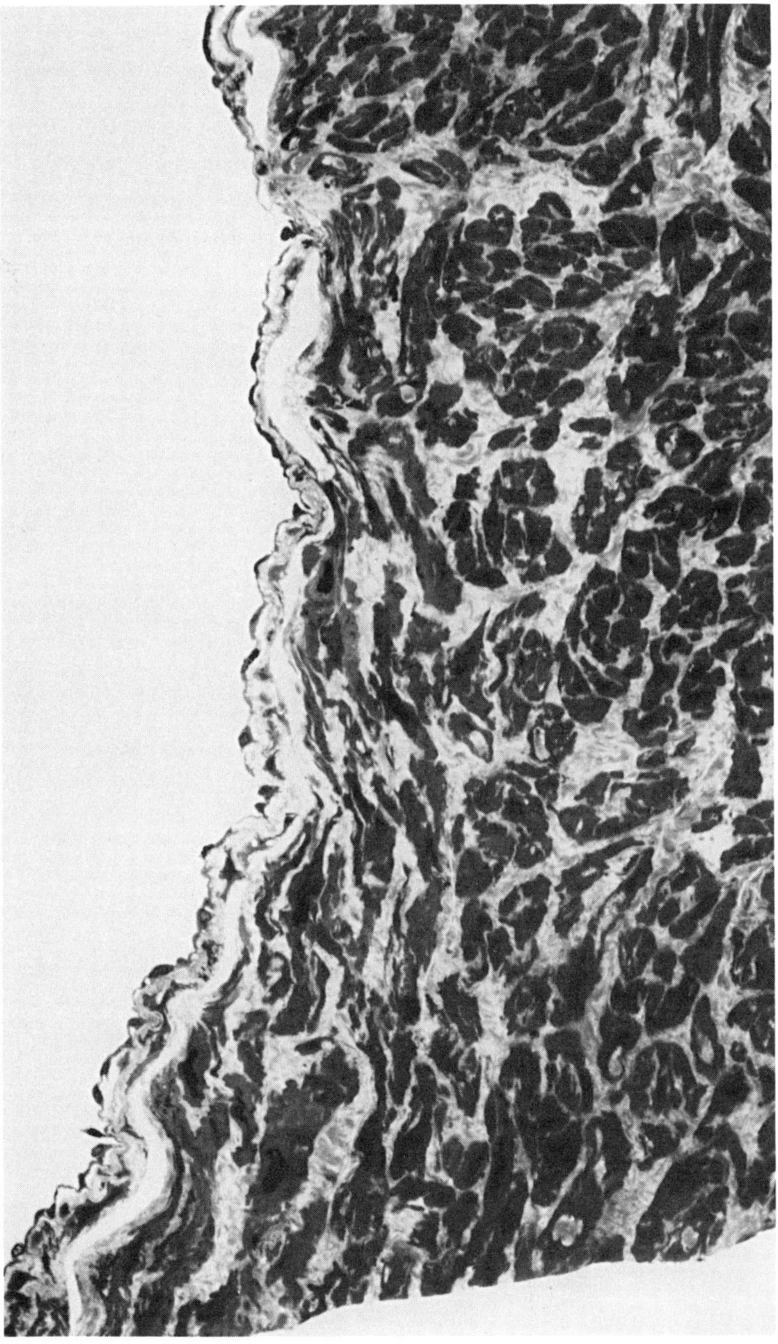

Fig. 1. Light micrographs showing transverse sections of dog coronary artery rings (A) with intact endothelium (original magnification, ×430) and (B) after mechanical removal of the endothelium (original magnification, ×1200). Note the extensive damage to the intima in (B), which contains no intact cells and is odematous. Also, the internal elastic lamina is unstained in (A) (from Cocks et al., 1985).

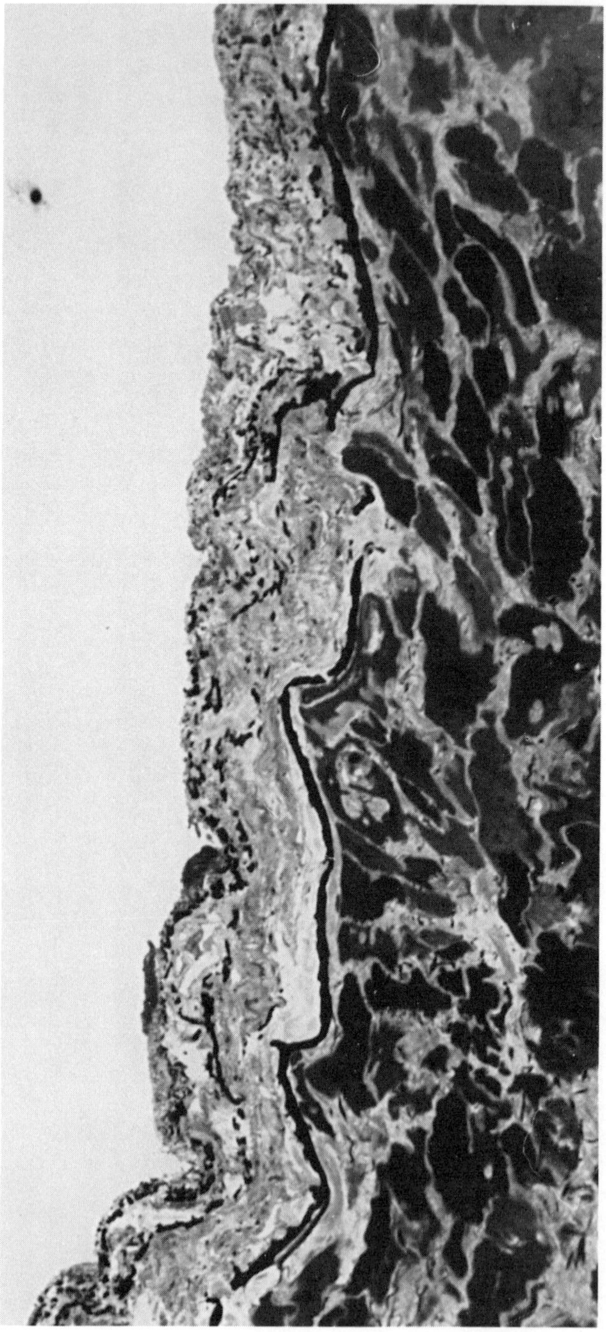

Fig. 1B.

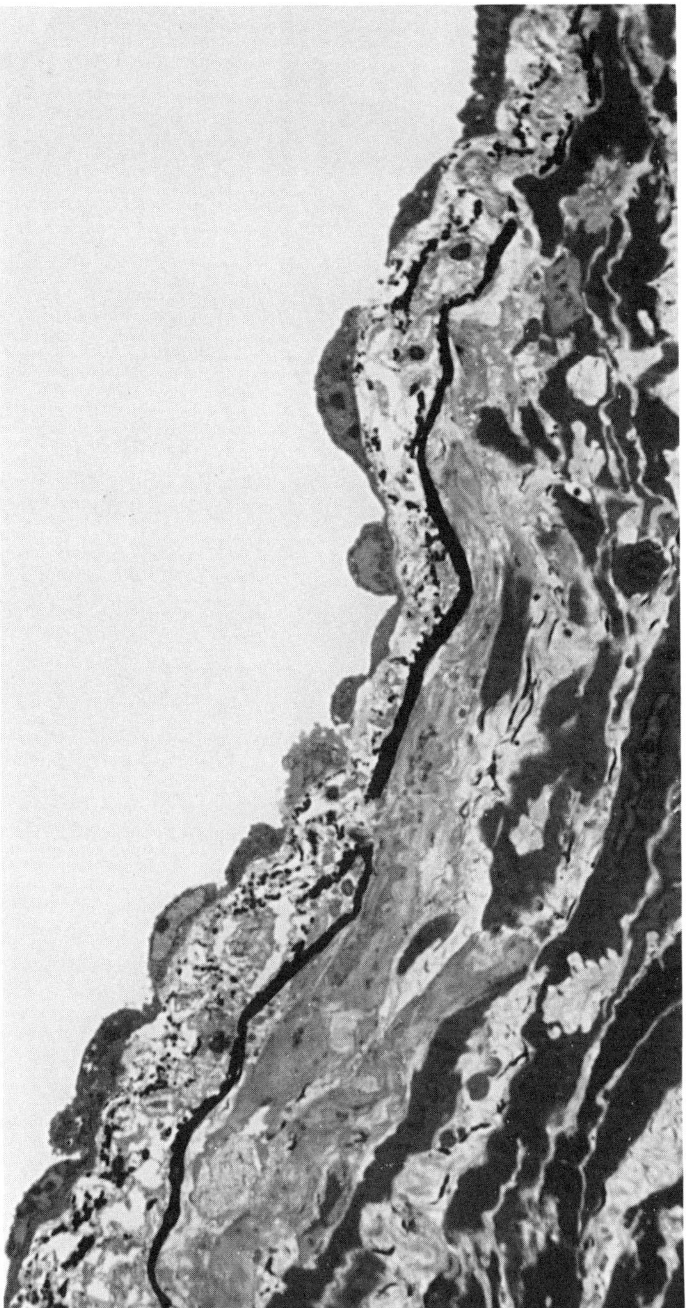

Fig. 2. Light micrographs showing transverse sections of endothelium-denuded dog coronary artery rings after regrowing bovine aortic endothelial cells onto the luminal surface. (A) The endothelial cells have started to flatten, forming a continuous monolayer over the luminal surface (original magnification, ×1200). (B) This section shows some areas of exposed subendothelium. Also, the regrown endothelial cells are either attenuated or enlarged (original magnification, ×1200). Note the extensive damage to both the intima and inner layers of the media caused by the denudation procedure (from Cocks et al., 1985).

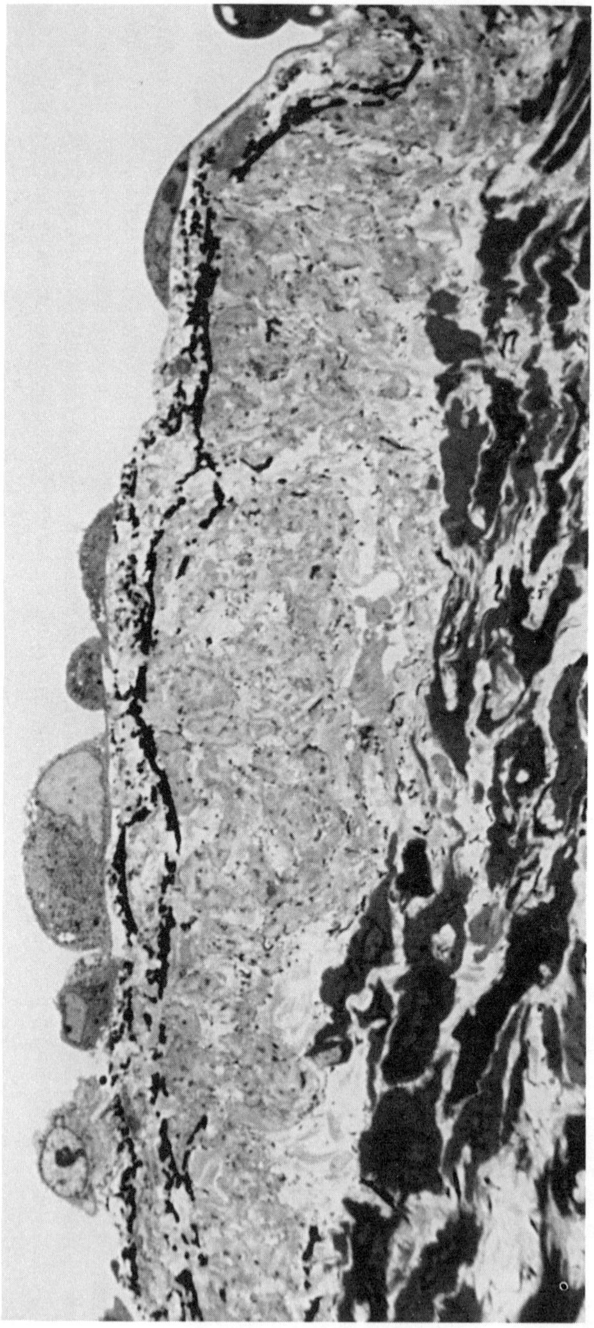

Fig. 2B.

Following the reseeding procedure outlined above, bovine endothelial cells were observed adhering to the internal surface of each arterial ring. Although some of the cells were flattened and formed a continuous monolayer like the normal artery, many were rounded and separate from their neighbors. Also no close contacts between endothelial cells and smooth muscle were observed (Fig. 2). In the organ bath, however, these rings relaxed in response to the Ca^{2+} ionophore A23187 and bradykinin, but not to acetylcholine or substance P, whereas the same rings prior to the readdition of cells failed to relax any of these agonists (Fig. 3). The shortest time tested for reseeding was 2 h, and the same results were obtained. Normal rings of artery with intact endothelium taken through the same procedure retained their responsiveness to all EDRF-releasing agonists, whereas endothelium-denuded rings incubated in medium only for 4 h remained unresponsive upon retesting. The lack of EDRF release in response to acetylcholine and substance P is discussed below in section 2.3.

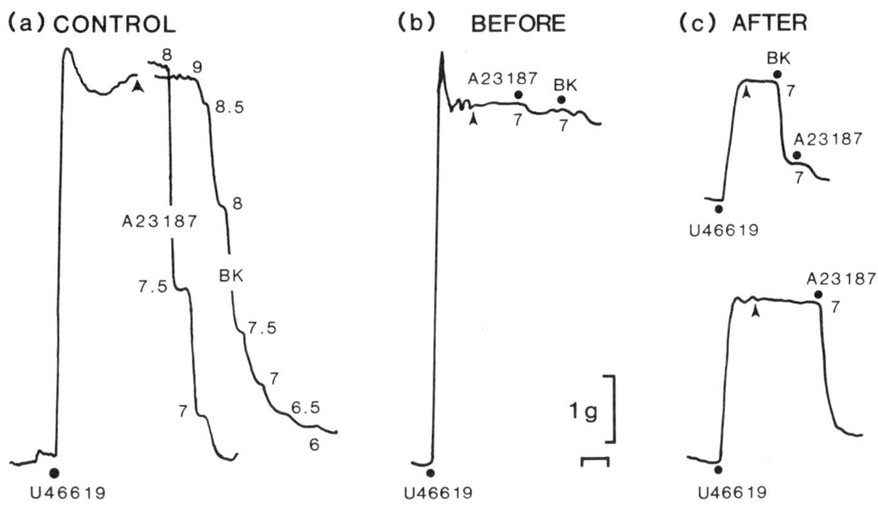

Fig. 3. Effect of regrowing cultured bovine endothelial cells onto previously denuded dog coronary artery rings (*see* Fig. 2) on their responsiveness to EDRF agonists. (a) Cumulative relaxation responses in two rings to bradykinin (BK) and the Ca^{2+} ionophore A23187 (concentrations of both given as $-\log M$). (b) Responses to maximal concentrations of BK and A23187 in the same rings (only one shown) after removal of their endothelium. (c) The same rings after the endothelial cell regrowing procedure. Indomethacin (30 μM) was present throughout. The time scale before and after the arrows represent 12 and 2 min, respectively (from Cocks et al., 1985).

The reconstitution-type experiment described here mainly serves to assess the suitability of various sources of endothelial cells with respect to their responsiveness to EDRF-releasing agonists. It also demonstrates that close contacts between endothelial and smooth muscle cells are probably not involved in the endothelium-dependent relaxation of blood vessels (*see* Peach et al., 1985). We observed relatively large distances between the endothelial cells and the underlying medial smooth muscle cells, which were largely the result of damage to the intima and media caused during removal of the endothelium (*see* Fig. 2). It should be mentioned, however, that endothelial cells and smooth muscle cells in coculture can rapidly (within 2 h) form intercellular junctions that permit the transfer of small molecules such as ions and nucleotides between each cell type (Davies et al., 1985). Such junctions may also have formed during the 2-h incubation period for the experiments with reconstituted arteries described above, as well as in the classic sandwich-type experiments first described by Furchgott and Zawadzki (1980).

2.2. Culture of Endothelial Cells on Microcarrier Beads

Once it has been established that an endothelial cell line in culture produces EDRF, the next step is to grow the cells in such a way that will allow for both the production and collection of large amounts of the factor. Microcarrier beads are ideal since they allow a variety of animal cell lines to be grown in suspension culture. For example, Cytodex beads (Pharmacia) are spherical, cross-linked dextran matrices (approximately 150–170 μm in diameter), with either charged groups or denatured collagen on their surface. The major advantage of growing endothelial cells in suspension culture is that large numbers can be readily perfused with a small volume of a physiological salt solution or culture medium, which thereby enables EDRF to be released in a concentrated form.

The procedure for growing endothelial cells on microcarriers relies on a continual supply of cells grown under normal culture conditions, as well as a suitable system for suspension culture (*see* Cocks et al., 1985). The density of initial seeding is also important in as much that, for rapid production of cells, it is best to have as many cells as possible initially adhering to as many beads as possible. Ideally the aim is for most beads to be confluent within 2–3 d. This can be achieved by using large number of cells/g of hydrated Cytodex beads.

2.3. Release of EDRF from Endothelial Cells Grown on Microcarrier Beads

Scanning electron micrographs of bovine endothelial cells growing on microcarrier beads are shown in Fig. 4. The cells form a confluent

Fig. 4. Scanning electron micrographs depicting bovine thoracic aortic endothelial cells growing on Cytodex microcarrier beads. (A) Low magnification (×520). (B) Higher magnification (×1980) showing a single bead covered with a confluent monolayer of cells. The nuclei of individual cells bulge out from the beads surface and are covered by microvilla (from Cocks et al., 1985).

monolayer over the surface of most beads in 2–4 d, at which time they are ready for use. About 1 × 10^6 beads (approximately 5–6 × 10^7 cells) are poured into a cut-down 30-mL plastic syringe barrel or other suitable column that can be readily perfused with a physiological salt solution (e.g., Krebs' solution). The choice of assay tissue should be one that is among the most sensitive to EDRF. Coronary arteries of the dog and the pig are ideal since they are the most sensitive of many arteries tested (*see* Angus et al., 1986; Cocks and Angus, 1984; Cocks et al., 1985).

Once the cell column has been equilibrated with Krebs' solution, the assay tissue, which has previously had its own endothelium removed, is

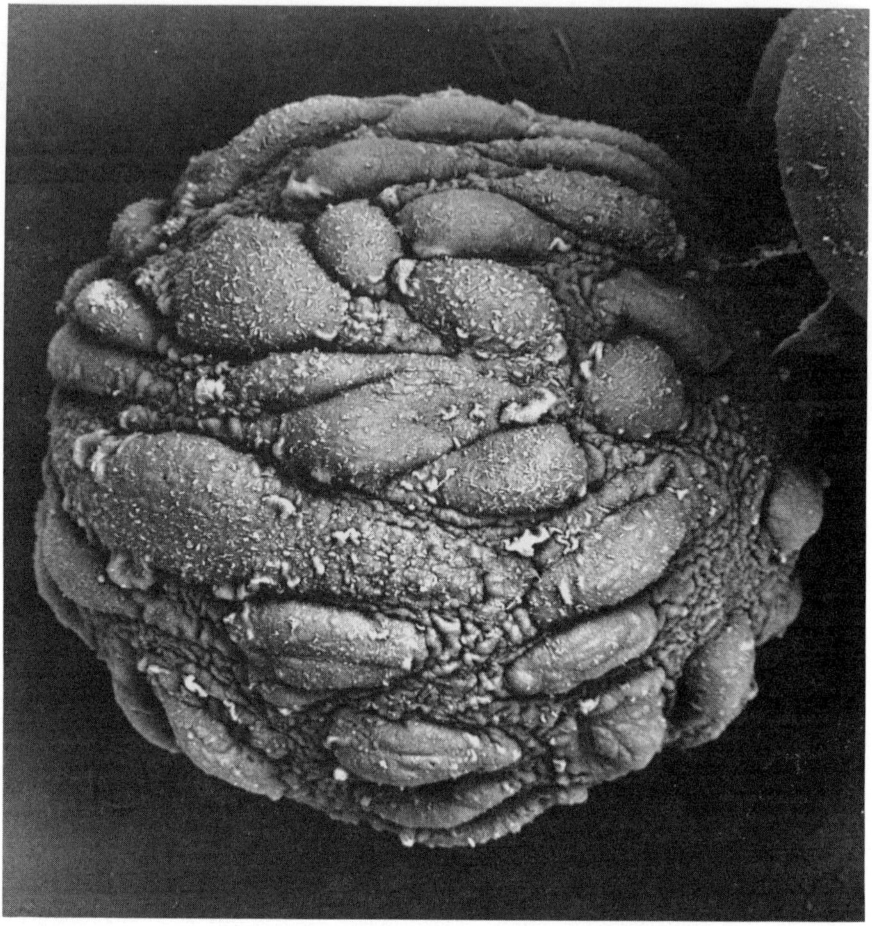

Fig. 4B.

suspended beneath the outlet of the column and superfused continuously by the column eluent. The artery is then contracted to a steady level of active force with a suitable vasoconstrictor. The concentration of the contracting agonist should be such that it puts the contraction at about 80% of its maximum, which allows the artery to relax sensitively to dilator substances. The column can then be challenged with a particular EDRF-releasing agonist by adding it to the reservoir containing Krebs' solution. If EDRF release ensues, it will be detected as a relaxation of the assay tissue.

A typical experiment using this type of system is described in Fig. 5. The assay tissue was a deendothelialized ring of porcine circumflex

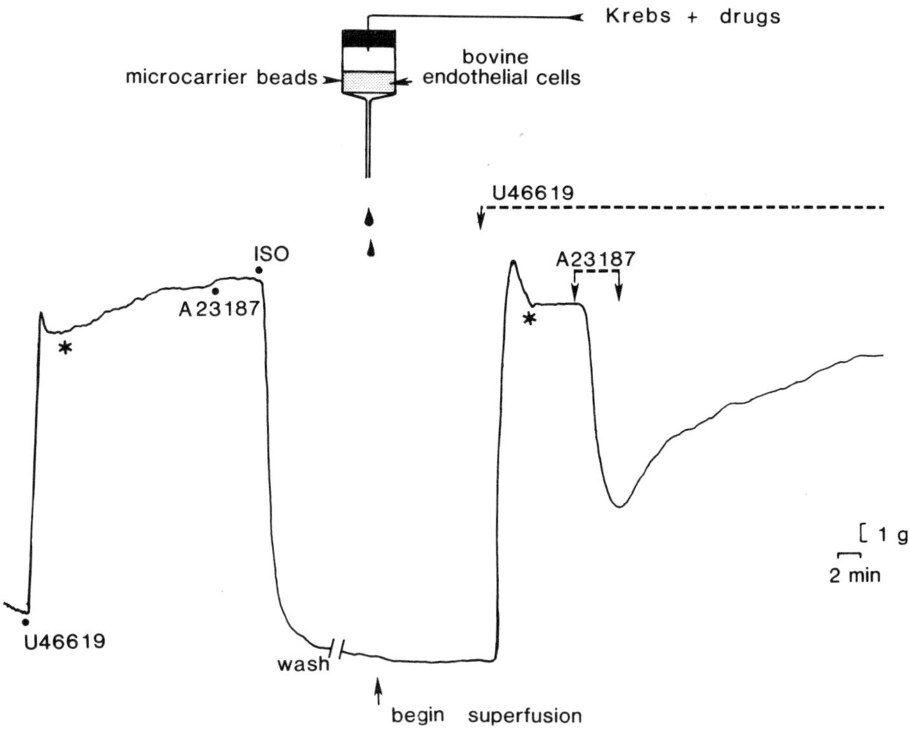

Fig. 5. Demonstration of the release of EDRF from a column of microcarrier beads containing bovine endothelial cells (*see* Fig. 4). After the assay tissue (a ring of de-endothelialized pig coronary artery) was ascertained to be nonresponsive to A23187 (0.1 μM) it was then removed from the organ bath and superfused with the eluent from the cell column. Challenge of the cells now with A23187 (0.1 μM) caused the artery to relax, indicating release of a humoral substance. The thromboxane mimetic U46619 (30 n*M*) was used to contract the tissue, and indomethacin (10 μM) was present throughout. Isoprenaline (ISO; 1 μM) was used to check that the assay tissue could maximally relax. The temperature was maintained at 30 $\pm$ 1 °C, and the time scale before the asterisk is 12 min (from Cocks et al., 1985).

coronary artery. Stimulation of the cell column with the Ca^{2+} ionophore A23187 and bradykinin caused a relaxation (greater than 50%). In contrast, little response was observed with acetylcholine and substance P. Indomethacin (10 μM) was present throughout to prevent the synthesis of prostanoids. Also, since the coronary artery of the pig does not relax to prostacyclin, the relaxations to bradykinin and A23187 can be considered not to be caused by a prostanoid, in particular prostacyclin (*see* Cocks et al., 1985). They are also unlikely to be caused by release of a purine nucleotide [e.g., adenosine triphosphate (ATP); *see* Pearson and Gordon, 1979] since with the canine splenic artery as the assay tissue, the same pattern of responses from the cell column were observed. This tissue does not relax in response to direct application of ATP, however, and only poorly to adenosine (*see* De Mey and Vanhoutte, 1982). Also, bovine endothelial cells in culture do not release significant amounts of ATP in response to bradykinin or A23187 (Cocks and White, unpublished data). Thus the relaxation signal generated from the cell column is most probably EDRF or an EDRF-like substance (*see* Cocks et al., 1985).

Using the same bioassay system, a variety of endothelial cells (grown on beads) qualitatively showed the same pattern of release of an EDRF-like substance. Cells derived from bovine aorta or vena cava, canine aorta and coronary artery, bovine and porcine epi- and endocardium, and human umbilical vein all released EDRF-like material in response to bradykinin, thrombin, trypsin, A23187, and ATP, as well as to mechanical vibration. All the cell lines failed, however, to release any inhibitory substance in response to substance P and acetylcholine. Other, non-endothelial-type cells, such as aortic smooth muscle, and fibroblasts from the pig aorta, also grown on microcarrier beads failed to respond to any of the above chemical stimuli. Mechanical vibration was not tested in these non-endothelial cells.

The reason for the failure of any of the endothelial cell lines to release EDRF in response to acetylcholine (and its stable analogs, methacholine and bethanechol) and substance P is unknown. Gordon and Martin (1983) similarly have shown that muscarinic agonists do not stimulate the release of Rb^+ efflux from porcine endothelial cells in culture, in contrast with other EDRF-releasing agonists. Perhaps under culture conditions these cells do not possess receptors for acetylcholine (*see* Peach et al., 1985) and substance P, or if they do these agonists are less efficacious than the other EDRF-releasing agonists (e.g., bradykinin) in producing EDRF. This is supported by the finding that fresh, luminal scrapings of endothelial cells respond to bethanechol and substance P, although to a much lesser extent than bradykinin and A23187 (Cocks et al., 1985). Also, it has been claimed that in cocultures of endothelial and smooth muscle cells, a muscarinic agonist such as methacholine is able to induce an increase in cyclic GMP, the proposed mediator of the EDRF-induced relaxation (Loeb et al., 1985). In separate cultures, both cell types were nonresponsive to methacholine, which suggests that the increase in cyclic GMP is likely to occur in the smooth muscle cell. This has been clearly demonstrated in a variation of the coculture technique with A23187 as the agonist (*see* Ganz et al., 1986). These preliminary results suggest that the smooth muscle cells are important for the phenotypic expression of muscarinic receptors.

To examine this phenomenon further, cocultures of endothelial and subcultured smooth muscle cells were tested for direct evidence of EDRF release (i.e., by relaxation of arterial smooth muscle). No response could be demonstrated to acetylcholine. The mixed cell preparation still responded maximally to bradykinin, however, although the culture was dominated by smooth muscle cells (*see* Fig. 6). The failure to observe release of EDRF in these experiments should be treated with caution. Perhaps the phenotype of the smooth muscle cells used is important (*see* Chamley-Campbell and Campbell, 1981). Smooth muscle of the synthetic type was used here, not the contractile type that is found in the media of arteries. While Loeb et al. (1985) did not precisely indicate the phenotype of their smooth muscle, however, it would most likely be the synthetic type, like that used by Ganz et al. (1986). Also, the response of EDRF to acetylcholine is not affected in arteries with severe, experimentally induced intimal thickening. This is predominantly made up of smooth muscle cells of the synthetic type (Cocks et al., 1987). Therefore if smooth muscle cells do play an important role in regulating the responsiveness of the endothelial cells to acetylcholine (and substance P), the phenotype (contractile or synthetic) does not appear to be important. It is interesting to note here that skeletal muscle cells in culture have been found to control the expression of acetylcholine receptors on parasympathetic neurons (Crean et al., 1982).

The negative result from the experiments described in Fig. 6 may in fact indicate that cyclic GMP is not the second messenger to EDRF resulting in relaxation of the smooth muscle. Whatever the reason, the apparent discrepancy between EDRF release and increases in cyclic GMP needs to be resolved if measurement of the latter is to be used as a means of detecting EDRF.

3. Properties of EDRF

3.1. Half-Life In Vitro: Its Estimation by the Analysis of Concentration–Relaxation Curves

Estimations of the half-life ($t_{1/2}$) of EDRF by bioassay range from 6 to 49 s (*see* Table 1). In all cases $t_{1/2}$ was equated with the transit time between the source of EDRF and the assay tissue that was required to reduce the relaxation response by half. Such estimates of $t_{1/2}$ are therefore estimates of "half-response times." Another different approach is to consider $t_{1/2}$ as the increase in transit time that will reduce the concentration of EDRF by half. Biological half-life is defined as the time interval within which the drug amount falls by half (Bacq, 1971). If the kinetics of elimina-

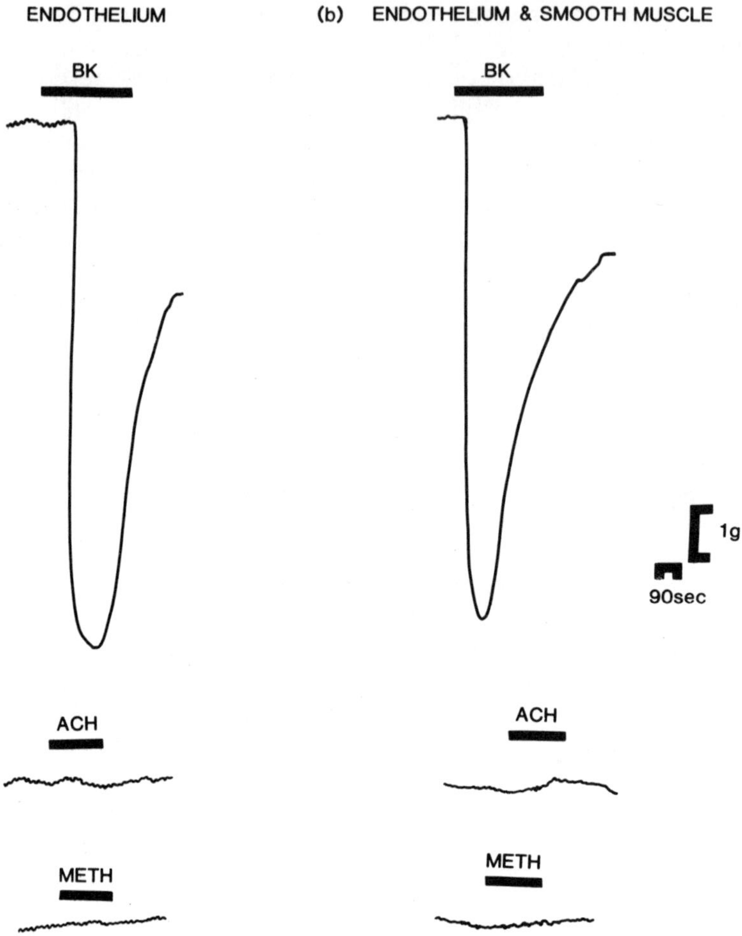

Fig. 6. The effect of coculturing endothelial and smooth muscle cells on the release of EDRF. (a) Bovine aortic endothelial cells (subcultured) only were scraped from a 4 × 90 mm culture dish and mixed with approximately 0.3 g Cytodex 3 microcarrier beads. The mixture was then prepared for bioassay as described in Fig. 5. The beads here acted as spacers. (b) As for (a), except scrapings were a mixture of rabbit aortic smooth muscle cells (subcultured) and bovine aortic endothelial cells (subcultured). The smooth muscle cells were first grown to confluence. Endothelial cells were then layered on top and allowed to grow for a further 3 d. The cultures appeared to be dominated by smooth muscle cells since the endothelial cells migrated underneath them (not shown). In both (a) and (b) only bradykinin (BK, 0.1 μM) caused relaxation. Acetylcholine (ACh, 1 μM) and a stable muscarinic analog, methacholine (Meth, 10 μM), were inactive. Note that the level of active force prior to stimulation with either Ach or Meth was the same as that prior to stimulation with BK for each bioassay.

Table 1
Types of Transferral Bioassays for EDRF and Half-Life ($t_{1/2}$) Estimates

EDRF source	Assay tissue	$t_{1/2}$, s	Agonist		Reference
Rabbit aorta	Rabbit coronary	6.3 ± 0.6	ACh	Single concentration	Griffith et al. (1984)
Dog femoral	Dog coronary	6.3 ± 1.2	ACh	Single concentration	Rubanyi et al. (1985)
Rabbit aorta	Rabbit femoral	24.0 ± 3.0	ACh	Single concentration	Fostermann et al. (1984)
Dog femoral	Dog femoral	49.0 ± 5.0	ACh	Single concentration	Fostermann et al. (1984)
Bovine cultured cells	Dog coronary	6.4 ± 0.4	BK	Single concentration	Cocks et al. (1985)
Porcine cultured cells	Rabbit aorta	<7	BK	Single concentration	Gryglewski et al. (1986)
Bovine cultured cells	Dog coronary	41	BK	(full curve analysis)	Angus and Cocks (1987)

tion of the drug (or hormone) are first order (exponential), then it is only necessary to measure relative concentrations to determine half-life. For a known substance this is relatively simple, since it can be compared with a given concentration, and the assay tissue calibrated accordingly. It is clearly more difficult, however, for an unknown substance such as EDRF. Its concentration will be determined by the concentration of agonist that stimulates the endothelial cells to release it. Thus, the challenge is to determine the half-life of EDRF by measuring changes in its concentration when only the concentration of the releasing agonist is known. This can be achieved by constructing a family of relaxation curves to the releasing agonist in the absence and presence of coils delaying transit-time. The bioassay described above (section 2) is ideal for this. The advantage of this system (i.e., large cell numbers and sensitive assay) is that it allows full relaxation curves to be constructed to the releasing agonist, in this case bradykinin. From the rightward shift in location of the curves with increasing delay time, the fall in potency of EDRF (as concentration ratio), and thereby the half-life of EDRF, can be determined, at least in Krebs' solution at 37°C.

Figure 7 depicts schematically the experimental design for this new approach to the estimation of the $t_{½}$ of EDRF. Duplicate cell columns from the same batch of endothelial cells and assay tissues from the same artery were perfused continuously with Krebs' solution. The assay tissues were then tested with substance P (10 n*M*) and bradykinin (0.1 μM), and only those that failed to relax were used in the experiment. Indomethacin (10 μM) was also present throughout to prevent activation of cyclooxygenase by bradykinin. The subsequent relaxation responses to increasing, sequential concentrations of bradykinin with and without a delay coil are shown in Fig. 8. The nominal zero delay curve (control) can be seen to be very steep since 0.3 n*M* bradykinin caused little relaxation compared to near maximal for only a fivefold increase in concentration. In the paired assay on the same day with 134 s delay there was a clear decrease in the maximal response to bradykinin, as well as an increase in the concentration causing 50% relaxation (EC_{50}).

Logistic fitting of the control curves (no delay) to bradykinin gave an EC_{50} value of 9.09 $\pm$ 0.05 ($-\log M$, mean $\pm$ SEM; n = 15), with a maximal relaxation of 96 $\pm$ 0.7% of the contraction to U46619. For 24- and 48-s delay times, the EC_{50} values were shifted to the right with no decrease in the maxima. Delays of 74 and 134 s further shifted the curves to the right, but the maxima were decreased to 78.8 $\pm$ 8.9% and 46.5 $\pm$ 13.6%, respectively (Fig. 9).

The concentration ratio (CR) for each pair of assays (control versus delay) was determined from the EC_{50} values, and the half-life of EDRF

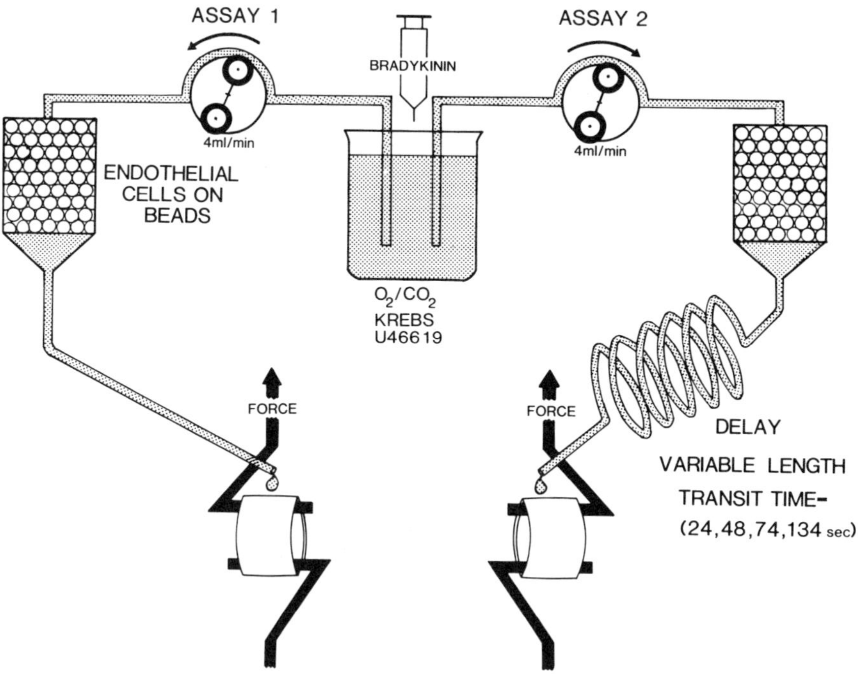

Fig. 7. Schema of the duplicate bioassay system for estimation of the half-life of EDRF. Assay 1 has a nominal zero-second delay between the cell column and assay tissue, and assay 2 has variable time delays of between 24 and 134 s.

estimated from the equation: ln CR = bT. This gave a mean half-life of 44.7 ± 8.5 s (n = 11) for 24–74 s delays (Fig. 9). Adding the data for 134 s delay gave a more shallow regression line and a longer estimate of half-life (51.1 ± 6.9 s; n = 15).

The strength of this new analytical approach to determine the half-life of EDRF is the ability to measure concentration ratios of this unstable substance by knowing the location of the concentration–response curve to the indirectly acting agonist (bradykinin) in the absence and presence of transit delay times. For ideal conditions, the EC_{50} value is the only parameter of the response curve that is altered by changing the delay. If the maximum is depressed under less than ideal conditions, however, modeling has been used to indicate how the fall in maximum can be used to adjust the concentration ratio (*see* Angus and Cocks, 1987).

An alternative method of measuring half-life of EDRF is to determine the increase in transit time that will reduce by half the relaxation response to a single concentration of an endothelium-dependent agonist (bradykinin). From the data shown in Fig. 9, it can be readily seen that

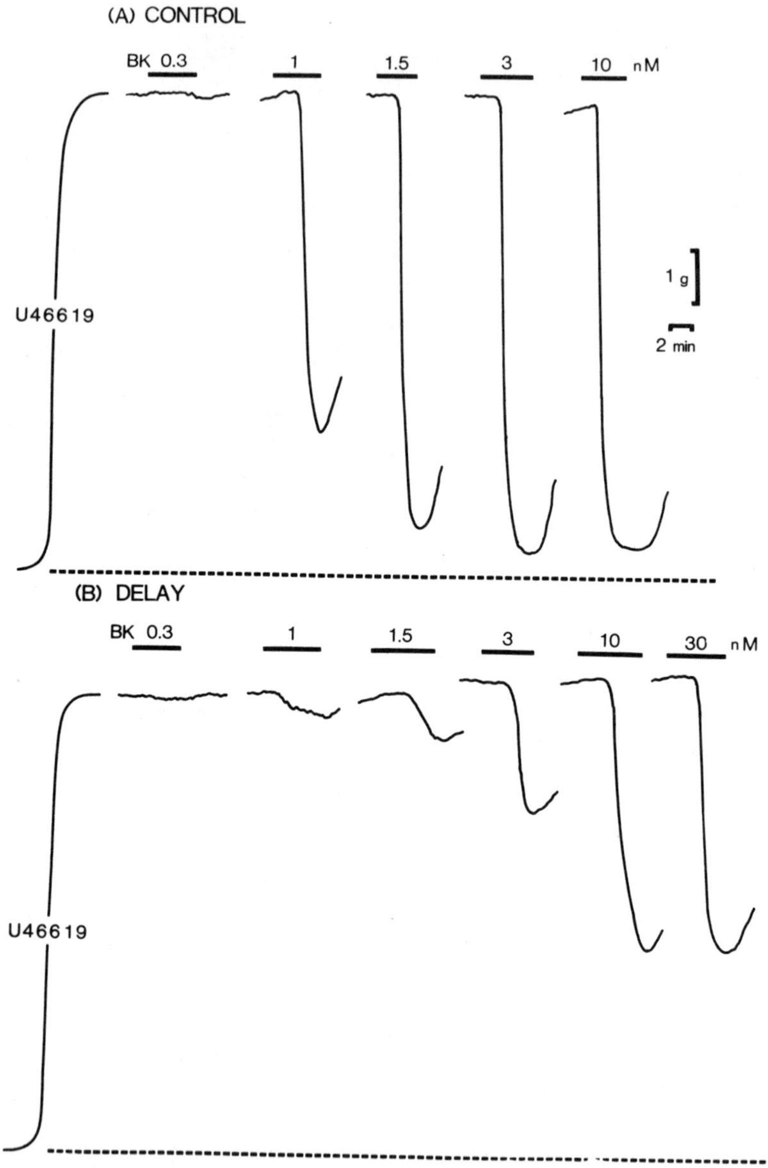

Fig. 8. Typical relaxation responses to EDRF from the paired bioassay system described in Fig. 7. Each assay artery was first contracted with U46619 (30 n*M*) and tested for endothelial cell removal with bradykinin (BK, 0.1 μM, not shown). The cell columns were then stimulated sequentially with increasing concentrations of BK for the times indicated by the solid bars. The breaks in the traces represent 20–40 min.

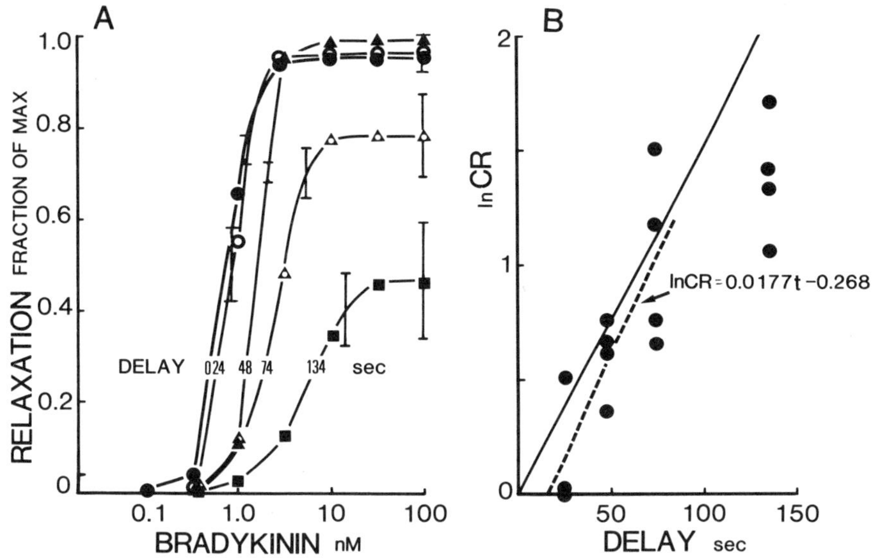

Fig. 9. (A) The effect of increasing transit time delays on the average concentration–relaxation curve for bradykinin-induced release of EDRF from endothelial cell columns described in Figs. 7 and 5. Error bars on each curve are average SEM from two-way analysis of variance. Those at maximal bradykinin response (100 n*M*) are 1 SEM for that concentration. (B) Linear regression of the rightward displacement of the bradykinin concentration–response curves as shown in (A) [expressed as a concentration ratio (CR) at the EC_{50} concentration] and delay time. The broken line is a plot of the regression equation for delays of 24, 48, and 74 s. The solid line is theoretically derived from the equation $\ln cr = 0.0154t$, where half-life is 45 s.

with this technique the half-life depends on the concentration of the releasing agonist. Thus, at the top of the curve, introducing delay times of up to 74 s does not alter the response. This would then indicate a half-life of more than 74 s, assuming that the response is equated with the concentration of the EDRF. Alternatively, the estimate of half-life could have been made at 1 n*M* bradykinin. There a value of less than 48 s would be obtained. Such an analysis is unsound because of the assumption that the relationship between the relaxation and the occupancy of the receptors for EDRF is linear, whereas it is most likely to be hyperbolic.

The advantages of the current bioassay for determining half-life are threefold. First, a large number of cultured endothelial cells is used compared to other sources of endothelium, usually *in situ*, such as in perfused whole arteries (see below). Second, the canine coronary artery is used as the assay tissue since it is one of the most sensitive when compared to many other arteries (Cocks and Angus, unpublished observations), and third, the relaxation evoked by bradykinin is measured under equilibrium conditions. That is, equilibrium is achieved between the concentration of bradykinin and all the endothelial cells in the column, as well as between the concentration of EDRF released and the recorded relaxation. This should have eliminated a potential source of error if the target tissue did not respond rapidly enough or if nonequilibrium kinetics hold for the reaction governing the release of EDRF by bradykinin.

The analysis of the half-life of EDRF as outlined here serves to illustrate two points. First, EDRF is, under in vitro conditions, more stable than previously believed (but, *see* Forstermann et al., 1984). Second, and perhaps more importantly, it clearly indicates the types of problems associated with measuring a labile substance (EDRF) on a single response basis (vertical analysis) without at least some knowledge of the location of that response on the full concentration–response curve to the EDRF-releasing agonist.

The same half-life of 40–50 s for EDRF may not hold in blood or when EDRF is released in close proximity to other endothelial and nonendothelial cells within the arterial wall in vitro. In blood, factors such as inactivation by hemoglobin (Martin et al., 1985a,b) and binding to plasma proteins may be important. Also, if EDRF release from endothelial cells is polar, directed predominantly toward the abluminal surface of the vessel, then little EDRF may in fact be released into the lumen. In blood-perfused canine femoral arteries, it was not possible to transfer EDRF in the arterial lumen from as close as 1–2 cm upstream (Angus et al., 1983).

3.2. Other Properties of EDRF Released from Cultured Endothelial Cells

EDRF released from cultured endothelial cells in response to bradykinin is hydrophilic and negatively charged at pH 7.4. These properties were ascertained by passing released EDRF through short transit-time hydrophobic (C-18 BondElut) and anion-exchange columns, respectively, prior to assay (*see* Figs. 10 and 11).

EDRF also binds or is inactivated by hemoglobin (Fig. 12). This is in addition to the apparent ability of hemoglobin to inhibit guanylate cyclase in smooth muscle and thus inhibit EDRF-mediated relaxations (*see* Martin et al., 1985a). Since EDRF appears to be highly susceptible to some oxidizing agents such as benzoquinone and superoxide anion, the ability of oxidized hemoglobin to reduce the EDRF-mediated response here may be caused, in part, by an oxidation–reduction reaction.

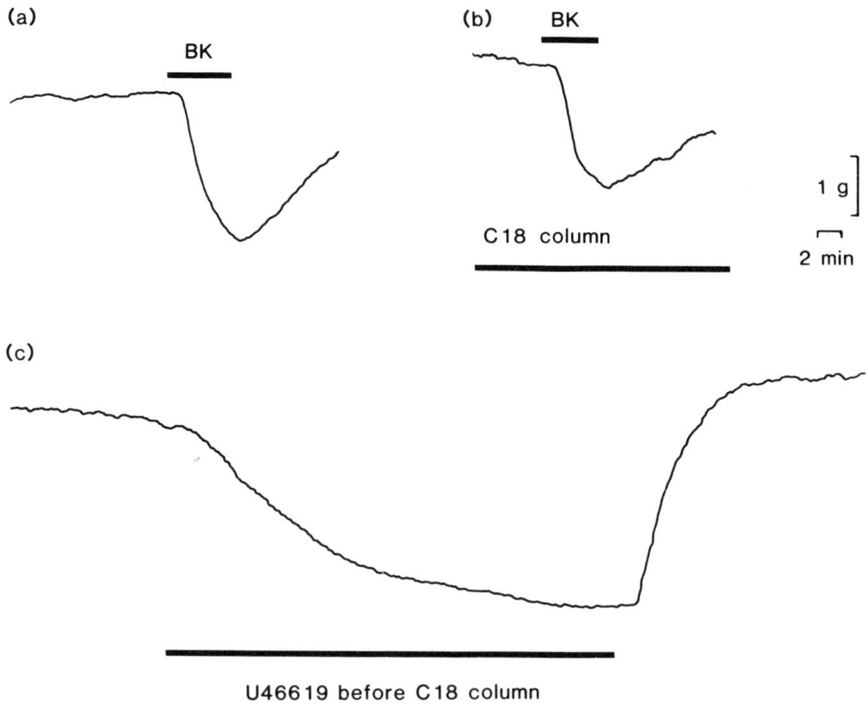

Fig. 10. The effect of passing EDRF released from an endothelial cell column through a short transit time (approximately 2 s) C-18 hydropholic chromatographic column (BondElut) prior to bioassay. The procedure was as described in Fig. 5, except canine, not porcine coronary artery was used as the assay tissue, and U46619 (the contracting agonist) was added via a sidearm just prior to the artery. (a) Control release of EDRF to bradykinin (BK, 0.1 μM). (b) As for (a), except the C-18 column was placed between the cell column and the assay tissue and before the U46619 sidearm. At least 10 min was allowed for the C-18 column to equilibrate prior to stimulation with bradykinin. (c) The C-18 column completely extracted U46619 when it was added to the Krebs' solution prior to it passing over the cells (from Cocks et al., 1985).

Benzoquinone has two properties that affect EDRF. First, it releases EDRF from endothelial cells (Fig. 13; *see* Cocks and Angus, 1984), and second, it inactivates EDRF once released presumably via an oxidation–reaction (Fig. 14). This is supported by the ability of the scavenger of superoxide anion, superoxide dismutase to potentiate the relaxation to EDRF released from endothelial cells of the canine femoral artery or in culture and to substantially stabilize it once released (Rubanyi and Vanhoutte, 1986; Gryglewski et al., 1986).

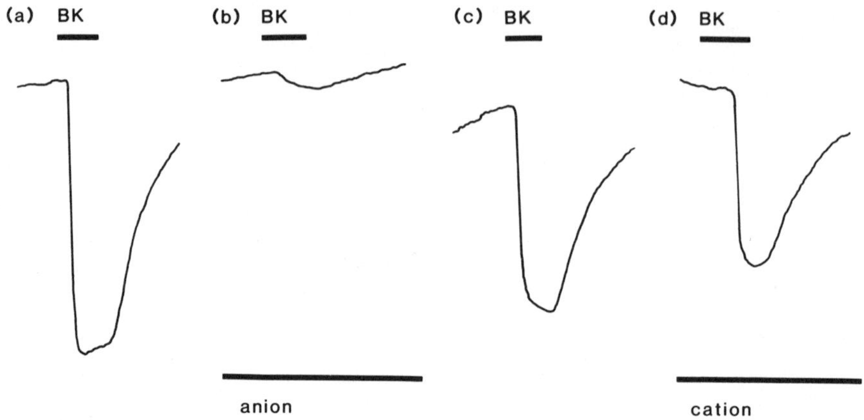

Fig. 11. Evidence that EDRF released from bovine endothelial cells is negatively charged at pH 7.4. Experimental conditions were as described in Fig. 10, except small, ion exchange chromatagraphic columns (short transit time) replaced the C-18 column. Also, U46619 was added to the Krebs' solution prior to it passing over the column of cells. (a) Control response to bradykinin (BK, 0.1 μM). (b) As for (a), except an anion exchange column (AGI × 8 resin) intervened between the cells and assay tissue. (c) Control response. (d) As for (b), except a cationic resin (AG 50w × 4 resin) was used in the intervening column. Thus only the anionic resin reduced the response to bradykinin (BK, 0.1 μM), indicating that EDRF has a net negative charge under these conditions (from Cocks et al., 1985).

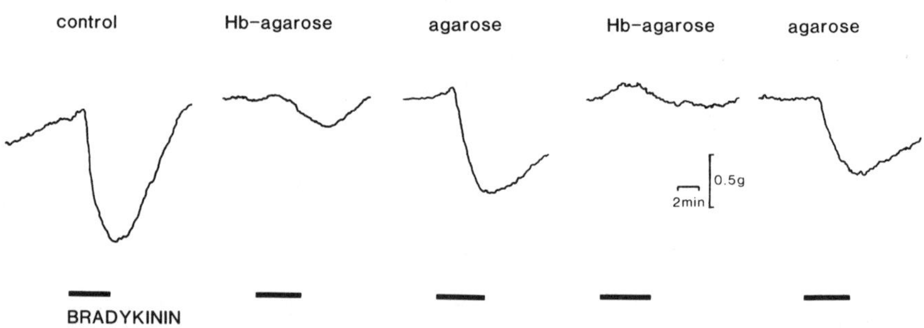

Fig. 12. Evidence that EDRF is inactivated and/or binds to hemoglobin (oxidized form). The experimental design was identical to that described in Fig. 11, except that a short delay (approximately 2 s) agarose-hemoglobin affinity column replaced the ion exchange columns. The responses to bradykinin (0.1 μM) are sequential and separated by 30–40 min. Note that agarose alone does not affect the relaxation response (from Cocks and Angus, 1985).

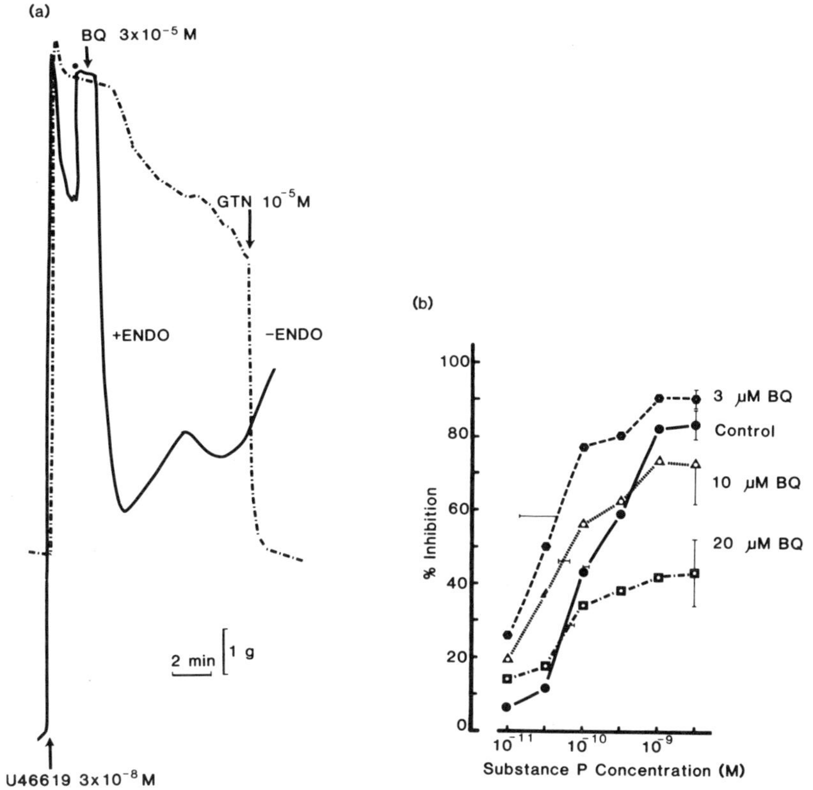

Fig. 13. (a) The effect of benzoquinone (BQ) on U46619-contracted rings of dog coronary artery with (+ENDO) and without (−ENDO) endothelium. The original chart records show that only the +ENDO ring of artery relaxed with a time course like that for other EDRF agonists. An endothelium-independent agent, glyceryl trinitrate (GTN), still relaxed the −ENDO artery. The horizontal bar represents 12 min before and 2 min after the bullet (●). (b) Cumulative concentration–relaxation curves to substance P in the absence and presence of increasing concentrations of BQ. Horizontal bars are SEM values of the EC_{50} values for each curve; vertical bars are SEM values of the mean maximal responses.

4. Whole Artery as a Source of EDRF—A Comparison with Cells in Culture

Many groups involved in the quest for EDRF's chemical nature use whole perfused arteries as their starting source of active material. The

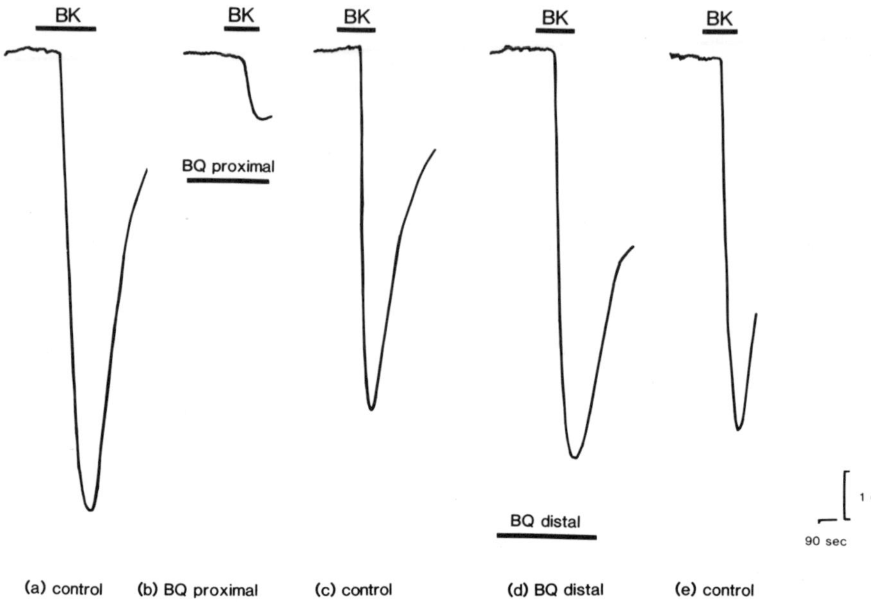

Fig. 14. The effect of benzoquinone (BQ, 10 μM) on the response to EDRF release from endothelial cells in culture by bradykinin (BK). The bioassay system was as described in Fig. 5, except that dog coronary artery was used as the assay tissue. Also, there was an appropriate 20-s delay coil of tubing inserted between the cell column and assay tissue. This allowed for infusion of BQ into the tubing either proximally or distally to the cell column. Times for interaction of EDRF and BQ were estimated to have been approximately 15 and 3 s for proximal and distal infusions, respectively. Responses to BK (10 nM) were sequential with the breaks in the trace representing 30–40 min.

major advantage of this approach is that the classic mediator of EDRF release, acetylcholine, is active, whereas as discussed above, both it and substance P do not release EDRF from cultured cells. The major disadvantage, however, is the small amount of EDRF released. There are obvious limits to the useful length and diameter of the donor artery. That is, like the cells in culture there are optimal conditions for maximizing EDRF transfer. The main problem with a perfused artery is the small number of endothelial cells compared to the column of microcarrier beads with endothelial cells described here. Then factors such as diameter and length of the vessel become important. If the vessel is too wide, a large dilution factor will occur; if too long, the rate of inactivation may mean EDRF released at the proximal end does not reach the assay tissue. The same considerations apply to the cell-bead column. Here the dead volume is drastically reduced, however, such that the large endothelial cell surface-to-volume ratio ensures that EDRF will be released from the column outlet in a concentrated form. This allows the assay tissue to respond to EDRF-releasing agonists with maximal sensitivity. To illustrate the significance of the magnification factor provided by cells in culture, three bioassays run in parallel are shown in Fig. 15, in which only the source of the EDRF differed. The first is an endothelium-intact ring of dog coronary artery, in which the responsiveness to bradykinin was compared to that in both types of transferal bioassay systems mentioned above. It can be seen that the endothelium-denuded ring of coronary artery (the assay tissue) detected EDRF released from the cell-bead column in a near identical manner as for the intact artery, whereas for the whole perfused artery, both the sensitivity and maximal response were greatly reduced. Thus, the concentrations of EDRF released by bradykinin from cell-bead column probably match closely those released into the microenvironment of the arterial wall in a normal artery.

Finally, another possible bioassay system for EDRF activity to emerge recently is the measurement of levels of cyclic GMP in smooth muscle. EDRF is believed to relax vascular smooth muscle by activating guanylate cyclase and thereby raising intracellular cyclic GMP (for review, *see* Ignarro and Kadowitz, 1985). Although this method has not been used in transferral bioassay systems like those described here, it has been successful in demonstrating that smooth muscle may modulate the responsiveness of endothelial cells to muscarinic agonists (*see* above and Loeb et al., 1985).

5. Concluding Remarks

The main problems to be solved in the search for the identity of EDRF are: (1) sufficient starting material; (2) a sensitive assay system; and (3) a means of stabilizing it once it is released. This chapter has considered the first two problems. Endothelial cells in suspension culture provide an ideal source of large amounts of EDRF that can be sensitively bioassayed using canine or porcine coronary artery rings. Other bioassays for EDRF in which the amount of EDRF available for release is limited simply by the number of cells and/or the assay tissue not being sensitive enough may provide some information as to the mechanism of the production and the breakdown of EDRF.

EDRF appears to be an unstable molecule with a half-life in Krebs' solution at 37 °C of 40–50 s. It is hydrophilic, negatively charged at pH 7.4, and susceptible to oxidation, particularly by benzoquinone and superoxide anion. Heme proteins also inactivate EDRF.

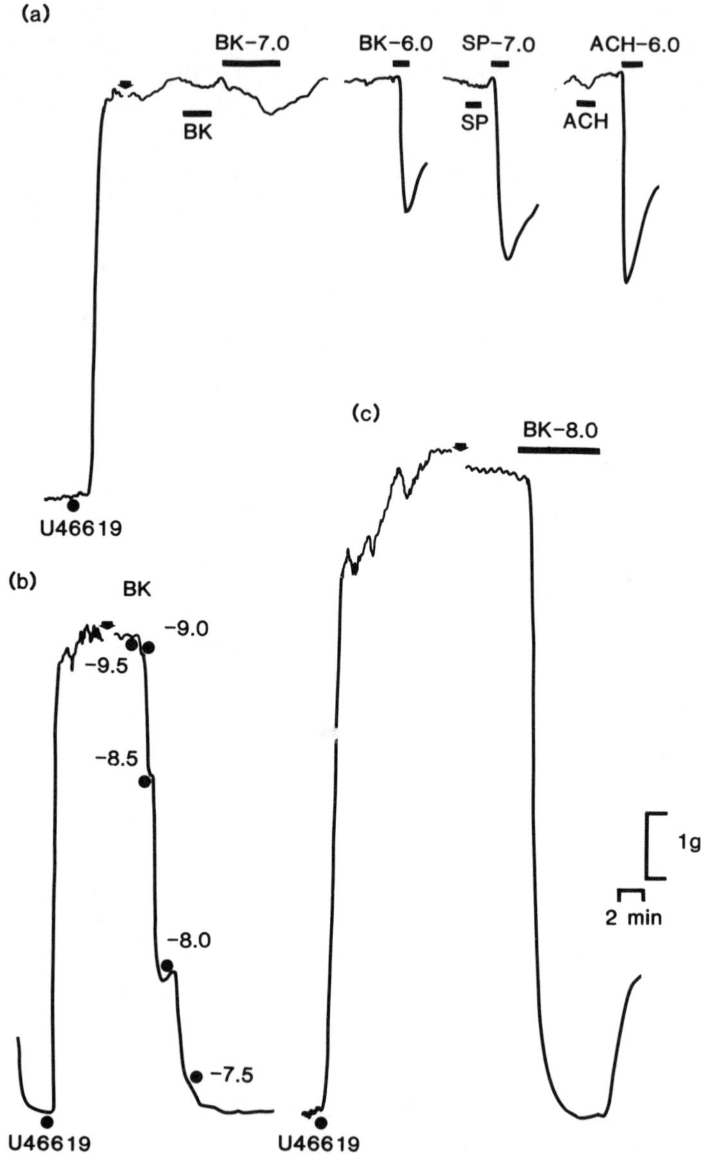

Fig. 15. Comparison of the relaxation responses to bradykinin (BK) in three types of bioassay system. (a) A transferral assay as described in Fig. 5, except that a perfused length (5 cm) of dog abdominal aorta was used as the source of EDRF. The aorta was filled with fine glass beads to reduce the dead space (EDRF is not inactivated by glass, Cocks, unpublished data). (b) An endothelium-intact ring of dog coronary artery. (c) A transferral assay similar to that described in

The instability of EDRF is now the major problem in this field. As mentioned earlier, experiments with scavengers of superoxide anions and benzoquinone indicate that one way EDRF can be degraded is by oxidation. Others have suggested that it is inactivated by a reduction reaction (Griffith et al., 1984). The data with superoxide dismutase are suggestive of the former mechanism. Oxidation by oxygen-derived free radicals is not the only means of inactivation, however, since EDRF collected in the presence of superoxide dismutase still rapidly loses its activity (Cocks and Angus, unpublished).

Once the problem of instability has been overcome, collection of large amounts of material from cells in suspension culture similar to those described here may enable chemists to unravel the secret of the structure of EDRF.

Acknowledgments

This work was supported by an Institute Grant from the National Health and Medical Research Council of Australia. We thank Nella Puglisi, Di Karamoshos, and Silvina Rainone for expert technical assistance, and Dr. P. Mosse for assistance with the scanning electron microscopy.

References

Angus, J. A. and Cocks, T. M: The half-life of endothelium-derived relaxing factor released from bovine aortic endothelial cells in culture. *J. Physiol.* (Lond.) **388:**71–81, 1987.

Angus, J. A., Campbell, G. R., Cocks, T. M., and Manderson J. A.: Vasodilation by acetylcholine is endothelium-dependent. A study by sonomicrometry in canine femoral artery in vivo. *J. Physiol.*(Lond.) **344:**209–222, 1983.

← Fig. 15 (*Continued*)

Fig. 5, with a column of cultured cells growing on microcarrier beads as the source of EDRF. It can be seen that responses to BK in the intact ring and the cell column assays are similar, whereas those for the perfused aorta are poor by comparison. The relaxation responses to both acetylcholine (ACh) and substance P (SP) in the perfused aorta assay are greater than for BK, although no more than 40–60% relaxation to maximal concentrations of each agonist was achieved. All assay tissues were de-endothelialized dog coronary artery rings, and each contracted with U46619 (30 n*M*). The bars under the trace indicate maximal concentrations of each agonist applied directly to the assay tissue. The time calibration bar represents 12 min before the arrow.

Angus, J. A., Cocks, T. M., and Satoh, K.: Alpha-adrenoceptors and endothelium dependent relaxation in canine large arteries. *Br. J. Pharmacol.* **88:**767–777, 1986.

Bacq, A. M.: *Fundamentals of Biochemical Pharmacology.* Pergamon, Oxford, p. 207, 1971.

Chamley-Campbell, J. H. and Campbell, G. R.: What controls smooth muscle phenotype? *Atherosclerosis* **40:**347–357, 1981.

Cocks, T. M. and Angus, J. A.: Endothelium-Dependent Modulation of Blood Vessel Reactivity, in *The Peripheral Circulation* Hunyor, S., Ludbrook, J., Shaw, J., and McGrath, M., eds. Elsevier, Amsterdam, New York, Oxford, 1984.

Cocks, T. M., Angus, J. A., Campbell, J. H., and Campbell, G. R.: Release and properties of endothelium-derived relaxing factor (EDRF) from endothelial cells in culture. *J. Cell Physiol.* **123:**310–320, 1985.

Cocks, T. M., Manderson, J. A., Mosse, P. R. L., Campbell, G. R., and Angus, J. A.: The effect of fibromuscular intimal thickening on endothelium-dependent relaxation in the rabbit carotid artery. *Blood Vess.* in press, 1987.

Crean, G., Pilar, G., Tuttle, J. B., and Vaca, K.: Enhanced chemosensitivity of chick parasympathetic neurones in co-culture with myotubes. *J. Physiol.* (Lond.) **331:**87–104, 1982.

Davies, P. F., Ganz, P., and Diehl, P.S.: Reversible microcarrier-mediated junctional communication between endothelial and smooth muscle cell monolayers: An in vitro model of vascular cell interactions. *Lab. Invest.* **85:**710–718, 1985.

DeMey, J. G. and Vanhoutte, P. M.: Heterogenous behavior of the canine arterial and venous wall. *Circ. Res.***51:**439–447, 1982.

Forstermann, U., Trogisch, G., and Busse, R.: Species-dependent differences in the nature of endothelium-derived vascular relaxing factor. *Eur. J. Pharmacol.* **106:**639–643, 1984.

Furchgott, R. F. and Zawadzki, J. V.: The obligatory role of endothelial cells in the relaxation of arterial smooth muscle by acetylcholine. *Nature* (Lond.) **288:**373–376, 1980.

Ganz, P., Davies, P. F., Leopold, J. A., Gimbrone, M. A., and Alexander, R. W.: Short and long-term interactions of endothelium and vascular smooth muscle in co-culture: Effects on cyclic GMP production. *Proc. Natl. Acad. Sci. USA* **83:**3552–3556, 1986.

Gordon, J. L. and Martin, W.: Endothelium-dependent relaxation of the pig aorta: Relationship to stimulation of ^{86}R6 efflux from isolated endothelial cells. *Br. J. Pharmacol.* **79:**531–541, 1983.

Griffith, T. M., Edwards. D. H., Lewis, M. J., Newby, A. C., and Henderson, A. H.: The nature of endothelium-derived vascular relaxant factor *Nature* (Lond.) **308:**645–647, 1984.

Gryglewski, R.J., Palmer, R.M.J., and Moncada, S.: Superoxide anion is involved in the breakdown of endothelium-derived vascular relaxing factor. *Nature* (Lond.) **320:**454–456, 1986.

Ignarro, L. J. and Kadowitz, P. J.: The pharmacological and physiological role of cyclic GMP in vascular smooth muscle relaxation. *Ann. Rev. Pharmacol. Toxicol.* **25:**171–191, 1985.

Loeb, A. L., Owens, G. K., and Peach, M. J.: Evidence for endothelium-derived relaxing factor in cultured cells. *Hypertension* **7:**804–807, 1985.

Martin, W., Villani, G. M., Jothianandan, D., and Furchgott, R. F.: Selective blockade of endothelium-dependent and glyceryl trinitrate-induced relaxation by hemoglobin and by methylene blue in the rabbit aorta. *J. Pharmacol. Exp. Ther.*, **232:**708–716, 1985a.

Martin, W., Villani, G.M., Jothianandan, D., and Furchgott, R. F.: Blockade of endothelium-dependent and glyceryl trinitrate induced relaxation of rabbit aorta by certain ferrous hemoproteins. *J. Pharmacol. Exp. Ther.* **233:**679–685, 1985b.

Peach, M. J., Singer, H. A., and Loeb, A. L.: Mechanisms of endothelium-dependent vascularsmooth muscle relaxation. *Biochem. Pharmacol.* **34:** 1867–1874, 1985.

Pearson, J. D. and Gordon, J. L.: Vascular endothelial and smooth muscle cells in culture selectively release adenine nucleotides. *Nature* (Lond.) **281:**384–386, 1979.

Rubanyi, G. M. and Vanhoutte, P. M.: Superoxide anions and hyperoxia inactivate endothelium-derived relaxing factor. *Am J. Physiol.* **250:**H822–H827, 1986.

Rubanyi, G. M., Lorenz, R. R., and Vanhoutte, P. M.: Bioassay of endothelium-derived relaxing factor(s): Inactivation by catecholamines. *Am. J. Physiol.* **249:**H95–H101, 1985.

Chapter 7

Endothelial Cells in Culture and Production of Endothelium-Derived Constricting Factors

Robert F. Highsmith, David Aichholz, Oliver FitzGerald, Richard Paul, Gabor M. Rubanyi, and Kristine Hickey

1. Introduction

It is now well established that the vascular endothelium possesses many biological functions in addition to its role in the microvascular transport of material between blood and tissues. Of major importance in the cardiovascular arena is our increasing knowledge of the role that the endothelial cell may play in modulating the reactivity and/or responsiveness of the underlying vascular smooth muscle. Furchgott and Zawadzki (1980) made the pioneering observation that the relaxation of isolated arteries to acetylcholine was dependent on the presence of an intact endothelial lining. They hypothesized that the vasodilatation was mediated by a secondary signal initiated in the intimal layer of the blood vessel wall. Since this discovery, a variety of endogenous and exogenous agents has been shown to elicit relaxations that are mediated by the release from the endothelium of a short-lived substance referred to as endothelium-derived relaxing factor (EDRF).

Initially, the authors' laboratory investigated the use of cell culture to study endothelium-derived vasoactive substances. It was not possible to detect significant relaxation of contracted coronary artery strips exposed to media conditioned by bovine aortic endothelial cells in culture. Contrary to expectations, however, a consistent tonic contraction of relaxed coronary artery rings was observed when treated with relatively small amounts of such media (Agricola et al., 1984). This potent endothelium-derived constricting factor(s) (EDCF) was characterized as being distinctly different from the commonly known endogenous vasoconstrictors (Hickey et al., 1985). This chapter will review the methodology used in the authors' laboratory, as well as current findings on EDCF, and examine studies of others who have confirmed and extended these initial observations.

2. Evidence for the Presence of Constricting Factors in Vascular Endothelium

The presence of EDCF in cultured endotheial cells has now been confirmed in several studies. Of importance, however, is the demonstration that vasoactive secretory products of cultured endothelial cells bear a functional relationship to the normal influence of endothelium in contact with vascular smooth muscle. Intact endothelium has been shown to facilitate the contraction of vascular smooth muscle under several conditions. For example, in isolated canine arteries and veins, anoxia augments the contractile response to several vasoconstrictor agents (Vanhoutte, 1976; Van Nueten and Vanhoutte, 1980). De Mey and Vanhoutte (1982), in comparing the responses of isolated arteries and veins, reported that mechanical removal of the endothelium reduced the maximal contractile response to norepinephrine and decreased the augmentation of the response to norepinephrine caused by anoxia. The same authors (De Mey and Vanhoutte, 1983) have additionally noted that cyclooxygenase inhibition did not prevent the potentiating response of anoxia on norepinephrine-induced tone in the canine femoral artery. Others (Holden and McCall, 1983) have reported preliminary observations that hypoxic or anoxic contractions of porcine pulmonary arteries were endothelium-dependent. The generation of a facilitatory EDCF-like substance by the endothelium would be a likely explanation of these results.

Perhaps the most convincing evidence to date that the intact endothelium may produce a potent constricting factor, at least under hypoxic or anoxic conditions, comes from the findings of Rubanyi and Vanhoutte (1985), who evaluated the role of the endothelium in the facilitation of the contractile responses of canine coronary arteries. The release of vasoactive material from the endothelium was assessed by comparing the responses of vessel strips with or without endothelium to those of denuded strips that had been layered ("sandwiched") with vessel segments containing an endothelial lining. In intact vessels and in layered preparations with endothelium, anoxia augmented contractile responses to prostaglandin $F_{2\alpha}$;

hypoxia elicited contractions in the presence of indomethacin. Removal of the endothelium abolished the anoxic facilitation as well as the contractions in response to hypoxia. Furthermore, inhibitors of cyclooxygenase, lipoxygenase, or phospholipase A_2 or antagonists of adrenergic, serotonergic, or histaminergic receptors failed to prevent the contractile response to anoxia. Likewise, methylene blue, an inhibitor of the action of EDRF, did not alter the anoxic facilitation. The authors concluded that hypoxia and anoxia caused the release of a diffusible vasoconstrictor substance(s) from the endothelium. The pharmacological properties of this vasoconstrictor are virtually identical to those which have been described by the authors' laboratory for EDCF produced by endothelial cells in culture (Agricola et al., 1984; Hickey et al., 1985). In particular, the release of the vasoconstrictor substance in the studies of Rubanyi and Vanhoutte (1985) was enhanced by hypoxia in a fashion analogous to the pattern observed for the stimulation of EDCF release by hypoxic endothelial cells in culture (Aichholz et al., 1986). Taken in conjunction, these results may suggest that the same EDCF molecule is released from the intact endothelium or from endothelial cells in culture when hypoxia is imposed. This then would indicate that cell culture may serve as a useful model for studying endothelium-dependent vasoconstrictor mechanisms.

3. Endothelium-Derived Constricting Factor(s) Produced by Endothelial Cells in Culture

3.1. Methodology for Studying EDCF

3.1.1. Control Cell Culture Techniques

Normal human skin fibroblasts were obtained as established cell lines (MC-31 and MC-41) from the Repository for Mutant Human Cell Strains, Montreal, Canada, and were cultured as described by Gelman et al. (1980). For any given experiment, all cell lines were at the same passage number (± 2), and all experiments were done on confluent monolayers of cells. Swiss mouse 3T3 fibroblasts were obtained from Dr. M. Lieberman and cultured as described by Whittenberger and Glaser (1977). Fresh 3T3 cultures were started from frozen stocks after six to eight serial transfers and allowed to reach confluency before the medium was harvested. Vascular smooth muscle cells were obtained fresh from rat aorta as described by Harder and Sperelakis (1979) and were cultured identically as described below for endothelial cell culture or were subjected to reaggregate culture as reported by Yoneda et al. (1981). After establishing confluent monolayers of each cell type or after primary culture of smooth muscle reaggregates, each of the cell types were placed into the fresh medium used for culturing endothelial cells. At 24-h intervals, aliquots of the conditioned medium were tested for biological activity.

3.1.2. Endothelial Cell Culture Techniques

3.1.2.1. Bovine Origin. Endothelial cell cultures derived from pooled, adult bovine thoracic aortas obtained at a local slaughterhouse were established in the following manner. Immediately following the death of the animal, the aortas were dissected free from surrounding tissues and placed in 100 m*M* phosphate-buffered saline for transport to the tissue culture laboratory. Under aspectic conditions in a laminar flow hood, the vessels were opened longitudinally and washed repeatedly with buffer to remove remaining blood clots. Endothelial cells were obtained by gentle scraping of the luminal aortic surface using a scalpel blade at an angle of 45°. The cells were then placed in a sterile tube and incubated for 20 min at 37°C with 0.3% trypsin in Ca^{2+}-free Eagle's minimum essential medium. Following this incubation, the cells were passed through successive filters (75 and 10 μm) to remove undigested tissue fragments, layered on a Percoll-Eagle's minimum essential medium (3:7, vol/vol) density gradient and centrifuged at 400*g* for 15 min. Endothelial cells were thus separated from remaining tissue fragments and from cells of different density; they formed a distinct band approximately 1 cm from the bottom of the centrifuge tube. This band was then collected, and the cells were suspended in medium containing 0.5 mg deoxyribonuclease/mL and 1.5 mg soybean trypsin inhibitor/mL. Final cell suspensions were obtained following further centrifugation and resuspension in alpha-modification of the medium containing 10% horse serum and 1% antibiotic/antimycotic. The cells were plated on 25 cm^2 tissue culture flasks at a density of 7.5 $\times 10^5$ cells/flask and then incubated with 5% CO_2, 95% humidified air at 37°C. Approximately 8 million endothelial cells were obtained per 20 cm segment of aorta.

3.1.2.2. Human Origin. Human endothelial cells were obtained from umbilical veins using a modification of the method first described by Jaffe et al. (1973). Umbilical cords were collected on the day of delivery and processed under aseptic conditions in a laminar flow hood. Using a pediatric nasogastric feeding tube, both ends of the umbilical vein were cannulated, and the tube was tied securely in place. A 60-mL syringe was then attached to the top end of the feeding tube, and the vein was flushed extensively with sterile saline and Ca^{2+}- and Mg^{2+}-free Hanks balanced salt solution containing 1% antibiotic/antimycotic. Following clamping of the distal end of the feeding tube, the vein was filled with 12–20 mL of 0.2% collagenase and incubated for 15 min at 37°C. The veins were then flushed with 12–20 mL of media 199 containing 20% (vol/vol) pooled AB human serum followed by 60 mL of salt solution. The media was collected and centrifuged at 400*g* for 10 min, and the resulting cell pellet was resuspended in 10 mL of complete media 199 containing 20% human serum and supplemented with 187 μg/mL L-glutamine, 100 units/mL penicillin G, 100 μg/mL streptomycin, 5 μg/mL gentamicin, 20 μg/mL endothelial cell growth factor, and 15 units/mL porcine, preservative-free heparin. The cells were plated at a density of 2.5×10^5 cells/25 cm^2 tissue culture flasks, which had been coated with 1% gelatin and incubated at 37°C with 5% CO_2, 95% humidified air.

3.1.2.3. Cell Passage and Endothelial Cell Microcarrier Culture. After an initial 24 h in culture, the media were replaced every 3 d. Upon reaching confluence, endothelial cells (bovine and human) were washed twice with salt solution and then incubated for 2 min with 1 mL of a mixture of 0.25 mg/mL trypsin and 0.1 mg/mL EDTA. Culture dishes were scraped with a cell scraper before adding 5 mL of medium containing 10 or 20% serum and transferring the mixture to a sterile centrifuge tube. Culture dishes were further washed with 20 mL of salt solution, and this solution was also added to the centrifuge tube. Following centrifugation at 400*g* for 10 min, the cell pellet was resuspended in the appropriate culture medium and counted using a hemocytometer. Endothelial cells were passaged at an approximate split ratio of 2:1.

Cells grown on gelatin-coated beads were used routinely for the microcarrier cultures. Prior to establishing the culture, the beads were incubated with the appropriate culture medium for 15 min at 37°C with 5% CO_2. Endothelial cells were then added at a cell-to-bead ratio of 10:1 and cultured in a 500-mL microcarrier culture vessel equipped with a flexible stir bar. Cells were initially cultured for 3 h without stirring in one-third the calculated final volume of media (calculated as 3 mg of beads/mL of media). After this period, the culture volume was brought to the final volume, and cultures were kept at 37°C, 5% CO_2, 95% air, and gently stirred. Additional media and microcarrier beads were added when the cultures reached confluency.

3.1.2.4. Characterization of Endothelial Cells. Endothelial cells were routinely used between the first and third passages. Viability was assessed by trypan blue exclusion, and cultures were characterized based on their morphological appearance and by positive immunofluorescent staining with an antifactor VIII/von Willebrand factor monoclonal antibody using the method described by Jaffe et al. (1973).

3.1.2.5. Bioassay. The biological activity of media from endothelial cell cultures was assessed on strips of porcine or bovine coronary arteries. Adult porcine or bovine hearts were collected from a local slaughterhouse immediately following death of the animal and transported to the laboratory where they were processed in an identical fashion. Segments of the left anterior descending coronary artery were dissected free from surrounding fat and connective tissue. Care was taken to keep the tissue moist during this procedure; following cleaning, the artery was cut into 5-mm segments. Depending upon the particular experiment, the vessel rings were either used with intact endothelium on the luminal side or everted and rubbed to remove endothelium as confirmed by electron microscopy (Hickey et al., 1985). Unless noted otherwise, all experiments reviewed below were completed with everted, deendothelialized rings.

The segments were suspended on a pair of stainless-steel hooks, which were mounted in a chamber filled with Krebs-bicarbonate solution perfused with 95% O_2, 5% CO_2 and maintained at 37°C. One of the two hooks on which the arterial ring was suspended was fixed to a plate outside the chamber, and the other was connected to a force-displacement transducer. The transducer in turn was mounted on a rack-and-pinion device above the chamber that could be precisely adjusted to alter resting muscle length. Isometric tension was then set at a level that resulted in optimal active tension development upon addition of KCl to the chamber. The arterial rings were allowed to equilibrate for 2 h before testing the response to KCl. The effect of nonconditioned media or media derived from endothelial cells on isometric force development was assessed by adding the media at various dilutions directly to the muscle chamber. The contraction elicited was measured as the increase in isometric tension above the initial base line.

3.2. Endothelial Cell Source, Culture Conditions, and Time Course of Production of EDCF

The origins of endothelial cells successfully maintained in primary culture or serially passaged include bovine aorta and pulmonary artery, human umbilical artery and vein, rat microvasculature (epididymal fat pad), and rat atrium. Production of EDCF was confirmed by cells derived from all of these sources. Only two sources of endothelial cells have resulted in an apparent lack of production of EDCF: Human endothelial cells isolated from omental fat (McMurty, 1985) and bovine corneal endothelial cells (McMurty, 1985) did not elicit constriction of the coronary artery. Bovine aortic endothelial cells have been used in the majority of experiments discussed below mainly because of their abundance and availability.

Endothelial cells obtained and treated as described above were routinely cultured in 25-cm^2 flasks containing α-modified Eagle's medium and 10% horse serum. The cells were cultured at high density for an initial 4–6 d period. Following the initial culture period, aliquots of the conditioned media were removed at 24-h intervals and tested for biological activity. At the first 24-h sampling interval, the conditioned media (with or without 10% serum) elicited a significant vasoconstriction in all coronary preparations tested. The potency of culture media in inducing coronary vasoconstriction increased significantly with time in culture, reaching a maximum on day 3–4 of incubation. Comparing the vasoactive responses of aliquots of media taken with or without volume replacement indicated that the increase in vasoconstrictor potency as a function of time in culture was caused by the progressive accumulation of the constrictor material in the medium. The production rate of EDCF basically paralleled the cell growth curve; maximal production during the log-phase growth period and contact inhibition of production as the cells approached confluency. This production cycle was repeatable for at least seven serial passages.

3.3. Physiological Response to EDCF

3.3.1. Effect of EDCF on Isolated Coronary Arteries

The vasoconstriction induced in porcine coronary arteries by 10% endothelial cell-conditioned media was of slow onset with a latency of 0.5–5 min (Fig. 1a). Maximal tension was usually achieved within 15 min, but on occasion required 30 min or more. In contrast to the KCl-induced contraction, the contractile response to the endothelial cell media was difficult to wash out. Ninety-one percent of the vessels tested did not relax completely following extensive washing, thus illustrating a substantial tonic component to the contractile response. Attainment of the original baseline tension in these vessels required approximately 1 h. The contractile response did not require an intact endothelium on the bioassay tissue. In parallel control experiments, nonconditioned medium was incubated in the absence of cells and then tested for biological activity. No response to the control media was obtained in the majority of the vessels, whereas the remainder demonstrated only a small contraction that averaged approximately 10% of that induced by the same concentration of conditioned media.

The cumulative addition of EC-conditioned media to a final concentration of 80% in the chamber elicited a contractile response of 100 m*N* over the initial baseline tension (Fig. 1b). The tonic component of the contraction was illustrated by the substantial tension (+55 m*N*) that remained after the second wash-out of the media.

A dose–response curve was obtained by recording the individual steady-state tensions of the coronary arteries in response to fixed concentrations of the endothelial cell-conditioned media. Any response to the same concentration of control media was subtracted. The isometric tension in

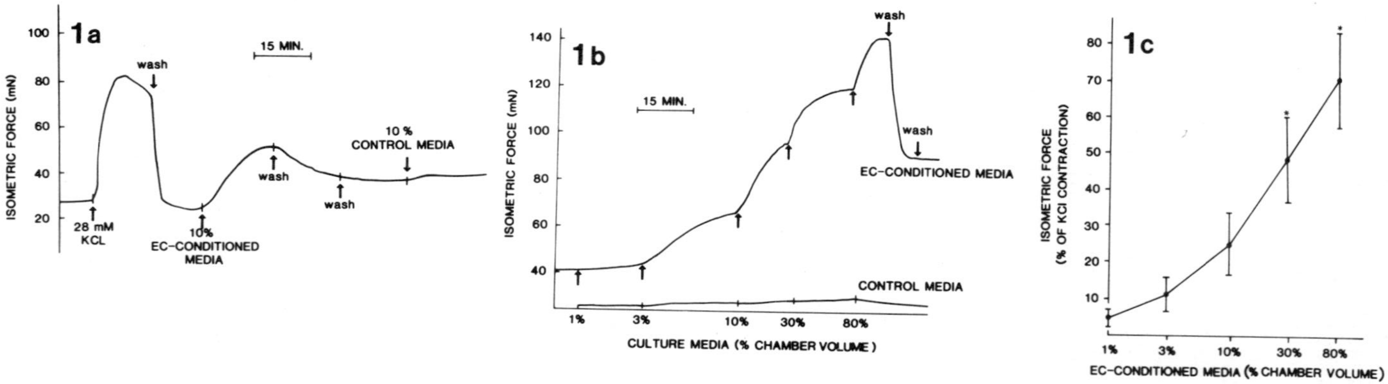

Fig. 1. Isometric contractions of deendothelialized porcine coronary arterial rings to endothelium-conditioned (EC) culture media. (a) A typical recording of the response obtained with a single porcine arterial segment. Isometric tension is expressed in m*N* as a function of time following the addition of 10% endothelium-conditioned or control culture media. (b) A cumulative tension recording from a single porcine coronary artery as a function of time following the addition of increasing volumes of endothelium-conditioned or control culture media. (c) The steady-state isometric tension obtained as a function of concentration of endothelium-conditioned culture media. Any responses to the same concentration of control media were subtracted. Values are expressed as percentage of a 28 m*M* KCl-induced contraction. Each point represents the mean ± SEM response of ten coronary arterial segments to different volumes of endothelium-conditioned media pooled from seven different cultures. An asterisk above a point indicates that the value is significantly different ($p < 0.05$) from the response obtained at the preceding lower dose (reproduced with permission).

the porcine coronary arteries increased in a dose-dependent fashion as the percentage of the conditioned media was increased from 1 to 80% in the organ chamber (Fig. 1c). Saturation of the response did not occur at higher (100%) concentrations of the media, suggesting that a maximal response was not yet obtained. The variability in the response of the vessels to a fixed concentration of media was most likely caused by differences in the concentrations of the vasoconstrictor substance in each particular pool of media. Repeated challenges to different arterial segments with the same media gave reproducible responses.

Specificity of the cell type involved in producing the vasoconstrictor substance was evaluated by testing the biological activity of several media obtained from control cultures of fibroblasts (human skin and mouse 3T3 cells) and vascular smooth muscle cells (rat aorta). Conditioned media obtained from each of these cell types after 4–6 d in culture had no effect on the coronary arteries when tested at concentrations ranging from 10 to 80%. The coronary arteries maintained their responsiveness, since the same contractile response to KCl and to the endothelium-conditioned media was observed before and after testing the control media.

3.3.2. Effect of EDCF on Coronary Artery Blood Flow

Myocardial ischemia and infarction may be causally related to inappropriate vasospasm of the major epicardial arteries. The qualitative nature of the response of isolated coronary arteries to EDCF suggests that this factor may be an important element in the etiology of coronary vasospasm. Thus, investigations of the effect of EDCF on coronary blood flow and myocardial performance are of obvious clinical importance.

Using an isolated Langendorff-perfused rat heart preparation, the effects of EDCF were evaluated on coronary blood flow and cardiac performance. Endothelium-conditioned media were diluted 1:10 and retrogradely perfused at 1 mL/min into the coronary circulation of four hearts paced at 225 bpm. The coronary blood flow was measured directly. Following attainment of a new steady state, coronary flow was markedly reduced (43.8%) when compared to infusions of nonconditioned, control media (8.1 $\pm$ 1.4 vs 14.2 $\pm$ 1.9 mL/min; $p < 0.005$). A modest, yet significant reduction (17.3%) in cardiac contractility (1310 $\pm$ 64 vs 1080 $\pm$ 37 mm Hg/s; $p < 0.005$) was also apparent. In addition, in nonpaced hearts, a slight negative chronotropic effect was seen (heart rate decreased approximately 15%). Nonconditioned control media or conditioned media obtained from control, nonendothelial cell cultures (fibroblasts and smooth muscle) had no effect on any of the above parameters when infused at the same dosage.

In a similar fashion, Gillespie et al. (1986) have recently investigated the effects of EDCF in isolated hearts of the rabbit in a nonrecirculating system. Coronary vascular resistance was estimated by measuring coronary perfusion pressure during a constant infusion of EDCF at 10 mL/min/kg body wt. EDCF was derived from a 20-fold concentrated solution of media conditioned by bovine aortic endothelial cells that had been passaged 5–8 times. Following bolus injections of 100–600 μL of EDCF, dose-dependent increases in coronary perfusion pressure were recorded. The injection of 600 μL of EDCF resulted in an increase of 20 mm Hg over control. The vasoconstriction induced by EDCF was remarkably tonic in nature and in some cases was maintained throughout the duration of the experiment. The contractile response to EDCF in Langendorff-perfused hearts is in marked contrast to that obtained with other pressor agents whose effects are characteristically transient. Gillespie et al. (1986) also found that the nonspecific vasodilator sodium nitroprusside ($3.8 \times 10^{-5}M$) blocked the initiation and reversed ongoing EDCF-induced vasoconstriction. In contrast to the authors' studies, Gillespie et al. (1986) did not observe any significant inotropic or chronotropic effect of EDCF. These differences could be attributable to differences in species or in experimental design, and/or to the greater responsiveness of the retrogradely perfused rat heart.

3.4. Biochemical and Pharmacological Characterization of EDCF

EDCF present in large volumes of media conditioned by bovine aortic endothelial cells cultured on microcarrier beds is very stable and does not lose significant activity upon storage at −70°C for at least 2 mo. This property has permitted the initial characterization studies (Agricola et al., 1984; Hickey et al., 1985). Most of these studies have utilized endothelium-conditioned culture media as a crude preparation of EDCF. More definitive biochemical characterization must await the purification of EDCF.

Initial dialysis and ultrafiltration experiments indicated that the molecular weight of the endothelium-derived vasoconstrictor was greater than 3500, but less than 12,000 dalton. In separate experiments, three different pools of conditioned media were gel filtered on a column of Biogel P-10, and the resulting fractions were bioassayed for vasoactivity using porcine coronary arteries. In each case a single predominant peak of vasoconstrictor activity eluted midway between pancreatic trypsin inhibitor and cytochrome C. A selectivity curve based on the calculated partition coefficients of standard marker proteins indicates that the average molecular weight of the isolated endothelium-derived vasoconstrictor is 8500 ± 1500 dalton (Fig. 2).

The gel filtration behavior and the molecular weight of the vasoconstrictor are consistent with the substance being a polypeptide with approxi-

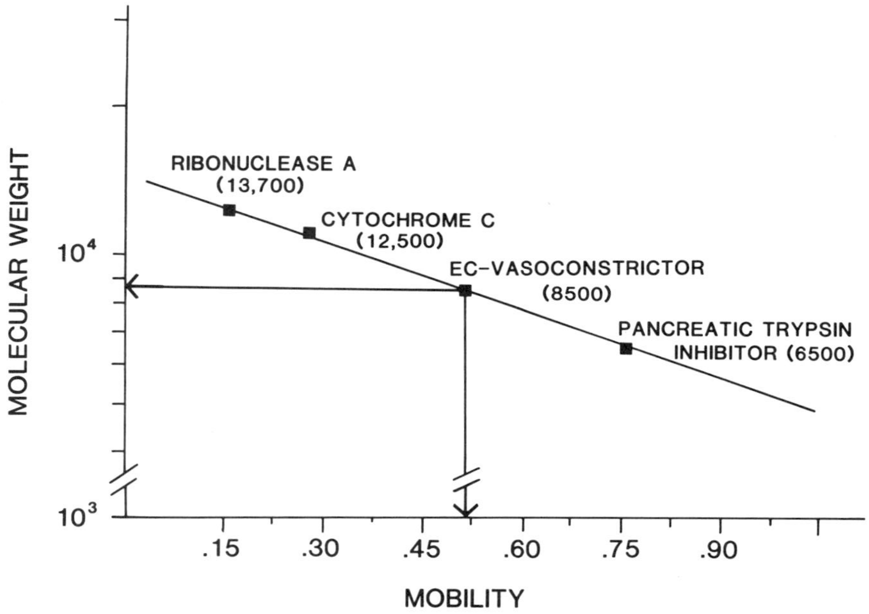

Fig. 2. A selectivity curve obtained on a column of Biogel P-10 indicating the calculated mobilities and molecular weights for standard marker proteins and for the endothelium-derived vasoconstrictor. The value obtained for the vasoconstrictor is indicated by arrows and represents the mean of three separate determinations. Each of the standards as well as the endothelium-conditioned media were run separately on the same column (1.0×30 cm) under the same conditions of flow rate and sample volume. The calculated molecular weight for the endothelium-derived vasoconstrictor was 8500 ± 1500 (reproduced with permission).

mately 60–70 amino acid residues. The protein nature of the vasoconstrictor is supported further by pharmacological experiments in which the effects of various antagonists on the activity of the vasoconstrictor were evaluated. The vasoconstriction induced by the endothelium-conditioned media was unaffected by the presence of inhibitors against various receptor systems, including alpha-adrenergic, beta-adrenergic, serotonergic, H_1-histaminergic, and cholinergic receptors.

EDCF is not likely to be a product derived from the metabolism of arachidonic acid by the endothelium or of vascular smooth muscle cells, for the mechanical response to the endothelium-derived vasoconstrictor was not affected by the inclusion of phenidone, indomethacin, or meclofenamate. Treatment of cultured endothelial cells with cyclohexamide or exposure of EDCF to either 100 °C, Na-dodecyl-SO_4, trypsin, alkali, or acid hydrolysis abolished the vasoconstrictor effect on porcine coronary arteries. These data, coupled with the results of the gel filtration experiments strongly suggest that EDCF is a polypeptide with a molecular weight of approximately 7–10 kdalton.

Independent confirmation of nearly all of the above properties of EDCF have been obtained in subsequent studies using arterial rings (O'Brien and McMurty, 1984), as well as isolated perfused hearts (Gillespie et al., 1986). Remarkably good agreement has been obtained in the data from the three laboratories involved, despite considerable differences in experimental design (species and vessels of origin for endothelial cells, subcultured vs. primary cultures of EC, bioassay preparations, and so on). For example, Gillespie et al. (1986), using the retrogradely perfused rabbit heart, reported that neither the cyclooxygenase inhibitor meclofenamate nor the putative lipoxygenase inhibitor diethylcarbamazine affected EDCF-induced increases in coronary perfusion pressure, thereby excluding a role for the metabolites of arachidonate. The action of EDCF was also unaffected by blockade of alpha-adrenergic, serotonergic and histaminergic receptors. The protein nature of EDCF was also suggested by their finding that EDCF-induced increases in coronary flow resistance were markedly attenuated by prior treatment with trypsin, and that the trypsin inhibitor aprotinin protected EDCF from inactivation. In addition to confirming the peptidergic nature of EDCF in an isolated heart preparation, these studies have apparently ruled out angiotensin II as a potential candidate for EDCF, since saralasin, at concentrations fully blocking the action of angiontensin II, failed to antagonize the increases in perfusion pressure elicited by EDCF. Thus, the studies of Gillespie et al. (1986) are in excellent agreement with the authors' initial characterization of EDCF and fully support the contention that EDCF is a polypeptide that can elicit profound increases in isometric tension in isolated coronary arteries as well as marked reductions in coronary blood flow in the intact beating heart. Further purification of EDCF will be an arduous task, primarily because of the difficulties inherent with bioassays and the increased lability and susceptibility of purified EDCF to proteolysis. Nevertheless, small quantities have been obtained of highly purified EDCF (800–1100-fold increase in specific activity) by a combination of HPLC procedures using reverse phase chromatography, ion exchange, and gel filtration. Present efforts are directed toward developing an antiserum to EDCF for potential use in quantification of EDCF by radioimmunoassay and purifying the polypeptide by immunoaffinity techniques. Immunofluorescent localization of EDCF in body fluids and tissues and confirmation of the identity of EDCF-like molecules reported by other laboratories will then be possible.

3.5. Potential Mechanism(s) of Action of EDCF on Vascular Smooth Muscle

Because of the lack of sufficient amounts of purified EDCF, experiments designed to elucidate the precise mechanism of action of the vasoconstrictor may be clouded by the presence of vasoactive contaminants or cofactors in the crude EDCF preparations from endothelium-conditioned media. Certain patterns have emerged, however, that are worthy of discussion. The vasoconstrictor action apparently requires the influx of extracellular Ca^{2+} since porcine coronary arteries are unresponsive to the endothelium-conditioned media when the vessels were incubated in Ca^{2+}-free solution containing EGTA (Hickey et al., 1985). Upon removal of EGTA and addition of 2.5 m*M* Ca^{2+} to the same arterial segment, full contractile responsiveness was restored. In addition, the contraction elicited by the endothelium-conditioned media was blocked by the Ca^{2+} antagonist verapamil. These data suggest that the influx of extracellular Ca^{2+} through voltage-dependent or receptor-operated slow channels may be an integral part of the action of EDCF on smooth muscle.

In collaborative preliminary studies with Dr. D. Pang evidence was obtained that supports the above hypothesis. Monolayer cultures of aortic smooth muscle cells of the rat were subjected to various concentrations of EDCF, and the resultant flux of extracellular Ca^{2+} across the membrane of the smooth muscle was determined by the cellular uptake of ^{45}Ca. Within 2 min following the addition of 30% endothelium-conditioned media, a significant increase in Ca^{2+} uptake was apparent that was approximately tenfold greater than that occuring under nonstimulated conditions. The maximal Ca^{2+} uptake occurred within 5 min and corresponded to 4.8 nmol Ca^{2+}/mg protein. This value was slightly greater than that obtained by depolarization with 50 m*M* K^+ (3.9 nmol Ca^{2+}/mg protein) and was 30-fold greater than that occurring in control smooth muscle cells treated with the same amount of nonconditioned media. These preliminary studies indicate that EDCF derived from endothelium-conditioned media elicits both a time- and concentration-dependent increase in Ca^{2+} uptake into monolayers of vascular smooth muscle. Thus, these data, coupled with the fact that the time course of Ca^{2+} uptake parallels the increase in isometric force development that was abolished by verapamil, suggest that EDCF's action may be mediated by the gating of the voltage-dependent or receptor-operated Ca^{2+} channels.

4. Effect of Hypoxia on EDCF Release

In recent years it has become evident that reduction in the oxygen supply to the myocardium can be ascribed to constriction or vasospasm of the larger coronary arteries. The evidence suggests that some cases of transient myocardial ischemia that occur at rest are the result of such coronary vasoconstriction. Furthermore, the acute reduction in blood flow accompanying coronary vasospasm may be instrumental in the development of myocardial infarction or fatal cardiac arrhythmias.

It has been widely held that hypoxia induces coronary vasodilatation and a resultant increase in coronary blood flow. Experiments with isolated large coronary arteries indicate, however, that a decrease of PO_2 in the vicinity of the vessel wall induces vasoconstriction. Although the immediate response to ischemia in the overall coronary circulation is vasodilatation, vascular resistance in the ischemic region increases because of active vasoconstriction if ischemia is maintained for a longer period of time. Collectively, these observations indicate that a local decrease in oxygen tension (hypoxia and ischemia) should be considered as an important etiological factor in the development of coronary vasospasm, and also that hypoxic vasospasm is a property of the vessel wall itself. Very little information is available, however, on the role of the endothelium in hypoxic vasospasm.

Since the in vitro and ex vivo contractile response of the coronary artery to EDCF is not transient, but tonic in nature and difficult to washout, one might hypothesize that the inappropriate release of EDCF may be involved in the etiology of coronary vasospasm occurring under hypoxic or ischemic conditions. To evaluate this possibility, the effect of hypoxia on production of EDCF by endothelial cells in culture was investigated (Aichholz et al., 1986). This approach allows for the assessment of the direct effect of hypoxia on the release of EDCF independently of smooth muscle cells or neural inputs.

Bovine aortic endothelial cells were cultured (Hickey et al., 1985), and each culture consisted of cells pooled from the aortas of 8–10 animals. For these studies, primary as well as passaged cells (2–5 times) were used as preconfluent monolayers in 25 cm^2 flasks after culturing 2–3 d in α-modified Eagles medium—10% horse serum at 5% CO_2 and 95% humidified air. Following a fresh change of media, equal numbers of viable cells were randomly divided into two groups and incubated for various lengths of time in either a normoxic or hypoxic environment. The final CO_2 and O_2 concentrations in the incubators and the gas partial pressures in the media for the two groups were as follows: normoxia, 5% CO_2, PCO_2 = 30 mm Hg; 20% O_2, PO_2 = 140 mm Hg; hypoxia, 5% CO_2, PCO_2 = 30 mm Hg; 2% O_2, PO_2 = 80 mm Hg. Conditions of temperature (37°C), pH (7.4), and evaporative fluid loss were identical for the two groups. Control, nonconditioned media were also incubated under the same conditions.

Aliquots of conditioned or control media were removed from each culture incubated under normoxic or hypoxic conditions at 0, 6, 12, 24, and 48 h. Samples were then pooled into appropriate groups and stored frozen at −70°C until analyzed. In experiments testing for reversibility after the initial 48-h collection period, all cultures received fresh media, and the hypoxic cells were placed into the normoxic environment; samples were then removed at a later time point and processed.

Isometric tension measurements were performed on everted and de-endothelialized rings of porcine coronary arteries. Prior to testing, pooled samples were thawed and the pH readjusted to 7.4. The effect of control and conditioned media on isometric force development was assessed by adding various concentrations of the samples directly to the muscle chamber. Contractions to KCl were obtained periodically to ensure that the responsiveness of the bioassay tissue was unchanged; myogenic tone was determined at the conclusion of each experiment by recording the response to papaverine.. For each determination, a minimum of five culture media harvested and pooled from five separate cultures (hypoxic or normoxic) was used. The data displayed in the figures as typical contractile responses were obtained on the same coronary arterial ring and were independent of the order of addition of the samples (data not shown). The contractile responses to control nonconditioned media was not different for hypoxic versus normoxic endothelial cells, and both averaged less than 10% of that observed in endothelium-conditioned media. No significant differences were noted in cell protein or viability between hypoxic and normoxic cultured endothelial cells under these conditions.

The response of the bioassay tissues to endothelium-conditioned media was considerably different for the cells cultured in a normoxic environ-

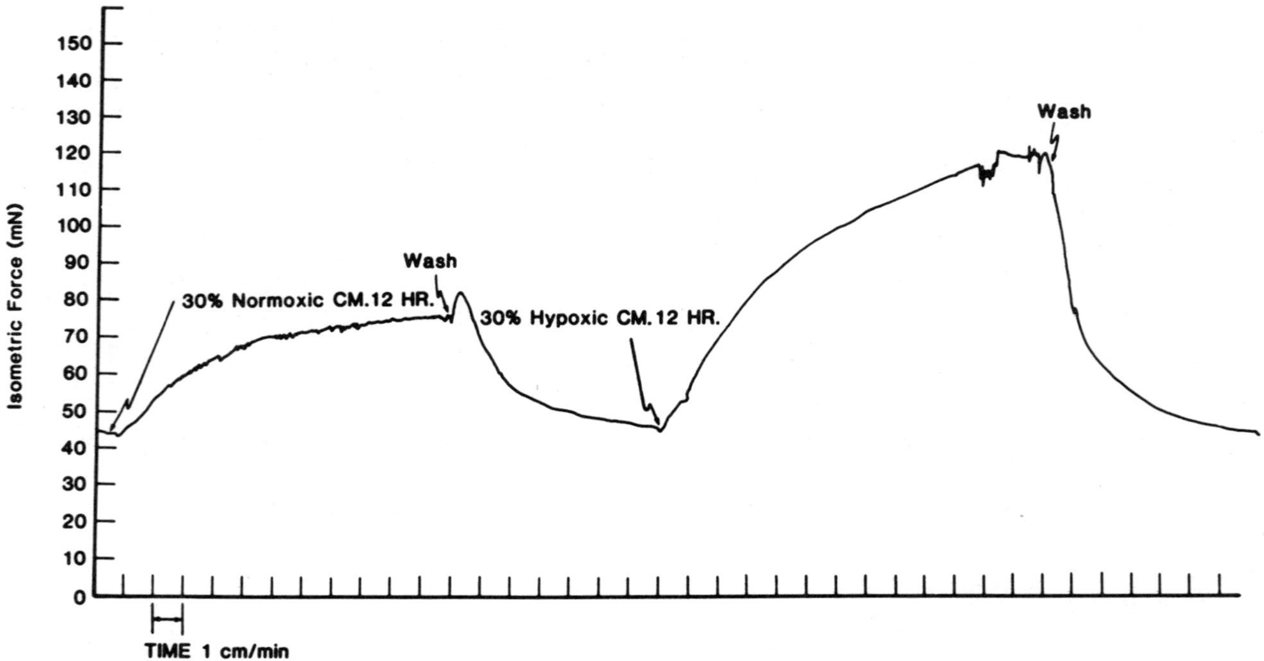

Fig. 3. Isometric tension in deendothelialized rings of porcine left anterior descending coronary artery. Vessels were challenged with equal dosages (30%, final chamber volume) of conditioned media (CM) obtained from monolayers of bovine aortic endothelial cells cultured for *12 h* under normoxic or hypoxic conditions. The EDCF activity in the media from cells exposed to hypoxia was approximately 200% of that found in the normoxic medium.

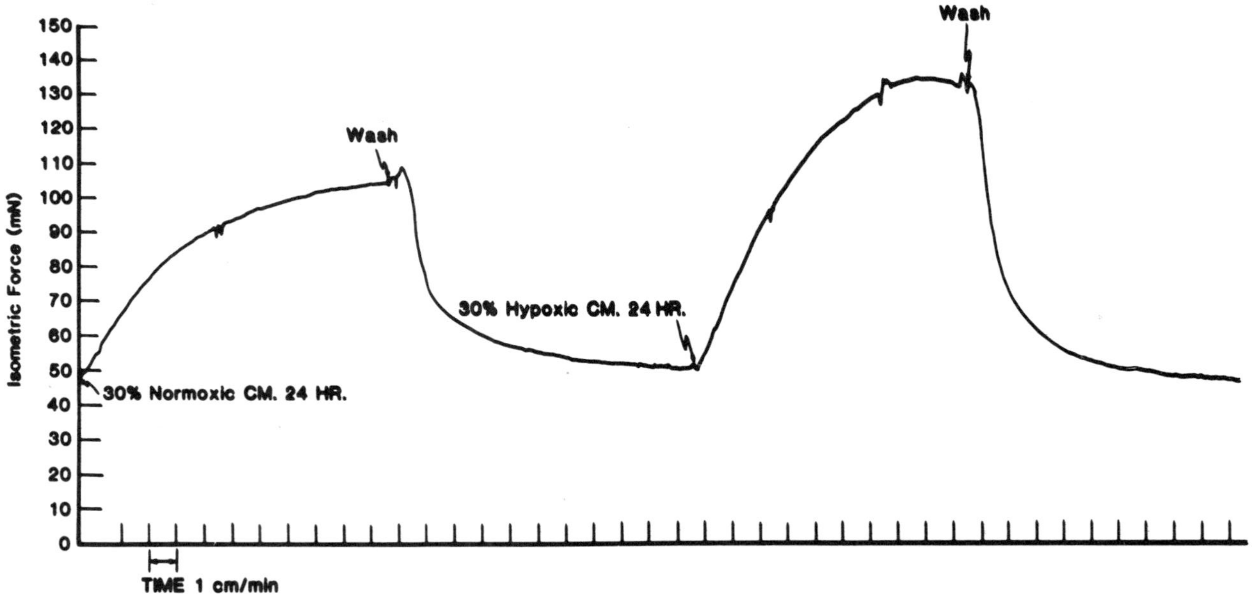

Fig. 4. Isometric tension in deendothelialized rings of porcine coronary artery. Vessels were challenged as described in Fig. 3, except that the endothelium-conditioned media were harvested after *24 h* in culture. The EDCF activity in the hypoxic culture media was approximately 150% of that present in the normoxic one.

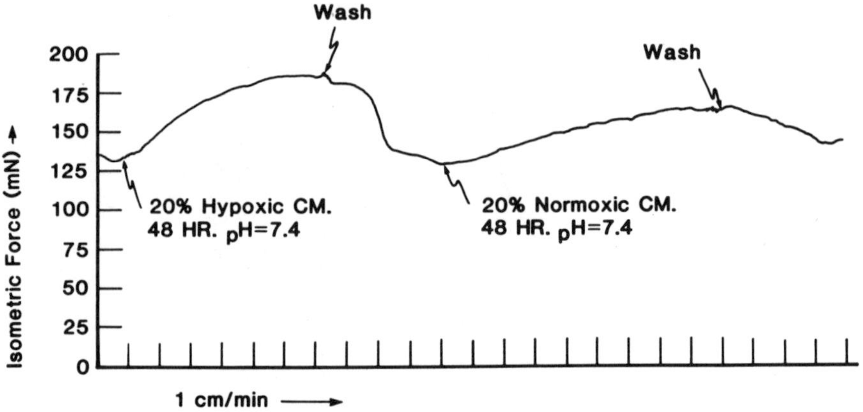

Fig. 5. Isometric tension in deendothelialized rings of porcine coronary artery. Vessels were challenged as described in Fig. 3, except that the endothelium-derived culture media was harvested after *48 h* in culture. The large difference in EDCF activity between hypoxic media and normoxic media that was apparent after 12 and 24 h in culture was attenuated at 48 h. The EDCF activity in the hypoxic culture media was approximately 1.3 times that present in the normoxic medium.

ment vs hypoxic conditions (Figs. 3–5). After only 12 h of culture, the EDCF activity in the media from cells exposed to hypoxia was approximately 2.0 times greater than the normoxic value in terms of maximal isometric force development. In addition, coronary arteries treated with hypoxic endothelium-derived media demonstrated a much greater initial rate of rise in tension. A similar pattern was observed after 24 h of culture with the magnitude of increase in maximal tension being somewhat less (1.5 times) than that observed at 12 h. The differences in maximal tension were further attenuated when the culture period was extended to 48 h or longer, at which time the endothelial cells in both groups had attained confluency.

Five additional experiments were performed and indicated that, in each case, the contractile response of the coronary artery to EDCF derived from hypoxic endothelial cell cultures significantly exceeded that obtained from normoxic cultures. The mean percentage increase in maximal isometric force elicited by hypoxic as compared to normoxic endothelium-derived media after 24 h of culture was 49.3 + 9.3% and was statistically different.

Although no significant differences in cell protein or viability were apparent, changes in morphology of the endothelial cells were evident following the mild hypoxic challenge. Compared to the normoxic cells, those exposed to hypoxic conditions for 48 h were larger in size and displayed what appeared to be numerous dense cytoplasmic granules and/or marked membrane blebbing—a characteristic usually associated with active secretion. This morphological pattern was reversible upon exposure of the endothelial cells to normoxic conditions. More intense degrees of hypoxia (PO_2 in the media lower than 50 mm Hg) resulted in the loss of cytoplasmic granularity and the appearance of poorly defined cell boundaries.

The reversibility of the effect of hypoxia on EDCF release was assessed after placing the hypoxic endothelial cell cultures back into a normoxic environment. After 48 h of additional culture in a normoxic environment, the difference between EDCF activity in the previously hypoxic endothelial cells, as compared to cells kept normoxic for an equal length of time, was abolished (Fig. 6).

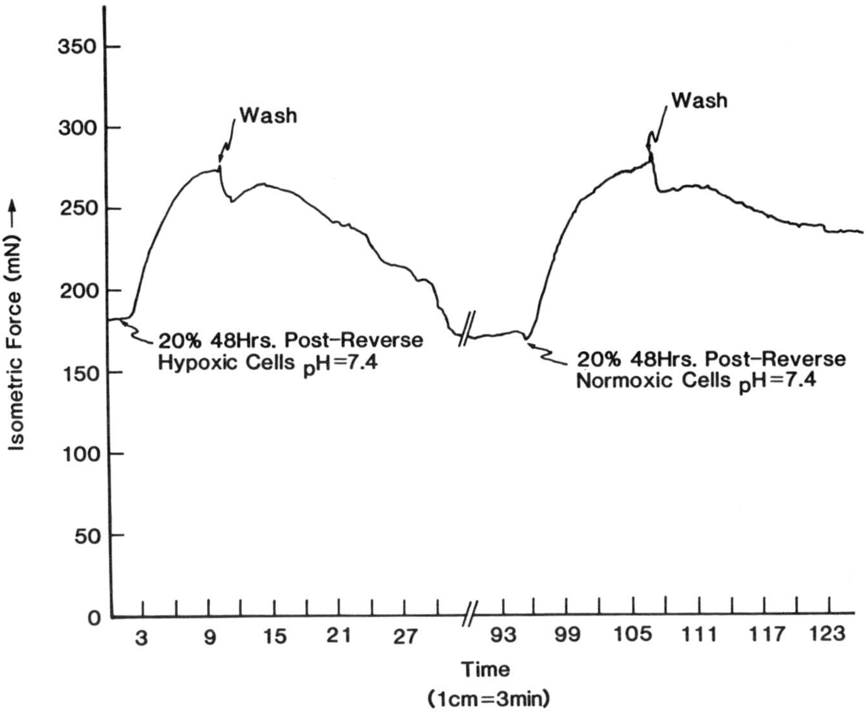

Fig. 6. Recording of isometric tension in deendothelialized rings of porcine coronary artery. Vessels were challenged as described in Fig. 3, except that endothelial culture media were harvested either from normoxic cultures or from previously hypoxic cultures that had been cultured for an additional 48 h under normoxic conditions. Reversibility is indicated in that the EDCF activity in the normoxic and hypoxic media were not significantly different after returning the hypoxic cells to a normoxic environment.

Thus, the release of EDCF by cultured bovine aortic endothelial cells is enhanced in a time-dependent fashion following a mild hypoxic challenge. In particular, the release of EDCF under hypoxic conditions was significantly enhanced within 12 h. After longer periods of hypoxia, however, EDCF derived from hypoxic endothelial cells was not significantly different from that present under normoxic culture. These results suggest that hypoxia stimulates the release of stored EDCF rather than *de novo* synthesis. Also, the morphological changes in endothelial cells exposed to hypoxia suggest that release of cellular or membrane-bound EDCF may occur under these conditions. Finally, the effect of mild hypoxia on EDCF release was fully reversible within 48 h after returning the endothelial cells to normoxic conditions. These results indicate that the synthesis and/or the release of EDCF may be linked to an O_2-sensing mechanism within the endothelial cells, whose perturbation could result in profound alterations in vasomotor tone mediated by the contracting factor.

5. Concluding Remarks

A rapidly expanding body of evidence now indicates that endothelial cells may play a critical role in modulating the responsiveness of vascular smooth muscle. Since the discovery of EDRF, significant progress has been obtained in the understanding of the physiology and pharmacology of this substance(s). As is often the case, the intense investigations spawned by the characterization of EDRF have resulted in the discovery of other vasoactive secretory products of endothelial cells. This chapter has been concerned with endothelium-dependent vasoconstriction mediated by EDCF.

The experimental evidence supporting the existence of a distinct EDCF molecule is accumulating from several laboratories using a variety of experimental approaches. EDCF activity has been detected in the unstimulated basal secretions of primary and subcultured endothelial cells derived from a variety of different animal vascular beds; its action is not vessel- or species-specific. Under the appropriate experimental conditions, EDCF can also be expressed by the intact endothelial lining in contact with the underlying vascular smooth muscle. The progressive vasoconstriction elicited in isolated coronary arteries by EDCF is not endothelium-dependent or transient, has a considerable tonic component, and is difficult to wash out. The time course of contraction is accompanied by a similar time-dependent influx of extracellular Ca^{2+}; the contractile response is abolished by removal of Ca^{2+} and by Ca^{2+} antagonists. Retrograde infusions of EDCF into hearts of rats or rabbits elicited a time- and dose-dependent decrease in coronary blood flow that was remarkably well-sustained. Although unconfirmed, a modest, yet significant negative inotropic and chronotropic effect of EDCF on cardiac performance was observed in one study. All of the available biochemical and pharmacological data are consistent with EDCF being a polypeptide with a molecular weight of approximately 7000–10,000. Although a physiological role for EDCF has not yet been demonstrated, two different experimental approaches have revealed that hypoxia and/or anoxia can enhance the release of EDCF(s) from either ECs in culture or from intact vessel preparations layered with endothelium. Although the identity of the EDCF(s) released under conditions of oxygen deprivation remains to be elucidated, these results do suggest a possible role for EDCF in the pathophysiology of hypoxic vasospasm.

Acknowledgments

The authors gratefully acknowledge the excellent technical assistance of Marcia Gallaher, Arlene Chance, Denise Koth, and Glenn Doerman and the skillful secretarial help of Anita Tolle.

The work was supported by research grants from the National Institutes of Health (HL-31543) and from the American Heart Association, Southwestern Ohio Affiliate.

References

Agricola, K., Rubanyi, G., Paul, R. J., and Highsmith, R. F.: Characterization of a potent coronary artery vasoconstrictor produced by endothelial cells in culture. *Fed. Proc.* **43:**899, 1984.

Aichholz, D., FitzGerald, O., and Highsmith, R. F.: Hypoxia enhances the release of an endothelial cell derived constricting factor (EDCF). *Fed. Proc.* **45:**532, 1986.

De Mey, J. G. and Vanhoutte, P. M.: Heterogeneous behavior of the canine arterial and venous wall. *Circ. Res.* **51:**439–447, 1982.

De Mey, J. G. and Vanhoutte, P. M.: Anoxia and endothelium-dependent reactivity of the canine femoral artery. *J. Physiol.* (Lond.) **335:**65–74, 1983.

Furchgott, R. F. and Zawadzki, J. V.: The obligatory role of endothelial cells in the relaxation of arterial smooth muscle by acetylcholine. *Nature* **228:**373–376, 1980.

Gelman,, B. B., Papa, L., Davis, M. H., and Gruenstein, E.: Decreased liposomal dipeptidyl aminopeptidase I activity in cultured human skin fibroblasts in Duchenne's muscular dystrophy. *J. Clin. Invest.* **65:**1398–1406, 1980.

Gillespie, M. N., Owasoyo, J. O., McMurtry, I. F., and O'Brien, R. F.: Sustained coronary vasoconstriction provoked by a peptidergic substance released from endothelial cells in culture. *J. Pharm. Exp. Ther.* **236:**339–343, 1986.

Harder, D. R. and Sperelakis, N.: Action potential generation in reaggregates of rat aortic smooth muscle cells in primary culture. *Blood Vess.* **16:**186–201, 1979.

Hickey, K. A., Rubanyi, G., Paul, R. J., and Highsmith, R. F.: Characterization of a coronary vasoconstrictor produced by cultured endothelial cells. *Am. J. Physiol.* **248:**C550–C556, 1985.

Holden, W. E. and McCall, E.: Hypoxic vasoconstriction of porcine pulmonary artery strips in vitro requires an intact endothelium. *Am. Rev. Respir. Dis.* **127:**301, 1983.

Jaffe, E. A., Nachmann, R. L., Becker, C. G., and Minick, C. R.: Culture of human endothelial cells derived from umbilical veins. Identification by morphological and immunological criteria. *J. Clin. Invest.* **52:**2745–2756, 1973.

McMurtry, I. F.: Personal communication, 1985.

O'Brien, R. F. and McMurtry, I. F.: Endothelial cell supernates contract bovine pulmonary artery rings. *Am. Rev. Respir. Dis.* **129:**337, 1984.

Rubanyi, G. and Vanhoutte, P. M.: Hypoxia releases a vasoconstrictor substance from the canine vascular endothelium. *J. Physiol.* (Lond.) **364:**45–56, 1985.

Vanhoutte, P. M.: Effects of anoxia and glucose depletion on isolated veins of the dog. *Am. J. Physiol.* **230:**1261–1268, 1976.

Van Nueten, J. M. and Vanhoutte, P. M.: Effect of Ca^{2+} antagonist lidoflazine on normoxic and anoxic contractions of canine coronary arterial smooth muscle. *Eur. J. Pharmacol.* **64:**173–176, 1980.

Whittenberger, B. and Glaser, L.: Inhibition of DNA synthesis in cultures of 3T3 cells by isolated surface membranes. *Proc. Natl. Acad. Sci. USA* **74:**2251–2255, 1977.

Yoneda, S., Forber, M. S., Zelcer, E., and Sperelakis, N.: Electron-microscope studies on reaggregate cultures of vascular smooth muscle cells from normotensive and spontaneously hypertensive rats. *Cell Tissue Res.* **217:**225-243, 1981.

Chapter 8

Basal Release of Endothelium-Derived Relaxing Factor

William Martin

1. Scope

This chapter reviews evidence that the endothelium-derived relaxing factor (EDRF) described by Furchgott and Zawadzki (1980) is released spontaneously and is active tonically in certain blood vessels. Release of EDRF elicited by chemical (e.g., acetylcholine) or physical (shear stress) stimuli is not discussed, nor is the release of prostacyclin, another endothelium-derived vasodilator, although this also occurs spontaneously (Gimbrone and Alexander, 1975; Gordon and Martin, 1983).

2. Detection of Basal EDRF Release

2.1. Cascade Bioassay Experiments

The most direct evidence that EDRF is released spontaneously from endothelial cells has been obtained in cascade bioassay experiments in which an endothelium-denuded vessel is used as a bioassay tissue for the detection of EDRF released from an endothelium-containing blood vessel. By using such cascade systems, basal release of EDRF has been detected from the rabbit aorta (Griffith et al., 1984a), the rat aorta (Long and Stone, 1985), and the femoral artery of the dog (Rubanyi et al., 1985). This spontaneously released EDRF appears to be similar to that liberated by acetylcholine: In both cases the relaxing action of EDRF is inhibited by a variety of carbonyl group reagents and antioxidants, including potassium borohydride, dithiothreitol, phenylhydrazine, phenidone, nordihydroguiaretic acid, and the soluble guanylate cyclase inhibitor methylene blue, and is potentiated by the phosphodiesterase inhibitor, MB 22948, which is selective for cyclic GMP (Griffith et al., 1984a; 1985).

An alternative cascade system used involves the use of endothelial cells grown on microcarrier beads and perfused in columns. No report has been made of basally released EDRF from bovine aortic endothelial cells in microcarrier columns (Cocks et al., 1985). Following protection of EDRF, however, from the destructive action of oxygen-derived free radicals by the use of superoxide dismutase (Vanhoutte and Rubanyi, 1985), spontaneous release of EDRF was detected in the perfusate from porcine aortic endothelial cells on microcarriers (Gryglewski et al., 1986a).

An unanswered question relevant to all the cascade bioassay techniques is: What proportion, if any, of the EDRF detected is truly basal rather than release elicited by flow-induced shear stress (Holtz et al., 1984)?

2.2. Basal EDRF Release and Cyclic GMP

EDRF-induced relaxation is associated with increased levels of cyclic GMP (Holzmann, 1982; Rapoport and Murad, 1983) resulting from activation of soluble guanylate cyclase (Busse et al., 1985). Resting levels of cyclic GMP are higher in endothelium-containing than in endothelium-denuded preparations of the rat aorta (Table 1; Rapoport and Murad, 1983; Martin et al., 1986a), the rabbit aorta (Diamond and Chu, 1983; Furchgott et al., 1984a; Martin et al., 1985a), and the bovine coronary (Holzmann, 1982) and intrapulmonary arteries (Ignarro et al., 1984). Furthermore, the EDRF-blocking agent hemoglobin (Martin et al., 1985a,b) and an inhibitor of soluble guanylate cyclase, methylene blue (Rapoport et al., 1985), each reduce the resting level of cyclic GMP in endothelium-containing, but not endothelium-denuded blood vessels (Table 1; Holzmann, 1982; Furchgott et al., 1984b; Martin et al., 1985a, 1986b). It would appear therefore that the higher resting level of cyclic GMP in endothelium-containing arteries is caused, at least in part, by the tonic action of spontaneously released EDRF.

Following introduction of soluble guanylate cyclase and ^{32}P-labeled GTP into the lumen of aortic segments of the rabbit, more conversion into cyclic GMP was seen if the endothelium was left intact (Busse et al., 1985). This finding provides additional evidence for the basal release of EDRF from the endothelium of the aorta of the rabbit.

Table 1
Phenylephrine or Clonidine Fail To Elicit EDRF Release in the Rat Aorta as Judged by Their Inability to Elevate Cyclic GMP Content, but Hemoglobin Reduces the Resting Cyclic GMP Content of Endothelium-Containing Rings by Abolishing the Tonic Action of Spontaneously Released EDRF[a]

Endothelium	Treatment	Cyclic GMP, pmol/g protein	Cyclic AMP, pmol/g protein
+	None (control)	309 ± 60 (6)	1669 ± 194
−	None	121 ± 29 (6)[b]	2156 ± 421
+	PE (0.1 μM, 1 min)	326 ± 54 (5)	1628 ± 213
+	Clon (1 μM, 1 min)	311 ± 50 (6)	1551 ± 114
+	Clon (1 μM, 10 min)	312 ± 123 (4)	1224 ± 281
+	Hb (10 μM, 10 min)	150 ± 20 (7)[b]	1112 ± 164

[a]Rings of rat aorta with (+) or without (−) endothelium present were suspended under 2 g of resting tension in Krebs solution at 37°C for at least 90 min. They were then exposed to phenylephrine (PE), clonidine (Clon), or hemoglobin (Hb) at the concentration and for the times indicated, prior to freezing in liquid N_2. Rings were then extracted in 6% TCA, and cyclic GMP and cyclic AMP measured by radioimmunoassay. Results are expressed as mean ± SEM, and the number of observations is shown in parentheses.

[b]$p < 0.001$; significantly different from control.

2.3. Calcium Dependency

Singer and Peach (1982) showed that endothelium-dependent relaxations of the rabbit aorta were inhibited by removal of extracellular Ca^{2+} and provided the first indication that stimulated release of EDRF was calcium-dependent. Application of more versatile cascade bioassay techniques in which Ca^{2+} was removed from the endothelium-containing donor vessel, but not the endothelium-denuded assay tissue, has confirmed the calcium-dependency of the evoked release of EDRF from the aorta of the rabbit (Griffith et al., 1986) and the rat (Long and Stone, 1985). Long and Stone (1985) presented additional evidence in the rat aorta that also the spontaneous release of EDRF is dependent on the presence of extracellular calcium. This finding is consistent with that of Miller et al. (1985), who found that calcium depletion reduced the higher resting level of cyclic GMP in endothelium-containing preparations of the rat aorta to that of endothelium-denuded preparations.

2.4. Potentiating Action of Phosphodiesterase Inhibitors

Since cyclic GMP mediates the vasodilator action of EDRF (Rapoport and Murad, 1983), it might be expected that inhibitors of phosphodiesterase would potentiate the vascular actions of EDRF. MB 22948, a phosphodiesterase inhibitor selective for cyclic GMP, relaxes endothelium-containing preparations of rat and rabbit aorta (Fig. 1; Griffith et al., 1985; Martin et al., 1986b). These relaxations are produced by concentrations of MB 22948 that induce only slight relaxation of endothelium-denuded preparations. The endothelium-dependent relaxation is accompanied by a rise in cyclic GMP content of endothelium-containing, but not endothelium-denuded, preparations (Table 2; Martin et al., 1986b). Furthermore, both the relaxation and the concomitant increase in cyclic GMP content induced by MB 22948, in the aorta of both rat and rabbit, are blocked by hemoglobin (Fig. 1, Table 2; Martin et al., 1986b). MB 22948 must induce endothelium-dependent relaxation by potentiating the actions of spontaneously released EDRF, since the rise in cyclic GMP is observed in otherwise unstimulated preparations, and MB 22948 itself has been shown not to release EDRF in a cascade bioassay system (Griffith et al., 1985).

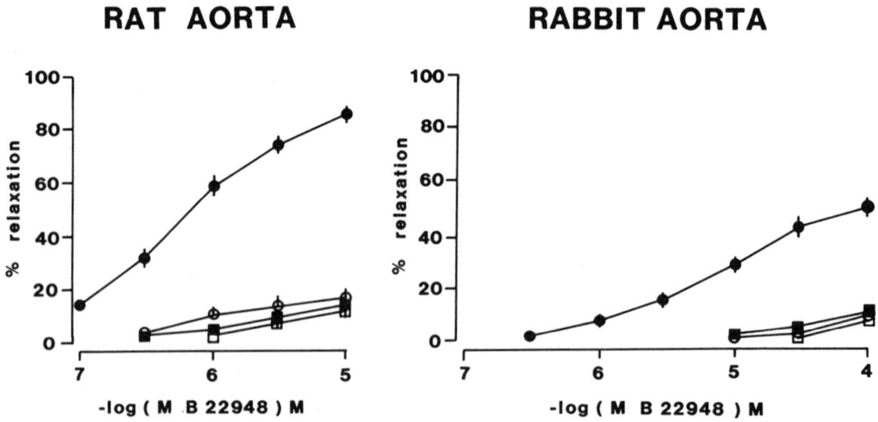

Fig. 1. MB 22948 induces endothelium-dependent relaxation of rat and rabbit aortas by potentiating the effects of spontaneously released EDRF. Hemoglobin abolishes the endothelium-dependent relaxation induced by MB 22948 in phenylephrine-contracted aortic rings of rat and rabbit, but does not affect the small component of relaxation that is endothelium-independent. Points show relaxation induced by MB 22948 in aortas of rats and rabbits with endothelium (●, $n = 12$ and $n = 7$, respectively) and without endothelium (■, $n = 9$ and $n = 5$, respectively), and relaxation induced following treatment for 5–10 min with hemoglobin (10 μM) in endothelium-containing (○, $n = 6$ and $n = 4$, respectively) and endothelium-denuded (□, $n = 4$, and $n = 4$, respectively) rings. Each point is the mean, and vertical bars indicate the SEM. Note that MB 22948 induces greater endothelium-dependent relaxations in rat than in rabbit aortas, which suggests higher basal EDRF activity in the aorta of the rat.

Table 2
MB 22948 Selectively Elevates the Cyclic GMP Content of Endothelium-Containing Rings of Rat and Rabbit Aortas by Potentiating the Action of Spontaneously Released EDRF[a]

Endo-thelium	Treatment	Cyclic GMP, pmol/g protein	Cyclic AMP, pmol/g protein	Relaxation, %
Rat				
+	PE (control)	624 ± 71 (7)	2381 ± 285	
+	PE + MB	2210 ± 353 (5)[b]	3313 ± 846	38.9 ± 1.9
+	PE + Hb	276 ± 55 (6)[c]	1573 ± 382	
+	PE + Hb + MB	579 ± 91 (4)	2545 ± 504	4.9 ± 2.0[b]
−	PE (control)	121 ± 29 (6)	2156 ± 421	
−	PE + MB	2630 ± 241 (6)	2162 ± 310	11.1 ± 3.7[b]
+	None (control)	518 ± 44 (5)	1582 ± 288	
+	MB	2630 ± 241 (4)[b]	2162 ± 310	
Rabbit				
+	PE (control)	145 ± 12 (4)	985 ± 104	
+	PE + MB	243 ± 34 (5)[d]	1081 ± 136	27.4 ± 4.7
+	PE + Hb	125 ± 16 (9)	966 ± 156	
+	PE + Hb + MB	114 ± 14 (4)	1128 ± 106	0.6 ± 0.6[c]

[a]Following induction of submaximal tone with phenylephrine (PE), rings of rat and rabbit aorta were exposed to MB 22948 (30 μM, for 1.5 min and 3 min, respectively). The effect of hemoglobin (10 μM, for 10 min) was examined on resting and MB 22948-stimulated cyclic GMP content. Rings were then frozen in liquid N_2, and cyclic GMP and cyclic AMP content were determined by radioimmunoassay. In one set of experiments, aortic rings of rats were not contracted with PE before exposure to MB 22948. The presence and absence of endothelium is indicated by + and −, respectively. Results are expressed as mean ± SEM, and the number of observations is shown in parentheses.

[b]$p < 0.001$; [c]$p < 0.005$; [d]$p < 0.05$: significantly different from cyclic nucleotide content of control or significantly different from relaxation induced in untreated endothelium-containing rings.

Nonselective phosphodiesterase inhibitors also induce endothelium-dependent relaxation: there is an endothelium-dependent component to the relaxation induced by papaverine or isobutylmethyl xanthine in the rat aorta and to papaverine in the rabbit aorta (Fig 2; Martin et al., 1986b). These two phosphodiesterase inhibitors induce a greater degree of relaxation than MB 22948 in endothelium-denuded preparations. As with MB 22948, however, the endothelium-dependent component of the relaxation is mediated by EDRF since both papaverine and isobutylmethyl xanthine selectively elevate the cyclic GMP content of endothelium-containing, but not endothelium-denuded, rings of rat or rabbit aorta (Martin et al., 1986b). Furthermore, in these two blood vessels, hemoglobin blocks the endothelium-dependent component of the relaxation and concomitant rise in

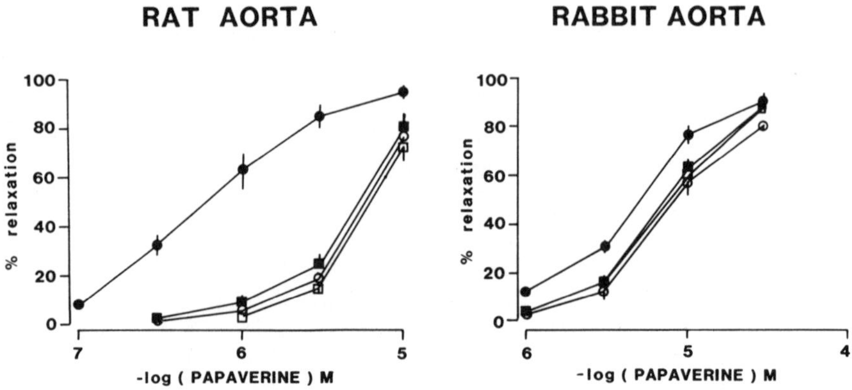

Fig. 2. Papaverine induces endothelium-dependent relaxation of rat and rabbit aorta by potentiating the effect of spontaneously released EDRF. Hemoglobin selectively abolishes the endothelium-dependent, but does not affect the endothelium-independent, components of the relaxation induced by papaverine. Points show the relaxation induced by papaverine in aortas of rats and rabbits with (●, n = 6 and n = 11, respectively) and without endothelium (■, n = 5 and n = 9, respectively), and that induced following treatment for 5–10 min with hemoglobin (10 μM) in rings with (○, n = 4 and n = 8, respectively), and without endothelium (□, n = 4 and n = 6, respectively). Each point is the mean, and vertical bars indicate the SEM. Note that the relaxation to papaverine has a greater endothelium-dependent component in the aorta of the rat than in that of the rabbit, suggesting a higher basal EDRF activity in the former.

cyclic GMP content, but has no effect on the endothelium-independent component of relaxation induced by papaverine or isobutylmethyl xanthine (Fig. 2; Martin et al., 1986b). Although the ability of papaverine or isobutylmethyl xanthine to elicit EDRF release has not been tested, it is likely that the endothelium-dependent relaxation induced by these phosphodiesterase inhibitors, like that induced by MB 22948 (Griffith et al., 1985), is mediated by potentiation of the vasodilator actions of spontaneously released EDRF.

3. Endothelium-Dependent Depression of Vasoconstrictor Responses

Many reports show that the sensitivity of endothelium-containing arteries and veins is depressed when compared with endothelium-denuded blood vessels. The endothelium depresses constrictor responses to norepinephrine or serotonin in the canine and porcine coronary arteries (Cocks and Angus, 1983; Cohen et al., 1983); to norepinephrine in the canine iliac artery

(Young and Vatner, 1984), pulmonary artery and vein, femoral artery and vein (Miller and Vanhoutte, 1985), and central ear artery of the rabbit (Verrecchia et al., 1985); to serotonin in the jugular vein of the chick (Imaizumi et al., 1984); to histamine, phenylephrine, or serotonin in the coronary artery of the rabbit (Griffith et al., 1984b), and to a range of alpha-adrenergic agonists, histamine, angiotensins I and II, serotonin, or prostaglandin $F_{2\alpha}$ in the aorta of the rat (Allan et al., 1983; Konishi and Su, 1983; Van de Voorde and Leusen, 1983; Egleme et al., 1984; Lues and Schumann, 1984; Miller et al., 1984; Bigaud et al., 1984; Carrier and White, 1985; Miller et al., 1985; Godfraind et al., 1985; Murakami et al., 1985; Martin et al., 1986a).

The endothelium-dependent depressions of contractile responses listed above are caused either by EDRF release elicited by the vasoconstrictor itself or by the spontaneous release of EDRF. In the cases of the endothelium-dependent depression of contractions induced by histamine, phenylephrine, or serotonin in the coronary artery of the rabbit (Griffith et al., 1984b) and by norepinephrine in the canine iliac artery (Young and Vatner, 1984) and the central ear artery of the rabbit (Verrecchia et al., 1985), it has not yet been established whether the depression is mediated by the evoked or spontaneous release of EDRF.

3.1. EDRF Release Elicited by Vasoconstrictor Agents

In both canine and porcine coronary arteries, yohimbine blocks and ketanserin potentiates, respectively, the endothelium-dependent relaxations induced by norepinephrine or serotonin (Cocks and Angus, 1983; Cohen et al., 1983), showing that EDRF release can be elicited from these tissues by activation of endothelial alpha$_2$-adrenoceptors and S_1-serotonergic receptors. There is no endothelium-dependent depression of the contractile responses induced by elevated potassium or by the endoperoxide analog U46619 in porcine coronaries, and by elevated potassium, U46619, prostaglandin $F_{2\alpha}$, or phenylephrine in canine coronary arteries (Cocks and Angus, 1983; Cohen et al., 1983), suggesting that the basal EDRF release is low in these tissues. In the jugular vein of the chick, also, serotonin induces EDRF release and endothelium-dependent relaxation, but the subtype of serotonergic receptor involved has not been characterized (Imaizumi et al., 1984).

In canine pulmonary and femoral arteries and veins, the selective alpha$_2$-adrenergic agonist UK 14,304 induces endothelium-dependent relaxations that are inhibited by the selective alpha$_2$-adrenergic antagonist rauwolscine (Miller and Vanhoutte, 1985). In a separate study, Matsuda et al. (1985) showed that luminal application of the alpha$_2$-adrenergic agonists UK 14,304 or azepexole, but not the alpha$_1$-adrenergic agonists phenylephrine or cirazoline, induce vasodilatation in rat tail artery and side branches of the canine femoral artery that is antagonized by rauwolscine.

Van de Voorde and Leusen (1983) have shown in the rat aorta that the direct contractile action of histamine is counteracted by EDRF released as a consequence of activation of endothelial H_1-histaminergic receptors.

3.2. Depression of Alpha-Adrenergic Contractions by Spontaneously Released EDRF in the Rat Aorta

In the rat aorta the ability of vasoconstrictor agents, other than histamine (Van de Voorde and Leusen, 1983), to elicit EDRF release is controversial. Allan and coworkers (1983) showed that the presence of the endothelium depressed contractions induced by norepinephrine, phenylephrine, or angiotensins I and II in this tissue. Egleme et al. (1984) showed subsequently that the endothelium reduced sensitivity to norepinephrine with no depression of the maximal contractile response, but the sensitivity to the selective $alpha_2$-adrenergic agonist clonidine was reduced more and was associated with a profound depression of the maximal response. These findings prompted the conclusion that stimulation of endothelial $alpha_2$-adrenoceptors elicited EDRF release, which counteracted the direct contractile effect of alpha-adrenergic agonists (Egleme et al., 1984). Others have confirmed that the presence of endothelium depresses the maximal response to clonidine more than to $alpha_1$-adrenergic agonists such as phenylephrine or norepinephrine (Figs. 3–5; Lues and Schumann, 1984; Godfraind et al., 1985; Martin et al., 1986a). However, other data do not support the presence of endothelial $alpha_2$-adrenoceptors in the rat aorta. For example, the endothelium greatly depresses the maximal contractions not only to the selective $alpha_2$-adrenoceptor agonists, B-HT 920, clonidine, UK 14,304, and oxymetazoline, but also to the $alpha_1$-adrenoceptor agonist St 587. Furthermore, maximal responses to the $alpha_2$-adrenergic agonist guanfacine are not depressed by the presence of endothelium (Lues and Schumann, 1984; Godfraind et al., 1985). In addition, $alpha_2$-adrenergic antagonists such as rauwolscine do not inhibit the endothelium-dependent depression of the response to alpha-adrenergic agonists (Lues and Schumann, 1984), and $alpha_2$-adrenergic agonists do not elicit endothelium-dependent relaxations (Godfraind et al., 1985; Martin et al., 1986a). The inability of $alpha_2$-adrenergic agonists to induce endothelium-dependent relaxation might be caused by a dominant vasoconstrictor action of these agents. This is improbable, however, since when the vasoconstrictor action of these substances, which is mediated by $alpha_1$-adrenoceptors in this tissue (Digges and Summers, 1983; Macia et al., 1984), is blocked

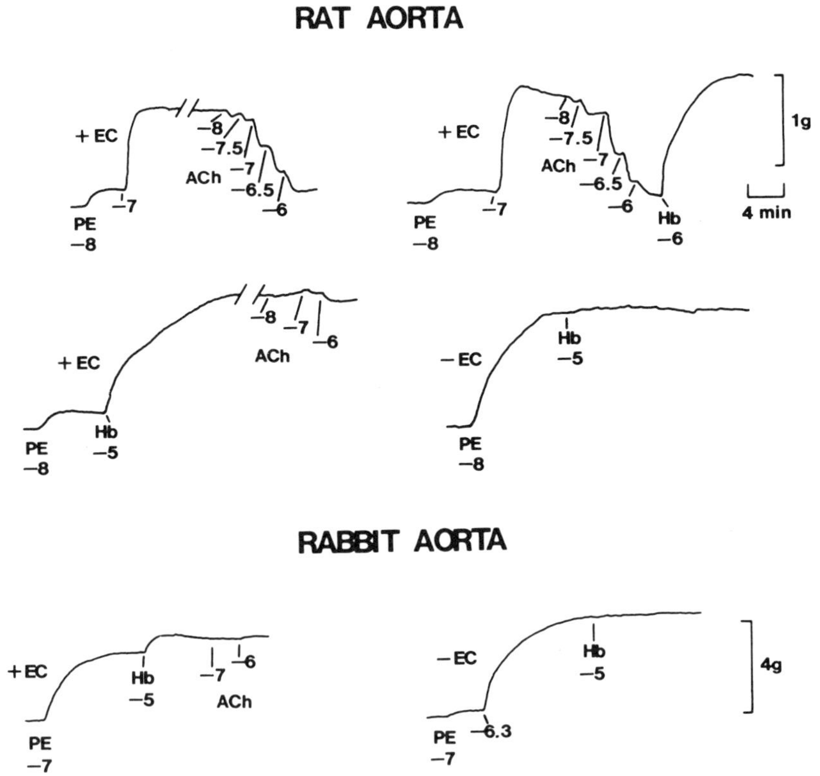

Fig. 3. Hemoglobin (Hb) selectively augments phenylephrine (PE)-induced contractions in endothelium-containing (+EC), but not endothelium-denuded (−EC), rings of rat and rabbit aorta by blocking the tonic inhibition exerted by spontaneously released EDRF. The top panels show the acetylcholine-induced, endothelium-dependent relaxation in the rat aorta and its blockade by Hb. The middle panels show that Hb markedly augments the PE-induced contraction in endothelium-containing, but not in endothelium-denuded, rings of rat aorta. The lower panels show that also in the aorta of the rabbit, Hb augments PE-induced contractions in endothelium-containing, but not endothelium-denuded, rings and subsequently blocks the acetylcholine-induced, endothelium-dependent relaxation. Note that the ability of Hb to augment the response to PE is greater in the rat than in the rabbit, suggesting higher basal EDRF activity in the former. Cumulative molar concentrations are expressed in log units.

by the alpha$_1$-adrenergic antagonist prazosin, alpha$_2$-adrenergic agonists such as clonidine still fail to elicit relaxation (Martin et al., 1986a).

Evidence that the selective alpha$_1$-adenergic agonists phenylephrine and methoxamine and the selective alpha$_2$-adrenergic agonists clonidine

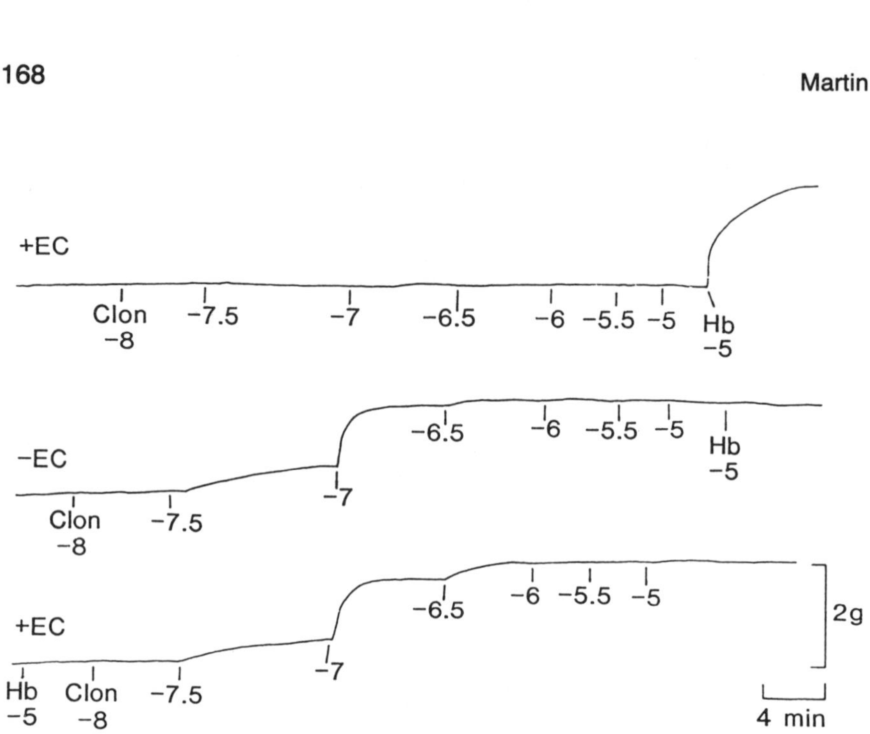

Fig. 4. Spontaneously released EDRF depresses the ability of aortic rings of the rat to contract to clonidine (Clon), and hemoglobin (Hb) abolishes this depression. Records were obtained simultaneously in rings from the same aorta. The presence and absence of endothelium is indicated by +EC and −EC, respectively. The top panel shows the depression by spontaneously released EDRF of clonidine-induced contractions and how Hb rapidly abolishes this depression. The middle panel shows the ability of clonidine to contract endothelium-denuded rings and the inability of Hb to augment tone in these rings. The bottom panel shows that Hb does not cause contraction by itself in endothelium-containing rings, but abolishes the endothelium-dependent depression of the response to clonidine. Cumulative molar concentrations are expressed in log units.

and B-HT 920 each elicit small (1.5–2-fold) increases in the cyclic GMP content of endothelium-containing rings of the rat aorta has been presented and taken as support for the existence of endothelial alpha$_2$-adrenoceptors (Bigaud et a., 1984; Miller et al., 1984, 1985). Other workers have failed, however, to detect increases in cyclic GMP in response to norepinephrine, phenylephrine, or clonidine (Table 1; Rapoport and Murad, 1983; Martin et al., 1986a) in endothelium-containing rings of this blood vessel. Until these differences in results are explained, one should use caution when implying the existence of endothelial alpha$_2$-adrenoceptors linked to EDRF release in the aorta of the rat from measurements of levels of cyclic GMP.

Thus, the weight of evidence is against the presence of such receptors.

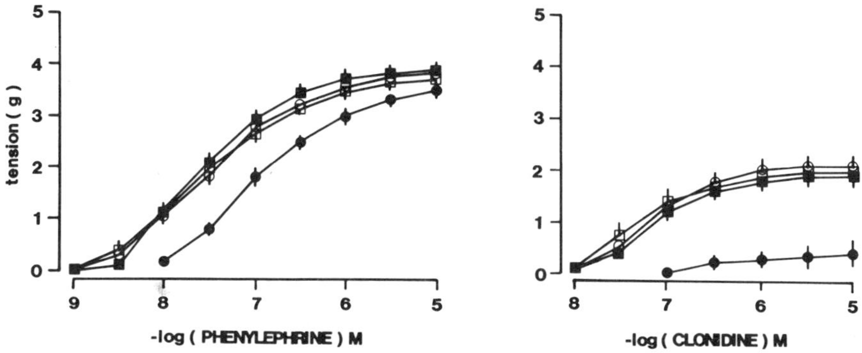

Fig. 5. Spontaneously released EDRF depresses contractions to the partial agonist clonidine more than those to the full agonist phenylephrine in aortic rings of the rat, and hemoglobin (Hb) abolishes these depressions. Points show responses induced by phenylephrine and clonidine in endothelium-containing (●, $n = 11$ and $n = 9$, respectively) and endothelium-denuded (■, $n = 6$ and $n = 10$, respectively) aortic rings and responses induced following treatment for 5–10 min with Hb (10 μM) in endothelium-containing (○, $n = 6$ and $n = 6$, respectively) and endothelium-denuded (□, $n = 4$ and $n = 4$, respectively) rings. Each point is the mean, and vertical bars indicate the SEM. Note that the maximal contraction induced by clonidine is about 50% of that induced by phenyleprine. In endothelium-containing rings, the reduction in sensitivity is associated with a depression in the maximal constrictor response to clonidine, but not to phenylephrine. Hemoglobin selectively abolishes the endothelium-dependent depression of phenylephrine- and clonidine-induced contractions but does not affect responses induced by these agents in endothelium-denuded rings.

3.3. Hemoglobin and Methylene Blue

Although the above results suggest that alpha-adrenergic agonists do not elicit EDRF release in the rat aorta, it is evident from experiments using selective blocking agents that the endothelium-dependent depression of alpha-adrenergic-induced contractions is mediated by EDRF. Hemoglobin, which binds and inactivates EDRF (Martin et al., 1985a,b), inhibits the endothelium-dependent depression of the contractions to norepinephrine, phenylephrine, or clonidine (Figs. 3–5; Martin et al., 1986a). Furthermore, methylene blue, which inhibits soluble guanylate cyclase (Rapoport et al., 1985), inhibits the endothelium-dependent depression of the response to B-HT 920 (Miller et al., 1984). These data, taken together with the inability of alpha-adrenergic agents to induce endothelium-dependent relaxation or to consistently elevate the levels of cyclic GMP, suggest that spontaneously released EDRF is responsible for the endothelium-dependent depression of the response of alpha-adenergic agonists.

3.4. Efficacy of Alpha-Adrenergic Agonists

The contractions of the aorta of the rat to different alpha-adrenergic agonists are reduced to different extents by the presence of the endothelium. Those whose constrictor responses are reduced most (i.e., clonidine, B-HT 920, St 587, and UK 14,304) are partial agonists in this blood vessel (Fig. 5; Ruffolo et al., 1979; Digges and Summers, 1983; Lues and Schumann, 1984; Godfraind et al., 1985). It follows from the receptor theory that if a sufficient fraction of alpha$_1$-adrenoceptors were inactivated irreversibly, the maximal response to a full agonist like phenylephrine could be reduced to that obtained with a partial agonist like clonidine in the absence of receptor inactivation. Treatment of rings of rat aorta with the irreversible alpha-adrenergic antagonist dibenamine (2 × 10^{-7}*M*, for 20 min) left only about 3% of adrenoceptors still active and reduced the maximal response to phenylephrine by about 50%. The presence of endothelium now greatly depressed the maximal contraction to phenylephrine (Fig. 6;

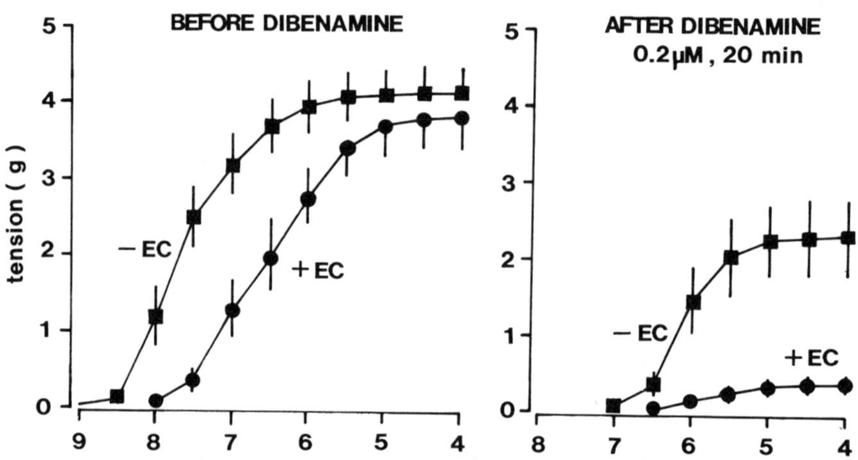

Fig. 6. Partial alpha-adrenergic blockade using dibenamine enhances the ability of spontaneously released EDRF to depress contractions to phenylephrine (PE) in the rat aorta. Concentration–response curves to phenylephrine were obtained in untreated (left panel) and dibenamine-treated (0.2 *μM* for 20 min, right panel) endothelium-containing (●) and endothelium-denuded (■) aortic rings. Each point is the mean, and vertical bars indicate the SEM of four observations. Note that before dibenamine the endothelium reduced the sensitivity to PE with no depression of the maximal response. Following treatment with dibenamine (which reduced the maximal response to PE in endothelium-denuded rings to approximately that of the partial agonist clonidine), the endothelium-dependent reduction in sensitivity to PE was associated with a depression of the maximal response.

Martin et al., 1986a). Therefore, the endothelium of the aorta of the rat, through spontaneous release of EDRF, which acts as a physiological antagonist to vasoconstrictor agents, can depress the contraction more profoundly in the case of an agonist with low efficacy, and consequently with a negligible receptor reserve. This explains why maximal responses induced by clonidine, B-HT 920, St 587, and UK 14,304 are depressed more than those induced by the full agonists phenylephrine and norepinephrine.

3.5. *Aorta of the Rabbit*

In the aorta of the rabbit as in that of the rat, alpha-adrenergic agonists do not stimulate the release of EDRF. They do not induce endothelium-dependent relaxations, and they do not induce endothelium-dependent increases in cyclic GMP content (Furchgott et al., 1984a). There is, however, ample evidence that the smooth muscle of the rabbit aorta, like that of the rat, is under the tonic influence of spontaneously released EDRF. In endothelium-containing, but not endothelium-denuded, preparations of rabbit aortas, the contractions to phenylephrine, norepinephrine, histamine, serotonin, or high concentrations of potassium are augmented by the EDRF inhibitors hemoglobin or hydroquinone; furthermore, in preparations displaying intrinsic tone, hydroquinone or hemoglobin induce small endothelium-dependent contractions (Fig. 3; Griffith et al., 1984b; Martin et al., 1985a).

4. Depression of Resting and Stimulated $^{45}Ca^{2+}$ Influx by Spontaneously Released EDRF

Collins et al. (1984, 1986a) found that in the rabbit aorta the release of EDRF by acetylcholine inhibits norepinephrine-induced accumulation of $^{45}Ca^{2+}$. This suggests that a major pathway by which EDRF induces vasodilatation is by inhibiting Ca^{2+} influx. These authors found no evidence in the rabbit aorta of inhibition of $^{45}Ca^{2+}$ accumulation by spontaneously released EDRF either in the resting state or following exposure to norepinephrine. In the rat aorta, however, the presence of endothelium reduced significantly both resting and clonidine-stimulated accumulation of $^{45}Ca^{2+}$ in a hemoglobin-sensitive manner (Collins et al., 1986b).

5. Hypoxia-Induced Endothelium-Dependent Responses

In general, hypoxia causes vasodilatation in systemic arteries, but vasoconstriction in pulmonary arteries (Fishman, 1976). De Mey and Vanhoutte

(1982, 1983) demonstrated that the hypoxic contractions seen in canine arteries and veins are endothelium-dependent. Hypoxia-induced vasoconstriction in the isolated pig pulmonary artery also is endothelium-dependent (Holden and McCall, 1984). Because anoxia blocks endothelium-dependent relaxations to acetylcholine (De Mey and Vanhoutte, 1980; Furchgott and Zawadzki, 1980), one explanation for hypoxia-induced contractions might be that it is caused by abolition of basal EDRF activity. Hypoxia induces endothelium-dependent contractions also in the canine coronary artery; in this blood vessel, the hypoxic contraction has been attributed to an unidentified vasoconstrictor substance released from the endothelium (Rubanyi and Vanhoutte, 1985).

The mechanisms underlying hypoxic vasodilatation in systemic arteries are also likely to be complex. In the tail artery of the rat, side branches of canine femoral arteries and in the femoral artery of the rabbit, it appears to be mediated jointly by EDRF and prostacyclin (Busse et al., 1983, 1985). A further possibility is that the activity of EDRF may be enhanced by the reduced production of oxygen-derived free radicals, which may normally contribute to its inactivation (Vanhoutte and Rubanyi, 1985; Rubanyi and Vanhoutte, 1986; Gryglewski et al., 1986b).

6. Differences in Basal EDRF Release

The basal EDRF activity is not similar in all arteries: If a comparison of the aortas of rat and rabbit is made, it is clear that spontaneously released EDRF is responsible for: (1) a greater depression of contractile responses in rat than in rabbit aorta (Fig. 3; Martin et al., 1986a); (2) a greater depression of alpha-adrenoceptor-induced $^{45}Ca^{2+}$ influx in the rat than in the rabbit aorta (Collins et al., 1986b); and (3) a greater endothelium-dependent relaxation and concomitant increase in cyclic GMP content induced by a range of phosphodiesterase inhibitors in the aorta of the rat than in that of the rabbit (Figs. 1 and 2; Martin et al., 1986b). Each of these criteria suggest that the effect of basally released EDRF is greater in the rat than in the rabbit. Whether this higher basal EDRF activity in the rat aorta is caused by greater amounts of EDRF being released, greater sensitivity of soluble guanylate cyclase, or lower phosphodiesterase activity is not known. The sensitivities of endothelium-denuded rings of aortas of the rat and the rabbit to glyceryl trinitrate, which like EDRF induces relaxation by activating soluble guanylate cyclase, are similar (Martin et al., 1986b). If sensitivity to glyceryl trinitrate can be taken as a rough estimate of the sensitivity to EDRF, this would suggest that more EDRF is released spontaneously from endothelial cells in the aorta of the rat than in that of the rabbit. More rigorous studies will be required, however, to determine precisely whether the differences in basal EDRF activity in different arteries reflect differences in the amount of EDRF released or in sensitivity of the smooth muscle.

7. Basal EDRF Release and Vasospasm

As with evoked release of EDRF, one can only speculate at this time on the possible involvement of spontaneously released EDRF in the etiology of vasospasm. Studies in the rat aorta show clearly that destruction of the endothelium enhances greatly the responsiveness to vasoconstrictor stimuli. It is possible therefore that damage to the endothelium in vivo at sites of high basal EDRF activity might similarly lead to augmented sensitivity to vasoconstrictor stimuli resulting in vasospasm.

Hemoglobin released from lysed erythrocytes is believed to play a role in the profound and sometimes intractable cerebral vasospasm that often follows subarachnoid hemorrhage (Echlin, 1971; Osaka, 1977; Tanishima, 1980). The ability of hemoglobin to block the vascular actions of EDRF in systemic (Martin et al., 1985a, 1986a), as well as in cerebral, arteries (Fujiwara et al., 1986) makes it possible that vasospasm following hemorrhage in the cerebral circulation, or elsewhere, may involve blockade of the actions of EDRF.

Despite being complexed with haptoglobin, hemoglobin in the plasma can inhibit endothelium-dependent relaxation (Edwards et al., 1985). The hemoglobin–haptoglobin complex does not normally cross the endothelial barrier, but following hemorrhage this complex will enter the interstitial space and may contribute to vasospasm.

8. Concluding Remarks

Although basal EDRF release has been studied extensively in only a few blood vessels, it is clear that a large difference exists in the extent to which the smooth muscle of these arteries is under the tonic influence of spontaneously released EDRF. It will be necessary to obtain much more information from different vascular sites and at different levels of the vascular tree before a unified hypothesis on the physiological role of spontaneously released EDRF can be constructed. The recent application of microangiographic techniques is already providing a new and exciting insight into the role of EDRF in the microcirculation (*see* chapter by Griffith, Edwards, Davies, Harrison, and Evans). The fact remains, however, that certain blood vessels are subject to the influence of spontaneously released EDRF to the extent that their responsiveness to vasoconstrictor stimuli is markedly depressed. This suggests an important role for spontaneously released EDRF in the control of vasomotor tone.

References

Allan, G., Brook, C. D., Cambridge, D., and Hladkiwiskyj, J.: Enhanced responsiveness of vascular smooth muscle to vasoactive agents after removal of endothelial cells. *Br. J. Pharmacol.* **79:**334P, 1983.

Bigaud, M., Schoeffter, P., Stocklet, J. C., and Miller, R. C.: Dissociation between endothelial mediated increases in tissue cGMP levels and modulation of aortic contractile responses. *Naunyn Schmiedebergs Arch. Pharmacol.* **328:**221–223, 1984.

Busse, P., Pohl, U., Kellner, C., and Klemm, U.: Endothelial cells are involved in the vasodilatory response to hypoxia. *Pflugers Arch.* **397:**78–80, 1983.

Busse, R., Trogisch, G., and Bassenge, E.: The role of endothelium in the control of vascular tone. *Basic Res. Cardiol.* **80:**475–490, 1985.

Carrier, G. O. and White, R. E.: Enhancement of alpha-1 and alpha-2 adrenergic agonist-induced vasoconstriction by removal of endothelium in rat aorta. *J. Pharmac. Exp. Ther.* **232:**682–687, 1985.

Cocks, T. M. and Angus, J. A: Endothelium-dependent relaxation of coronary arteries by noradrenaline and serotonin. *Nature* **305:**627–630, 1983.

Cocks, T. M., Angus, J. A., Campbell, J. H., and Campbell, G. R.: Release and properties of endothelium-derived relaxing factor (EDRF) from endothelial cells in culture. *J. Cell. Physiol.* **123:**310–320, 1985.

Cohen, R. A., Shepherd, J. T., and Vanhoutte, P. M.: 5-Hydroxytryptamine can mediate endothelium-dependent relaxation of coronary arteries. *Am. J. Physiol.* **245:**H1077–H1180, 1983.

Collins, P., Griffith, T. M., Henderson, A. H., and Lewis, M. J.: Effect of endothelium-derived relaxing factor on calcium fluxes in aortic ring preparations of the rabbit. *J. Physiol.* **381:**427–437, 1986a.

Collins, P., Chappell, S. P., Griffith, T. M., Lewis, M. J., and Henderson, A. H.: Differences in basal endothelium-derived relaxing factor activity in difference artery types. *J. Cardiovasc. Pharmacol.* **8:**1158–1162, 1986b.

Collins, P., Griffith, T. M., Henderson, A. H., and Lewis, M. J.: Endothelium and calcium flux in rabbit aortic preparations. *J. Physiol.* **360:**63P, 1984.

De Mey, J. and Vanhoutte, P. M.: Interaction between Na^+, K^+ exchanges and the direct inhibitory effect of acetylcholine on canine femoral arteries. *Circ. Res.* **46:**826–836, 1980.

De Mey, J. G. and Vanhoutte, P. M.: Heterogeneous behavior of the canine arterial and venous wall: Importance of the endothelium. *Circ. Res.* **51:**439–447, 1982.

De Mey, J. G. and Vanhoutte, P. M.: Anoxia and endothelium-dependent reactivity of the canine femoral artery. *J. Physiol.* **335:**65–74, 1983.

Diamond, J. and Chu, E. B.: Possible role for cyclic GMP in endothelium-dependent relaxation of rabbit aorta by acetylcholine. Comparison with nitroglycerine. *Res. Commun. Path. Pharmacol.* **41**:369–372, 1983.

Digges, K. G. and Summers, R. J.: Characterisation of postsynaptic alpha-adrenoceptors in rat aortic strips and portal veins. *Br. J. Pharmacol.* **79**:655–665, 1983.

Echlin, F.: Experimental vasospasm, acute and chronic, due to blood in the subarachnoid space. *J. Neurosurg.* **35**:646–656, 1971.

Edwards, D. H., Griffith, T. M., Henderson, A. H., Lewis, M. J., and Ryley, H. C.: Endothelium-dependent relaxation is inhibited by a high molecular weight protein fraction of whole human plasma. *Br. J. Pharmacol.* **85**:266P, 1985.

Egleme, C., Godfraind, T., and Miller, R. C.: Enhanced responsiveness of rat isolated aorta to clinidine after removal of the endothelial cells. *Br. J. Pharmacol.* **81**:16–18, 1984.

Fishman, A. P.: Hypoxia on the pulmonary circulation. *Circ. Res.* **38**:221–231, 1976.

Fujiwara, S., Kassell, N. F., Sasaki, T., Nakagomi, T., and Lehman, R. M.: Selective hemoglobin inhibition of endothelium-dependent vasodilatation of rabbit basilar artery. *J. Neurosurg.* **64**:445–452, 1986.

Furchgott, R. F. and Zawadzki, J. V.: The obligatory role of endothelial cells in the relaxation of arterial smooth muscle by acetylcholine. *Nature* **288**:373–376, 1980.

Furchgott, R. F., Cherry, P. D., Zawadzki, J. V., and Jothianandan, D.: Endothelial cells as mediators of vasodilatation of arteries. *J. Cardiovasc. Pharmacol.* **6**:S336–S343, 1984a.

Furchgott, R. F., Martin, W., Jothianandan, D., and Villani, G. M.: Comparison of Endothelium-Dependent Relaxation by Acetylcholine and Endothelium-Independent Relaxation by Light in the Rabbit Aorta, in *Proceedings IUPHAR 9th International Congress of Pharmacology* (Paton, W., Mitchell, J., and Turner, P., eds.) vol. 1, Macmillan, London, 1984b.

Gimbrone, M. A., Jr. and Alexander, R. W.: Angiotensin II stimulation of prostacyclin production in cultured human vascular endothelium. *Science* **189**:219–220, 1975.

Godfraind, T., Egleme, C., and Osachie, A. L.: Role of endothelium in the contractile response of rat aorta to alpha-adrenoceptor agonists. *Clin. Sci.* **68** (suppl. 10):65S–71S, 1985.

Gordon, J. L. and Martin, W.: Stimulation of endothelial prostacyclin production plays no role in endothelium-dependent relaxation of the pig aorta. *Br. J. Pharmacol.* **80**:179–186, 1983.

Griffith, T. M., Edwards, D. H., Lewis, M. J., and Henderson, A. H.: Evidence that cyclic guanosine monophosphate (cGMP) mediates endothelium-dependent relaxation. *Eur. J. Pharmacol.* **112**:195–202, 1985.

Griffith, T. M., Edwards, D. H., Lewis, M. J., Newby, A. C., and Henderson, A. H.: The nature of endothelium-derived vascular relaxant factor. *Nature* **308**:645–647, 1984a.

Griffith, T. M., Henderson, A. H., Huges Edwards, D., and Lewis, M. J.: Isolated perfused rabbit coronary artery and aortic strip preparations: The role of endothelium-derived relaxant factor. *J. Physiol.* **351:**13–24, 1984b.

Griffith, T. M., Edwards, D. H., Newby, A. C., Lewis, M. J., and Henderson, A. H.: Production of endothelium-derived relaxant factor is dependent on oxidative phosphorylation and extracellular calcium. *Cardiovasc. Res.* **20:**7–12, 1986.

Gryglewski, R. J., Moncada, S. and Palmer, R. M. J.: Bioassay of prostacyclin and endothelium-derived relaxing factor (EDRF) from porcine aortic endothelial cells. *Br. J. Pharmacol.* **87:** 685–694, 1986a.

Gryglewski, R. J., Palmer, R. M. J., and Moncada, S.: Superoxide anion is involved in the breakdown of endothelium-derived vascular relaxing factor. *Nature* **320:** 454–456, 1986b.

Holden, W. E. and McCall, E.: Hypoxia-induced contractions of porcine pulmonary artery strips depend on intact endothelium. *Exp. Lung. Res.* **7:** 101–112, 1984.

Holtz, J., Forstermann, U., Pohl, U., Giesler, M., and Bassenge, E.: Flow-dependent, endothelium-mediated dilatation of epicardial coronary arteries in conscious dogs: Effects of cyclooxygenase inhibition. *J. Cardiovasc. Pharmacol.* **6,** 1161–1169, 1984.

Holzmann, S.: Endothelium-induced relaxation by acetylcholine associated with larger rises in cyclic GMP in coronary arterial strips. *J. Cyclic Nucleotide Res.* **8:**409–419, 1982.

Ignarro, L. J., Burke, T. M., Wood, K. S., Wolin, M. S., and Kadowitz, P. J.: Association between cyclic GMP accumulation and acetylcholine-elicited relaxation of bovine intrapulmonary artery. *J. Pharmacol. Exp. Ther.* **228:**682–690, 1984.

Imaizumi, Y., Baba, M., Imaizumi, Y., and Watanabe, M.: Involvement of endothelium in the relaxation of isolated chick jugular vein by 5-hydroxytryptamine. *Eur. J. Pharmacol.* **97:**335–336, 1984.

Konishi, M. and Su, C.: Role of endothelium in dilator responses of spontaneously hypertensive rat arteries. *Hypertension* **5:**881–886, 1983.

Long, C. J. and Stone, T.W.: The release of endothelium-derived relaxant factor is calcium-dependent. *Blood Vess.* **22:**205–208, 1985.

Lues, I and Schumann, H-J.: Effect of removing the endothelial cells on the reactivity of rat aortic segments to different alpha-adrenoceptor agonists. *Naunyn Schmiedebergs Arch. Pharmacol.* **328:**160–163, 1984.

Macia, R. A., Matthews, W. D., Lafferty, J., and De Marinis, R. M.: Assessment of alpha-adrenergic receptor subtypes in isolated aortic segments. *Naunyn Schmiedebergs Arch. Pharmacol.* **325:**306–309, 1984.

Martin, W., Villani, G. M., Jothianandan, D., and Furchgott, R. F.: Selective blockade of endothelium-dependent and glyceryl trinitrate-induced relaxation by hemoglobin and by methylene blue in the rabbit aorta. *J. Pharmacol. Exp. Ther.* **232:**708–716, 1985a.

Martin, W., Villani, G. M., Jothianandan, D., and Furchgott, R. F.: Blockade of endothelium-dependent and glyceryl trinitrate-induced relaxation of rabbit aorta by certain ferrous hemoproteins. *J. Pharmacol. Exp. Ther.* **233:**679–685, 1985b.

Martin, W., Furchgott, R. F., Villani, G. M., and Jothianandan, D.: Depression of contractile responses in rat aorta by spontaneously released endothelium-derived relaxing factor (EDRF). *J. Pharmacol. Exp. Ther.* **237:**529–538, 1986a.

Martin, W., Furchgott, R. F., Villani, G. M., and Jothianandan, D.: Phosphodiesterase inhibitors induce endothelium-dependent relaxation of rat and rabbit aorta by potentiating the effects of spontaneously released endothelium-derived relaxing factor. *J. Pharmacol. Exp. Ther.* **237:**539–547, 1986b.

Matsuda, H., Kuon, E., Holtz, J., and Busse, R.: Endothelium-mediated dilations contribute to the polarity of the arterial wall in vasomotion induced by α_2-adrenergic agonists. *J Cardiovasc. Pharmacol.* **7:**680–688, 1985.

Miller, V. M., and Vanhoutte, P. M.: Endothelial α_2-adrenoceptors in canine pulmonary and systemic blood vessels *Eur. J. Pharmacol.* **118:**123–129, 1985.

Miller, R. C., Mony, M. C., Schini, V., Schoeffter, P., and Stoclet, J. C.:Endothelial mediated inhibition of contraction and increase in cGMP levels evoked by the alpha-adrenoceptor agonist B-HT 920 in rat isolated aorta. *Br. J. Pharmacol.* **83:**903–908, 1984.

Miller, R. C., Schoeffter, P., and Stoclet, J. C.: Insensitivity of calcium-dependent endothelial stimulation in rat isolated aorta to the calcium entry blocker, flunarizine. *Br. J. Pharmacol.* **85:**481–487, 1985.

Murakami, K., Karaki, H., and Urakawa, N.: Role of endothelium in the contractions induced by norepinephrine and clonidine in rat aorta. *Jap. J. Pharmacol.* **39:**357–364, 1985.

Osaka, K.: Prolonged vasospasm produced by the breakdown products of erythrocytes. *J. Neurosurg.* **47:**403–410, 1977.

Rapoport, R. M. and Murad, F.: Agonist-induced endothelium-dependent relaxation in rat thoracic aorta may be mediated through cGMP. *Circ. Res.* **52:**352–357, 1983.

Rapoport, R. M., Waldman, S. A., Schwartz, K., Winquist, R. J., and Murad, F.: Effects of atrial natriuretic factor, sodium nitroprusside, and acetylcholine on cyclic GMP levels and relaxation in rat aorta. *Eur. J. Pharmacol.* **115:**219–229, 1985.

Rubanyi, G. M. and Vanhoutte, P. M.: Superoxide anions and hyperoxia inactivate endothelium-derived relaxing factor(s). *Am. J. Physiol.* **250:**H822–H827, 1986.

Rubanyi, G. M. and Vanhoutte, P. M.: Hypoxia releases a vasoconstrictor substance from the canine vascular endothelium. *J. Physiol.* **364:**45–56, 1985.

Rubanyi, G. M., Lorenz, R. R., and Vanhoutte, P. M.: Bioassay of endothelium-derived relaxing factor(s): inactivation by catecholamines. *Am. J. Physiol.* **249:**H95–H101, 1985.

Ruffolo, R. R. Jr., Rosling, E. L, and Waddell, J. E.: Receptor interactions of imidazolines. I. Affinity and efficacy for *alpha* adrenergic receptors in rat aorta. *J. Pharmacol. Exp. Ther.* **209:**429–436, 1979.

Singer, H, A. and Peach, M. J.: Calcium- and endothelial-mediated vascular smooth muscle relaxation in rabbit aorta. *Hypertension* **4** (suppl. II):19–25, 1982.

Tanishima, T.: Cerebral vasospasm: Contractile activity of hemoglobin in isolated canine basilar arteries. *J. Neurosurg.* **53:**787–793, 1980.

Van De Voorde, J. and Leusen, I.: Role of endothelium in the vasodilator response of rat thoracic aorta to histamine. *Eur. J. Pharmacol.* **87:**113–120, 1983.

Vanhoutte, P. M. and Rubanyi, G. M.: Superoxide dismutase prolongs the half-life of endothelium-derived relaxing factor(s). *Clin. Res.* **33:**522A, 1985.

Verrecchia, C., Sercombe, R., and Seylaz, J.: Influence of endothelium on noradrenaline-induced vasoconstriction in rabbit central ear artery. *Clin. Exp. Pharmacol. Physiol.* **12:**169–179, 1985.

Young, M. A. and Vatner, S. F.: Removal of endothelium reduces β -adrenergic dilatation and enhances α -adrenergic constriction in iliac arteries of conscious dogs. *Circulation* **70** (suppl. II):122, 1984.

Chapter 9

Calcium Transport Mechanisms in Endothelial Cells Regulating the Synthesis and Release of Endothelium-Derived Relaxing Factor

Gabor M. Rubanyi, Arnold Schwartz, and Paul M. Vanhoutte

1. Introduction

Calcium ions (Ca^{2+}) play an important role in excitation–contraction coupling in muscle (including vascular smooth muscle) and also in excitation–secretion coupling in neurons and endocrine cells. Endothelium-dependent relaxation of blood vessels in response to numerous pharmacological [e.g., acetylcholine, adenosine diphosphate (ADP), and so on] and physiological (e.g., increase in flow) stimuli is mediated by a still-unidentified substance ("endothelium-derived relaxing factor" or EDRF; Furchgott, 1983) released from endothelial cells (Furchgott, 1983; Peach et al., 1985; Vanhoutte et al., 1986). The sequence of events leading to endothelium-dependent relaxation is initiated by an excitation–secretion coupling mechanism in endothelial cells leading to the release of endothelium-derived relaxing factor, resembling events occurring in neurons (initiating the release of neurotransmitters) or in endocrine cells (initiating the release of a hormone). That Ca^{2+} plays an important role in the release of endothelium-derived relaxing factor from endothelial cells was indicated by the finding that the Ca^{2+} ionophore A23187 evokes endothelium-dependent relaxation in a variety of blood vessels isolated from different species (for ref., *see* Furchgott, 1983) and stimulates the release of endothelium-derived relaxing factor from perfused rabbit aorta (Griffith et al., 1984) and cultured endothelial cells (Cocks et al., 1985).

The purpose of this chapter is to summarize the available data regarding the possible sources and mechanisms of translocation of activator Ca^{2+} in endothelial cells, which underlies the synthesis and release of endothelium-derived relaxing factor.

2. Influx of Extracellular Ca^{2+}

2.1. Removal of Extracellular Ca^{2+}

Evidence that influx of extracellular Ca^{2+} into endothelium cells is an essential step in the synthesis and/or release of endothelium-derived relaxing factor was provided by experiments in which extracellular Ca^{2+} had been removed. Exposure of the aorta of the rat (Winquist et al., 1985) or the rabbit (Singer and Peach, 1982) to Ca^{2+}-free medium abolished or depressed the endothelium-dependent relaxations in response to muscarinic agonists (acetylcholine and metacholine) and to the Ca^{2+} ionophore A23187. Exposure of perfused aortic segments (with endothelium) of the rat (Long and Stone, 1985) or the rabbit (Griffith et al., 1986) to Ca^{2+}-free solution, while restoring a normal Ca^{2+} concentration in the perfusate before it reached the bioassay tissue, abolished basal and acetylcholine-induced release of endothelium-derived relaxing factor from the endothelium of the perfused segments within seconds. These bioassay studies suggest that removal of extracellular Ca^{2+} prevented the synthesis or release of endothelium-derived relaxing factor rather than its action on vascular smooth muscle. Similarly, basal or acetylcholine-induced increases in cyclic GMP levels in blood vessels with endothelium (compared to the levels in the absence of endothelium) were abolished in Ca^{2+}-free medium, whereas increases caused by sodium nitroprusside were not affected (Holzmann, 1982; Rapoport and Murad, 1983; Miller et al., 1985).

2.2. Ca^{2+} Channel Agonists

Although extracellular Ca^{2+} is essential for sustained release of endothelium-derived relaxing factor in response to most agonists in several blood vessels (*see* section 2.1), the involvement of Ca^{2+} channels in the entry of extracellular Ca^{2+} into endothelial cells is still controversial. The 1,4-dihydropyridine derivative Bay K 8644 is a "calcium channel agonist" that acts presumably at voltage-dependent Ca^{2+} channels in cardiac and vascular smooth muscle by increasing the time that the Ca^{2+} channel remains open (Schramm et al., 1983; Nilius et al., 1985; Nowycky et al., 1985). Bay K 8644 did not evoke relaxations or increases in cyclic GMP levels in aortic strips (with endothelium) of the rat (Spedding et al., 1986). Similarly, Bay K 8644 did not evoke endothelium-dependent relaxations in canine femoral arteries (unpublished observation). In contrast to Bay K 8644 [which is a racemic mixture of (+) and (−) enantiomers and can be considered as a partial agonist (Hof et al., 1985)], a more selective Ca^{2+} channel agonist, the 1,4-dihydropyridine derivative compound (+)-202,791 (Hof et al., 1985) evoked endothelium-dependent relaxations in the canine femoral artery in a concentration range (10^{-11} to $10^{-9}M$; unpublished observations), which is significantly lower than the concentrations (10^{-8} to $10^{-6}M$) required to induce contraction of vascular smooth muscle cells (Dube et al., 1985; Spedding et al., 1986). Concentration-dependent contractions of the aorta of the rat, evoked by Bay K 8644 in the presence of 12–15 mM K^+, were shifted to the right by a factor of two and the maximal response was inhibited by 50% in the presence of endothelium (Spedding et al., 1986). Since Bay K 8644 did not evoke relaxation or an increase in cyclic GMP content, these studies suggested that the depression of Bay K 8644-induced contractions in the presence of the endothelium was caused by the inhibitory effect of endothelium-derived relaxing factor, released continuously under basal conditions, on the action of the Ca^{2+} channel agonist on vascular smooth muscle, rather than on the release of endothelium-derived relaxing factor triggered by Bay K 8644 (Spedding et al., 1986). Recent studies in the porcine coronary artery demonstrated that the Ca^{2+} channel antagonist, (−)-202,791 augmented the contractions in response to Bay K 8644 only in the presence of endothelium (Williams et al., 1986). The difference between the concentration–response curves obtained in rings with or without endothelium was abolished by the antagonist. These findings favor the hypothesis that as in the canine femoral artery (Rubanyi et al., 1986), the Ca^{2+} channel agonist stimulates the release of endothelium-derived relaxing factor in the porcine coronary artery. These observations on intact vascular rings were confirmed by bioassay studies that demonstrated that Bay K 8644 triggers the release of endothelium-derived relaxing factor from perfused canine femoral artery segments (Rubanyi et al., 1985a) and from cultured bovine aortic endothelial cells (R. Busse, personal communication). Thus, endothelial cells must possess voltage sensitive Ca^{2+} channels, and selective Ca^{2+} agonists can trigger the release of endothelium-derived relaxing factor by acting on these channels. Even if a voltage-operated Ca^{2+} channel is involved in the entry of Ca^{2+} into endothelial cells, it may not be identical with the Ca^{2+} channels of the underlying vascular smooth muscle, since much lower concentrations of the Ca^{2+} agonists seem to activate it, and they are activated by the more selective agonist (+)-202,791, but not, or only to a small extent, by the partial agonist Bay K 8644 (Rubanyi et al., 1985a).

2.3. Ca^{2+} Channel Antagonists

The possibility that translocation of extracellular Ca^{2+} into endothelial cells may occur via Ca^{2+} channels has been analyzed using Ca^{2+} channel antagonists. In aortic preparations of the rabbit, the Ca^{2+} antagonists nifedipine (5×10^{-7} *M*) and verapamil (10^{-5} *M*) inhibited the maximal endothelium-dependent responses to metacholine and the Ca^{2+} ionophore A23187 by approximately 40%, without changing the EC_{50} values of these agonists or the direct relaxant responses of smooth muscle to sodium nitroprusside (Singer and Peach, 1982). Although these observations may suggest that Ca^{2+} channels may be involved in translocation of extracellular Ca^{2+} in response to metacholine, the relative high concentration of the antagonists needed and the inhibition of the response induced by the Ca^{2+} ionophore indicated nonspecific rather than specific inhibition of the Ca^{2+} channels. Verapamil (10^{-6} to 10^{-5}*M*) and nifedipine (10^{-8} to 5×10^{-7}*M*) had no or only very minor effects on the EC_{50} values and maximal responses elicited by acetylcholine and A23187 in the aorta of the rat (Winquist et al., 1985). Similar negative results were obtained with verapamil in canine femoral arteries (De Mey and Vanhoutte, 1980) and with flunarizine in rat aorta (Miller et al., 1985). Endothelium-mediated increases in tissue content of cyclic GMP in the aorta of the rat evoked by acetylcholine were not affected by Ca^{2+} channel antagonists (Miller et al., 1985). In contrast, the release of endothelium-derived relaxing factor from perfused canine femoral arteries evoked by the Ca^{2+} channel agonists Bay K 8644 and (+)-202,791 were prevented by nitrendipine (5×10^{-9}*M*) (Rubanyi et al., 1985a). These observations indicate that potential-operated Ca^{2+} channels are not involved in the entry of Ca^{2+} required to sustain the release of endothelium-derived relaxing factor in response to stimulation by muscarinic agonists. *d-cis*-Diltiazem, but not verapamil, nitrendipine, or the pure antagonist enantiomer (−)-202,791 causes stereoselective inhibition of acetylcholine-induced release of endothelium-derived relaxing factor from perfused canine femoral arteries (Rubanyi et al., 1985b; unpublished observation). These findings confirm the possibility that if a Ca^{2+} channel is involved in Ca^{2+} translocation essential to the synthesis or release of endothelium-derived relaxing factor, it is not identical with the potential operated Ca^{2+} channels in vascular smooth muscle.

2.4. Na^+–Ca^{2+} Exchange

Increases in intracellular sodium ion (Na^+) concentration can lead to a rise in free cytoplasmic Ca^{2+} concentration either by facilitation of Ca^{2+} influx via a Na^+–Ca^{2+} exchange mechanism at the plasma membrane (Siegl et al., 1984) or via Na^+-induced Ca^{2+} release from intracellular binding sites (Van Breemen et al., 1979). Several lines of experimental evidence suggest that Na^+–Ca^{2+} exchange may play an important role in the synthesis or release of endothelium-derived relaxing factor from endothelial cells. Replacement of Na^+ by lithium or sucrose in the Krebs-Ringer solution inhibited the acetylcholine-induced, endothelium-dependent, relaxations of canine femoral arteries (De Mey and Vanhoutte, 1980). Winquist et al. (1985) have demonstrated that endothelium-mediated relaxations of rat and rabbit aortic preparations in response to acetylcholine and A23187 are inhibited by dichlorobenzamil, an amiloride analog and a potent inhibitor of Na^+-dependent Ca^{2+} influx (Siegl et al., 1984); more recent observations confirmed these findings (Schoeffler and Miller, 1986). Similar inhibition by dichlorobenzamil of acetylcholine and A23187-induced relaxations were found in canine femoral arteries (unpublished observations). Dichlorobenzamil did not inhibit direct inhibition of smooth muscle by atrial natriuretic factor or sodium nitroprusside in aorta of the rat (Winquist et al., 1985) and selectively prevented the release of endothelium-derived relaxing factor (but not its action on bioassay tissue) from perfused canine femoral arteries (unpublished observation). Thus, a Na^+–Ca^{2+} exchange may play an important role in the synthesis or release of endothelium-derived relaxing factor. An interaction with Na^+–Ca^{2+} exchange mechanisms may also explain the inhibitory effect of *d-cis*-diltiazem on acetylcholine-induced release of endothelium-derived relaxing factor (Rubanyi et al., 1985b), since in contrast to other Ca^{2+} antagonists, low concentrations of *d-cis*-diltiazem inhibit the Na^+-induced Ca^{2+} release exchange process in isolated cardiac mitochondria (Vaghy et al., 1982; Matlib et al., 1983). Although the primary action of amiloride and its analogs occur at the plasma membrane, recent observations by Jurkowitz et al. (1983) demonstrated that dichlorobenzamil produced a diltiazem-like inhibition of the Na^+–Ca^{2+} exchange of cardiac mitochondria. This illustrates that dichlorobenzamil, and other analogs of amiloride, have multiple effects (Winquist et al., 1985). The similarity of the effects of diltiazem and dichlorobenzamil may imply that a mitochondrial (or other intracellular) site may be involved in the release of endothelium-derived relaxing factor.

3. Liberation of Ca^{2+} from Intracellular Pools

Several lines of experimental evidence suggest that influx of extracellular Ca^{2+} is not an essential requirement for all endothelium-dependent vascular responses. Although removal of extracellular Ca^{2+} depressed endothelium-dependent responses in the aorta of the rabbit, metacholine was able to evoke transient endothelium-dependent relaxations several minutes after exposure to Ca^{2+}-free solution (Singer and Peach, 1982). Similarly, endothelium-dependent relaxations of the rabbit aorta in response to the bee venom polypeptide melittin were not affected by incubation for 10 min in Ca^{2+}-free buffer solution (Loeb et al., 1986). Indirect assessment of changes in intracellular Ca^{2+} concentration by measuring Ca^{2+}-activated potassium transport (using ^{86}Rb) in cultured porcine aortic endothelial cells revealed a close correlation between endothelium-dependent relaxations of aortic strips and stimulation of ^{86}Rb efflux in response to bradykinin, ATP, and the Ca^{2+} ionophore A23187 (Gordon and Martin, 1983). During progressive prevention of transmembrane Ca^{2+} movements with lanthanum, bradykinin and ATP augmented the ^{86}RB efflux before a complete block was established (Gordon and Martin, 1983); this is analogous to the effect of lanthanum on norepinephrine-induced contractions of the aorta of the rabbit (Deth and Van Breemen, 1974). These findings suggest that endothelial cells have an intracellular Ca^{2+} store, which can be utilized by bradykinin and ATP. Direct assessment of changes in free cytoplasmic Ca^{2+} concentration in monolayers of bovine aortic endothelial cells by measuring changes in fluorescence of the Ca^{2+} chelator quin 2 demonstrated that activation of purinergic receptors [by adenosine diphosphate (ADP) and ATP] causes an increase in free intracellular Ca^{2+}, which may be essential for the synthesis of prostacyclin and endothelium-derived relaxing factor (Luckhoff and Busse, 1986). In the absence of extracellular Ca^{2+}, the ability of ATP to stimulate increases in intracellular Ca^{2+} concentration was preserved. Repetitive challenges were followed by decreasing responses, however. These findings confirm that ATP is able to mobilize Ca^{2+} from intracellular stores in endothelial cells. The Ca^{2+} entry blocker verapamil had no effect on the stimulation of the ^{86}Rb efflux (Gordon and Martin, 1983) or the increase in quin 2-fluorescence (Luckhoff and Busse, 1986) evoked by ATP, indicating that Ca^{2+} channels, similar to those present in vascular smooth muscle cells, do not play a role in transmembrane Ca^{2+} translocation in endothelial cells during stimulation by bradykinin or ATP.

4. Concluding Remarks

Both indirect and direct experimental evidences indicate that, as postulated earlier (Furchgott et al., 1981), an increase in free cytoplasmic calcium concentration in endothelial cells seems to be an important initial step in the synthesis and/or the release of endothelium-derived relaxing factor(s). A variety of Ca^{2+} transport mechanisms may be involved in the regulation of cytoplasmic calcium in endothelial cells (Fig. 1).

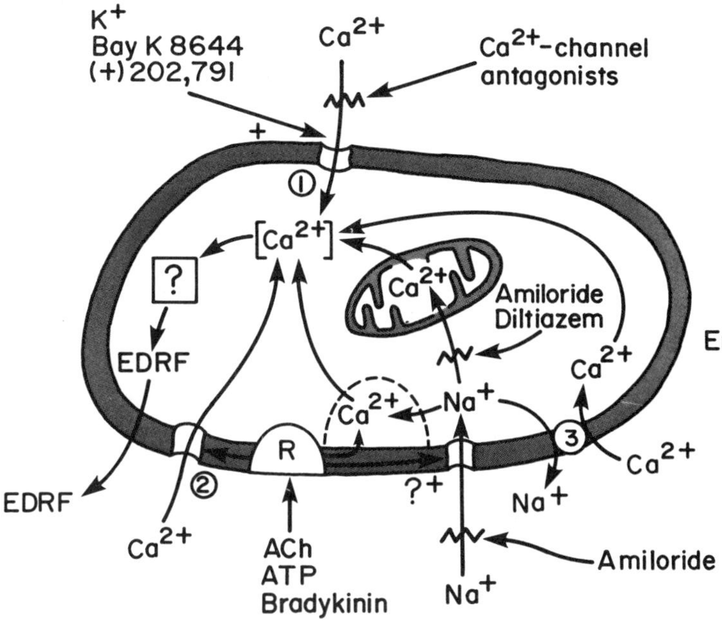

Fig. 1. Ca^{2+} transport mechanisms that could be involved in the regulation of cytoplasmic calcium ([Ca^{2+}]) in endothelial cells, leading to the synthesis and/or the release of endothelium-derived relaxing factor (EDRF). Translocation of extracellular Ca^{2+} may occur at voltage-operated (1; VOC) and receptor-operated (2; ROC) Ca^{2+} channels and via a Na^+–Ca^{2+} exchange mechanism at the plasma membrane (3). VOC can be activated by high K^+ and by Ca^{2+} channel agonists [Bay K 8644 and (+)-202,791] and inhibited by various Ca^{2+} channel antagonists. Stimulation of receptors (R) by acetylcholine (ACh), adenine nucleotides, or bradykinin can lead to Ca^{2+} influx via ROC or by enhancing Na^+ infux and thereby stimulating Na^+–Ca^{2+} exchange; it is insensitive to Ca^{2+} channel antagonists. Liberation of Ca^{2+} from intracellular stores (e.g., sarcoplasmic reticulum or mitochondria) or binding sites (e.g., inner surface of the plasma membrane) may be triggered by an increase in intracellular Na^+ concentration or via other (still-unknown) mechanisms. It is still uncertain how an increase in [Ca^{2+}] stimulates the synthesis or release of the unidentified mediator endothelium-derived relaxing factor(s).

References

Cocks, T. M., Angus, J. A., Campbell, J. H., and Campbell, G. R.: Release and properties of endothelium-derived relaxing factor (EDRF) from endothelial cells in culture. *J. Cell. Physiol.* **123:**310–320, 1985.

De Mey, J. G. and Vanhoutte, P. M.: Interaction between Na^+, K^+ exchanges and the direct inhibitory effect of acetylcholine on canine femoral arteries. *Circ. Res.* **46:**826–836, 1980.

Deth, R. and Van Breemen, C.: Relative contributions of Ca^{2+}-influx and cellular Ca^{2+} released during drug-induced activation of the rabbit aorta. *Pflugers. Arch. Gen. Physiol.* **348:**13–22, 1974.

Dube, G. P., Baik, Y. H., Vaghy, P. L., and Schwartz, A.: Nitrendipine potentiates Bay K 8644-induced contraction of isolated porcine coronary artery: Evidence for functionally distinct dihydropyridine receptor subtypes. *Biochem. Biophys. Res. Comm.* **128:**1295–1302, 1985.

Furchgott, R. F.: Role of the endothelium in responses of vascular smooth muscle. *Circ. Res.* **53:**557–573, 1983.

Furchgott, R. F., Zawadzki, J. V., and Cherry, P. D.: Role of Endothelium in the Vasodilator Response to Acetylcholine, in *Vasodilatation* (Vanhoutte, P. M. and Leusen, I., eds.) Raven, New York, 1981.

Gordon, J. L. and Martin, W.: Endothelium-dependent relaxation of the pig aorta: Relationship to stimulation of ^{86}Rb efflux from isolated endothelial cells. *Br. J. Pharmacol.* **79:**531–541, 1983.

Griffith, T., Edwards, D., Lewis, M., Newby, A., and Henderson, A.: The nature of endothelium-derived relaxing factor. *Nature* **308:**645–647, 1984.

Griffith, T. M., Edwards, D. H., Newby, A. C., Lewis, M. J., and Henderson, A. H.: Production of endothelium-derived relaxing factor is dependent on oxidative phosphorylation and extracellular calcium. *Cardiovas. Res.* **20:**7–12, 1986.

Hof, R. P., Ruegg, U. T., Hof, A., and Vogel, A.: Stereoselectivity at the calcium channel; opposite action of the enantiomers of a 1,4-dihydropyridine. *J. Cardiovasc. Pharmacol.* **7:**689, 1985.

Holzmann, S.: Endothelium-induced relaxation by acetylcholine associated with larger rises in cyclic GMP in coronary arterial strips. *J. Cyclic Nucleotide Res.* **8:**409–419, 1982.

Jurkowitz, M. S., Altschuld, R. A., Brierley, G. P., and Crague, E. L., Jr.: Inhibition of Na^+-dependent Ca^{2+} efflux from heart mitochondria by amiloride analogues. *Fed. Eur. Biochem. Soc.* **162:**262–265, 1983.

Loeb, A. L., Johns, R. A., and Peach, M. J.: Phospholipase activation by melittin releases prostacyclin and an endothelium-derived relaxing factor from rabbit aorta. *Blood Vessels* **23:**86, 1986.

Long, C. J. and Stone, T. W.: The release of endothelium-derived relaxing factor is calcium dependent. *Blood Vessels* **22:**205–208, 1985.

Luckhoff, A. and Busse, R.: Increased free calcium in endothelial cells under stimulation with adenine nucleotides. *J. Cell. Physiol.* **126:**414–420, 1986.

Matlib, M. A., Lee, S. W., Depover, A., and Schwartz, A.: A specific inhibitory action of certain benzothiazepines and benzodiazepines on the sodium-calcium exchange process of heart and brain mitochondria. *Eur. J. Pharmacol.* **89:**327–328, 1983.
Miller, R. C., Schoeffter, P., and Stoclet, J. C.: Insensitivity of calcium-dependent endothelium stimulation in rat isolated aorta to the calcium entry blocker, flunarizine. *Br. J. Pharmacol.* **85:**481–487, 1985.
Nilius, B., Hess, P., Lansman, J. B., and Tsien, R. W.: A novel type of cardiac channel in ventricular cells. *Nature* **316:**443–446, 1985.
Nowycky, M. C., Fox, A. P., and Tsien, R. W.: Three types of neuronal calcium channel with different calcium agonist sensitivity. *Nature* **316:**440–443, 1985.
Peach, M. J., Loeb, A. L., Singer, H. A., and Saye, J. A.: Endothelium-derived vascular relaxing factor. *Hypertension* **7** (suppl. I):I94–I100, 1985.
Rapoport, R. M. and Murad, F.: Agonist-induced endothelium-dependent relaxation in rat thoracic aorta may be mediated through cGMP. *Circ. Res.* **52:**352–357, 1983.
Rubanyi, G. M., Schwartz, A., and Vanhoutte, P. M.: The calcium agonists Bay K 8644 and (+)202,791 stimulate the release of endothelium relaxing factor from canine femoral arteries. *Eur. J. Pharmacol.* **117:**143–144, 1985a.
Rubanyi, G. M., Schwartz, A., and Vanhoutte, P. M.: The effect of diltiazem and verapamil on endothelium-dependent responses in canine blood vessels. *Pharmacologist* **27:**290, 1985.
Rubanyi, G. M., Schwartz, A., and Vanhoutte, P. M.: Calcium-channel agonists stimulate the release of endothelial relaxing factor from perfused canine femoral arteries. FASEB, St. Louis, Missouri, 1986.
Schoeffler, P. and Miller, R. C.: Role of sodium-calcium exchange and effects of calcium entry blockers on endothelial-mediated responses in rat isolated aorta. *Mol. Pharmacol.* **30:**53–57, 1986.
Schramm, M., Thomas, G., Towart, R., and Franchowiak, G.: Novel dihydropyridine with positive inotropic action through activation of Ca^{++} channels. *Nature* **303:**535–537, 1983.
Siegl, P. K. S., Cragoe, E. J., Jr., Trumble, M. J., and Kaczorowski, G. J.: Inhibition of Na^+/Ca^{2+} exchange in membrane vesicle and papillary muscle preparations from guinea pig heart by analogs of amiloride. *Proc. Natl. Acad. Sci. USA* **81:**3238–3242, 1984.
Singer, H. A. and Peach, M. J.: Calcium- and endothelial-mediated vascular smooth muscle relaxation in rabbit aorta. *Hypertension* **4** (suppl. II)**:**19–25, 1982.
Spedding, M., Schini, V., Schoeffter, P., and Miller, R. C.: Calcium channel activation does not increase release of endothelium-derived relaxing factors (EDRF) in rat aorta although tonic release of EDRF may modulate calcium channel activity in the smooth musccle. *J. Cardiovasc. Pharmacol.* **8:**1130–1137, 1986.

Vaghy, P. L., Johnson, J. D., Matlib, M. A., Wang, T., and Schwartz, A.: Selective inhibition of Na^+-induced Ca^{2+} release from heart mitochondria by diltiazem and certain other Ca^{2+} antagonist drugs. *J. Biol. Chem.* **257:**6000, 1982.

Van Breemen, C., Aaronson, P., and Loutzenhiser, R.: Sodium-calcium interactions in mammalian smooth muscle. *Pharmacol. Rev.* **30:**167–208, 1979.

Vanhoutte, P. M., Rubanyi, G. M., Miller, V. M., and Houston, D. S.: Modulation of vascular smooth muscle contraction by the endothelium. *Ann. Rev. Physiol.* **48:**307–320, 1986.

Williams, J. S., Baik, Y. H., and Schwartz, A.: Interactions between calcium agonists and antagonists in coronary smooth muscle: Role of the endothelium. *Blood Vessels* **23:**No. 2, abstract, 1986.

Winquist, R. J., Bunting, P. B., and Schofield, T. L.: Blockade of endothelium-dependent relaxation by the amiloride analog dichlorobenzamil: Possible role of Na^+/Ca^{++} exchange in the release of endothelium-derived relaxing factor. *J. Pharmacol. Exp. Ther.* **235:**644–650, 1985.

Chapter 10

Release of Endothelium-Derived Relaxing Factor(s) by Physicochemical Stimuli

Eberhard Bassenge, Rudi Busse, and Ulrich Pohl

1. Introduction

When Furchgott and Zawadzki (1980) published their pioneering observations on endothelium-dependent vascular responses, their results looked somewhat like a biological curiosity. Most of the endothelial stimulators do not circulate and therefore do not reach the endothelium from the luminal side in significant concentrations. The principal stimulator, acetylcholine, is rapidly inactivated by cholinesterases. Obviously, the biological concept and physiological relevance of endothelium-stimulated vasomotion by serotonin, norepinephrine, thrombin, substance P, bradykinin, adenosine triphosphate (ATP), and diphosphate (ADP) is not well understood so far. From this uncertainty, the question arises as to whether endothelium-dependent vasomotion plays a significant role in the regulation of blood flow to the organs under physiological or pathophysiological conditions. If such regulation plays a significant part, one would expect a continuous basal release of endothelium-derived vasoactive compounds, which would be effectively modulated or amplified by a number of physiological activators like metabolites, changes in PO_2 and PCO_2 (or pH), and similar parameters affected by increased activity of the tissues. Such basal release has indeed been demonstrated, and various physiological factors associated with increased activity of the organs have a modulatory role on the continuous liberation of endothelium-derived relaxing factor (EDRF). These comprise changes in PO_2, changes in blood flow-induced shear stress acting on the endothelial surface, and pulsatile stretching of the blood vessel wall. Indeed, these physiological factors can elicit endothelium-dependent responses.

Such endothelium-dependent vasomotor responses so far have been demonstrated only in segments from large conduit arteries. In these, presence or absence of endothelium-dependent vasomotion seems more important under pathophysiological conditions, when the endothelial function is impaired or absent and cannot counterbalance the direct vasomotor effects of various circulating agonists on the vasculature, which are predominantly constrictor (serotonin, histamine, norepinephrine, angiotensin II, vasopressin, oxytocin, ATP). Under physiological conditions, it would be even more important to analyze the existence of these endothelium-dependent effects in resistance vessels, which determine blood flow rates in various organs. A flow-dependent EDRF release at the level of resistance vessels could initiate a positive feedback mechanism for vasodilatation, resulting in an uncontrolled excess perfusion if not adequately counterbalanced by the metabolic control mechanisms (exaggerated washout). There are some hints that such effects on resistance vessels exist (Pohl et al., 1987b; Griffith et al., 1986a), but an exact analysis and quantification is still lacking. In particular, the stimulating effect of lowered PO_2, pulsation-induced stretching, and augmented shear stress (flow-induced) in arterioles can only be derived from extrapolations; these effects can be convincingly demonstrated in large conduit arteries. Although a continuous basal release of EDRF has been demonstrated (Griffith et al., 1984; Forstermann et al., 1986b), a potential contribution to vasomotor control is only tenable if a release of the factor toward the adjacent smooth muscle can be demonstrated. Such a release, penetrating the entire vascular wall toward the adventitia, occurs (Bassenge et al., 1987). It could even be elicited, when the EDRF-releasing agonists were applied extraluminally by adventitial superfusion. That this EDRF-release is truly able to stimulate guanylate cyclase activity, which leads to relaxation of the vascular smooth muscle, has been demonstrated by Forstermann et al. (1986b), who described a substantial increase in activity of the enzyme by stimulation of EDRF release in arterial segments.

2. EDRF Release and Vascular Response: Asymmetric Behavior of the Vascular Wall

2.1. Bipolar Release

The endothelium exhibits a pronounced morphological and functional polarity. Various receptor and effector mechanisms of endothelial cells are located exclusively at the luminal side. This is the case for several ecto-enzymes like nucleotidases, angiotensin-converting enzyme, lipoprotein lipase, peptidases, and carboanhydrase (Ryan and Ryan, 1984), which are found only at the luminal surface. Likewise, various compounds involved in the hemostatic system, (e.g., von Willebrand factor, antithrombin III, plasminogen activator; [*see* Giddings, 1983]) are secreted predominantly toward the vascular lumen. Also, an array of receptors for vasoactive substances seems to be situated mainly on the luminal surface (Buonassisi and Venter, 1976; Leitman and Murad, 1986). A particular feature of the abluminal side is the secretion of a variety of glycoproteins, which are constituents of the basement membrane (Duance and Bailey, 1983). Furthermore, myoendothelial junctions (i.e., close membrane appositions of endothelial and smooth muscle cells endowed with gap junctions) are characteristically located at the basal side of the endothelial cells (Rhodin, 1967). Although the number of myoendothelial junctions varies considerably depending on the vessel type, in general, it increases toward the microcirculatory level (Hammersen and Hammersen, 1985). On the basis of this morphological and biochemical polarity of the endothelium, the detection of an EDRF release at the luminal side cannot be taken as evidence for an EDRF release at the abluminal side. Therefore, it has been suggested that the signal transmission from the endothelium to the underlying vascular smooth muscle is realized via myoendothelial junctions (Chand and Altura, 1981; Gerlach et al., 1985).

To demonstrate the existence of an abluminal EDRF release, the intraluminal perfusate as well as the adventitial superfusate of endothelium-intact aortic segments of the rabbit were assayed for dilator activities during stimulation with acetylcholine. The EDRF releaser was administered to the intraluminal perfusate. A dilator response indicating EDRF activity was observed not only in the endothelium-denuded segment assaying the intraluminal perfusate, but also in the segment assaying the adventitial superfusate (Fig. 1). The latter dilatation was delayed in onset by about 240 s compared to the onset of dilatation in the luminal assay segment. In the adventitial superfusate, a dilator activity could be detected only at concentrations of acetylcholine, which were about tenfold higher than those needed to detect dilator activity in the luminal perfusate (Fig. 1). The vasodilator activity that is detected in the adventitial superfusate during stimulation with acetylcholine is probably caused by abluminal EDRF release. Indeed, the appearance of the dilating signal in the adventitial superfusate strictly depended on the presence of an intact endothelium. Furthermore, the dilator activity in both perfusates (intraluminal and abluminal) was inhibited in the presence of either hemoglobin (1 μM), the antioxidant dithiothreitol (0.2 mM), or the antioxidant and lipoxygenase inhibitor gossypol (5 μM).

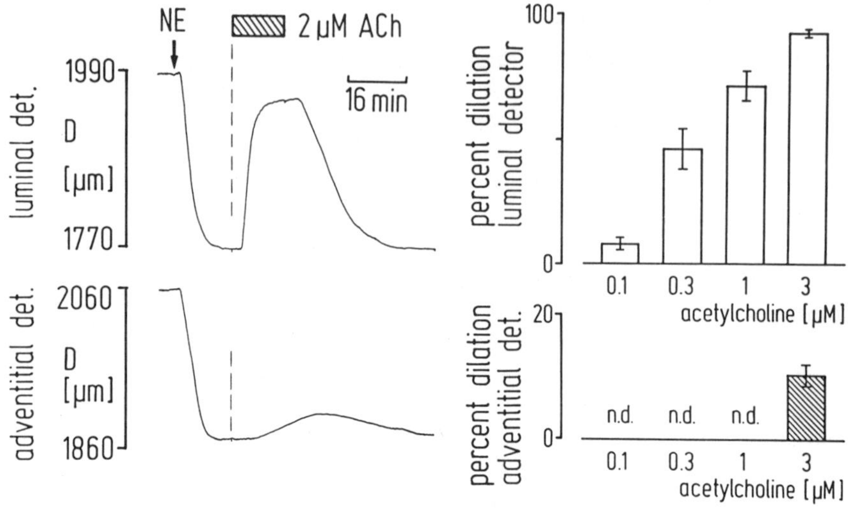

Fig. 1. (Left) Representative recordings of the diameter (D) of two endothelium-denuded femoral arteries of the rabbit used to bioassay dilator activity (EDRF) in the luminal perfusate (luminal det.) and the adventitial superfusate (adventitial det.) from an endothelium-intact donor segment (rabbit aorta). During stimulation of the donor by acetylcholine, a dilator response in both bioassay vessels [contracted with norepinephrine (NE, 0.1 *μM*)] is observed. The presence of acetylcholine (ACh) in the intraluminal perfusate of the donor segment is indicated by a dashed bar. (Right) Dilator response of luminal and adventitial detectors ($n = 14$) during stimulation of the EDRF donor with increasing doses of acetylcholine. No dilator activity could be detected in the adventitial superfusate of the donor during stimulation with acetylcholine at concentrations below 3 *μM* (data from Bassenge et al., 1987).

Experiments using EDRF released from cultured endothelial cells demonstrate that penetration of the arterial wall by EDRF occurs only at very high concentrations of the factor. The intraluminal perfusion of de-endothelialized aortic segments of the rabbit with the EDRF-containing effluent from stimulated endothelial cells in culture elicited a maximal vasodilatation of the detector assaying the luminal perfusate, whereas no dilator activity could be detected in the adventitial superfusate.

It must be emphasized that the dilator activity assayed in the intraluminal perfusate in these experiments was in the same order of magnitude as obtained during acetylcholine-induced stimulation of the rabbit aorta with endothelium. Therefore, the fact that adventitial vasodilator activity could only be detected when endothelial cells *in situ* were stimulated (but not during intraluminal perfusion of a denuded vessel with EDRF from *cultured* cells in comparable luminal concentrations), indicates that in the intact blood vessel there was a higher abluminal than luminal release of EDRF. A rough estimate of the expected amount of EDRF at the adventitial side can strengthen this view. Assuming a diffusion time across the aortic wall of about 4 min and a half-life of about 50 s (Cocks et al., 1986; Forstermann et al., 1986b), the dilator activity of EDRF detected in the adventitial superfusate should amount to only 3% of its initial value. Since the dilator response of the adventitial detector approximated 10% of that of the luminal detector, the amount of abluminally released EDRF must have exceeded substantially that of luminally released EDRF.

2.2. Differential Response of Inner and Outer Smooth Muscle to EDRF

This higher abluminal release of EDRF is associated with a preferential sensitivity of subintimal smooth muscle layers to the factor. Indeed, EDRF, released from cultured endothelial cells, induced a significant greater dilation when administered at the intimal side than when administered adventitially (Fig. 2). A similar asymmetry, with a higher sensitivity of inner compared to outer smooth muscle layers, has been described for the vascular response to contractile agonists (Keatinge and Torrie, 1976; Graham and Keatinge, 1972; Cohen et al., 1984). By contrast, sodium nitroprusside elicited virtually identical dilatations following intimal or adventitial application (Fig. 2). The asymmetry in EDRF sensitivity may have important functional implications. The signal transmission of endothelium-mediated dilatation across the smooth muscle layers of the media does not appear to be solely caused by diffusion of EDRF. Direct relaxant effects of EDRF on the smooth muscle cells in the outer media should only play a minimal role because of the low sensitivity to EDRF and the decreased concentration of the factor in these layers. It is conceivable, however, that an electrical signal [hyperpolarization (Bolton et al., 1984; Feletou and Vanhoutte, 1988)] generated by EDRF in the inner smooth muscle layer is propagated by myogenic coupling into the outer layers of the wall. Propagation of this signal is likely to be much faster than the diffusion of the relaxant factor across the wall. Although only suggestive, two observations support this view. First, the initial phase of dilatation induced by EDRF is, compared with the response characteristics of other dilators, too fast to fit the concept of a pure diffusive propagation of dilator response. Second, in endothelium-denuded arteries, dilatation induced by EDRF released from cultured endothelial cells was significantly smaller in the presence of depolarizing concentrations of potassium than in the presence of norepinephrine or prostaglandin (Luckoff et al., 1987).

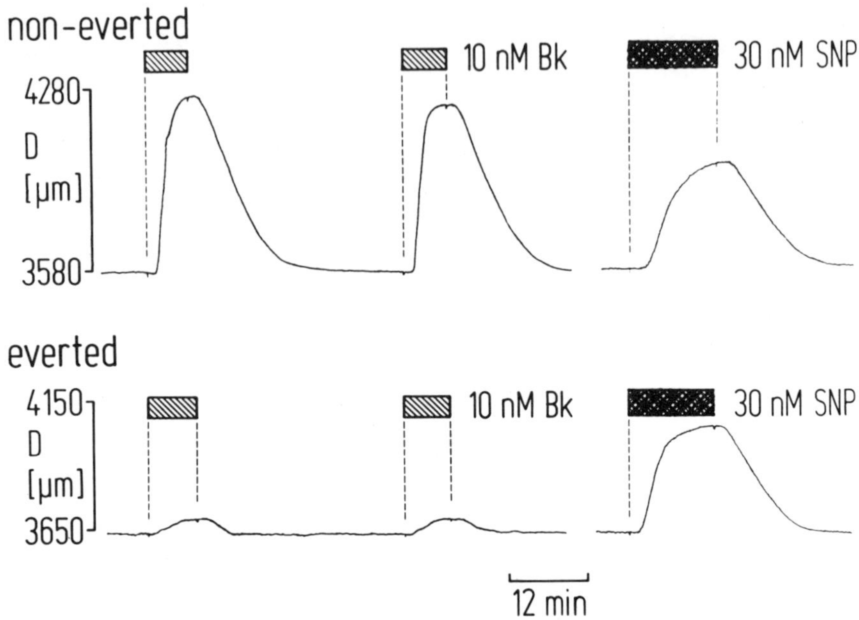

Fig. 2. Dilator responses of paired segments of endothelium-denuded rabbit aorta to EDRF released from cultured endothelial cells by bradykinin (BK, 10 n*M*) or to sodium nitroprusside (SNP, 30 n*M*). Presence of the compounds in the perfusate is indicated by bars. Either the intimal surface (noneverted segment) or the adventitial surface (everted segment; inside out) was exposed to the perfusate containing EDRF or SNP (D, diameter of the segment).

2.3. Transmural Endothelial Stimulation

Some of the endogenous compounds that elicit endothelium-dependent dilatations, (e.g., acetylcholine, substance P, and ATP) are not present in vasoactive concentrations in the circulating blood. All these compounds, however, are released from perivascular nerve endings (Burnstock and Kennedy, 1985). It is conceivable that if these compounds were released in sufficient concentrations and low extracellular degradation and/or cellular uptake prevailed, they could reach endothelial receptors by diffusion through the media. No data are available concerning the local concentrations of these compounds following neuronal release. For acetylcholine it can be demonstrated that there is a substantial inactivation in thick-walled arteries. In side branches of canine femoral arteries, the effects of luminal and adventitial application of acetylcholine were compared. With adventitial application, a 10-fold higher concentration had to be used to elicit comparable dilator responses (Fig. 3a). The same holds true in canine cor-

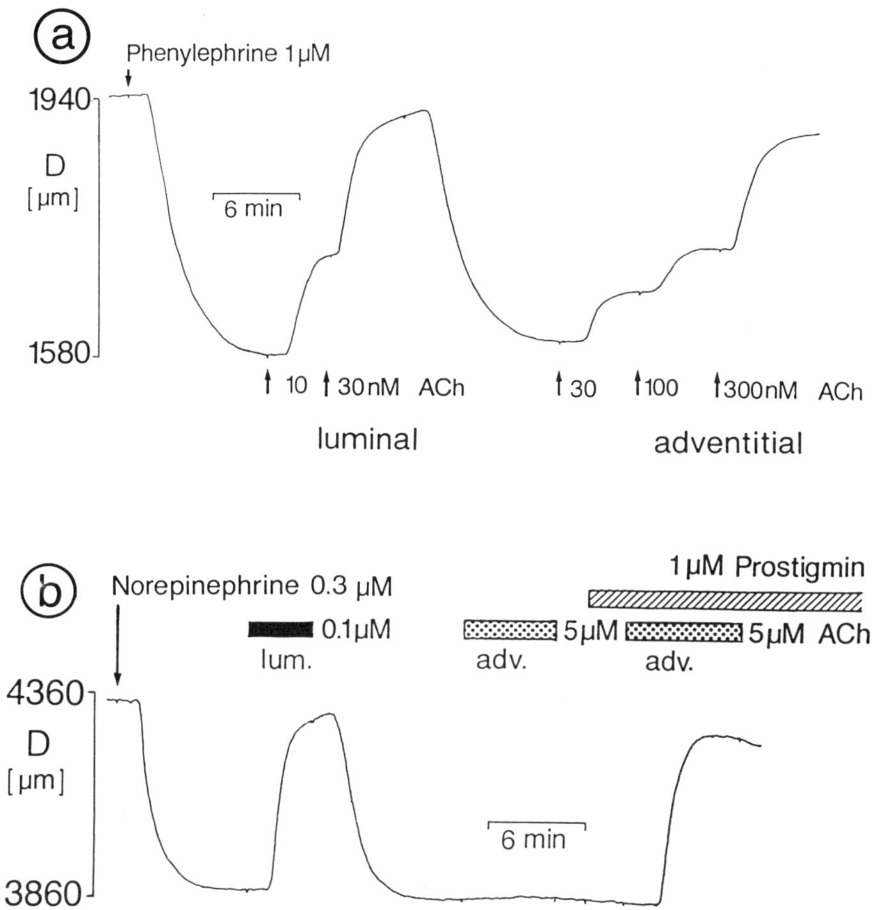

Fig. 3. (a) Diameter recordings of a side branch of the canine femoral artery, contracted with phenylephrine (1 μM). Acetylcholine (ACh) was applied either to the luminal (left) or to the adventitial side (right) of the segment. At application from the adventitial side a 10-fold higher concentration of ACh was required to induce maximal dilation of the segment. b) Diameter recordings (D) of a perfused segment of thoracic aorta of the rabbit, contracted with norepinephrine (0.3 μM). Luminal application (1 μM) of acetylcholine (ACh, 0.1 μM) elicited a virtually complete dilatation. Adventitial application of acetylcholine (ACh, 5 μM) only elicited a dilatation in the presence of the cholinesterase-inhibitor prostigmin.

onary arteries. In the rabbit aorta, however, dilator responses to adventitially administered acetylcholine can be evoked only following inhibition of cholinesterase activity (Fig. 3b). By contrast, in small arteries of the rat, rabbit, and dog with outer diameters smaller than 600 μm, the endothelium-dependent dilatation elicited by adventitial application of acetylcholine is only 10–20% less than that induced by the same concentration upon luminal application (Busse et al., 1985b).

Unlike acetylcholine, ATP applied from the adventitial side did not reach the endothelial cells in concentrations sufficient to induce endothelium-dependent vasodilatation (Bassenge and Busse, 1987). This holds true for all large- and medium-sized arteries studied so far.

Thus, the available data suggest that acetylcholine (and maybe other transmitters) can reach the endothelial receptors by diffusion. This mechanism may be important in small arteries and arterioles. These blood vessels contain only few layers of smooth muscle, which may not represent important barriers for diffusion or sites of degradation.

3. Flow Rate and Regulation of Vascular Tone

3.1. Vasomotion of Arteries in Response to Changes in Flow Rate

An important stimulator of endothelium-mediated vasomotion appears to be the flow rate. Changes in flow are associated with parallel changes in shear stress acting as viscous drag on the intimal surface. Besides chronic effects on vascular growth and regression, changes in viscous drag initiate acute adjustments of the caliber of the blood vessels in a moment-to-moment fashion. Increases in flow cause enlargements of the caliber, flow decreases cause reductions (Fig. 4). Originally it was thought that an activated tissue (e.g., working muscle) generates a vasoactive signal affecting primarily the local blood vessels, but that is then conducted upstream to cause "ascending dilatation" of the feed conduit arteries (Fleisch, 1935; Hilton, 1959). The transsection of the conduit blood vessel does not impede the dilatation, however, which argues in favor of a local dilator mechanism and against an ascending conduction (Lie et al., 1970).

3.2. Endothelial Cells as Flow Sensors

To initiate appropriate adjustments of the caliber of the blood vessel, endothelial cells must act as flow sensors and be able to transduce mechanical signals into biochemical events, which in turn initiate the release of vasoactive compounds and induce functional and structural changes (Davies et al., 1984; Franke et al., 1984; Kamiya and Togawa, 1980; Langille and O'Donell, 1986). The obligatory role of endothelial cells in transduc-

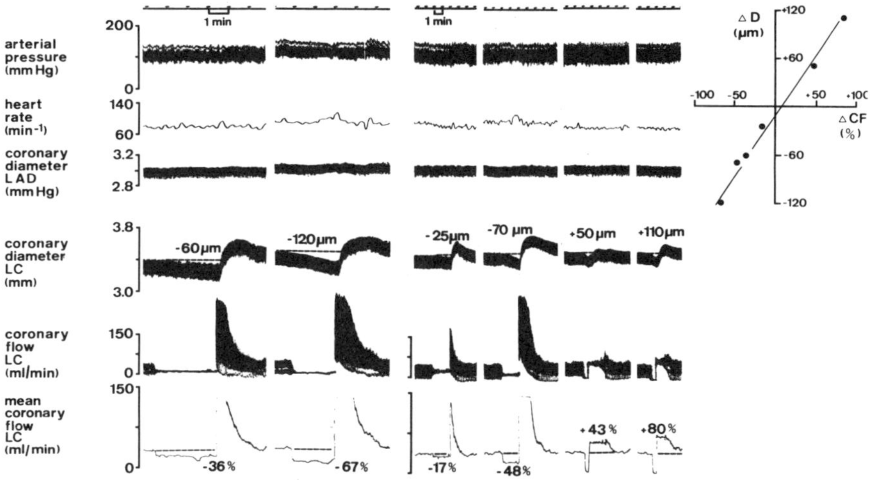

Fig. 4. Flow-dependent dilatation in coronary arteries of conscious dogs. Flow changes were induced either by partially inflating an occlusion cuff so that mean coronary flow was kept constant at a lower level for at least 2 min, or by limiting postocclusion (30 s) reactive hyperemic flow to a constant level above the preocclusion flow by only partially deflating the cuff (*see* the two panels on the right). Ninety seconds after lowering or elevating mean coronary flow, the percent change in mean coronary flow and the change in (μm) in mean coronary diameter were read and yielded the flow-diameter relation depicted in the right-hand insert. Note that reactive hyperemia following the temporary reductions in coronary flow induced the typical flow-dependent dilatations of the epicardial artery. ΔD, change in mean diameter; ΔCF, change in mean coronary flow; LAD, left anterior descending; LC, left circumflex (from Holtz et al., 1984, by permission).

ing flow signals (shear stress) into vasomotor events was demonstrated by Holtz et al. (1983a,b), Smiesko et al. (1983, 1985), and Pohl et al. (1986a). When the endothelium is removed or its function impaired, however, the vasomotor response to changes in flow is suppressed, though the response to endothelium-independent agonists is fully preserved (Smiesko et al., 1985; Pohl et al., 1986a). For example, in the canine femoral artery, stimulation of endothelium-dependent dilatation by acetylcholine resulted in a marked dilatation in the control state, as did the sudden increase in flow rate (Fig. 5). Nitroglycerin and norepinephrine caused dilatation and constriction, respectively. In the absence of endothelium, however, when the responses to acetylcholine are abolished, sudden increases in flow do not bring about a dilator response, although the effects of nitroglycerin and norepinephrine were preserved fully. In a canine coronary artery studied in vivo, the flow-dependent adjustments of the caliber amount to roughly

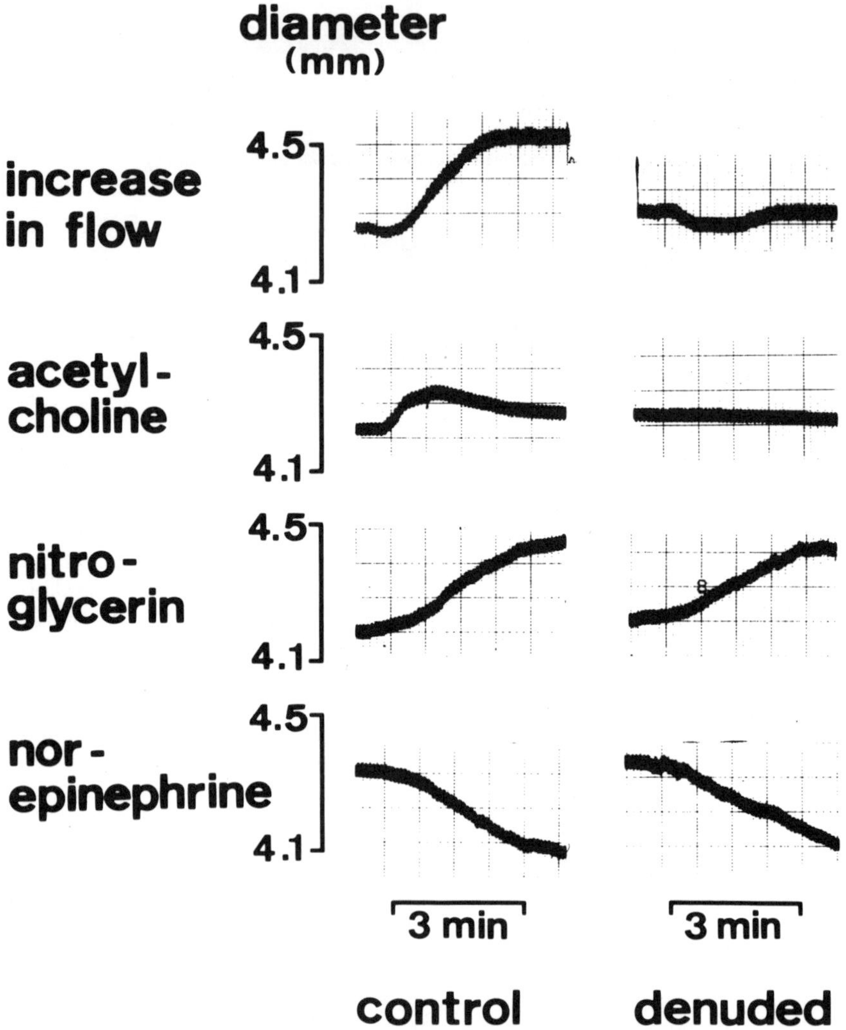

Fig. 5. Endothelial cells are crucial for flow-dependent dilatation. *In situ* recordings of the diameter of the canine femoral artery demonstrate the vasomotor responses to different stimuli before (control) and after intimal denudation. An increase in flow elicits vasodilatation, which is suppressed in the denuded state leaving a slight, passive, pressure-dependent reduction in diameter. Application of acetylcholine (ACh) elicits similar responses. The effects of the endothelium-independent agonists norepinephrine ($10^{-10}M$) and nitroglycerin ($10^{-7}M$), however, are not affected by denudation (four lower panels) (from Pohl et al., 1986a, by permission).

100 μm per 100% increase in blood flow (Fig. 4). The reductions in caliber show a similar magnitude (*see* insert of Fig. 4). These caliber changes were recorded using implanted piezoelectric crystals, which induce scar formation at the site of implantation. Therefore, the physiological range of the changes in caliber will probably be larger. Similar observations on the importance of fully preserved endothelial function in the flow-induced vasomotor responses were made in isolated perfused segments of arteries (Holtz et al., 1983a; Rubanyi et al., 1986). Pulsatile flow (pulsatile stretching) and increases in steady state perfusion rates caused release of an EDRF (Pohl et al., 1986b; Rubanyi et al., 1986b). The findings of Kaiser et al. (1986) are consistent with the increased EDRF release during flow-induced dilatation in vivo. They found a suppression of flow-induced dilatation using the inhibitor of guanylate cyclase methylene blue and the inhibitor of lipoxygenase 5,8,11,14-eicosatetraynoic acid (ETYA). With regard to the mediator role played by EDRF, however, these findings may have to be interpreted with caution. Indeed, both methylene blue and ETYA are rather nonspecific compounds that interfer with many cellular processes, often in an irreversible manner.

3.3. Physiological Relevance of Flow-Dependent Dilation

The significance of flow-dependent dilatation in the regulation of large artery tone seems well established. This may be of particular relevance under pathophysiological conditions, when endothelial function is reduced or impaired at sites of atheromas or plaque formation.

The potential role of flow-dependent, endothelium-mediated dilatation for the regulation of organ blood flow and tissue perfusion is not yet established. This is because of the technical difficulties that one encounters in the process of recording diameters in small resistance vessels. Hitherto such effects of flow on vasomotion could be demonstrated only in large arteries (Schretzenmayr, 1933; Fleisch, 1935; Hilton, 1959; Lie et al., 1970; Gerova et al., 1981; Holtz et al., 1983b, 1984; Smiesko et al., 1983; Hintze and Vatner, 1984; Kaiser et al., 1986), which exert only a minimal role in the regulation of perfusion rates to the tissues. Endothelium-dependent vasomotion, however, may significantly affect small arteries and resistance vessels that determine the perfusion of the tissues (Owen and Bevan, 1985; Pohl et al., 1987b). By extrapolation, one could speculate that flow-induced, endothelium-mediated vasomotion may also play a substantial role in resistance, as it does in conductance vessels. Direct observation of the arterioles in the isolated, saline-perfused rabbit ear support this view (Griffith et al., 1986a). The flow-dependent vasodilatation would lead to a positive feedback amplification of the augmentation in blood flow.

This flow-augmenting mechanism may counteract the consequences of the reduction of tissue-derived vasodilator metabolites that follows the increase in perfusion and the resulting accelerated washout.

3.4. Effect of Long-Term Alterations in Flow Rate on Vascular Caliber

Alterations in shear stress induced by changes in flow rate acting upon the luminal vascular surface not only cause acute but also chronic adjustments in caliber (vessel growth or regression; Fig. 6). This may be triggered by endothelial factors released when luminal shear stress is augmented and/or by the increased circumferential wall stress once the vessels are substantially dilated subsequent to increases in flow rate. As early as 1893, Thoma observed an increased vessel growth with increased flow velocities,

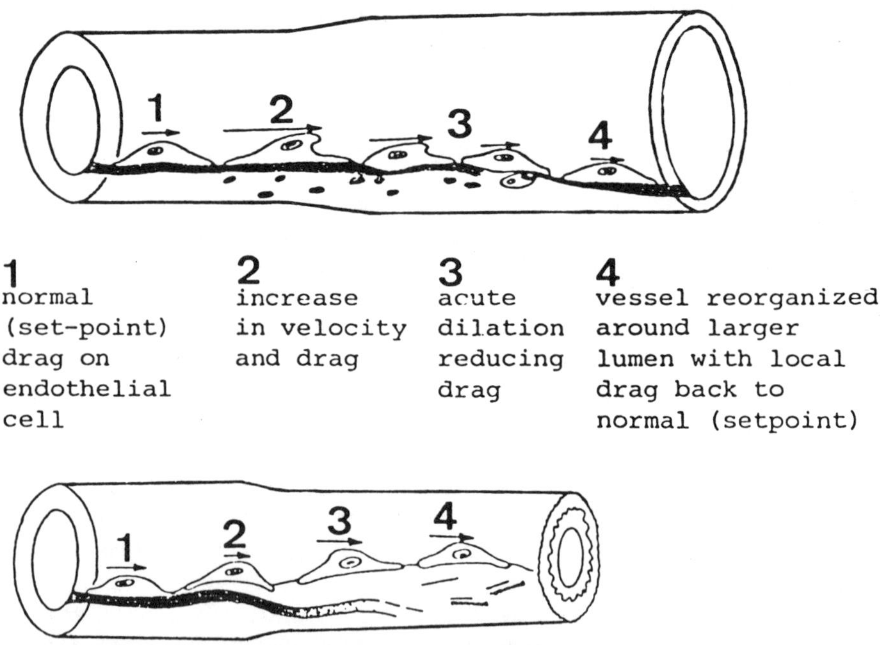

Fig. 6. Prediction of endothelial "transducer function" by Dr. S. Rodbard. Length of arrows represents actual size of hydrodynamic drag on the intimal surface. (Top) Increases in caliber, growth in response to increases in flow. (Bottom) Reductions in caliber, regression by reductions in flow. (Step 2) Sudden flow increase (top) or decrease (bottom) involves the augmented release or reduced basal release of EDRF. (Steps 3 and 4) Acute adjustments in caliber are transferred into chronic growth (regression) processes (modified from Rodbard, 1975).

and regression with reduced velocities. This was studied in more detail by Kamiya and Togawa (1980). They found that a chronically increased shear stress (flow increase caused by a carotid arteriovenous fistula) could be normalized within 6 mo when the step increase in flow did not exceed 400%. Langille and O'Donnell (1986) were able to demonstrate that these flow-induced increases and decreases in caliber are strictly endothelium dependent. In the absence of a functional endothelial lining, the growth and its regression could not be detected. These findings point to a potential key role of endothelial cells in the formation of blood vessels and collaterals.

4. Pulsatile Flow and Release of Endothelium-Derived Vasodilators

4.1. Prostacyclin

In various tissues, mechanical stimuli are able to elicit specific metabolic events. Ferreira and Vane (1967) observed in isolated splenic and gastric muscle an augmented release of prostacyclin in response to stretch. The enhanced protein synthesis in skeletal muscle that is induced by rhythmic stretching also appears to be associated with cyclooxygenase-dependent steps (Smith et al., 1983). In the vascular wall, an increased release of histamine (DeForrest and Hollis, 1978) and of prostacyclin (Quadt et al., 1982) was observed during pulsations. In these studies, however, no attempt was made to investigate the specific effects of pulsatile perfusion on the release of vasoactive compounds by the endothelial cells. In cultured endothelial cells, the production of prostacyclin and endocytosis (Frangos et al., 1985; Dewey et al., 1983) is increased in response to oscillating shear stress. In isolated segments of the canine femoral artery, the induction of pulsations (1–1.4 Hz; diameter amplitudes 1–15% of the resting outer diameter) elicited a significant augmentation of the basal release of prostacyclin (Pohl et al., 1986b). Removal of the endothelium abolished both the basal and the stimulated release of prostacyclin, indicating that endothelial cells are the only (or main) source of the prostanoid released in response to pulsatile flow. Similar results were reported by Rubanyi et al. (1986). The amount of prostacyclin release increases with augmented pulsations (Fig. 7). Although in the experiment depicted it cannot be differentiated whether it is the pressure or the diameter oscillations that act as the stimulus, systematic variations of either parameter yielded a correlation of the augmentation of the release of prostacyclin with the amplitude of the changes in diameter during the pulsations. Therefore it appears that rhythmic stretching of the endothelial cells is responsible for an augmented production and release of prostacyclin during pulsatile flow.

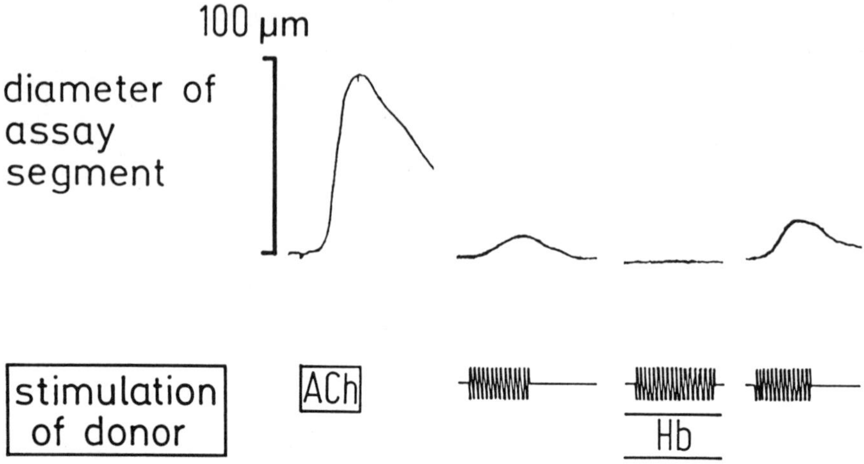

Fig. 7. Diameter recordings of an endothelium-denuded femoral artery used to assay EDRF in the effluent from a segment of rabbit aorta. Stimulation of this donor segment (see bottom) by acetylcholine (ACh, 2 μM) or by pulsation (jagged line) elicited a dilatation of the bioassay vessel, indicating transfer of EDRF. In the presence of hemoglobin (Hb, 0.5 μM) administered distal to the donor, the dilatation of the bioassay vessel during pulsatile stimulation of the donor was abolished reversibly (from Pohl et al., 1986b, by permission).

4.2. Endothelium-Derived Relaxant Factor

Unstimulated isolated arterial segments exhibit a basal release of EDRF (Griffith et al., 1984; Rubanyi et al., 1985; Forstermann et al., 1986b). Little is known, however, about the basal release in vivo and the factors that maintain the continuous release of EDRF. Pulsatile flow, which is observed even in capillaries, may represent an important stimulus in vivo. Bioassay experiments examined whether a nonprostanoid endothelium-derived vasodilator compound was released from native endothelium in the rabbit aorta in response to pulsatile flow. Nonpulsatile perfusion of endothelium-denuded assay segments (femoral artery of the rabbit) was performed to exclude direct mechanical influences on vascular tone in these segments. Pulsatile perfusion of the donor elicited a vasodilatation in the assay segments that could be inhibited by hemoglobin (0.5 μM) or dithiothreitol (0.2 mM). After removal of the endothelium from donor segments, a transfer of dilator compound was no longer observed (Pohl et al., 1986b). These findings suggest that EDRF (or another compound with similar properties) was released during pulsatile perfusion. Rubanyi et al. (1986) reported findings in perfused segments from canine arteries that confirm this interpretation.

Thus pulsatile perfusion of arteries represents an important endothelial stimulus for the release of both prostacyclin and EDRF. Therefore, mechanical effects of pulsatile blood flow *per se* may significantly contribute to a continuous basal release of these autacoids. This basal release may help to maintain the balance between neural and local constrictor mechanisms, on the one hand, and local dilator mechanisms on the other. Since EDRF inhibits platelet aggregation (Busse et al., 1987), its pulsation-induced release may contribute to the antiaggregatory properties of the luminal surface.

5. Oxygen

5.1. Hypoxia Vs Anoxia

When studying the effects of oxygen lack on the vascular wall, one must differentiate between "anoxia" and "hypoxia." Anoxia refers to a complete absence of oxygen resulting in cessation of cellular respiration. On the contrary, hypoxia comprises a broader range of partial pressures of oxygen. It may be defined as the range wherein the actual oxygen demand of some oxygen-consuming processes in cells is not fulfilled completely, but mitochondrial respiration is still maintained. The threshold values for hypoxia may differ depending on the experimental conditions. In isolated thick-walled vessels, because of the omission of the oxygen supply through the vasa vasorum as well as the occurrence of unstirred layers (Hellstrand, 1978), high environmental PO_2 values are required to bring about a sufficient diffusive transport of oxygen (Pittman and Duling, 1973). In vivo, the perfusion through the vasa vasorum substantially reduces PO_2 gradients across thick-walled vessels (Moss et al., 1968), and the oxygenation of the vascular wall is well maintained even at lower PO_2 values than in vitro. Especially in resistance vessels the PO_2 level within the wall is close to that of the blood. Indeed, direct intra- and paravascular measurements of the PO_2 in the microcirculation yield values between 60 and 30 mm Hg (Duling and Berne, 1970). In large conduit arteries, these values may be considered hypoxic. It remains to be established, however, whether the cellular constituents of the wall in these small arteries are hypoxic according to the above definition. True hypoxia may only occur in these blood vessels following reduction of the PO_2 below the physiological range. At present, however, it cannot be excluded that the low (physiological)

PO_2 impairs O_2-dependent processes to some extent in small blood vessels. The vasoconstriction in response to hyperoxia that is observed in these vessels (Jackson and Duling, 1983) would then result from activation of processes that require a higher PO_2 than exists normally.

5.2. Oxygen Sensitivity of the Vascular Wall

Mild to moderate hypoxia is a potent vasodilator stimulus in most arteries in vivo and in vitro, whereas hyperoxia results in constriction. The mechanisms, underlying these oxygen-dependent responses are still not completely understood. In vivo, there is convincing evidence that tissue-derived vasoactive metabolites contribute to the vasodilatation in response to prolonged hypoxia or anoxia (Berne and Rubio, 1977). In vitro as well as in vivo studies also suggest, however, that the lack of oxygen affects the contractile state of vascular smooth muscle, even in the absence of tissue-derived dilator metabolites. This indicates that components of the vascular wall act as oxygen sensors. The capability to sense oxygen tension seems to result from specific cellular processes and does not reflect cell damage caused by oxygen lack as seen in experiments in isolated blood vessels. Furthermore, relaxation of isolated arteries apparently is not the result of impaired mitochondrial respiration. Coburn et al. (1979) described a PO_2-sensitive dilatation of the rabbit aorta even after pharmacological blockade of the respiratory chain. In vivo studies yielded results that also pointed toward a direct sensitivity to oxygen of the vascular wall. In arterioles of the cheek pouch of the hamster, Jackson and Duling (1983) demonstrated a direct effect of oxygen on vascular diameter that was independent of flow conditions or the presence or absence of adjacent parenchymal tissue. This led to the conclusion that the oxygen sensitivity of these vessels resulted from the direct local action of oxygen on some not yet identified component of the vascular wall. In canine skeletal muscle, vasodilatation in response to a sudden onset of hypoxemia occurred as soon as the hypoxic blood reached the organ before the oxygenation of the tissues was affected. Thus, the observed dilatation probably resulted from a direct action of oxygen on the vasculature of the skeletal muscle (Pohl et al., 1982). These findings, altogether, demonstrate that direct effects of oxygen on some component(s) of the vascular wall contribute to the local control of vascular tone. In blood-perfused systems, changes in PO_2 are associated with changes in oxygen content of the blood. The effects of lowering the PO_2 seem not, however, to be only secondary to changes in oxygen content. In isolated blood-perfused rat hearts a sudden decrease in oxygen content induced augmentation of the coronary blood flow. This flow increase was much smaller when the same sudden decrease of oxygen content was induced in the presence of an simultaneously increased PO_2 (Durable et al., 1985).

5.3. Oxygen and Release/Transfer of Vasoactive Compounds from Endothelial Cells

5.3.1. Anoxia

During anoxia ($PO_2 < 1$ mm Hg), rings of canine femoral artery with intact endothelium exhibited an enhanced response to norepinephrine (De Mey and Vanhoutte, 1978), whereas the relaxant effects of acetylcholine were abolished (De Mey and Vanhoutte, 1981). Recently, Griffith et al. (1986b) demonstrated, by using several blockers of the respiratory chain, that the production and/or release of EDRF was abolished by inhibiting mitochondrial respiration, whereas blockade of glycolysis by deoxyglucose exerted only minor effects. Therefore, processes involved in the production or release of EDRF apparently depend on adequate energy production through mitochondrial respiration, which is certainly not maintained during anoxia. In addition, anoxia facilitated contractions in (endothelium-intact) rings of various arteries and veins (De Mey and Vanhoutte, 1982, 1983; Rubanyi and Vanhoutte, 1985; Peach et al., 1985). At present, it is unclear whether these anoxic contractions may be elicited by the enhanced release of a vasoactive peptide (clearly different from EDRF) that was identified in cultures of endothelial cells (Hickey et al., 1985; Gillespie et al., 1986). Other mechanisms may also be involved, since anoxic contractions can be elicited by increased K^+ concentrations (anoxia-induced cellular potassium loss) and norepinephrine release (Borda et al., 1980). In general, in all the experiments hitherto described, no distinction is possible between direct effects on endothelium and potential concomitant changes of the sensitivity of the smooth muscle to certain compounds released by the anoxia.

5.3.2. Hypoxia

In contrast to anoxia, which occurs only under pathologic conditions, hypoxia may be a physiological state, associated with fluctuations of arterial PO_2, as well as with cyclic diameter changes in the microcirculation. In addition, provided that in vivo and in vitro PO_2 effects are comparable, in the microcirculation a physiological hypoxic range may exist because of longitudinal PO_2 gradients in the arterial blood (Duling and Berne, 1970). This may contribute to the continuous balance of vasoconstrictor and dilator influences on vascular smooth muscle.

To investigate effects of hypoxia on endothelial cells and their modulator role on vascular smooth muscle, a model was developed that allows for induction of selective intimal hypoxia while the smooth muscle layers of the media are well supplied with oxygen (Busse et al., 1984). This is achieved by perfusing and superfusing intact vascular segments via separate routes. Whereas the PO_2 in the organ bath superfusing the adventitial side of the vessel is maintained continuously above 350 mm Hg, that of the intraluminal perfusate of the vessels is varied from 130 mm Hg (normoxia) to hypoxic values (60–20 mm Hg). Intimal hypoxia is associated with a vasodilatation of contracted segments with intact endothelium, whereas after endothelial denudation, the dilator response is abolished (Busse et al., 1983). Originally, these investigations were performed in the tail artery of the rat and in canine femoral and coronary arteries (Busse et al., 1984). These vessels are characterized by a high endothelial production of prostacyclin as well as a high sensitivity to this compound. The measurements of prostacyclin (determined as 6-keto-prostagland H in F_α by radioimmunoassay), showed an enhanced release during intimal hypoxia. Since indomethacin inhibited the hypoxia-induced vasodilatation, it was concluded that endothelium-derived prostacyclin may be an important mediator of the endothelium-dependent hypoxic dilatation (Busse et al., 1984). The endothelium also contributes to the relaxation of aortic rings of the rabbit in response to hypoxia (Coburn et al., 1986). Aortic smooth muscle of that species, however, is virtually insensitive to the vasodilator effects of prostacyclin (Forstermann et al., 1984). In addition, compared to the other prostanoids, prostacyclin is not released in significant amounts from the microvascular endothelium (Charo et al., 1984). Accordingly, one may anticipate that prostacyclin is not the only endothelium-derived mediator released during hypoxia. In fact, the reactivity to oxygen of small arterioles (<45 μm) of the cheek pouch and the cremaster muscle was not affected by inhibitors of cyclooxygenase (Jackson, 1986). If the endothelium plays a role in the oxygen reactivity of these blood vessels, other endothelial dilators must be involved. Since prostacyclin and EDRF are released in parallel in response to many endothelial stimuli such as ATP, bradykinin, or acetylcholine (Busse et al., 1985b; Peach et al., 1985), hypoxia may induce not only the release of prostacyclin, but also that of EDRF.

In the rabbit aorta, selective intraluminal hypoxia caused dilatations of up to 40% of the contractile response to norepinephrine, phenylephrine, or $PGF_{2\alpha}$ in the presence of indomethacin (10 μM) (Pohl et al., 1987a). No hypoxia-induced dilatations were observed in endothelium-denuded segments (Fig. 8). This indicates the release of a nonprostanoid vasodilator

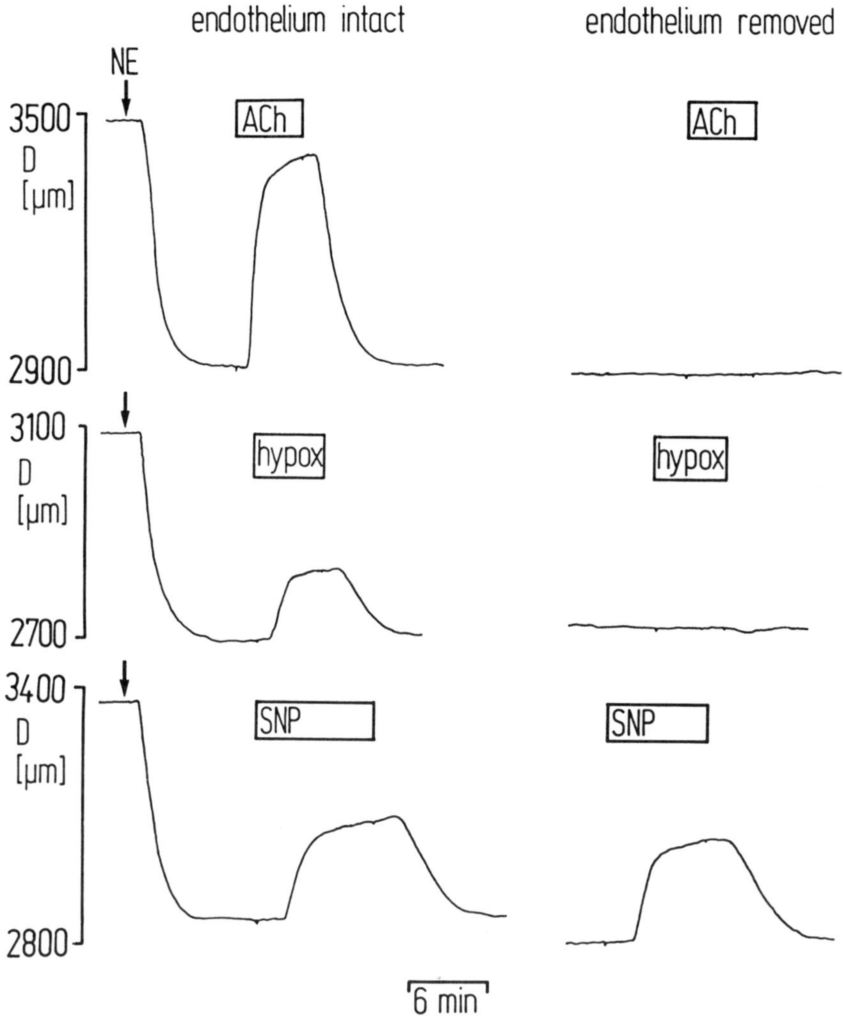

Fig. 8. Changes of outer diameter (D) of segments of rabbit aorta with and without endothelium. In segments with intact endothelium [contracted with norepinephrine (NE)], stimulation by acetylcholine (ACh), selective endothelial hypoxia (28 mm Hg), or sodium nitroprusside (SNP) induced dilatations (recordings from top to bottom). After removal of the endothelium (rubbing the intimal surface over a steel cannula), the vasodilator responses to acetylcholine and to endothelial hypoxia were abolished, indicating the endothelial nature of both reactions.

signal from hypoxic endothelial cells in this blood vessel. The magnitude of the dilator response was inversely related to the PO_2 of the intraluminal perfusate in a range between 80 and 15 mm Hg. True anoxic values were not reached. Exposure of the aortas to dithiothreitol (0.2 m*M*) or hemoglobin (5 μM) significantly attenuated or abolished the hypoxia-induced dilatations. The effects of these inhibitors (Griffith et al., 1984; Martin et al., 1985) indicate that the hypoxia-induced endothelium-dependent vasodilation in the aorta of the rabbit may be mediated by EDRF.

Bioassay experiments (endothelium-intact rabbit aorta as donor combined with endothelium-denuded femoral artery of the same species as detector) investigated the transfer (and the inhibition) of a humoral endothelium-derived vasodilator compound following induction of intraluminal hypoxia in the donor segment (Fig. 9). During the intraluminal hypoxia of the donor segment normoxic conditions were maintained in the detector preparation. The vasodilatation of the detector indicated the release of a humoral compound during hypoxic stimulation of the donor. The biological activity of the factor was inhibited by hemoglobin (Fig. 9). In addition, increasing the transit time of the effluent between donor and detector reduced the dilatation of the detector (not shown). These findings are consistent with an enhanced EDRF release during hypoxic stimulation of the endothelium.

In cultured endothelial cells derived from the bovine aorta, the oxygen consumption rate was in the range of 50–70 nL/min/10^6 cells, which is about 10-fold higher than previously reported for freshly isolated endothelium of venous origin (Dobrina and Rossi, 1983). Lowering the PO_2 in the perfusate from 130 to 40 mm Hg did not change the oxygen consumption of the cells (Pohl et al., 1987a). Inhibition of cyclooxygenase [indomethacin (10 μM) or lipoxygenase (gossypol (10 μM)] did not reduce the oxygen consumption measurably, whereas cyanide (0.1 m*M*) abolished it (Pohl et al., unpublished observations). This suggests that oxygen is nearly exclusively consumed by mitochondrial respiration [which seems not to be affected by hypoxic perfusion (see above)], whereas the amount of oxygen needed for the oxygenation of the metabolites of arachidonic acid is rather low. Moreover the available data suggest that hypoxia (to PO_2 values of 10–15 mm Hg) does not impair the activity of the oxidases (cyclooxygenase, lipoxygenase, and cytochrome P-450 monoxygenase) involved in the metabolism of the fatty acid.

The mechanism whereby hypoxia induces the enhanced release of the endothelial autacoids remains to be clarified. The cellular levels of arachidonic acid increase under hypoxic conditions (Lands et al., 1978). Whether this is caused by an impaired reacylation of lysophosphatides or stimulation of phospholipase A_2 is not clear. Since the latter enzyme is Ca^{2+} depen-

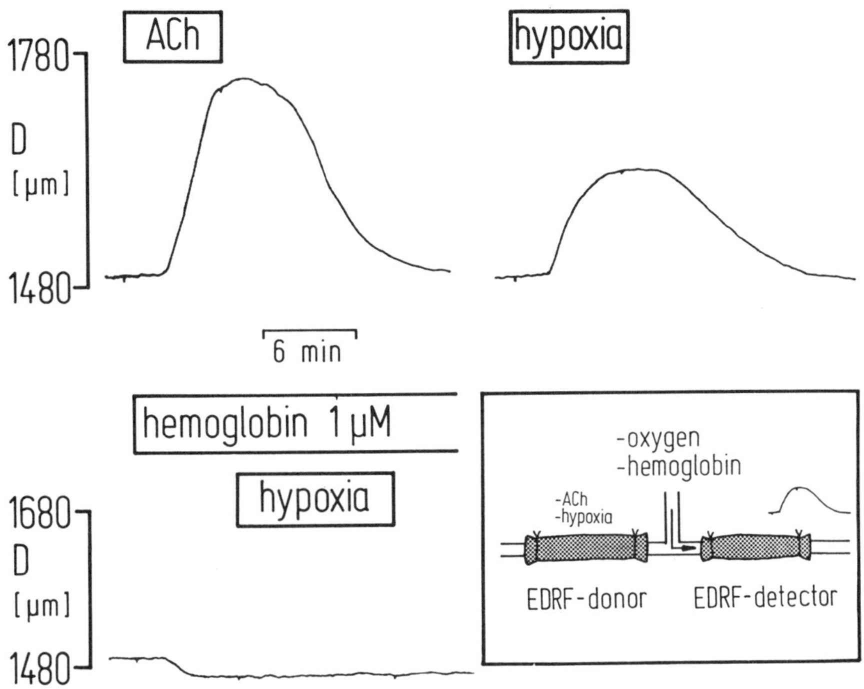

Fig. 9. Recordings of the outer diameter of a femoral artery (without endothelium) used to bioassay EDRF in the effluent from a rabbit aorta (see insert). Stimulation of the donor by acetylcholine (ACh) or by hypoxia (PO_2 = 35 mm Hg) elicited the release of a humoral dilator from the donor. In the presence of hemoglobin (administered distal to the donor), the dilatation of the bioassay vessel during hypoxic stimulation of the donor (bottom recording) was abolished.

dent (Lapetina, 1982), it is interesting to note that hypoxic dilatations in endothelium-intact segments are abolished in the presence of a calcium entry blocker (verapamil or nifedipine) or during intraluminal perfusion of the segments with Ca^{2+}-free solution (Busse et al., 1984; Pohl et al., 1987a). In hypoxic (PO_2, 30–50 mm Hg) feline pulmonary arteries, an increased calcium conductance was also observed (Harder et al., 1985). An enhanced Ca^{2+} influx into the endothelial cells may represent an important trigger for the hypoxia-induced enhancement of the release of endothelial autacoids.

Changes of oxygen tension not only affect the release, but also the stability and the biological activity of EDRF. Under hyperoxia (PO_2 > 300 mm Hg; induced by administration of high PO_2 solution distal to the donor), the dilatation of the detector is significantly smaller than under normoxia (Busse et al., 1985) (Fig. 10). The inhibitor effects of high oxygen tensions may be caused by an enhanced generation of free radicals. Application of superoxide dismutase to the effluent from donor segments stabilized the biological activity of EDRF (Rubanyi and Vanhoutte, 1986; Gryglewski et al., 1986). The effect of superoxide dismutase was potentiated by switching from hyperoxia to a lower PO_2 (Rubanyi and Vanhoutte, 1986). These findings indicate that active oxygen species, mostly superoxide anions, may be involved in inactivation of EDRF, at least in vitro.

Thus, the release as well as the biological half-life of EDRF are influenced by the partial pressure of oxygen. The low levels of oxygen that occur physiologically in the microcirculation may represent a continuous stimulus for the endothelial cells. These findings imply a functional role of endothelial cells as oxygen sensors in the vascular wall.

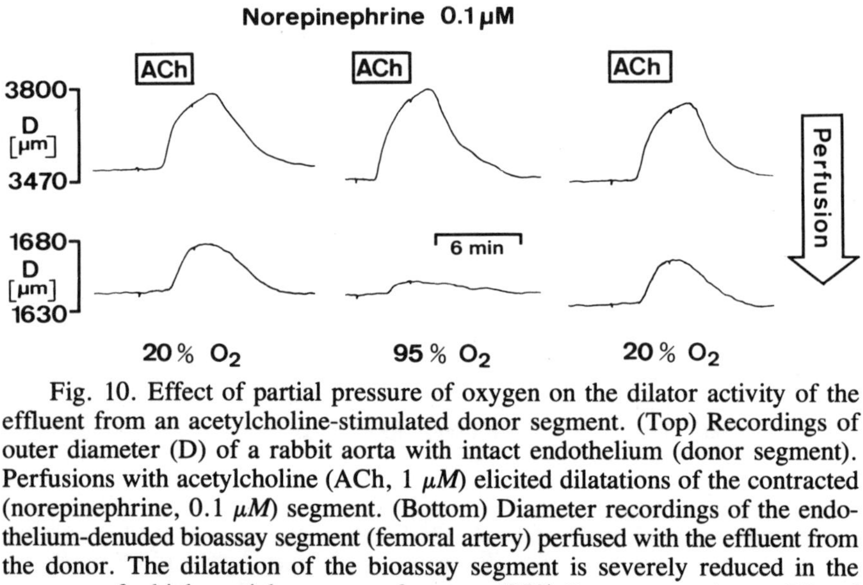

Fig. 10. Effect of partial pressure of oxygen on the dilator activity of the effluent from an acetylcholine-stimulated donor segment. (Top) Recordings of outer diameter (D) of a rabbit aorta with intact endothelium (donor segment). Perfusions with acetylcholine (ACh, 1 μM) elicited dilatations of the contracted (norepinephrine, 0.1 μM) segment. (Bottom) Diameter recordings of the endothelium-denuded bioassay segment (femoral artery) perfused with the effluent from the donor. The dilatation of the bioassay segment is severely reduced in the presence of a high partial pressure of oxygen (95% O_2 corresponds to a PO_2 of 658 mm Hg; 20% O_2 corresponds to 139 mm Hg) indicating reduced biological activity of EDRF (from Busse et al., 1985, by permission).

6. Concluding Remarks

Agonists such as acetylcholine and substance P, when administered only at the adventitial surface are able to penetrate the vascular wall and to cause EDRF release. EDRF is released both at the luminal surface and toward the underlying smooth muscle. It can be detected at the luminal and adventitial surface of the arteries under appropriate bioassay conditions. Physicochemical stimuli such as flow-induced increases in shear stress, pulsatile stretching, and low levels of PO_2 act on the endothelial lining to cause EDRF-mediated relaxations in arteries, and probably in arterioles. Not only native endothelium but also cultured endothelial cells can be stimulated by low PO_2 levels (20–60 mm Hg; comparable to levels in arterioles) to release EDRF resulting in gradual relaxation. In anoxia, however, no EDRF release and relaxation can be observed. Under physiological conditions, all these stimuli interact, thus exerting a continuous tonic influence on endothelial cell EDRF production and release.

Acknowledgment

The work of the authors cited in this chapter was generously supported by the Dr. Karl Wilder-Stiftung of the German Life Insurances.

References

Bassenge, E. and Busse, R.: Endothelial modulation of coronary tone. *Prog. Cardiovasc. Dis.* **30**:000, 1987.

Bassenge, E., Busse, R., and Pohl, U.: Abluminal release and asymmetrical response of the rabbit arterial wall to endothelium-derived relaxant factor. *Circ. Res.* in press, 1987.

Berne, R. M. and Rubio, R.: Circulatory Effects of Tissue Oxygen Tension Sensors, in *Tissue Hypoxia and Ischemia* (Reivich, M., Coburn, R., Lahiri, S., and Chance, B., eds.) Plenum, New York, London, 1977.

Boulton, T. B., Lang, R. J., and Takewaki, T.: Mechanisms of action of noradrenaline and carbachol on smooth muscle of guinea-pig anterior mesenteric artery. *J. Physiol.* **351:549**–572, 1984.

Borda, L. J., Shuchleib, R., and Henry, P. D.: Hypoxic contraction of isolated canine coronary artery. Mediation by potassium-dependent exocytosis of norepinephrine. *Circ. Res.* **46**:870–879, 1980.

Buonassisi, V. and Venter, J. C: Hormone and neurotransmitter receptors in an established vascular endothelial cell line. *Proc. Natl. Acad. Sci. USA* **73:** 1612–1616, 1976.

Burnstock, G. and Kennedy, C.: A dual function for adenosine 5′-triphosphate in the regulation of vascular tone. Excitatory cotransmitter with noradrenaline from perivascular nerves and locally released inhibitory intravascular agent. *Circ. Res.* **58:**319–330, 1985.

Busse, R., Pohl, U., Kellner, C., and Klemm, U.: Endothelial cells are involved in the vasodilatory response to hypoxia. *Pflugers Arch.* **397:**78–80, 1983.

Busse, R., Forstermann, U., Matsuda, H., and Pohl, U.: The role of prostaglandins in the endothelium-mediated vasodilatory response to hypoxia. *Pflugers Arch.* **401:**77–83, 1984.

Busse, R., Trogisch, G., and Bassenge, E.: The role of endothelium in the control of vascular tone. *Basic Res. Cardiol.* **80:**475–490, 1985.

Busse, R., Luckhoff, A., and Bassenge, E.: Endothelium-derived relaxant factor inhibits platelet activation. *Naunyn Schmiedebergs Arch. Pharmacol.* 1987, in press.

Chand, N. and Altura, B. M.: Acetylcholine and bradykinin relax intrapulmonary arteries by acting on endothelial cells: Role in lung vascular disease. *Science* **213:**1376–1379, 1981.

Charo, I. F., Shak, S., Karasek, M. A., Davison, P. M., and Goldstein, I. M.: Prostaglandin I_2 is not a major metabolite of arachidonic acid in cultured endothelial cells from human foreskin microvessels. *J. Clin. Invest.* **74:** 914–919, 1984.

Coburn, R. F., Grubb, B., and Aronson, R. D.: Effect of cyanide on oxygen tension-dependent mechanical tension in rabbit aorta. *Circ. Res.* **44:**368–378, 1979.

Coburn, R. F., Eppinger, R., and Scott, D. P.: Oxygen-dependent tension in vascular smooth muscle. Does the endothelium play a role? *Circ. Res.* **58:**341–347, 1986.

Cocks, T. M., Angus, J. A., Campbell, J. H., and Campbell, G. R.: Nature and role of endothelium-derived relaxing factor (EDRF). *J. Mol. Cell. Cardiol.* **18:**(suppl. I):17, 1986.

Cohen, R. A., Shepherd, J. T., and Vanhoutte, P. M.: Endothelium and asymmetrical responses of the coronary arterial wall. *Am. J. Physiol.* **247:**H403–H408, 1984.

Davies, P. F., Dewey, C. F., Bussolari, S. R., Gordon, E. J., and Gimbrone, M. A.: Influence of hemodynamic forces on vascular endothelial function. In vitro studies of shear stress and pinocytosis in bovine aortic cells. *J. Clin. Invest.* **73:**1121–1129, 1984.

DeForrest, J. M. and Hollis, D. M.: Shear stress and aortic histamine synthesis. *Am. J. Physio.* **234:**H701–H705, 1978.

De Mey, J. G. and Vanhoutte, P. M.: Oxygen-dependency of the acetylcholine induced relaxation in vascular smooth muscle. *Arch. Int. Pharmcodyn.* **234:**339, 1978.

De Mey, J. G and Vanhoutte, P. M.: Contribution of the endothelium to the response to anoxia in the canine femoral artery. *Arch. Int. Pharmacodyn.* **253:**325–326, 1981.

De Mey, J. G. and Vanhoutte, P. M.: Anoxia and endothelium dependent reactivity of the canine femoral artery. *J. Physiol.* **335:**65–74, 1983.

De Mey, J. G. and Vanhoutte, P. M.: Heterogeneous behavior of the canine arterial and venous wall: Importance of the endothelium. *Circ. Res.* **51:**439–447, 1982.

Dewey, C. F., Gimbrone, M. A., Bussolari, S. R., White, G. W., and Davies, P. F.: Response of Vascular Endothelium to Unsteady Fluid Shear Stress In Vitro, in *Fluid Dynamics as as Localizing Factor for Atherosclerosis*

(Schettler, G., Nerem, R. M., Schmid-Schonbein, H., Morl, H., and Diehm, C., eds.) Springer, Berlin, Heidelberg, New York, Tokyo, 1983.
Dobrina, A. and Rossi, F.: Metabolic properties of freshly isolated bovine endothelial cells. *Biochim. Biophys. Acta* **762:**295–301, 1983.
Duance, V. C. and Bailey, A. J.: The Nature Structure and Function of the Vascular Basement Membrane, in *Biochemical Interactions at the Endothelium* (Cryer, A., ed.) Elsevier, Amsterdam, New York, Oxford, 1983.
Duling, B. R. and Berne, R. M.: Longitudinal gradients in periarteriolar oxygen tension. A possible mechanism for the participation of oxygen in local regulation of blood flow. *Circ. Res.* **27:**669–678, 1970.
Duruble, M., Duvelleroy, M., Gauduel, Y., Martin, J. L., and Teisseire, B.: Transient responses of coronary flow in the blood-perfused isolated rat heart submitted to changes in oxygen content. *J. Physiol.* **358:**321–334, 1985.
Feletou, M. and Vanhoutte, P. M.: Endothelium-derived factor(s) and membrane potential of vascular smooth muscle. *Blood Vess.* **23:**68, 1986.
Feletou, M. and Vanhoutte, P. M.: Endothelium-dependent hyperpolarization of canine coronary smooth muscle. *Br. J. Pharmacol.* **93:**515–524, 1988.
Ferreira, S. H. and Vane, J. R.: Prostaglandins: Their disappearance from and release into the circulation. *Nature* **216:**868–873, 1967.
Fleisch, A.: Les reflexes nutritifs ascendants producteurs de dilatation arterielle. *Arch. Int. Physiol.* **41:**141–167, 1935.
Forstermann, U., Hertting, G., and Neufang, B.: The importance of endogenous prostaglandins other than prostacyclin for the modulation of contractility of some rabbit blood vessels. *Br. J. Pharmacol.* **81:**623–630, 1984.
Forstermann, U., Goppelt-Strube, M., Frolich, J. C., and Busse, R.: Inhibitors of acyl-coenzyme A: lysolecithin acyltransferase activate the production of endothelium-derived vascular relaxing factor (EDRF). *J. Pharmacol. Exp. Ther.* **238:**352–359, 1986a.
Forstermann, U., Mulsch, A., Bohme, E., and Busse, R.: Stimulation of soluble guanylate cyclase by an acetylcholine-induced endothelium-derived factor from rabbit and canine arteries. *Circ. Res.* **58:**531–538, 1986b.
Frangos, J. A., Eskin, S. G., McIntire, L. V., and Ives, C. L.: Flow effects on prostacyclin production by cultured human endothelial cells. *Science* **227:**1477–1479, 1985.
Franke, R. -P., Grafe, M., Schnittler, H., Seiffge, D., and Mittermayer, C.: Induction of human vascular endothelial stress fibres by fluid shear stress. *Nature* **307:**648–649, 1984.
Furchgott, R. F. and Zawadzki, J. V.: The obligatory role of endothelial cells in the relaxation of arterial smooth muscle by acetylcholine. *Nature* **288:**373–376, 1980.
Gerlach, E., Nees, S., and Becker, B. F.: The vascular endothelium: A survey of some newly evolving biochemical and physiological features. *Basic Res. Cardiol.* **80:**459–474, 1985.
Gerova, M., Gero, J., Barta, E., Dolezel, S., Smiesko, V., and Levicky, V.: Neurogenic and myogenic control of conduit coronary artery: A possible interference. *Basic Res. Cardiol.* **76:**503–507, 1981.

Giddings, J. C.: The Control of Intravascular Blood Coagulation and Haemostasis at Endothelial Surface, in *Biochemical Interactions at the Endothelium* (Cryer, A., ed.) Elsevier, Amsterdam, New York, Oxford, 1983.

Gillespie, M. N., Owasoyo, J. O., McMurtry, I. F., and O'Brien, R. F.: Sustained coronary vasoconstriction provoked by a peptidergic substance released from endothelial cells in culture. *J. Pharmacol. Exp. Ther.* **236:** 339–343, 1986.

Graham, J. M. and Keatinge, W. R.: Differences in sensitivity to vasoconstrictor drugs within the wall of the sheep carotid artery. *J. Physiol.* **221:** 477–492, 1972.

Griffith, T. M., Edwards, D. H., Lewis, M. J., Newby, A. C., and Henderson, A. H.: The nature of endothelium-derived vascular relaxant factor. *Nature* **308:**645–647, 1984.

Griffith, T. M., Edwards, D. H., Davies, R. L., Harrison, T. J., and Evans, K. T.: The role of EDRF in the microcirculation. *Int. J. Microcirc. Clin. Exp.* **5:**168, 1986a.

Griffith, T. M., Edwards, D. H., Newby, A. C., Lewis, M. J., and Henderson, A. H.: Production of endothelium derived relaxant factor is dependent on oxidative phosphorylation and extracellular calcium. *Cardiovasc. Res.* **20:** 7–12, 1986b.

Gryglewski, R. J., Palmer, R. M. J., and Moncada, S.: Superoxide anion is involved in the breakdown of endothelium-derived vascular relaxing factor. *Nature* **320:**454–456, 1986.

Hammersen, F. and Hammersen, E.: Some structural and functional aspects of endothelial cells. *Basic Res. Cardiol.* **80:**491–501, 1985.

Harder, D. R., Madden, J. A., and Dawson, C.: Hypoxic induction of Ca^{2+}-dependent action potentials in small pulmonary arteries of the cat. *J. Appl. Physiol.* **59:**1389–1393, 1985.

Hellstrand, P.: Effects of hypoxia on the rat portal vein in vitro: pO_2 gradients in tissue and surrounding fluid. *Acta Physiol. Scand.* **103:**472–474, 1978.

Hickey, K. A., Rubanyi, G., Paul, R. J., and Highsmith, R. F.: Characterization of a coronary vasoconstrictor produced by cultured endothelial cells. *Am. J. Physiol.* **248:**C550–C556, 1985.

Hilton, S. M.: A peripheral arterial conducting mechanism underlying dilatation of the femoral artery and concerned in functional vasodilatation in skeletal muscle. *J. Physiol.* (Lond.) **149:**93–111, 1959.

Hintze, T. H. and Vatner, S. F.: Reactive dilation of large coronary arteries in conscious dogs. *Circ. Res.* **54:**50–57, 1984.

Holtz, J., Busse, R., and Giesler, M.: Flow-dependent dilation of canine epicardial coronary arteries in vivo and in vitro: Mediated by the endothelium. *Naunyn Schmiedebergs Arch. Pharmacol.* **322**(suppl.):R44, 1983a.

Holtz, J., Giesler, M., and Bassenge, E.: Two dilatory mechanisms of antianginal drugs on epicardial coronary arteries in vivo: Indirect, flow-dependent, endothelium-mediated dilation and direct smooth muscle relaxation. *Z. Kardiol.* **72** (suppl. 3):98–106, 1983b.

Holtz, J., Forstermann, U., Pohl, U., Giesler, M., and Bassenge, E.: Flow-dependent, endothelium-mediated dilation of epicardial coronary arteries in conscious dogs: Effects of cyclooxygenase inhibition. *J. Cardiovasc. Pharmacol.* **6:**1161–1169, 1984.

Jackson, W. F.: Prostaglandins do not mediate arteriolar oxygen reactivity. *Am. J. Physiol. 250:*H1102–H1108, 1986.

Jackson, W. F. and Duling, B. R.: The oxygen sensitivity of hamster cheek pouch arterioles. In vitro and in situ studies. *Circ. Res.* **53:**515–525, 1983.

Kaiser, L., Hull, S. S., and Sparks, H. V.: Methylene blue and ETYA block flow-dependent dilation in canine femoral artery. *Am. J. Physiol.* **251:** H974–H981, 1986.

Kamiya, A. and Togawa, T.: Adaptive regulation of wall shear stress to flow change in the canine carotid artery. *Am. J. Physiol.* **239:**H14–H21, 1980.

Keatinge, W. R. and Torrie, M. C.: Action of sympathetic nerves on inner and outer muscle of sheep carotid artery and effect of pressure on nerve distribution. *J. Physiol.* **257:**699–712, 1976.

Lands, W. E. M., Sauter, J., and Stone, G. W.: Oxygen requirement for prostaglandin biosynthesis. *Prostaglandins Med.* **1:**117–120, 1978.

Langille, B. L. and O'Donnell, F.: Reductions in arterial diameter produced by chronic decreases in blood flow are endothelium-dependent. *Science* **231:** 405–407, 1986.

Lapetina, E. G.: Regulation of arachidonic acid production: Role of phospholipases C and A_2. *Trends Pharmacol. Sci.* **3:**115–118, 1982.

Leitman, D. C. and Murad, F.: Comparison of binding and cyclic GMP accumulation by atrial natriuretic peptides in endothelial cells. *Biochim. Biophys. Acta* **885:**74–79, 1986.

Lie, M., Sejersted, O. M., and Kiil, F.: Local regulation of vascular cross section during changes in femoral arterial blood flow in dogs. *Circ. Res.* **27:**727–737, 1970.

Luckhoff, A., Busse, R., Winter, I., and Bassenge, E.: Characterization of vascular relaxant factor released from cultured endothelial cells. *Hypertension* **9:**295–303, 1987.

Martin, M., Villani, G. M., Jothianandan, D., and Furchgott, R. F.: Selective blockade of endothelium-dependent and glyceryl trinitrate-induced relaxation by hemoglobin and by methylene blue in the rabbit aorta. *J. Pharmacol. Exp. Ther.* **232:**708–716, 1985.

Moss, A. J., Samuelson, P., Angell, C., and Minken, S. L.: Polarographic evaluation of transmural oxygen availability in intact muscular arteries. *J. Atheroscler. Res.* **8:**803–810, 1968.

Owen, M. P. and Bevan, J. A.: Acetylcholine induced endothelial-dependent vasodilation increases as artery diameter decreases in the rabbit ear. *Experientia* **41:**1057–1058, 1985.

Peach, M. J., Singer, H. A., and Loeb, A. L.: Mechanisms of endothelium-dependent vascular smooth muscle relaxation. *Biochem. Pharmacol.* **34:**1867–1874, 1985.

Pittman, R. N. and Duling, B. R.: Oxygen sensitivity of vascular smooth muscle. I. In vitro studies. *Microvasc. Res.* **6**:202–211, 1973.

Pohl, U., Busse, R., and Kessler, M.: Vascular Resistance and Tissue pO_2 in Skeletal Muscle During Perfusion with Hypoxic Blood, in *Cardiovascular System Dynamics* (Kenner, T., Busse, R., and Hinghofer-Szalkay, H., eds.) Plenum, New York, London, 1982.

Pohl, U., Busse, R. Kuon, E., and Bassenge, E.: Pulsatile perfusion stimulates the release of endothelial autacoids. *J. Appl. Cardiol.* **1**:235–255, 1986b.

Pohl, U., Busse, R., and Bassenge, E: Endothelial Cells as Oxygen Sensors, in *Mechanisms of Vasodilatation IV* (Vanhoutte, P. M., ed.) Raven, New York, in press, 1987a.

Pohl, U., Dezsi, L., Simon, B., and Busse, R.: Selective inhibition of endothelium-dependent dilation in resistance-sized vessels in vivo. *Am. J. Physiol.* **253**:H000–H000, 1987b.

Pohl, U., Holtz, J., Busse, R., and Bassenge, E.: Crucial role of endothelium in the vasodilator response to increased flow in vivo. *Hypertension* **7**:37–44, 1986a.

Quadt, J. F. A., Voss, R., and TenHoor, F.: Prostacyclin production of the isolated pulsatingly perfused rat aorta. *J. Pharmacol. Meth.* **7**:263–270, 1982.

Rhodin, J. A. G.: The ultrastructure of mammalian arterioles and precapillary sphincters. *J. Ultrastruct. Res.* **18**:181–223, 1967.

Rodbard, S.: Vascular caliber. *Cardiology* **60**:4–49, 1975.

Rubanyi, G. M. and Vanhoute, P. M.: Hypoxia releases a vasoconstrictor substance from the canine vascular endothelium. *J. Physiol.* **364**:45–56, 1985.

Rubanyi, G. M. and Vanhoutte, P. M.: Superoxide anions and hyperoxia inactivate endothelium-derived relaxing factor. *Am. J. Physiol.* **250**:H822–H827, 1986.

Rubanyi, G. M., Lorenz, R. R., and Vanhoutte, P. M.: Bioassay of endothelium-derived relaxing factor(s). Inactivation by catecholamines. *Am. J. Physiol.* **249**:H95–H101, 1985.

Rubanyi, G. M., Romero, J. C., and Vanhoutte, P. M.: Flow-induced release of endothelium-derived relaxing factor. *Am. J. Physiol. 250:*H1145–H1149, 1986.

Ryan, U. S. and Ryan, J. W.: Cell biology of pulmonary endothelium. *Circulation* **70**(suppl. III):III46–62, 1984.

Schretzenmayr, A.: uber kreislaufregulatorische Vorgange an den großen Arterien bei der Muskelarbeit. *Pflugers Arch. Ges. Physiol.* **232**:743–748, 1933.

Smiesko, V., Kozik, J., and Dolezel, S.: The control of arterial diameter by blood flow velocity is dependent upon intact endothelium. *Physiol. Bohemoslow* **32**:558A, 1983.

Smiesko, V., Kozik, J., and Dolezel, S.: Role of endothelium in the control of arterial diameter by blood flow. *Blood Vess.* **22**:247–251, 1985.

Smith, R. H., Palmer, R. M., and Reeds, P. J/: Protein synthesis in isolated rabbit forelimb muscles. The possible role of metabolites of arachidonic acid in the response to intermittent stretching. *Biochem. J.* **214:**153–161, 1983.
Thoma, R.: Untersuchungen Uber die Histogenese und Histomechanik des Gefaβ systems. Enke, Stuttgart, 1893.

Chapter 11

Role of Cyclic GMP in Endothelium-Dependent Relaxation of Vascular Smooth Muscle

Robert M. Rapoport and Ferid Murad

1. Introduction

There are four classes of agents that elevate cyclic GMP levels in smooth muscle. These are: (1) the agents generally known as the nitrovasodilators, such as sodium nitroprusside, sodium azide, and nitroglycerin (Diamond and Blisard, 1976; Katsuki and Murad, 1977; Schultz et al., 1977); (2) contractile agents (Diamond and Blisard, 1976; Katsuki and Murad, 1977; Schultz et al., 1977); (3) atrial natriuretic factor (Winquist et al., 1984; Rapoport et al., 1985a); and (4) the endothelium-dependent vasodilators (Rapoport and Murad, 1983a). Initial studies suggested that cyclic GMP formation was associated with contraction (Lee et al., 1972; Dunham et al., 1974). This hypothesis was seriously questioned, however, when it was found that the elevated levels of cyclic GMP occurred after the initiation of contraction (Diamond and Blisard, 1976; Katsuki and Murad, 1977).

It was then hypothesized that cyclic GMP formation may be associated with relaxation of smooth muscle, since the nitrovasodilator-induced relaxation was associated with accumulation of cyclic GMP (Katsuki et al., 1977a; Schultz et al., 1977). In support of this hypothesis are the fairly recent discoveries that vascular relaxation induced by atrial natriuretic factor (Winquist et al., 1984; Rapoport et al., 1985a) and the endothelium-dependent vasodilators (Furchgott and Zawadzki, 1980) were also associiated with cyclic GMP formation (Rapoport and Murad, 1983a). The purpose of the present review is to evaluate in more detail the evidence that suggests that relaxation induced by the endothelium-dependent vasodilators may be mediated through formation of cyclic GMP. Attention will be focused on those studies that utilized agents that suggested the involvement of cyclic GMP or in fact measured cyclic GMP levels. Thus, many studies that demonstrated only changes in contractility have not been referred to.

2. Elevation of Cyclic GMP Levels in Smooth Muscle by Agents that Act on the Endothelium

The initial observations that the endothelium-dependent relaxants elevated cyclic GMP levels within the smooth muscle were performed in the thoracic aorta of the rat (Rapoport and Murad, 1983a). Relaxations induced by acetylcholine, histamine, and the Ca^{2+} ionophore A23187 were associated in time- and concentration-dependent manners with formation of cyclic GMP (Rapoport and Murad, 1983a; Rapoport et al., 1984). Removal of the endothelium abolished the relaxations and the associated elevations in cyclic GMP levels. Cyclic AMP levels remained unchanged by these endothelium-dependent relaxants.

Acetylcholine- or carbamylcholine-induced relaxations of the aorta of the rabbit (Diamond and Chu, 1983; Furchgott et al., 1984; Coburn et al., 1986), bovine intrapulmonary (Ignarro et al., 1984) and coronary arteries (Holzmann, 1982) and calcium ionophore A23187-induced relaxations of aorta of the rabbit (Furchgott et al., 1984), and human coronary artery (Rapoport et al., 1987) and melittin-induced relaxations of the aorta of the rat (Rapoport and Murad, unpublished observation) were also associated with elevated levels of cyclic GMP, in time- and/or concentration-dependent manners. Furthermore, it appears that the elevated cyclic GMP levels caused by the endothelium-dependent vasodilators occur within the smooth muscle, since levels of cyclic GMP remained elevated following exposure to acetylcholine and rapid removal of the endothelium (Rapoport and Murad, 1983a; Furchgott et al., 1985). The serine proteases trypsin and thrombin also induced endothelium-dependent relaxation of the aorta of the rat that was associated with increased formation of cyclic GMP (Rapoport et al., 1984). The elevated cyclic GMP levels and relaxations were dependent on the presence of the endothelium. When examined, cyclic AMP levels remained unaltered, and atropine inhibited the acetylcholine-induced relaxations and the elevated levels of cyclic GMP. Recently, methacholine and adenosine triphosphate (ATP) were demonstrated to increase the levels of cyclic GMP in mixed cultures of endothelial cells and smooth muscle (Loeb et al., 1985).

ATP also induced endothelium-dependent relaxation of the aorta of the rat that was associated with increased levels of cyclic GMP and with a small, but significant, increase in those of cyclic AMP (Rapoport et al., 1984). Removal of the endothelium prevented the increased levels of both cyclic nucleotides, whereas exposure to the inhibitor of cyclooxygenase indomethacin did not alter the level of either. The mechanism by which ATP increases levels of cyclic AMP is not clear, but may be the release of a factor from the endothelium.

Arachidonic acid has also been shown to induce endothelium-dependent relaxation (Furchgott, 1983). The relaxation was associated with increased levels of cyclic AMP and cyclic GMP in the bovine intrapulmonary artery (Ignarro et al., 1985). The increase in cyclic AMP levels, but not cyclic GMP levels, was abolished by indomethacin. We have been unable to observe increased levels of cyclic GMP in response to arachidonic acid in the thoracic aorta of the rat (Rapoport and Murad, unpublished observation).

Relaxations to the nitrovasodilators, in contrast to relaxations to the endothelium-dependent vasodilators, are thought to be independent of the endothelium (Rapoport and Murad, 1983b; Furchgott et al., 1985). Others have recently observed, however, that removal of the endothelium from the aorta of the rat significantly augmented relaxations to sodium nitroprusside and sodium nitrite (Shirasaki and Su, 1985), whereas still others have observed that removal of the endothelium from the aorta of the rabbit inhibited relaxations to nitroglycerin and isoproterenol (White et al., 1986). This is an area that needs to be investigated further in order to clarify these apparently contrasting results. In any case, levels of cyclic GMP are still elevated by the nitrovasodilators in tissues denuded of endothelium (Rapoport and Murad, 1983b; Furchgott et al., 1985; Winquist et al., 1984). The magnitude of the nitrovasodilator-induced elevations in cyclic GMP in the presence and absence of the endothelium are difficult to compare since basal levels of cyclic GMP are decreased in tissues denuded of endothelium (*see* section 9).

In some vascular tissues with damaged endothelium, exposure to acetylcholine resulted in contraction and elevated levels of cyclic GMP (Holzmann, 1982; Ignarro et al., 1984). The elevated levels of cyclic GMP may be caused by direct effect of acetylcholine on the smooth muscle and thus may function as a feedback inhibitor of the contraction (Schultz et al., 1977; Kukovetz et al., 1982). One might also consider that some functional endothelium may not have been destroyed, however, and/or that release of an endothelium-derived relaxant factor (EDRF) from the vaso vasorum might also account for some of the increase in cyclic GMP. The inhibitor of guanylate cyclase, methylene blue, and removal of extracellular Ca^{2+} have both been shown to inhibit the increased levels of cyclic GMP that accompany the contraction to acetylcholine (Kukovetz et al., 1982; Rapoport and Murad, 1983b; Ignarro et al., 1984). Exposure to methylene blue and removal of extracellular Ca^{2+} have also been shown to inhibit endothelium-dependent relaxations, although relaxations to the nitrovasodilators are also inhibited by methylene blue (*see* section 3). Thus it might be necessary to document complete removal of the endothelium from the vasculature prior to evaluating the relationship between the changes in magnitude of cyclic GMP levels and the degree of contraction.

3. Effects of Inhibitors of Guanylate Cyclase on Endothelium-Dependent Relaxation and Formation of Cyclic GMP

3.1. Effects of Free Radical Scavengers and Reducing Agents

The use of guanylate cyclase inhibitors to investigate the functional role of cyclic GMP in the regulation of smooth muscle contractility originated from the serendipitous observation that nitrovasodilators activate guanylate cyclase (Kimura et al., 1975). From these and other studies it was proposed that the nitric oxide free radical, formed from the nitrovasodilators (Arnold et al., 1977; Katsuki et al., 1977b; Murad et al., 1978), or other pharmacologically and/or physiologically formed free radicals, such as those containing oxygen (Mittal and Murad, 1977a,b; Murad et al., 1978) may be proximal activators of guanylate cyclase. Therefore, reducing agents and free radical scavengers, such as butylated hydroxyanisole, hydroquinone, methylene blue, hemoglobin, methemoglobin, and myoglobin, were successfully used to inhibit guanylate cyclase activation (Kaksuki et al., 1977b; Murad et al., 1978; Mittal et al., 1978). Subsequently, these agents were demonstrated to inhibit the relaxation and the associated increased levels of cyclic GMP caused by the nitrovasodilators (Rapoport and Murad, 1983b).

Based on the effects of the inhibitors of guanylate cyclase on nitrovasodilator-induced relaxation and elevated levels of cyclic GMP, the effects of these inhibitors on endothelium-dependent relaxation were investigated. Methylene blue inhibited acetylcholine-induced relaxation and/or the elevation in cyclic GMP it causes in coronary arteries of rabbits (Griffith et al., 1985) and cattle (Holzmann, 1982), in bovine intrapulmonary arteries (Ignarro et al., 1984, 1986), and in aortas of the rat (Rapoport et al., 1985a; Huang and Lee, 1985) and the rabbit (Furchgott et al., 1984, 1985; Martin et al., 1985a; Bowman et al., 1986). Methylene blue also inhibited the endothelium-dependent relaxation and the associated increase in cyclic GMP caused by arachidonic acid in bovine intrapulmonary arteries (Ignarro et al., 1985). In experiments designed to bioassay EDRF, methylene blue acted at the level of the smooth muscle and not through inhibition of the release or through chemical inactivation of EDRF (Griffith et al., 1985). Hydroquinone, another free radical scavenger, inhibited acetylcholine-induced, endothelium-dependent relaxation in the coronary artery and the aorta of the rabbit (Griffith et al., 1984a). It should be noted, however, that other oxygen-derived free radical scavengers, superoxide dismutase, catalase, and mannitol, have no effect on acetylcholine-induced relaxations in the intact aorta of the rabbit (Silin et al., 1985). In apparent contrast to the results of Silin et al. (1985), others have recently reported that superoxide dismutase and Cu^{2+}, but not catalase, protect EDRF from breakdown as tested in a bioassay system (Rubanyi and Vanhoutte, 1986; Gryglewski et al., 1986). These authors suggested that superoxide dismutase and Cu^{2+} may prevent the degradation of EDRF by superoxide anions.

Myoglobin was as effective as hemoglobin at inhibiting acetylcholine-induced relaxation and the elevation in cyclic GMP in the aorta of the rabbit, whereas the ferric hemoproteins methemoglobin and metmyoglobin were only slightly inhibitory and reduced cytochrome was without effect (Martin et al., 1985b). Similar results were obtained with the nitrovasodilator nitroglycerin. This suggested that the inhibition required ferrous hemoprotein containing ligand binding sites, possibly by preventing the interaction between the activators of guanylate cyclase and an endogenous heme moiety contained within guanylate cyclase (Rapoport and Murad, 1983b).

It is interesting to note that methylene blue, hemoglobin, and myoglobin had greater inhibitory effects on endothelium-dependent relaxation and/or the associated increase in cyclic GMP levels than on those induced by the nitrovasodilators (Furchgott et al., 1984; Martin et al., 1985a,b; Rapoport et al., 1985a; Bowman et al., 1986). The explanation for the greater inhibitor effects of these agents on endothelium-dependent responses is not clear, but they are probably not caused by an inhibitory effect on the release of EDRF or on EDRF itself (Griffith et al., 1985). Alternative explanations are that methylene blue, hemoglobin, and myoglobin interfere with the interaction of the endothelium-derived relaxant factor with the smooth muscle or the possible step(s) between this interaction and activation of guanylate cyclase. It must also be considered that the mechanisms of stimulation of guanylate cyclase by the activators of the enzyme, which result from the nitrovasodilators and the endothelium-derived relaxant factor, may not be identical. Thus methylene blue, hemoglobin, and myoglobin may be more potent inhibitors of activation of guanylate cyclase caused by EDRF than that caused by the nitrovasodilators. This concept is not supported by the nitroglycerin-desensitization studies reviewed below, however. The correct interpretation of these studies awaits further experimentation on the mechanisms of activation of guanylate cyclase.

Other reducing agents, including dithiothreitol, potassium borohydride, sucrose, butylated hydroxytoluene, and α-tocopherol, have also been shown to inhibit endothelium-dependent relaxations and/or the associated increase in cyclic GMP levels in the aorta of the rat (Rapoport et al., 1985a) and the rabbit (Griffith et al., 1984b, 1985; Forstermann and Neufang, 1984). The mechanism by which the reducing agents inhibit the relaxation may be through chemical inactivation of EDRF (Griffith et al., 1984b, 1985). The exact chemical nature of the inactivation is not clear, however, since the structure of EDRF is unknown. Furthermore, ascorbic acid did not inhibit endothelium-dependent relaxation in rat (Rapoport et al., 1985a) and rabbit aorta (Forstermann and Neufang, 1984). Relaxation of the canine coronary artery, however, without endothelium induced by the perfusate from an intact canine femoral artery exposed to acetylcholine was inhibited by ascorbic acid (Rubanyi et al., 1985).

The explanation for these apparently contrasting results is not clear, but it has been suggested that the time of diffusion between the endothelial and smooth muscle cells in vascular preparations with intact endothelium is too rapid for the inactivation of EDRF (Rubanyi et al., 1985). This explanation is not consistent, however, with the observations that other reducing agents were effective at inhibiting endothelium-dependent relaxation in intact vascular preparations (Forstermann and Neufang, 1984; Griffith et al., 1984b, 1985; Rapoport et al., 1985a), although considerable nonspecific effects of the reducing agents on the contraction of the tissues have been reported (Rapoport et al., 1985a).

Under bioassay conditions, norepinephrine and epinephrine also inhibited endothelium-dependent relaxation in the presence of alpha- and beta-adrenergic antagonists (Rubanyi et al., 1985). Sodium nitroprusside-induced relaxations were inhibited by the catecholamines to a much smaller magnitude. This suggested that the inhibition was caused by direct inactivation of EDRF by the catecholamines (Rubanyi et al., 1985). Consistent with these observations are the findings that norepinephrine inhibited the elevated cyclic GMP levels caused by acetylcholine to a much greater magnitude than those elevated in response to sodium nitroprusside (Rapoport et al., 1985a). Isoproterenol also inhibited the increase in cyclic GMP levels caused by acetylcholine or sodium nitroprusside in the presence of alpha- or beta-adrenergic blockers (Rapoport et al., 1985a). These inhibitory effects were reversed by ascorbic acid (Rapoport et al., 1985a). A slight inhibitory effect of ascorbic acid on the acetylcholine-induced increase in cyclic GMP levels was also reported (Rapoport et al., 1985a). Thus, the inhibitory effects of catecholamines on endothelium-dependent relaxation may be caused by their properties as reducing agents.

3.2. Effects of Cyanide

Cyanide inhibits guanylate cyclase activation by the nitrovasodilators (Kimura et al., 1975). Therefore, the effects of cyanide on endothelium-dependent relaxation and/or the associated increase in cyclic GMP levels have been investigated in a number of vascular tissues. Cyanide inhibited endothelium-dependent relaxations and the formation of cyclic GMP evoked by acetylcholine in the aorta of the rat (Rapoport and Murad, 1984) and, more recently, by acetylcholine and the Ca^{2+} ionophore A23187 in the aorta of the rabbit (Zawadzki and Furchgott, 1986). Cyanide was without effect, however, on the acetylcholine-induced relaxations and the elevations of cyclic GMP in bovine intrapulmonary arteries (Ignarro et al., 1986).

The mechanism by which cyanide inhibits endothelium-dependent relaxations is not clear. The production of EDRF may be dependent on oxidative phosphorylation (Griffith et al., 1986). Thus, cyanide may inhibit endothelium-dependent relaxations through inhibition of that process, although an inhibitory effect of cyanide on guanylate cyclase activation cannot be eliminated. This latter explanation is complicated by the observations in the aorta of the rat that relaxation and/or the elevation in cyclic GMP indued by sodium nitroprusside were inhibited to a much greater magnitude by cyanide than those caused by other nitrovasodilators (Rapoport and Murad, 1984). Furthermore, cyanide had no effect on relaxations induced by a variety of nitrovasodilators in aortas of rabbits and in bovine intrapulmonary arteries (Grayling et al., 1978; Kruszyna et al., 1982; Ignarro et al., 1986). It was also suggested that cyanide may have nonspecific effects that inhibit relaxation, since inhibitory responses of the aorta of the rat to isoproterenol, which are mediated through formation of cyclic AMP, were inhibited by cyanide (Rapoport and Murad, 1984).

4. Effects of Inhibitors of Phosphodiesterase on Endothelium-Dependent Relaxation and Formation of Cyclic GMP

The use of phosphodiesterase inhibitors and, in particular, inhibitors of cyclic GMP phosphodiesterase to investigate the role of cyclic GMP in endothelium-dependent relaxation is based on observations with the nitrovasodilators. Nitrovasodilator-induced relaxation and the associated elevation in cyclic GMP levels were potentiated by the phosphodiesterase inhibitor MB 22948 (Kukovetz et al., 1982; Rapoport and Murad, 1983b). MB 22948 also potentiated the acetylcholine-induced, endothelium-dependent relaxation and/or the associated increase in cyclic GMP levels in coronary arteries of cattle (Holzmann, 1982) and rabbits (Griffith et al., 1985). This potentiation was not caused by increased release of EDRF or interaction of MB 22948 with the relaxant factor, as demonstrated using a bioassay system (Griffith et al., 1985).

5. Effects of Inhibitors of Phospholipase A_2, Cyclooxygenase, and Lipoxygenase on Endothelium-Dependent Relaxation and Levels of Cyclic GMP

The effects of inhibitors of phospholipase A_2, cyclooxygenase, and lipoxygenase on endothelium-dependent relaxations in a variety of blood vessels have been previously reviewed (Furchgott, 1983; Rapoport and Murad, 1983b). This section will focus on reports that determined the effects of these inhibitors on cyclic GMP levels and changes in contractility. Because of the known, nonspecific nature of many of these inhibitors, caution should be used in interpretating the results discussed.

It was proposed that the endothelium-dependent inhibitory agonists activate phospholipase A_2, which causes the release of a free fatty acid (Furchgott, 1983). Consistent with this hypothesis were the observations that the inhibitor of phospholipase A_2, quinacrine, inhibited the endothelium-dependent relaxation and the associated elevation in cyclic GMP in the aorta of the rat (Rapoport and Murad, 1983a). However, 10 μM quinacrine partially inhibited acetylcholine-induced relaxation of bovine intrapulmonary arteries without altering the increase in cyclic GMP (Ignarro et al., 1984). Higher concentrations of quinacrine inhibited the latter, but this may be because of nonspecific, inhibitory effects of quinacrine (Rapoport and Murad, 1983a; Ignarro et al., 1984).

The inhibitor of phospholipase A_2, bromo phenacylbromide (*see* Furchgott, 1983), also inhibited the endothelium-dependent relaxations and the increase in cyclic GMP caused by ATP and thrombin in rat aorta, but not those caused by sodium nitroprusside (Rapoport et al., 1984). Exposure of the aorta of the rat to phospholipase A_2 caused an endothelium-independent relaxation that was inhibited by methylene blue (Huang and Lee, 1985). These results may suggest that activation of endogenous phospholipase A_2 could result in the release of product(s) that induce relaxation through formation of cyclic GMP.

Since the activity of phospholipase A_2 requires Ca^{2+} (*see* Furchgott, 1983), the effects of removal of extracellular Ca^{2+} on endothelium-dependent relaxation and the associated increase in cyclic GMP were investigated. In the aorta of the rat, removal of extracellular Ca^{2+} prevented the inhibitory effects of ATP and trypsin on contractions caused by norepinephrine and, in addition, decreased the magnitudes of the elevation in cyclic GMP caused by these endothelium-dependent dilators. In contrast, the magnitude of elevation of cyclic GMP caused by sodium nitroprusside and the inhibitory effect of sodium nitroprusside on the contraction caused by norepinephrine were not altered by the removal of extracellular Ca^{2+} (Rapoport et al., 1984). The endothelium-dependent relaxations of the aorta of the rat to acetylcholine, as well as the endothelium-dependent increase in the levels of cyclic GMP caused by acetylcholine and phenylephrine, depended on extracellular Ca^{2+}, but were not antagonized by the Ca^{2+} channel antagonist flunarizine (Miller et al., 1985). Of course, Ca^{2+} channels of endothelial cells activated by acetylcholine and phenylephrine may be different from Ca^{2+} channels in smooth muscle (Miller et al., 1985).

It has been proposed that endothelium-dependent relaxation was dependent on oxidation of a fatty acid by lipoxygenase (Furchgott, 1983). An inhibitor of lipoxygenase and cyclooxygenase, eicosatetraynoic acid, inhibited acetylcholine-induced endothelium-dependent relaxations and the associated increase in cyclic GMP in the thoracic aorta of the rat (Rapoport and Murad, 1983a). The elevated levels of cyclic GMP caused by sodium nitroprusside, however, were also inhibited by eicosatetraynoic acid in the aorta (Rapoport and Murad, 1983a) and the ductus deferens (Spies et al., 1980) of the rat. Thus, a nonspecific inhibitory effect of eicosatetraynoic acid on the activation of guanylate cyclase cannot be excluded. Indomethacin, an inhibitor of cyclooxygenase, did not alter the magnitudes of cyclic GMP elevation caused by ATP in the aorta of the rat (Rapoport et al., 1984) or by arachidonic acid in the bovine intrapulmonary artery (Ignarro et al., 1985).

6. Effects of Inhibitors of the Na^+,K^+ Pump and Membrane Depolarizing Agents on Endothelium-Dependent Relaxation and Formation of Cyclic GMP

The Na^+,K^+ pump inhibitor ouabain inhibits acetylcholine-induced relaxation and the elevated levels of cyclic GMP in the aorta of the rat (Rapoport et al., 1985b). Other procedures that inhibit the Na^+,K^+ pump, including exposure to K^+-free or Mg^{2+}-free Krebs-Ringer bicarbonate solution, also inhibited acetylcholine-induced relaxations and the associated increase in cyclic GMP (Rapoport et al., 1985b). These agents and procedures also inhibited relaxations and increases in cyclic GMP evoked by sodium nitroprusside (Rapoport and Murad, 1983c; Rapoport et al., 1985c).

The mechanism(s) underlying the inhibitory effects of these agents and procedures on the responses to acetylcholine levels are apparently complex. The inhibition of the relaxation may result from functional antagonism, although an effect of cyclic GMP on the Na^+,K^+ pump cannot be completely ruled out (Rapoport and Murad, 1983c; Rapoport et al., 1985b,c). The observations that the membrane-depolarizing agents KCl and tetraethylammonium also inhibited acetylcholine- and sodium nitroprusside-induced relaxations, as well as the accumulation of cyclic GMP, suggest that the inhibitory effects of the inhibitors of the Na^+,K^+ pump may also be caused by the membrane depolarization associated with inhibition of the Na^+,K^+ pump (Rapoport et al., 1985b,c).

The inhibitory effects of the membrane depolarizing agents and inhibitors of the Na^+,K^+ pump on endothelium-dependent relaxations to acetylcholine and the production of cyclic GMP were greater than those on the response to sodium nitroprusside (Rapoport and Murad, 1983c; Rapoport et al., 1985b,c). Therefore, it is possible that these agents and procedures also inhibit the release and/or formation of EDRF. It has recently been suggested that a Na^+/Ca^{2+} exchange mechanism may play a role in the release of EDRF (Winquist et al., 1985). The greater inhibitory effects of the inhibitors of the Na^+,K^+ pump and the membrane-depolarizing agents on endothelium-dependent relaxation and the production of cyclic GMP may be caused by decreased release of the relaxant factor through inhibition of Na^+/Ca^{2+} exchange of the endothelial cells.

The mechanism by which ouabain decreased the formation of cyclic GMP by acetylcholine and sodium nitroprusside is not simple, since the inhibition was dependent upon the presence of norepinephrine (Rapoport et al., 1985b,c). The regulation of the synthesis of cyclic GMP under these conditions may involve the activation of both the Na^+,K^+ pump and alpha-adrenergic receptors.

7. Effects of Nitroglycerin-Induced Desensitization on Endothelium-Dependent Relaxation and Formation of Cyclic GMP

Prior treatment of vascular tissue in vitro or of intact animals with nitroglycerin induces subsequent inhibition of vascular relaxation to nitroglycerin and of the associated formation of cyclic GMP (Axelsson and Andersson, 1983; Rapoport and Murad, 1983b). The mechanism for the desensitization may be an inhibitory effect on activation of guanylate cyclase and/or an increased activity of phosphodiesterase within the smooth muscle, since both of these phenomena were observed following treatment of rats with nitroglycerin (Axelsson and Andersson, 1983). We have investigated the effects of nitroglycerin treatment both in vitro and in vivo on the acetylcholine-induced relaxation of the aorta of the rat. Prior exposure to nitroglycerin either in vivo or in vitro inhibited both (Rapoport et al., 1987; Molina et al., 1987). The magnitude of the inhibition to the response to acetylcholine was less than that to nitroglycerin. Furthermore, the activation of guanylate cyclase by nitroglycerin, sodium nitroprusside, nitric oxide, and arachidonic acid were reduced in rat aorta desensitized with nitroglycerin (Waldman et al., 1986). These observations are consistent with the hypothesis that nitroglycerin-induced densensitization may be caused, at least in part, by an inhibitory effect on activation of guanylate cyclase.

The smaller magnitude of inhibition of the acetylcholine-induced relaxation may be caused by the possibility that EDRF and nitroglycerin activate guanylate cyclase through different mechanisms. An alternative explanation is that the tolerance to nitroglycerin involves two components: (1) metabolism and conversion of nitroglycerin to nitric oxide and (2) a stable modification in guanylate cyclase with decreased activatability. Presumably the decreased relaxation to endothelium-dependent vasodilators in nitroglycerin-tolerant vessels would only involve the latter mechanism, and this may explain the differences observed. Thus relaxations induced by different vasodilators may be differentially affected in tissues desensitized with nitroglycerin. It must also be considered that treatment with nitroglycerin may inhibit the release and/or formation of EDRF.

8. Effects of Contractile Agents on Cyclic GMP Levels and the Role of the Endothelium

Certain constrictor agonists elevate the levels of cyclic GMP in a variety of preparations containing smooth muscle (Rapaport and Murad, 1983b). These observations were made prior to the knowledge concerning the role of the endothelium. It may be reasonable to predict that at least part of the elevation in cyclic GMP levels in response to the constrictors were caused by the release of a factor(s) from cell types other than smooth muscle. For example, it has recently been demonstrated that removal of the epithelial cells and/or connective tissue from tracheal preparations increased the magnitudes of contraction to a number of agents (Flavahan et al., 1985; Barnes et al., 1985; Mitchell and Yu, 1985; Goldie et al., 1986; Holroyde, 1986).

Adrenergic agonists induce endothelium-dependent increases in the levels of cyclic GMP. Thus the alpha$_2$-adrenergic agonists B-HT 920 and clonidine elevated the levels of cyclic GMP twofold in the aorta of the rat with intact endothelium and induced relatively small contraction (Bigaud et al., 1984; Miller et al., 1984). Removal of the endothelium abolished the increases in cyclic GMP levels and potentiated the maximal contractions and/or increased the sensitivity to the alpha$_2$-adrenergic agonists. In contrast, the alpha$_1$-adrenergic agonist methoxamine induced an endothelium-dependent increase in cyclic GMP of approximately 1.5-fold in the aorta of the rat, whereas the contractile response to methoxamine was not potentiated upon removal of the endothelium (Bigaud et al., 1984). The authors suggested that there may be a dissociation between endothelium-mediated inhibition of maximal contractile responses and magnitude of the increase in cyclic GMP. This suggestion awaits further investigation; it should be noted, however, that it is difficult to establish whether elevations of 1.5-fold and twofold are or are not significantly different. Furthermore, the potential quantitative relationship between the levels of cyclic GMP and the degree of inhibition of maximal contraction is not clear.

Exposure of certain vascular preparations with intact endothelium to inhibitors of guanylate cyclase, including methylene blue, hydroquinone, hemoglobin, and myoglobin, potentiate contractile responses to agonists. These agonists include histamine and serotonin in the coronary artery of the rabbit (Griffith et al., 1984a), acetylcholine in bovine coronary arteries (Kukovetz et al., 1982), norepinephrine (Huang and Lee, 1985) and B-HT 920 in the aorta of the rat (Miller et al., 1984), norepinephrine, phenylephrine, KCl, and serotonin in the aorta of the rabbit (Griffith et al., 1984a; Martin et al., 1985a,b), and acetylcholine, KCl, phenylephrine, and U46619 in bovine intrapulmonary arteries (Ignarro et al., 1984, 1986). Cyanide potentiates the contractile responses to norepinephrine in the aorta of the rabbit with endothelium (Zawadzki and Furchgott, 1986), but not in intact bovine intrapulmonary arteries (Ignarro et al., 1986). In addition, MB 22948, the inhibitor of phosphodiesterase, inhibited the serotonin-induced contractions in the aorta of the rabbit, which were reversed by methylene blue (Griffith et al., 1985).

The mechanism(s) responsible for the endothelium-dependent potentiation by inhibitors of guanylate cyclase of the contractile responses to various agonists presumably involves several factors. Many of the potentiated agonists induce little, if any, endothelium-dependent relaxation in the tissue where the potentiation is observed. Other agonists induce considerable amounts of endothelium-dependent relaxation. Thus, the relative potency of an agonist and its affinity for receptors located on the smooth muscle cells and on the endothelial cells may determine the magnitude of the potentiation of the contractile response by the inhibitors of guanylate cyclase. Inhibitors of guanylate cyclase or the removal of the endothelium increases the basal tone of the smooth muscle (*see* section 9). Thus this magnitude may also be influenced by the amount of increase in basal tone caused by the inhibitors (Griffith et al., 1984a; White et al., 1986).

9. Role of the Endothelium and Cyclic GMP in the Regulation of Basal Tone

It appears that the endothelium plays an important role in the regulation of basal tone. We hypothesized that the endothelial cells continually release a factor that regulates basal levels of cyclic GMP and smooth muscle tone (Rapoport and Murad, 1983a). This hypothesis was based on observations that eicosatetraynoic acid decreased the basal levels of cyclic GMP in the aorta of the rat. Cyclic GMP levels were also decreased in the same blood vessel after removal of endothelium (Rapoport and Murad, 1983a); this has been observed by others in a variety of blood vessels (Holzmann, 1982; Diamond and Chu, 1983; Furchgott et al., 1984; Winquist et al., 1984; Coburn et al., 1986), but not by all (Ignarro et al., 1984).

Also consistent with the hypothesis that the endothelium regulates basal tone through continual release of a factor that alters cyclic GMP levels (Rapoport and Murad, 1983a) are the observations that the inhibitors of guanylate cyclase, methylene blue, hemoglobin, hemolysate, and hydroquinone, decreased basal levels of cyclic GMP and/or induced contraction in vascular tissue with intact endothelium (Holzmann, 1982; Griffith et al., 1984b, 1985; Ignarro et al., 1984, 1986; Martin et al., 1985a; Rapoport et al., 1985a). The contractile response of the aorta to methylene blue was abolished, however, by pretreating the rabbits with reserpine, which suggests that the contraction induced by methylene blue was caused by the release of norepinephrine from intramural nerves (Martin et al., 1985a). The observations that the MB 22948 induced endothelium-dependent relaxation in the aorta of the rabbit, which was inhibited by methylene blue (Griffith et al., 1985a), also support the hypothesis that the endothelium alters basal tone, at least in part, through regulation of the metabolism of cyclic GMP. These observations are further supported by our findings that aortas with an intact endothelium exhibited an increase in cyclic GMP-dependent protein kinase activity (Fiscus et al., 1985).

10. Conclusions

It is apparent that the endothelium exerts a major form of control over vascular contractility through the regulation the metabolism of cyclic GMP within smooth muscle. The regulation occurs in the form of a factor(s) that is released from the endothelium spontaneously and in response to a variety of agonists. The agonist-induced release of endothelium-derived relaxant factor(s) can cause relaxation or inhibit the magnitude of contraction. The spontaneously released factor(s) apparently functions to decrease basal tone. Thus, removal of this factor(s) increase tone, which, in turn, enhances vasoconstrictor responses and, possibly, inhibits vasodilatation. Therefore vascular disorders can be predicted in a variety of disease states in which the function of the endothelial cell is comprised.

There are a number of important basic questions that remain unanswered. The exact chemical nature of the endothelium-derived relaxing factor needs to be established. Following this, the mechanism by which this factor activates guanylate cyclase will be more readily elucidated. Preliminary steps in this direction have been recently accomplished through the demonstration that a factor released from the endothelium can activate guanylate cyclase (Forstermann et al., 1986).

A major question that needs to be answered concerns the mechanism by which cyclic GMP induces relaxation. We have demonstrated that agents that elevate levels of cyclic GMP, the endothelium-dependent vasodilators, nitrovasodilators, and atrial natriuretic factor, activate cyclic GMP-dependent protein kinase in rat aorta (Fiscus et al., 1984, 1985). Furthermore, removal of the endothelium decreased the basal activity of cyclic GMP-dependent protein kinase (Fiscus et al., 1984). Thus, cyclic GMP-dependent protein kinase may induce the phosphorylation/dephosphorylation of proteins that regulate the contractility of smooth muscle in a primary or secondary manner. In this regard, we have demonstrated that the endothelium-dependent and nitrovasodilators induce similar patterns of protein phosphorylation in intact aorta of the rat which were mimicked by 8-bromo cyclic GMP (Rapoport et al., 1982, 1983a,b, 1984).

One of the proteins dephosphorylated by the agents that elevate cyclic GMP was myosin light chain, which suggests that cyclic GMP may inhibit the development of tone by inhibiting phosphorylation of myosin light (Rapoport et al., 1983a,b; Draznin et al., 1986). The mechanism underlying this dephosphorylation of myosin light chain is unknown. It could result from decreased myosin light chain kinase activity and/or increased phosphatase activity; cyclic GMP-dependent kinase had no effect on activation of myosin light chain kinase (Nishikawa et al., 1984; Hathaway et al., 1985). In cultured vascular smooth muscle cells, nitroglycerin reduces

Ca^{2+} flux, which could be caused by the ability of cyclic GMP to decrease intracellular Ca^{2+} levels (Kobayashi et al., 1985). The lower levels of intracellular Ca^{2+} would decrease myosin light chain kinase activity and, thus, phosphorylation of myosin light chain. Other studies, however, have demonstrated that agents elevating the levels of cyclic GMP induced relaxation when myosin light chain phosphorylation had returned to basal levels (Gerthoffer et al., 1984).

Acknowledgments

This work was supported by grants to RMR from the National Institutes of Health (HL 34895), the University of Cincinnati (BRSG 507 RROS-408-24), the American Heart Association, Southwestern Ohio Chapter, and the Veterans Administration and to FM from the National Institutes of Health (AM 30787 and HL 28474), the Council for Tobacco Research, USA, Inc., and the Veterans Administration. We would like to thank Liz Wendelmoot and Kathryn Siefert for help in the preparation of this manuscript.

References

Arnold, W. P., Mittal, C. K., Katsuki, S., and Murad, F.: Nitric oxide activates guanylate cyclase and increases guanosine 3′:5′-cyclic monophosphate levels in various tissue preparations. *Proc. Natl. Acad. Sci. USA* **74:**3203–3207, 1977.

Axelsson, K. L. and Andersson, R. G. G.: Tolerance towards nitroglycerin, induced in vivo, is correlated to a reduced cGMP response and an alteration in cGMP turnover. *Eur. J. Pharmacol.* **88:**71–79, 1983.

Barnes, P. J., Cuss, F. M., and Palmer, J. B.: The effect of airway epithelium on smooth muscle contractility in bovine trachea. *Br. J. Pharmacol.* **86:**685–691, 1985.

Bigaud, M., Schoeffter, P., Stoclet, J. C., and Miller, R. C.: Dissociation between endothelium-mediated increases in tissue cGMP levels and modulation of aortic contractile responses. *Naunyn Schmiedebergs Arch. Pharmacol.* **328:**221–223, 1984.

Bowman, A., Gillespie, J. S., and Soarres-da-Silva, P.: A comparison of the action of the endothelium-derived relaxant factor and the inhibitory factor from the bovine retractor penis on rabbit aortic smooth muscle. *Br. J. Pharmacol.* **87:**125–181, 1986.

Coburn, R. F., Eppinger, R., and Scott, D. P.: Oxygen-dependent tension in vascular smooth muscle. Does the endothelium play a role? *Circ. Res.* **58:**341–347, 1986.

Diamond, J. and Blisard, K. S.: Effects of stimulant and relaxant drugs on tension and cyclic nucleotide levels in canine femoral artery. *Mol. Pharmacol.* **12:**688–692, 1976.

Diamond, J. and Chu, E. B.: Possible role for cyclic GMP in endothelium-dependent relaxation of rabbit aorta by acetylcholine. Comparison with nitroglycerin. *Res. Comm. Chem. Path. Pharmacol.* **41:**369–381, 1983.

Draznin, M. B., Rapoport, R. M., and Murad, F.: Myosin light chain dephosphorylation by sodium nitroprusside treatment of intact rat thoracic aorta: Role of cyclic GMP. *Intl. J. Biochem.* **18:**917–928, 1986.

Dunham, E. W., Haddox, M. K., and Goldberg, N. D.: Alteration of vein cyclic 3′:5′ nucleotide concentrations during changes in contractility. *Proc. Natl. Acad. Sci. USA* **71:**815–819, 1974.

Fiscus, R. R., Rapoport, R. M., and Murad, F.: Endothelium-dependent and nitrovasodilator-induced activation of cyclic GMP-dependent protein kinase in rat aorta. *J. Cyclic Nucleotide Protein Phosphor. Res.* **9:**415–425, 1984.

Fiscus, R. R., Rapoport, R. M., Waldman, S. A., and Murad, F.: Atriopeptin II elevates cyclic GMP, activates cyclic GMP-dependent protein kinase and causes relaxation in rat thoracic aorta. *Biochim. Biophys. Acta.* **846:**179–184, 1985.

Flavahan, N. A., Aarhus, L. L., Rimele, T. J., and Vanhoutte, P. M.: The respiratory epithelium inhibits bronchial smooth muscle tone. *J. Appl. Physiol. 58:*834–838, 1985.

Forstermann, U. and Neufang, B.: The endothelium-dependent relaxation of rabbit aorta: Effects of antioxidants and hydroxylated eicosatetraenoic acids. *Br. J. Pharmacol.* **82:**765–767, 1984.

Forstermann, U., Mulsch, A., Bohme, E., and Busse, R.: Stimulation of soluble guanylate cyclase by an acetylcholine-induced endothelium-derived factor from rabbit and canine arteries. *Circ. Res.* **58:**531–538, 1986.

Furchgott, R. F.: Role of endothelium in response of vascular smooth muscle: Role of cyclic GMP. *Circ. Res.* **53:**557–573, 1983.

Furchgott, R. F. and Zawadzki, J. V.: The obligatory role of endothelial cells in the relaxation of arterial smooth muscle by acetylcholine. *Nature* (Lond.) **288:**373–376, 1980.

Furchgott, R. F., Cherry, P. D., Zawadzaki, J. V., and Jothianandan, D.: Endothelial cells as mediators of vasodilatation of arteries. *J. Cardiovas. Pharmacol.* **6** (suppl 2):S336–S343, 1984.

Furchgott, R. F., Jothianandan, D., and Cherry, P. D.: Endothelium-Dependent Responses: The Last Three Years, in *Vasodilatation Mechanisms* (Vanhoutte, P. and Vatner, S. F., eds.) Karger, Basel, New York, 1985.

Gerthoffer, W. T., Trevethick, M A., and Murphy, R. A.: Myosin phosphorylation and cyclic adenosine 3′,5′-monophosphate in relaxation of arterial smooth muscle by vasodilators. *Circ. Res.* **54:**83–89, 1984.

Goldie, R. G., Papadimitriou, J. M., Paterson, J. W., Rigby, P. J., Self, H. M., and Spina, D.: Influence of the epithelium on responsiveness of guinea-pig isolated trachea to contractile and relaxant agonists. *Br. J. Pharmacol.* **87:**5–14, 1986.

Grayling, G. W., Miller, E. F., and Peach, M. J.: Sodium cyanide antagonism of the vasodilator action of sodium nitroprusside in the isolated rabbit aortic strip. *Anesthesiology* **49:**21–25, 1978.
Griffith, T. M., Edwards, D. H., Lewis, M. J., and Henderson, A. H.: Evidence that cyclic guanosine monophosphate (cGMP) mediates endothelium-dependent relaxation. *Eur. J. Pharmacol.* **112:**195–202, 1985.
Griffith, T. M., Henderson, A. H., Edwards, D. H., and Lewis, M. J.: Isolated perfused rabbit coronary artery and aortic strip preparations: The role of the endothelium-derived relaxant factor. *J. Physiol.* (Lond.) **351:**13–24, 1984a.
Griffith, T. M., Edwards, D. H., Lewis, M. J., Newby, A. C., and Henderson, A. H.: The nature of endothelium-derived vascular relaxant. *Nature* (Lond.) **308:**645–647, 1984b.
Griffith, T. M., Edwards, D. H., Newby, A. C., Lewis, M. J., and Henderson, A. H.: Production of endothelium-derived relaxant factor is dependent on oxidative phosphorylation and extracellular Ca^{2+}. *Cardiovas. Res.* **20:**7–12, 1986.
Gryglewski, R. J., Palmer, R. M. J., and Moncada, S.: Superoxide anion is involved in the breakdown of endothelium-derived vascular relaxing factor. *Nature* (Lond.) **320:**454–456, 1986.
Hathaway, D. R., Konick, M. V., and Codican, S. A.: Phosphorylation of myosin light chain kinase from vascular smooth muscle by cAMP- and cGMP-dependent protein kinases. *J. Mol. Cell. Cardiol.* **17:**841–850, 1985.
Holroyde, M. D.: The influence of epithelium on the responsiveness of guinea-pig isolated trachea. *Br. J. Pharmacol.* **87:**501–507, 1986.
Holzmann, S.: Endothelium-induced relaxation by acetylcholine associated with larger rises in cyclic GMP in coronary arterial strips. *J. Cyclic Nucleotide Res.* **8:**409–419, 1982.
Huang, H. -C. and Lee, C. Y.: Relaxant effect of phospholipase A_2 from *Vipera russeli* snake venom on rat aorta. *Eur. J. Pharmacol.* **118:**139–146, 1985.
Ignarro, L. J., Burke, T. M., Wood, K. S., Wolin, M. S., and Kadowitz, P. J.: Association between cyclic GMP accumulation and acetylcholine-elicited relaxation of bovine intrapulmonary artery. *J. Pharmacol. Exp. Ther.* **288:**682–690, 1984.
Ignarro, L. J., Harbison, R. G., Wood, K. S., and Kadowitz, P. J.: Dissimilarities between methylene blue and cyanide on relaxation and cyclic GMP formation in endothelium-intact intrapulmonary artery caused by nitrogen-oxide containing vasodilators and acetylcholine. *J. Pharmacol. Exp. Ther.* **236:**30–36, 1986.
Ignarro, L. J., Harbison, R. G., Wood, K. S., Wolin, M. S., McNamara, D. B., Hyman, A. L., and Kadowitz, P. J.: Differences in responsiveness of intrapulmonary artery and vein to arachidonic acid: Mechanism of arterial relaxation involves cyclic guanosine 3′:5′monophosphate and cyclic adenosine 3′:5′-monophosphate. *J. Pharmacol. Exp. Ther.* **233:**560–569, 1985.

Katsuki, S. and Murad, F.: Regulation of adenosine cyclic 3′,5′-monophosphate and guanosine cyclic 3′,5′-monophosphate levels and contractility in bovine tracheal smooth muscle. *Mol. Pharmacol.* **13:**330–341, 1977.

Katsuki, S., Arnold, W. P., and Murad, F.: Effects of sodium nitroprusside, nitroglycerin, and sodium azide on levels of cyclic nucleotides and mechanical activity of various tissues. *J. Cyclic Nucleotide Res.* **3:**239–247, 1977a.

Katsuki, S., Arnold, W. P., Mittal, C. K., and Murad, F.: Stimulation of guanylate cyclase by sodium nitroprusside, nitroglycerin and nitric oxide in various tissue preparations and comparison to the effects of sodium azide and hydroxylamine. *J. Cyclic Nucleotide Res.* **3:**23–25, 1977b.

Kimura, H., Mittal, C. K., and Murad, F.: Activation of guanylate cyclase from rat liver and other tissues by sodium azide. *J. Biol. Chem.* **250:**8016–8022, 1975.

Kobayashi, S., Kanaide, H., and Nakamura, M.: Cytosolic-free calcium transients in cultured vascular smooth muscle cells: Microfluorometric measurements. *Science* **229:**553–556, 1985.

Kruszyna, H., Kruszyna, M. A., and Smith, R. P.: Nitroprusside increases cyclic guanylate monophosphate concentrations during relaxation of rabbit aortic strips and both effects are antagonized by cyanide. *Anesthesiology* **57:**303–308, 1982.

Kukovetz, W. R., Holzmann, S., and Poch, G.: Function of cyclic GMP in acetylcholine-induced contraction of coronary smooth muscle. *Naunyn Schmiedebergs Arch. Pharmacol.* **319:**29–33, 1982.

Lee, T. P., Kuo, J. F., and Greengard, P.: Role of muscarinic cholinergic receptors in regulation of guanosine 3′,5′-cyclic monophosphate content in mammalian brain, heart muscle and intestinal smooth muscle. *Proc. Natl. Acad. Sci. USA* **69:**3287–3291, 1972.

Loeb, A. L., Owens, G. K., and Peach, M. J.: Evidence for endothelium-derived relaxing factor in cultured cells. *Hypertension* **7:**804–807, 1985.

Martin, W., Villani, G. M., Jothianandan, D., and Furchgott, R. F.: Selective blockade of endothelium-dependent and glyceryl trinitrate-induced relaxation by hemoglobin and by methylene blue in the rabbit aorta. *J. Pharmacol. Exp. Ther.* **232:**708–716, 1985a.

Martin, W., Villani, G. M., Jothianandan, D., and Furchgott, R. F.: Blockade of endothelium-dependent and glyceryl trinitrate-induced relaxation of rabbit aorta by certain ferrous hemoproteins. *J. Pharmacol. Exp. Ther.* **233:**679–685, 1985b.

Miller, R. C., Schoeffter, P., and Stoclet, J. C.: Insensitivity of calcium-dependent endothelial stimulation in rat isolated aorta to the calcium entry blocker, flunarizine. *Br. J. Pharmacol.* **85:**481–487, 1985.

Miller, R. C., Mony, M., Schini, V., Schoeffter, P., and Stoclet, J. C.: Endothelial mediated inhibition of contraction and increase in cyclic GMP levels evoked by the α-adrenoceptor agonist B-HT 920 in rat isolated aorta. *Br. J. Pharmacol.* **83:**903–908, 1984.

Mitchell, H. W. and Yu, L. L.: Attenuation of tracheal smooth muscle contraction by connective tissue. *Eur. J. Pharmacol.* **118:**171–174, 1985.

Mittal, C. K. and Murad, F.: Activation of guanylate cyclase by superoxide dismutase and hydroxyl radical: A physiological regulator of guanosine 3′,5′-monophosphate formation. *Proc. Natl. Acad. Sci. USA* **74:**4360–4364, 1977a.

Mittal, C. K. and Murad, F.: Properties and oxidative regulation of guanylate cyclase. *J. Cyclic Nucleotide Res.* **3:**381–391, 1977b.

Mittal, C. K., Arnold, W.P, and Murad, F.: Characterization of protein inhibitors of guanylate cyclase activation from rat heart and bovine lung. *J. Biol. Chem.* **253:**1266–1271, 1978.

Molina, C. R., Andresen, J. W., Rapoport, R. M., Waldman, S. A., and Murad, F.: The effect of in vivo nitroglycerin therapy on endothelium-dependent and independent relaxation and cyclic GMP accumulation in rat aorta. *J. Cardiovasc. Pharmacol.*, in press, 1987.

Murad, F., Mittal, C. K., Arnold, W. P., Katsuki, S., and Kimura, H.: Guanylate cyclase: Activation by azide, nitro compounds, nitric oxide, and hydroxyl radical and inhibition by hemoglobin and myoglobin. *Adv. Cyclic Nucl. Res.* **9:**145–158, 1978.

Nishikawa, M., de Lanerolle, P., Lincoln, T. M., and Adelstein, R. S.: Phosphorylation of mammalian myosin light chain kinase by the catalytic subunit of cyclic AMP-dependent protein kinase and by cyclic GMP-dependent protein kinase. *J. Biol. Chem.* **259:**8429–8436, 1984.

Rapoport, R. M. and Murad, F.: Effect of cyanide on nitrovasodilator-induced relaxation, cyclic GMP accumulation and guanylate cyclase activation in rat aorta. *Eur. J. Pharmacol.* **104:**61–70, 1984.

Rapoport, R. M. and Murad, F.: Agonist-induced endothelium-dependent relaxation in rat thoracic aorta may be mediated through cyclic GMP. *Circ. Res.* **52:**352–357, 1983a.

Rapoport, R. M. and Murad, F.: Endothelium-dependent and nitrovasodilator-induced relaxation of vascular smooth muscle: Role of cyclic GMP. *J. Cyclic Nucleotide Protein Phosphor. Res.* **9:**281–296, 1983b.

Rapoport, R. M. and Murad, F.: Effect of ouabain and alterations in potassium concentration on relaxation induced by sodium nitroprusside. *Blood Vess.* **20:**255–264, 1983c.

Rapoport, R. M., Draznin, M. B., and Murad, F.: Mechanism of adenosine triphosphate-, thrombin- and trypsin-induced relaxation in rat thoracic aorta. *Circ. Res.* **55:**468–471, 1984.

Rapoport, R. M., Waldman, S. A., Schwartz, K., Winquist, R. J., and Murad, F.: Effects of atrial natriuretic factor, sodium nitroprusside, and acetylcholine on cyclic GMP levels and relaxation in rat aorta. *Eur. J. Pharmacol.* **115:**219–229, 1985a.

Rapoport, R. M., Schwartz, K., and Murad, F.: Effects of Na+, K+-pump inhibitors and membrane depolarizing agents on acetylcholine-induced endothelium-dependent relaxation and cyclic GMP accumulation in rat aorta. *Eur. J. Pharmacol.* **110:**203–209, 1985b.

Rapoport, R. M., Schwartz, K. and Murad, F.: Effect of Na+K+-pump inhibitors and membrane depolarizing agents on sodium nitroprusside-induced relaxation and cyclic GMP accumulation in rat aorta. *Circ. Res.* **57:**164–170, 1985c.

Rapoport, R. M., Draznin, M. B., and Murad, F. : Endothelium-dependent relaxation in rat thoracic aorta may be mediated through cyclic GMP-dependent protein phosphorylation. *Nature* (Lond.) **306:**174–176, 1983a.

Rapoport, R. M., Draznin, M. B., and Murad, F.: Nitrovasodilator- and endothelium-dependent vasodilator-induced relaxation may be mediated through cyclic GMP formation and cyclic GMP-dependent protein phosphorylation. *Trans. Assoc. Am. Phys.* **96:**19–30, 1983b.

Rapoport, R. M., Draznin, M. B., and Murad, F.: Sodium nitroprusside-induced protein phosphorylation in intact rat aorta is mimicked by 8-bromo cyclic GMP. *Proc. Natl. Acad. Sci. USA* **79:**6470–6474, 1982.

Rapoport, R. M., Waldman, S. A., Ginsburg, R., Molina, C., and Murad, F.: Effect of nitroglycerin on endothelium-dependent and independent relaxation and cyclic GMP levels in rat aorta and human coronary artery. *J. Cardiovascular Pharmacol.* **10:**82–89, 1987.

Rubanyi, G. M. and Vanhoutte, P. M.: Superoxide anions and hyperoxia inactivate endothelium-derived relaxing factor(s). *Am. J. Physiol.* **250:**H822–H827, 1986.

Rubanyi, G. M., Lorenz, R. R., and Vanhoutte, P. M.: Bioassay of endothelium-derived relaxing factor(s): Inactivation by catecholamines. *Am. J. Physiol.* **249:**H95–H101, 1985.

Schultz, K.-D., Schultz, K., and Schultz, G.: Sodium nitroprusside and other smooth muscle relaxants. *Nature* (Lond.) **265:**750–751, 1977.

Shirasaki, Y. and Su, C.: Endothelium removal augments vasodilation by sodium nitroprusside and sodium nitrite. *Eur. J. Pharmacol.* **114:**93–96, 1985.

Silin, P. J., Strulowitz, J. A., Wolin, M. S., and Belloni, F. L.: Absence of a role for superoxide anion, hydrogen peroxide and hydroxyl radical in endothelium-mediated relaxation of rabbit aorta. *Blood Vess.* **22:**65–73, 1985.

Spies, C., Schultz, K.-D., and Schultz, G.: Inhibitory effects of mepacrine and eicosatetraynoic acid on cyclic GMP levels caused by calcium and hormonal factors in rat ductus deferens. *Naunyn Schmiedebergs Arch. Pharmacol.* **306:**1–9, 1980.

Waldman, S. A., Rapoport, R. M., Ginsburg, R., and Murad, F.: Molecular mechanisms regulating nitroglycerin-induced densensitization in vascular smooth muscle in rat and human. *Biochem. Pharmacol.* **35:**3525–3531, 1986.

White, D. G., Lewis, M. J., Griffith, T. M., Edwards, T. M., and Henderson, A. H.: Influence of endothelium on drug-induced relaxation of the rabbit aorta. *Eur. J. Pharmacol.* **121:**19–23, 1986.

Winquist, R. J., Bunting, P. B., and Schofield, T. L.: Blockade of endothelium-dependent relaxation by the amiloride analog dichlorobenzamil: Possible role of Na+/Ca++ exchange in the release of endothelium-derived relaxant factor. *J. Pharmacol. Exp. Ther.* **235:**644–650, 1985.

Winquist, R. J., Faison, E. P., Waldman, S. A., Schwartz, K., Murad, F., and Rapoport, R. M.: Atrial natriuretic factor elicits an endothelium independent relaxation and activates particulate guanylate cyclase in vascular smooth muscle. *Proc. Natl. Acad. Sci. USA* **81:**7661–7664, 1984.

Zawadzki, J. V. and Furchgott, R. F.: Endothelium-dependent effects of cyanide on mechanical tension in rabbit aorta. *Fed. Proc.* **45:**669, 1986.

Chapter 12

Modulation by the Endothelium of Agonist-Induced Contractions of Vascular Smooth Muscle

R. C. Miller, V. Schini, and P. Schoeffter

1. Introduction

Since the first report of the importance of endothelium for the relaxing effects of acetylcholine in blood vessels (Furchgott and Zawadzki, 1980), the role played by endothelial cells in the pharmacological and physiological responses of the vascular smooth muscle of all types of blood vessels has come under close scrutiny.

The object of this review is to summarize, first, the known modulatory effects of endothelium on agonist-induced contractile responses, particularly alpha-adrenergic agonists, and second, to examine the possible influence of endothelium-derived factor(s) (EDRF; Furchgott, 1984) on agonist-stimulated mobilization of Ca^{2+} necessary for the contraction of smooth muscle.

2. Modulatory Effect of Endothelium on Agonist-Induced Contractile Responses

As well as the endothelium-dependent relaxations to a variety of agonists (Furchgott, 1984), it has been shown that agonist-induced contractions of many isolated blood vessels are modulated by the presence of endothelium (Table 1). In the aorta of the rat, concentration–effect curves evoked

Table 1a
Modulation of Agonist-Induced Contractile Responses in Rat Aorta by the Endothelium[a]

Agonist	EC_{50}, n*M*		+ENDO/−ENDO	Increase in max resp −ENDO, %	Reference
	−ENDO	+ENDO			
Norepinephrine	0.65	31.8	48.9	NS	Allan et al., 1983
Norepinephrine	5.48	53.1	9.7	NS	Carrier and White, 1985
Norepinephrine	6.7	32.0	4.8	NS	Egleme et al., 1984a
Norepinephrine	0.28	4.8	17.1	29	Fortes et al., 1983
Norepinephrine	12.0	26.3	2.2	NS	Konishi and Su, 1983
Norepinephrine	2.3	20.0	8.7	NS	Lues and Schumann, 1984
Norepinephrine	6.7	8.7	1.3	30	Murakami et al., 1985
Guanfacine	87.0	340.0	3.9	33	Lues and Schumann, 1984
Guanfacine	210.0	700.0	3.3	20	Godfraind et al., 1985
Phenylephrine	7.48	16.6	2.2	NS	Allan et al., 1983
Phenylephrine	15.7	76.5	4.9	99	Carrier and White, 1985
Phenylephrine	24.0	118.0	4.9	NS	Miller and Stoclet, 1985
Phenylephrine	12.0	47.0	3.9	NS	Furchgott et al., 1985
Phenylephrine	10.0	41.0	4.1	52	Godfraind et al., 1985
Methoxamine	440.0	2460.0	5.6	108	Carrier and White, 1985
Methoxamine	260.0	2200.0	10.5	41	Lues and Schumann, 1984
Methoxamine	1700.0	8100.0	4.8	−27	Bigaud et al., 1984
Cirazoline	3.7	24.0	6.5	81	Lues and Schumann, 1984
ST 587	18.2	850.0	47.0	967	Lues and Schumann, 1984
Sgd 101/75	130.0	890.0	6.8	125	Lues and Schumann, 1984
Oxymetazoline	470.0	1000.0	2.1	633	Godfraind et al., 1985
UK 14,304	260.0	850.0	3.3	1015	Godfraind et al., 1985
B-HT 920	4600.0	9010.0	1.95	1639	Carrier and White, 1985
B-HT 920	1300.0				Lues and Schumann, 1984

B-HT 920	1300.0	1100.0	NS	1100	Miller et al., 1984
Clonidine	19.0	110.0	5.8	900	Egleme et al., 1984a
Clonidine	50.4	530.0	10.5	517	Carrier and White, 1985
Clonidine	40.0	170.0	4.3	280	Furchgott et al., 1985
Clonidine	10.7	110.0	10.0	580	Lues and Schumann, 1984
Clonidine	20.0			800	Bigaud et al., 1984
Clonidine	292.0				Murakami et al., 1985
Idazoxan	130.0			Sig	Lues and Schumann, 1985
Angiotensin I	0.96	4.84	5.0	Sig	Allan et al., 1983
Angiotensin II	0.43	2.14	5.0	Sig	Allan et al., 1983
Prostaglandin $F_{2\alpha}$			4.0	23	Egleme et al., 1984b

[a]Abbreviations: max resp, maximum response; NS, not significant; Sig, significant.

Table 1b
Modulation of Agonist-Induced Contractile Responses in Blood Vessels by the Endothelium[a]

Agonist	EC_{50}, n*M*				
	−ENDO	+ENDO	+ENDO/−ENDO	Increase in max resp −ENDO, %	Reference
Rabbit aorta					
Norepinephrine	97.3	1100.0	11.30	NS	Murakami et al., 1985
Norepinephrine			NS		Spokas et al., 1983
Phenylephrine			NS		Singer and Peach, 1983a
Phenylephrine	200.0		NS	NS	Singer and Peach, 1983b
Phenylephrine	64.0	56.0	NS	−14	Griffith et al., 1984
Serotonin	130.0	160.0	NS	−15	Griffith et al., 1984
Rat mesenteric artery					
Norepinephrine	9.81	118.0	12.0	203	White and Carrier, 1986
Methoxamine	2680.0	9300.0	3.5	300	White and Carrier, 1986
Clonidine	114.0	297.0	2.6	1200	White and Carrier, 1986
Rabbit ear artery					
Norepinephrine	150.0	280.0	1.9	≈100	Sercombe et al., 1985
Norepinephrine			NS	35	Verrecchia et al., 1985
Rat femoral artery					
Norepinephrine	100.0	450.0	4.5	NS	Konishi and Su, 1983
Dog femoral vein					
Norepinephrine	36.3	87.1	2.4	NS	Miller and Vanhoutte, 1985
Dog pulmonary artery					
Norepinephrine	67.6	186.2	2.75	NS	Miller and Vanhoutte, 1985
Dog pulmonary vein					
Norepinephrine	24.0	186.2	7.76	68	Miller and Vanhoutte, 1985

Dog carotid artery					
Norepinephrine	310.0	640.0	2.1		D'Orleans-Juste et al., 1985
Cat cerebral artery					
Norepinephrine	770.0	1240.0	NS	≈340	Sercombe et al., 1985
Pig coronary artery					
Norepinephrine			>3	233	Cocks and Angus, 1983
Dog coronary artery					
Norepinephrine			>2	130–300	Cocks and Angus, 1983
Rabbit coronary artery					
Phenylephrine	(3.6 m*M*)		>1	≈275	Griffith et al., 1984
Serotonin	400.0	0			Griffith et al., 1984

[a]See Table 1a for abbreviations.

by norepinephrine, for example, are displaced to the left in the absence of endothelium in comparison to responses obtained in the presence of endothelium, often without a major change in the magnitude of the maximal responses (Fortes et al., 1983; Allan et al., 1983; Egleme et al., 1984a; Carrier and White, 1985; Lues and Schumann, 1984). The contractile response to clonidine, on the other hand, is virtually absent in the presence of endothelium, whereas in the absence of endothelium, the drug elicits contractions comparable to those evoked by norepinephrine (Egleme et al., 1984a; Carrier and White, 1985). In the aorta of the rat, contractile responses induced by agonists, such as prostaglandin $F_{2\alpha}$, serotonin, angiotensin II, and other alpha-adrenoceptor agonists usually classified as selective $alpha_1$ or as nonselective (phenylephrine, methoxamine, epinephrine), are modulated by the endothelium somewhat like responses to norepinephrine rather than responses to clonidine (Egleme et al., 1984b; Bigaud et al., 1984; Miller and Stoclet, 1985; Godfraind et al., 1985; Oshiro et al., 1985). It must be noted, however, that although all investigators have described a shift to the left of concentration–response curves evoked by alpha-adrenoceptor agonists in this tissue, reports on modulation of maximal contractions vary, and maximal contractions elicited by all agonists listed in Table 1a have been noted to increase dramatically under some circumstances in the absence of endothelium. These contractile studies were usually performed in parallel on two segments of the same blood vessel, one with and one without endothelium, and the comparative variability of the relative maximal contractions to agonists may be caused by the differing amounts of endothelium in the tissues from which it was not deliberately removed.

Modulation of contractions induced by agonists have also been described in other isolated blood vessels (Table 1b). For example, maximal responses to norepinephrine and serotonin in the rat mesenteric artery (Criscione et al., 1984) and in coronary arteries of the pig and the dog (Cocks and Angus, 1983) were markedly enhanced by the removal of the endothelium. In the coronary arteries, however, maximal responses evoked by potassium chloride and the thromboxane A_2 mimetic U46619 were either not altered or reduced to a small extent (maximum reduction of about 20%) by the removal of the endothelium. In the saphenous vein and femoral artery of the dog, sensitivity to norepinephrine-induced contractions was not changed by the removal of endothelium, and maximal responses remained unchanged in some experiments (Rimele and Vanhoutte, 1983) and were depressed on removal by about 10–20%, respectively, in others (De Mey and Vanhoutte, 1981). These norepinephrine-induced contractions were inhibited by acetylcholine in the presence of endothelium. Contractions of rabbit aorta elicited by norepinephrine have generally been reported to be unchanged by the removal of endothelium (Table 1b), although there is one report of a shift of concentration–response curves to the left as seen with the rat aorta (Murakami et al., 1985).

In the aorta of the latter species, the contractile responses to alpha-adrenergic agonists, usually subclassified as $alpha_2$-selective (B-HT 920, oxymetazoline, UK 14,304, idazoxan), are modulated in a similar manner to clonidine-induced responses (Lues and Schumann, 1984; Miller et al., 1984; Godfraind et al., 1985). This modulatory effect of the endothelium on agonist-induced contractions has also been shown for serotonin in vivo in the canine coronary artery (Lamping et al., 1985; Brum et al., 1984), although responses to angiotensin II and phenylephrine were not altered by endothelial removal (Lamping et al., 1985); constrictor responses of the canine femoral artery were not affected either (Angus et al., 1983).

Observations in the isolated aorta of the rat suggest that it might be possible to divide agonists into two groups: one whose concentration–effect curves are displaced to the left by removal of endothelium (norepinephrine, phenylephrine, and prostaglandin $F_{2\alpha}$, for example), and a second whose maximal responses are greatly enhanced, typified by clonidine (Miller and Stoclet, 1985). These two groups might also be distinguished by their relative dependence on extracellular Ca^{2+}. Agonists in the first group evoke contractions that are in part independent of extracellular Ca^{2+}, whereas contractions elicited by the second group are almost abolished in the absence of the activator ion (Godfraind et al., 1982; Nghiem et al., 1982). Another feature distinguishing the two groups of agonists is their relative efficacy. Norepinephrine and the other $alpha_1$-adrenoceptor agonists generally are full agonists and have a receptor reserve. The agonists that behave in a similar manner to clonidine are classified as partial agonists and have a lower efficacy than the first group and no receptor reserve. Sgd 101/75 and cirazoline, two alpha-adrenoceptor agonists usually classed as $alpha_1$-selective seem to be affected by the removal of the endothelium to a degree intermediate between norepinephrine and clonidine: The maximal responses evoked by these two agonists are reduced in the presence of endothelium (but not to the extent of responses to clonidine), and concentration–effect curves are shifted to the right (Lues and Schumann, 1984). Sgd 101/75 has been described as a partial agonist and therefore presumably has no spare receptors. A possible explanation then for the observed effects of EDRF could be a physiological antagonism of agonist-induced responses, since it would be expected that any physiological antagonism of agonist-induced contractions would be more marked with partial than with full agonists.

Another possibility is that EDRF inhibits receptors, and it has been shown that endothelium reduced the number of binding sites for prazosin, but did not change their affinity in the bovine aorta (Carman-Krzan, 1985). Such an effect, however, to antagonize receptors would have to be nonspecific, since many different receptor types would have to be antagonized, and modulation of K^+- or Bay K 8644-induced contractions (Spedding et al., 1986) could not be explained in this way.

The observed modulatory effects of endothelium on agonist-induced contractions are consistent with a reduced efficacy of agonists in the presence of endothelium and an analysis of concentration–effect curves for phenylephrine, clonidine, and several other agonists in the aorta of the rat has shown that the effect of endothelium could indeed be accounted for by a similar degree of reduction of efficacy (of about sixfold) of all alpha-adrenoreceptor agonists (Malta et al., 1986a). This analysis also predicts that maximal responses to full agonists with a large receptor reserve, such as norepinephrine and phenylephrine, would be reduced by a small, but not dramatic, amount in the presence of endothelium, but, as already mentioned, this has not always been true of all published concentration–effect curves (Table 1a). Efficacy is undefined in biochemical terms, but among the many possible modulatory effects of EDRF on smooth muscle cellular metabolism that might contribute to a reduction in efficacy are changes in cellular handling of Ca^{2+}; for example, a reduction in influx of extracellular Ca^{2+} or of the mobilization of intracellular Ca^{2+} subsequent to receptor stimulation. Other possibilities are an increase in the activity of the Ca^{2+}-ATPase of the sarcolemma associated with Ca^{2+} extrusion from the smooth muscle or of that associated with the pumping of Ca^{2+} into sarcoplasmic reticulum, or an increased sequestration of Ca^{2+} into other intracellular storage sites. These all would reduce the free intracellular Ca^{2+} concentration subsequent to receptor stimulation and so reduce contractile responses.

These possible effects of EDRF on cellular Ca^{2+} handling are particularly attractive explanations for the observed modulation by endothelium of agonist-induced contractions because: (1) It has been reported that Ca^{2+}-induced contractions are enhanced in the absence of endothelium in canine mesenteric artery (Chiba and Tsukada, 1984), as are contractile responses to low ($\leqslant 15$ m*M*) concentrations of K^+ in the aorta of the rat and the coronary artery of the dog (Spedding et al., 1986; Cohen et al., 1984); (2) The presence of endothelium changes the basal tissue levels of cyclic GMP in many blood vessels (Holzmann, 1982; Rapoport and Murad, 1983; Miller et al., 1984), and some agonists, particularly alpha-adrenoceptor agonists, stimulate a modest (about two-fold) increase in vascular levels of cyclic GMP in the presence, but not in the absence, of endothelium (Miller et al., 1984, 1985; Bigaud et al., 1984; Spedding et al., 1986). Cyclic GMP may be involved in calcium gating in the cell membrane (Bkaily and Sperelakis, 1985) and in the activation of the calcium pump of the sarcoplasmic reticulum (Lincoln, 1983).

There are two well-studied sources of calcium that can be mobilized to induce contraction in vascular smooth muscle, extracellular calcium, and intracellularly stored calcium. The importance of extracellular calcium is often assessed by altering—usually reducing—its concentration in the bathing medium; the liberation of intracellular calcium by agonists is often studied in the absence of extracellular calcium. Therefore, before studying the effect of endothelium on smooth muscle it is necessary to first establish the degree of dependence of the release of EDRF on extracellular calcium.

3. Dependence of Endothelium-Modulated Responses on Calcium

The essential requirement of extracellular Ca^{2+} for both the basal and stimulated release of EDRF has been demonstrated by several means. First, relaxant effects of EDRF have been shown to be abolished (Winquist et al., 1985) or markedly reduced (Singer and Peach, 1982) in the absence of extracellular Ca^{2+} in tissues such as the aorta of the rat and rabbit in which there is a sufficient stock of intracellular Ca^{2+} to support an agonist-induced contraction (van Breemen et al., 1980; Godfraind et al., 1982). Second, release of EDRF from arterial segments with endothelium is abolished in the absence of external Ca^{2+}, as assayed on a second tissue from which the endothelium had been removed by a bioassay technique that allowed Ca^{2+} (and a contracting agonist) to be added to the perfusion fluid before it reached the second tissue (Long and Stone, 1985; Griffith et al., 1986). Third, by using the increased tissue levels of cyclic GMP evoked by EDRF as a measure of its presence or absence, it was shown that increases in basal tissue levels of cyclic GMP in the presence of endothelium compared to those in its absence, and increases in cyclic GMP levels evoked by acetylcholine in the presence of endothelium, are abolished in a Ca^{2+}-free medium, whereas endothelium independent increases caused by sodium nitroprusside are not affected (Holzmann, 1982; Spedding et al., 1986; Schoeffter and Miller, 1986; Rapoport et al., 1985b).

Using this third criterion, the Ca^{2+} agonist Bay K 8644 (Schramm et al., 1983) at concentrations (micromolar) usually associated with increased open time of Ca^{2+} channels (Nilius et al., 1985; Nowycky et al., 1985) did not increase tissue cyclic GMP levels in the presence of endothelium (Spedding et al., 1986), although very low concentrations of this compound, in the range of 5 p*M* to 2.5 n*M*, increase the release of EDRF under bioassay conditions (Rubanyi et al., 1985), but with Bay K 8644, this release was not obviously concentration-dependent, and the relaxation produced was less than that caused by basal liberation of EDRF. It can be concluded that extracellular Ca^{2+} is essential for sustained release of EDRF, but the relative lack of effect of the calcium agonist Bay K 8644 poses a question concerning the implication of calcium channels in the translocation of this extracellular Ca^{2+}.

If extracellular calcium is essential to EDRF release, then it is impossible to study possible effects of EDRF on the mobilization of intracellular calcium by reducing the extracellular concentration of the ion to zero. A possible aid, however, in studying the modulation of EDRF on mobilization of calcium for contraction in smooth muscle is the use of calcium entry blockers. With these compounds it should be possible to work in a calcium-containing solution, but to prevent the entry of extracellular calcium into smooth muscle cells via calcium channels in response to agonist stimulation. This would allow an action of EDRF on the liberation of intracellular calcium to be studied, provided that the blocker exerts a selective effect on the smooth muscle cells.

4. Calcium Channels and Endothelial Cells

In rabbit aortic preparations the calcium entry blockers nifedipine (0.5 μM) and verapamil (10 μM) antagonized concentration–relaxation curves to methacholine and the calcium ionophore A 23187, inhibiting maximal responses by about 40% without much change of EC_{50} values of the agonists (Singer and Peach, 1982). In these experiments verapamil did not affect sodium nitroprusside-induced inhibitory responses that were used as a control response independent of endothelium. Verapamil (1–10 μM) and nifedipine (10–500 n*M*) had no, or only slight, effects on EC_{50} values and maximal responses elicited by acetylcholine and A 23187 in the aorta of the rat. The small effects observed could be explained by an effect on the smooth muscle and not the endothelium, since in the presence of high concentrations of both blockers EC_{50} values of sodium nitroprusside were increased by about twofold (Winquist et al., 1985). Also, the rather high concentrations of calcium entry blockers used in these experiments would be expected to antagonize agonist-induced responses in a noncompetitive manner if the activation of a calcium channel was implicated and would not be expected to antagonize responses to the calcium ionophore that do not depend on the activation of these channels. The antagonist effects of the blockers therefore imply a nonspecific effect, unrelated to blockade of calcium channels. A shift to the right of the concentration–effect curves might imply an interaction of the blockers with muscarinic receptors that has been demonstrated for verapamil in this concentration range (Karlinger et al., 1982; Waelbroeck et al., 1984). Flunarizine also did not obviously affect maximal acetylcholine-induced relaxations in rat aorta (Miller et al., 1985), but it is difficult to be certain that acetylcholine responses were entirely unchanged because of the effect of the calcium entry blocker reducing the degree of agonist-induced contraction of the smooth muscle.

The calcium agonist Bay K 8644 contracts the aorta of the rat in the presence or absence of endothelium when tissues are maintained in a Krebs solution containing 12–15 m*M* K^+. Relative to concentration–effect curves elicited in the absence of endothelium, concentration–effect curves to Bay K 8644 in the presence of endothelium were displaced to the right by a factor of about two, and the maximal response was inhibited by up to 50% (Spedding et al., 1986). Neither cinnarizine nor verapamil at a concentration of 1 μM affected this endothelial modulation of Bay K 8644 contractions. A 10-fold higher concentration of cinnarizine did have inhibitory effects, although 10 μM of verapamil was inactive, perhaps indicating nonspecific depressent actions of cinnarizine at this concentration (Spedding, 1983a). Nifedipine (10 n*M*) antagonized Bay K 8644-induced contractions in the presence of endothelium in an apparently competitive manner. At all concentrations tested, the calcium entry blockers inhibited Bay K 8644-induced contractions of the aorta of the rat in the absence of endothelium (Spedding et al., 1986). The results of these contractile experiments therefore indicate that calcium entry blockers do not have a potent inhibitory effect on the release of EDRF by endothelial cells. Further, endothelium-dependent acetylcholine-induced increases in tissue cyclic GMP levels in rat aorta in the presence of endothelium were not antagonized by the known Ca^{2+} entry blockers verapamil, flunarizine, nifedipine, diltiazem, and bepridil (Miller et al., 1985; Schoeffter and Miller, 1986).

Taken together, all these observations suggest that if a Ca^{2+} channel is involved in the entry of Ca^{2+} essential to the liberation of EDRF, it is not identical with the calcium channels of the adjacent smooth muscle (Miller et al., 1985; Schoeffter and Miller, 1986; Spedding et al., 1986). In this regard it is interesting to note that recently calcium channels insensitive to derivatives of dihydropyridine that reduce or enhance calcium channel open time in various tissues have been described (Nowycky et al., 1985; Nilius et al., 1985).

5. Possible Importance of Na^+-Ca^{2+} Exchange for Release of EDRF

Replacement by Li^+ or sucrose of Na^+ in the Krebs-Ringer solution bathing canine femoral artery preparations inhibited relaxant responses to acetylcholine when the Na^+ concentration was less than about 50 m*M* (De Mey and Vanhoutte, 1980). This observation suggests that Na^+ is important for release of EDRF, and more recently Winquist et al. (1985) have demonstrated that dichlorobenzamil, an analog of amiloride and an inhibitor of Na^+-Ca^{2+} exchange, inhibits A 23187- and acetylcholine-induced relaxation of aortas of the rat and rabbit. Dichlorobenzamil does not block relaxation by atrial natriuretic factor or by sodium nitroprusside, two agents that exert direct effects on the smooth muscle, so these authors suggested that Na^+-Ca^{2+} exchange might play an important role in the release of EDRF from endothelium.

Some additional support for this idea is provided from experiments in the aorta of the rat, in which replacement of Na^+ by Li^+ or choline, or treatment with a high concentration of amiloride (1 m*M*), inhibited increases in cyclic GMP levels induced by 1-min exposure to acetylcholine. In these Na^+-replacement experiments, the Na^+ concentration of the bathing solution was reduced to 20 m*M*, and under these conditions relaxing responses to acetylcholine were abolished if Na^+ was replaced with choline. If Li^+ was used as replacement, then responses were not abolished, but the half-time of acetylcholine-induced relaxations was increased from about 8 s to about 10 min, an increase of about 70-fold. Choline replacement had no significant effect on sodium nitroprusside-induced relaxations, although with Li^+ replacement the time to half relaxation of sodium nitroprusside was increased by about fourfold. Li^+ can therefore apparently partially replace Na^+ to some extent in the endothelium and has some effects on the functioning of the smooth muscle (Schoeffter and Miller, 1986).

Even though the selectivity of action of the drugs used in these studies is not entirely certain, it can be concluded that Na^+ is essential for agonist activation of endothelium leading to the liberation of EDRF, and that a Na^+-Ca^{2+} exchange process may be involved.

6. Effect of Endothelium on Mobilization of Intracellular Calcium for Contraction

Comparing equieffective concentrations of alpha-adrenergic agonists or prostaglandin $F_{2\alpha}$ in the presence and absence of endothelium, it has been observed that responses induced in tissues with endothelium have a greater phasic component of contraction (phenylephrine) and are more resistant to inhibition by flunarizine (both phenylephrine and prostaglandin $F_{2\alpha}$), suggesting that these contractions are more dependent on the mobilization of intracellular Ca^{2+} stores (Miller and Stoclet, 1985). This could be interpreted to mean that one of the effects of EDRF is to enhance the intracellular store of Ca^{2+} releasable by agonists, perhaps by increasing the uptake of Ca^{2+} into these stores, or perhaps by increasing the amount of Ca^{2+} released from these stores by any given agonist stimulus. Regarding this second possibility, it is known that although caffeine releases only 20–30% of sarcoplasmic reticulum Ca^{2+} stores (Weber, 1968), pretreatment with caffeine almost abolishes agonist-induced contractions dependent on intracellular Ca^{2+} and ^{45}Ca efflux (Leitjen et al., 1985). There is thus a large reserve of sarcoplasmic Ca^{2+} whose availability for liberation by agonists could conceivably be altered by EDRF.

Oshiro et al. (1985) observed that after exposure to Ca^{2+}-free medium for 5 min, both angiotensin II and epinephrine could still evoke contractions in rabbit mesenteric and coeliac arteries, but that on readdition of Ca^{2+} in the presence of these agonists, there was an enhancement of contractions only in the absence of endothelium. This was interpreted as a possible reduction in Ca^{2+} influx in the presence of endothelium. Contractions evoked by these agonists in the absence of extracellular Ca^{2+} were similar in the presence and absence of endothelium, and this suggested that an endothelial factor did not affect intracellular Ca^{2+} stores available for contraction. Liberation of EDRF is abolished, however, within seconds of removal of extracellular Ca^{2+} (Long and Stone, 1985; Griffith et al., 1986), which might mean that effects on intracellular Ca^{2+} stores (that are activated very quickly on addition of agonists) would appear to be unaffected on re-addition of Ca^{2+}, whereas the much slower influx of extracellular Ca^{2+} could still be subject to the effects of the reestablished release of endothelial factor (see below).

7. Effect of Endothelium on Calcium Uptake by Vascular Tissues

7.1. Basal Uptake

In rat aorta the basal initial rate of uptake of Ca^{2+} (uptake over the first 7 min of exposure to ^{45}Ca; Table 2), measured as uptake of ^{45}Ca into the lanthanum-resistant tissue compartment, is reduced in tissues with endothelium compared to those without endothelium (Malta et al., 1986b).

Table 2
Influences of Endothelium on Basal Uptake and Efflux of Ca^{2+} with Time in Rat Isolated Aorta[a]

	P/mol calcium/mg wet wt					
	2 min	5 min	7 min	10 min	20 min	60 min
Influx of Ca^{2+} measured as uptake of ^{45}Ca resistant to displacement by La^{3+}						
+Endothelium	14.4[b]	25.8[b]	39.0[b]	47.9	70.8	110.2
−Endothelium	20.7	31.2	44.0	50.1	75.6	107.3
Efflux of Ca^{2+} measured as tissue retention of ^{45}Ca resistant to displacement by La^{3+}						
+Endothelium		105.0[b]	94.8[b]	87.5[b]	71.4	29.0
−Endothelium		81.6	77.1	73.7	65.3	30.8

[a]Data from Malta et al. (1986b).
[b]Significantly different from corresponding tissue content of calcium in absence of endothelium.

This reduction does not seem to be caused by an effect of the endothelium in reducing diffusion of ^{45}Ca into the tissue since it is abolished by treatment with methylene blue, a dye that antagonizes the effects of EDRF on tissue cyclic GMP (Holzmann, 1982; Rapoport and Murad, 1983; Diamond and Chu, 1983), probably by inhibiting guanylate cyclase, but that does not remove endothelial cells. This reduced net influx could be the result of an inhibition of influx or of an increased efflux (or both), but not of an effect of an endothelium-derived factor to increase Ca^{2+} uptake into intracellular stores alone. It does not rule out the possibility that there is increased intracellular sequestration of calcium, associated with reduced influx and/or increased efflux. In the aorta of the rabbit, the presence of endothelium has been reported to enhance the tissue uptake of ^{45}Ca, but this is an effect apparently associated with a product of the cyclooxygenase system and was not affected by 10 μM verapamil (Collins et al., 1985a).

7.2. Stimulated Uptake

^{45}Ca influx stimulated by B-HT 920 in the aorta of the rat was reduced or abolished in the presence of endothelium (Table 3), and this effect was also antagonized by methylene blue (Malta et al., 1986b). With a short exposure to B-HT 920 (2 min), however, there is not a good correlation between the inhibition of Ca^{2+} influx and that of contraction. An influx of about 6.7 pmol/mg tissue wet wt was associated with a small but significant increase in tension in the absence of endothelium, but a similar influx of 5.3 pmol/mg wet wt was not associated with any contractile activity in the presence of endothelium (Table 3). This is a small difference, but the influx of ^{45}Ca stimulated by a 2-min exposure to different concentrations of phenylephrine was not affected by the presence of endothelium although contractile responses were significantly different when measured after the same contact time (Table 3).

Table 3
Correlation Between Ca^{2+} Influx and Contraction Attained in Rat Isolated Aorta on Stimulation by Phenylephrine (PE) and B-HT 920 for 2 min[a]

Agonist	ENDO	Time, min	Ca^{2+} influx pmol/mg	Ca^{2+} influx (normalized)	Contraction mg	Contraction (normalized)
PE, 30 n*M*	+	2	10.6	(41)	120	(11)
	−	2	7.7	(30)	750	(69)
PE, 1 μ*M*	+	2	25.3	(98)	770	(71)
	−	2	25.7	(100)	1090	(100)
B-HT 920, 1 μ*M*	+	2	0.0	(0)	0	(0)
	−	2	6.7	(26)	70	(7)
B-HT 920, 10 μ*M*	+	2	5.3	(21)	0	(0)
	−	2	13.0	(51)	440	(40)

[a]Data from Malta et al., 1986b.

Influx of ^{45}Ca and tension development, induced by a submaximal concentration of norepinephrine were, however, enhanced in the absence of endothelium as compared to responses in its presence (Godfraind, 1986). In the aorta of the rabbit, the influx of ^{45}Ca stimulated by serotonin was greater in the absence than in the presence of endothelium (Collins et al., 1985c), but the reverse was the case with norepinephrine, which stimulated greater influx in the presence of the endothelial cells (Collins et al., 1985a). This norepinephrine-stimulated influx of ^{45}Ca in the aorta of the rabbit and of the rat was reduced in the presence of 8-bromo-cyclic GMP (Collins et al., 1985b; Godfraind, 1986), an observation compatible with the idea that the known increase in cyclic GMP is implicated in the mechanism of action of EDRF. The difference in stimulated uptake of ^{45}Ca in the rabbit aorta by serotonin and norepinephrine may be caused by differences in the ability of these agonists to stimulate the endothelium.

If one of the properties of EDRF is to reduce Ca^{2+} influx by a direct mechanism, such as reducing membrane permeability to Ca^{2+}, then it might be expected that calcium entry blockers would be more effective or that their effects would be unchanged (Spedding, 1983b) in the presence of a functional endothelium. In fact they have been found to be less effective (Spedding et al., 1986; Miller and Stoclet, 1985), also suggesting that an effect on membrane permeability to Ca^{2+} (an effect on Ca^{2+} channels) is not a very important part of the mechanism of action of EDRF.

Therefore, although there is an effect of endothelium to inhibit calcium influx into the lanthanum-resistant calcium compartment of aorta (which is probably caused by an effect of EDRF) that is marked in unstimulated tissues, it cannot account for the inhibitory effect of endothelium on alpha-adrenoceptor agonist-induced contractions. From these experiments it is difficult to know whether the effect of endothelium on B-HT 920-induced Ca^{2+} influx is caused by basal or by a stimulated release of EDRF.

8. Effect of Endothelium on Efflux of Calcium

8.1. Basal Efflux

The initial rate of efflux of ^{45}Ca (over the first 10 min) from the aorta of the rat loaded with the isotope is significantly reduced in the presence of endothelium (by about 28–18%; Malta et al., 1986b). That is, there was an increased tissue retention of ^{45}Ca apparent over the first 10 min of washout in the presence of endothelium, as compared to tissues without endothelium (Table 2). This retention was abolished by methylene blue and also in the absence of extracellular Ca^{2+}, treatments that abolish the effects of EDRF on the smooth muscle or abolish the liberation of EDRF, respectively. This tissue retention of Ca^{2+} in the presence of endothelium could be simply caused by a reduction of Ca^{2+} entry in the presence of endothelium, since in equilibrium (basal) conditions influx and efflux must be equal and does not necessarily imply an increased activity of uptake systems into sarcoplasmic reticulum or a reduced activity of the sarcolemmal Ca^{2+}-ATPase.

8.2. Efflux of Calcium from Stimulated Tissues

Efflux stimulated by low concentrations of B-HT 920 and phenylephrine (1 μM and 30 nM, respectively) in the presence of extracellular Ca^{2+} was also reduced in the presence of endothelium in the aorta of the rat (Malta et al., 1986b). B-HT 920 and other alpha$_2$-adrenoceptor agonists do not induce a liberation of intracellular Ca^{2+} that can be detected in the absence of extracellular Ca^{2+} (Godfraind et al., 1982; Malta et al., 1986b), and a stimulated efflux of ^{45}Ca by B-HT 920 (and other alpha$_2$-adrenoceptor agonists) can be detected in the presence of extracellular Ca^{2+} because of an exchange of entering ^{40}Ca for ^{45}Ca. Therefore, this reduction in stimulated efflux could also possibly be explained by reduced influx associated with the presence of endothelium. In the presence, however, of a calcium entry blocker such as flunarizine, for example, it is possible to study the liberation of intracellular Ca^{2+} stores labeled with ^{45}Ca by an agonist such as phenylephrine without the complication of the entry of ^{40}Ca. In these conditions it has been found that ^{45}Ca efflux from the aorta of the rat was significantly reduced by the presence of the endothelium. In the absence of endothelium, the tissue content fell from about 70 to about 43 pmol/mg wet wt, and in its presence from about 70 to about 53 pmol/mg wet wt after a 2-min exposure to 1 μM phenylephrine (Malta et al., 1986b).

This latter result clearly indicates an effect of endothelium to either antagonize the release of intracellular calcium by the agonist or to reduce the activity of the sarcolemmal Ca^{2+}-ATPase responsible for pumping Ca^{2+} out of the cell. Alternatively, there could be an increase in the activity of the Ca^{2+}-ATPase associated with the sarcoplasmic reticulum. A combination of any of these possibilities might also be envisaged. Regarding the first possibility, an increased agonist mobilization of intracellular calcium in the presence of endothelium (Miller and Stoclet, 1985) and an antagonism of phosphatidylinositol hydrolysis, that is likely to be associated with a decreased mobilization of intracellular Ca^{2+} (Rapoport, 1986), has been reported. The effect of EDRF on Ca^{2+}-ATPases is a possible mechanism of action that might merit investigation. This mechanism could involve cyclic GMP, since norepinephrine-stimulated efflux of ^{45}Ca from preparations of the aorta of the rabbit is reduced in the presence of 8-bromo-cyclic GMP (Collins et al., 1985b).

9. Effects of Muscarinic Agonists on Ca^{2+} Influx and Efflux in Vessels With and Without Endothelium

Further evidence against the primary mechanism of EDRF effects being caused by changes in membranal permeability to Ca^{2+} is provided by studying the effects of stimulating endothelial release of EDRF by muscarinic agonists. If the increase in tissue cyclic GMP levels is taken as a criterion of endothelial stimulation, then muscarinic stimulation by acetylcholine or carbachol in rat aorta is about 10–15-fold more effective than alpha-adrenergic stimulation (Rapoport and Murad, 1983; Schoeffter and Miller, 1986; Bigaud et al., 1984). Carbachol did not enhance the effect of the endothelium to reduce the basal influx or efflux of ^{45}Ca in the aorta of the rat. But, exposure to carbachol did inhibit alpha-adrenoceptor stimulated influx, and there was some correlation between this inhibition and the inhibition of phenylephrine-induced contractions produced by the same treatment (Schini et al., 1987). These observations are consistent with the idea that stimulated release of EDRF is associated with an inhibition of agonist stimulated Ca^{2+} influx into the smooth muscle, but the complete lack of effect of stimulated release of EDRF on basal Ca^{2+} influx is puzzling. In the aorta of the rabbit acetylcholine at 1 μM, a concentration that produces maximal relaxation (Furchgott and Zawadzki, 1980), reduced but did not abolish norepinephrine-stimulated ^{45}Ca efflux in the presence of endothelium (Collins et al., 1985a), and 8-bromo-cyclic GMP reduced norepinephrine-stimulated efflux in the absence of endothelium (Collins et al., 1985b). These observations are consistent with either a reduced liberation of intracellular Ca^{2+} or a reduced activity of sarcolemmal Ca^{2+}-ATPase, or both, that are dependent on endothelium and perhaps mediated by cyclic GMP.

10. Importance of the Modulatory Effect of Endothelium on Cellular Ca^{2+} Movements

Overall, it can be suggested that EDRF exerts effects on the membrane permeability to Ca^{2+} and increases the quantity of intracellularly stored Ca^{2+} available for liberation by agonists. This increased availability of intracellular Ca^{2+} may be caused by an increased amount of Ca^{2+} associated with intracellular sites of release or with a change in the availability of calcium already stored in these sites. EDRF also reduces significantly the amount of Ca^{2+} extruded from the smooth muscle by the sarcolemmal Ca^{2+}-ATPase. Thus, there may be multiple effects of EDRF on calcium movements in vascular smooth muscle, or EDRF may be a mixture of compounds having different sites of action.

11. Do Contractile Agonists Stimulate the Release of EDRF or Are Modulatory Effects Dependent on a Basal Release of EDRF?

Cocks and Angus (1983) were the first to demonstrate an endothelium-dependent relaxing effect of the alpha-adrenoceptor agonists norepinephrine and clonidine in contracted coronary arteries of pig and dog. Methoxamine did not relax the porcine coronary arteries, even in the presence of prazosin, and these authors concluded that endothelial alpha$_2$-adrenoceptors might be implicated in these responses. Egleme et al. (1984a) also thought that the marked inhibition by endothelium of clonidine-mediated contractions compared to norepinephrine-mediated contractions of rat aorta might be explained by the presence of endothelial receptors more sensitive to alpha$_2$- than to alpha$_1$-adrenoceptor agonists. This difference in agonist sensitivity can probably be explained, however, by a similar reduction of efficacy of the two agonists having different intrinsic activities (Malta et al., 1986a) and does not constitute a basis for supposing the presence of endothelial alpha$_2$-adrenoceptors.

In the presence of endothelium, mesenteric arteries of the rabbit contracted by either histamine or K^+ were relaxed by epinephrine when propranolol and prazosin were present (Oshiro et al., 1985). Norepinephrine-contracted aortic preparations of the rat can be relaxed by clonidine (Murakami et al., 1985). In the aorta of the rat and pulmonary and femoral arteries and veins of the dog treated with prazosin, both norepinephrine and the alpha$_2$-adrenoceptor agonist UK 14,304 relaxed contractions induced by the thromboxane analog U 46619 (Bullock et al., 1986; Miller and Vanhoutte, 1985). These relaxant responses were antagonized by phentolamine in the aorta of the rat and by rauwolscine in the canine blood vessels. Dilator responses in isolated perfused vessels from rat and dog in the presence of endothelium could be elicited by alpha$_2$-adrenergic (UK 14,304 and azepexole), but not alpha$_1$-adrenoceptor (phenylephrine and cirazoline) selective agonists (Matsuda et al., 1985). Other contractile agonists such as serotonin and angiotensin II have also been shown to relax contracted vascular preparations in the presence of endothelium (Cocks and Angus, 1983; Cohen et al., 1983; Imaizumi et al., 1984; Houston et al., 1985; Oshiro et al., 1985).

These observations seem to imply that, like acetylcholine (Furchgott and Zawadzki, 1980), many contractile agonists, including agonists that activate alpha$_2$-adrenoceptors, stimulate the liberation of an endothelial-derived factor that inhibits contractions of adjacent smooth muscle. Lues and Schumann (1984) could demonstrate, however, no significant influence of rauwolscine on alpha-adrenoceptor agonist-induced contractions of rat aorta in the presence of endothelium and concluded that the endothelial alpha-adrenoceptor was not identical with either alpha$_1$- or alpha$_2$-adrenoceptors. Concentrations of both alpha$_1$-adrenergic (methoxamine, phenylephrine) and alpha$_2$-adrenergic (clonidine, B-HT 920) agonists that stimulate maximal contractions in rat aorta increase the content of cyclic GMP in the presence, but not in the absence, of endothelium by about twofold (Miller et al., 1984, 1985; Bigaud et al., 1984) as does norepinephrine, although no increase was seen using submaximal concentrations of the agonists (Schoeffter and Miller, unpublished observations; Rapoport and Murad, 1983; Rapoport et al., 1985a,b). Such an increase in levels of cyclic GMP would be expected if the agonists stimulated the release of EDRF and imply that the endothelium is responsive to both alpha$_1$- and alpha$_2$-adrenoceptor agonists. No such increase, however, in tissue cyclic GMP levels in the aorta of the rat was seen by Furchgott and colleagues (1985) using a maximal concentration of clonidine, although in this series of experiments acetylcholine increased cyclic GMP levels by about four-fold, whereas increases of about 15–37-fold have more usually been reported in this tissue (Rapoport and Murad, 1983; Miller et al., 1985). Bullock et al. (1986) also detected no increase in cyclic GMP levels in rat aorta evoked by equi-relaxant concentrations of acetylcholine and UK 14,304.

Overall, there is sufficient evidence to suppose that many contractile agents stimulate endothelial cells to liberate a factor, possibly identical with the EDRF whose liberation is stimulated by acetylcholine, in many vascular tissues. It is not known if this is a general property of a particular agonist in all vessels, and interpretation of experimental results should allow for the possibility that the basal release of an endothelial factor may be sufficient to explain the observed modulatory effects of endothelium (Malta et al., 1986a).

References

Allan, G., Brook, C. D., Cambridge, D., and Hladkiwskyj, J.: Enhanced responsiveness of vascular smooth muscle to vasoconstrictor agents after removal of endothelial cells. *Br. J. Pharmacol.*, **79**:334P, 1983.

Angus, J. A., Cocks, T. M., Le Duc, M., and Campbell, G. R.: Endothelial-dependent vasodilation demonstrated in vivo. *Blood Vess.* **20**:185–212, 1983.

Bigaud, M., Schoeffter, P., Stoclet, J. C., and Miller, R. C.: Dissociation between endothelium-mediated increases in tissue cGMP levels and modulation of aortic contractile responses. *Naunyn Schmiedebergs Arch. Pharmacol.* **328**:221–223, 1984.

Bkaily, G. and Sperelakis, N.: Injection of guanosine 5′-cyclic monophosphate into heart cells blocks calcium slow channels. *Am. J. Physiol.* **248**:H745–H749, 1985.

Brum, J. M., Sufan, Q., Lane, G., and Bove, A. A.: Increased vasoconstrictor activity of proximal coronary arteries with endothelial damage in intact dogs. *Circulation* **70**:1066–1073, 1984.

Bullock, G. R., Taylor, S. G., and Weston, A. H.: Influence of the vascular endothelium on agonist-induced contractions and relaxations in rat aorta. *Br. J. Pharmacol.* **89**:819–830, 1986.

Carman-Krzan, M.: Role of endothelium in the binding of [^{3}H]prazosin to the aortic α_1-adrenergic receptors. *Eur. J. Pharmacol.* **111**:137–138, 1985.

Carrier, G. O. and White, R. E.: Enhancement of alpha-1 and alpha-2 adrenergic agonist-induced vasoconstriction by removal of endothelium in rat aorta. *J. Pharmacol. Exp. Ther.* **232**:682–687, 1985.

Chiba, S. and Tsukada, M.: Vasoconstrictor responses induced by alpha-adrenoceptor agonists before and after removal of the endothelial cells of dog mesenteric arteries. *J. Auton. Pharmacol.* **4**:257–260, 1984.

Cocks, T. M. and Angus, J. A.: Endothelium-dependent relaxation of coronary arteries by noradrenaline and serotonin. *Nature* **305:**627–630, 1983.

Cohen, R. A., Shepherd, J. T., and Vanhoutte, P. M.: 5-Hydroxytryptamine can mediate endothelium-dependent relaxation of coronary arteries. *Am. J. Physiol* **245:**H1077–H1080, 1983.

Cohen, R. A., Shepherd, J. T., and Vanhoutte, P. M.: Endothelium and asymmetrical responses of the coronary arterial wall. *Am. J. Physiol.* **247:**H403–H408, 1984.

Collins, P., Griffith, T. M., Henderson, A. H., and Lewis, M. J.: Endothelium and calcium flux in rabbit aortic preparations. *Br. J. Pharmacol.*, **85:**342P,, 1985a.

Collins, P., Griffith, T. M., Henderson, A. H., and Lewis, M. J.: 8-Bromo-cGMP and calcium flux in arterial smooth muscle. *Br. J. Pharmacol.* **85:**279P, 1985b.

Collins, P., Griffith, T. M., Henderson, A. H., and Lewis, M. J.: Endothelium and calcium fluxes in rabbit aortic preparations. *J. Physiol.* (Lond.) **360:**63P, 1985c.

Criscione, L., Muller, K., and Prescott, M. F.: Endothelial cell loss enhances the pressor response in resistance vessels. *J. Hypertension* **2** (Suppl. 3):441–444, 1984.

De Mey, J. G. and Vanhoutte, P. M.: Interaction between Na^+, K^+ exchanges and the direct inhibitory effect of acetylcholine on canine femoral arteries. *Circulation Res.* **46:**826–836, 1980.

De Mey, J. G. and Vanhoutte, P. M.: Role of the intima in cholinergic and purinergic relaxation of isolated canine femoral arteries. *J. Physiol.* (Lond.) **316:**347–355, 1981.

Diamond, J. and Chu, E. B.: Possible role for cyclic GMP in endothelium-dependent relaxation of rabbit aorta by acetylcholine. Comparison with nitroglycerin. *Res. Commun. Chem. Path. Pharmacol.* **41:**369–381, 1983.

D'Orleans-Juste, P., Dion, S., Mizrahi, J., and Regoli, D: Effects of peptides and non-peptides on isolated arterial muscles; role of endothelium. *Eur. J. Pharmacol.* **114:**9–21, 1985.

Egleme, C., Godfraind, T., and Miller, R. C.: Enhanced responsiveness of rat isolated aorta to clonidine after removal of the endothelial cells. *Br. J. Pharmacol.* **81:**16–18, 1984a.

Egleme, C., Godfraind, T., and Miller, R. C.: Role of endothelium in the rat aorta response to vasoconstrictors. *Arch. Int. Pharmacodyn.* **268:**116, 1984b.

Fortes, Z. B., Garcia Leme, J., and Scivoletto, R.: Vascular reactivity in diabetes mellitus: Role of the endothelial cell. *Br. J. Pharmacol.* **79:**771–781, 1983.

Furchgott, R. F.: The role of endothelium in the responses of vascular smooth muscle to drugs. *Ann. Rev. Pharmacol. Toxicol.* **24:**175–197, 1984.

Furchgott, R. F. and Zawadzki, J. V.: The obligatory role of endothelial cells in the relaxation of arterial smooth muscle by acetylcholine. *Nature* **288:**373–376, 1980.

Furchgott, R. F., Jothianandan, D., Martin, W., and Villani, G. M.: Depression of contractile responses by spontaneously released EDRF in rat aorta. *Br. J. Pharmacol.* **86:**565P,, 1985.

Godfraind, T.: EDRF and cyclic GMP control gating of receptor-operated calcium channels in vascular smooth muscle. *Eur. J. Pharmacol.* **126:**341–343, 1986.

Godfraind, T., Miller, R. C., and Socrates Lima, J.: Selective α_1- and α_2-adrenoceptor agonist induced contractions and ^{45}Ca fluxes in the rat isolated aorta. *Br. J. Pharmacol.* **77:**597–604, 1982.

Godfraind, T., Egleme, C., and Al Osachie, I.: Role of endothelium in the contractile response of rat aorta to alpha-adrenoceptor agonists. *Clin. Sci.* **68** (suppl. 10)**:**65s–71s, 1985.

Griffith, T. M., Edwards, D. H., Newby, A. C., Lewis, M. J., and Henderson, A. H.: Production of endothelium derived relaxant factor is dependent on oxidative phosphorylation and extracellular calcium. *Cardiovasc. Res.* **20:**7–12, 1986.

Griffith, T. M., Henderson, A. H., Edwards, D. H., and Lewis, M. J.: Isolated perfused rabbit coronary artery and aortic strip preparations: The role of endothelium-derived relaxant factor. *J. Physiol.* **351:**13–24, 1984.

Holzmann, S.: Endothelial-induced relaxation by acetylcholine associated with larger rises in cyclic GMP in coronary arterial strips. *J. Cyclic Nucleotide Res.* **8:**409–419, 1982.

Houston, D. S., Shepherd, J. T., and Vanhoutte, P. M.: Adenine nucleotides, serotonin, and endothelium-dependent relaxations to platelets. *Am. J. Physiol.* **248:**H389–H395, 1985.

Imaizumi, Y., Baba, M., Imaizumi, Y., and Watanabe, M.: Involvement of endothelium in the relaxation of isolated chick jugular vein by 5-hydroxytryptamine. *Eur. J. Pharmacol.* **97:**335–336, 1984.

Karliner, J. S., Motulsky, H. J., Dunlap, J., Brown, J. H., and Insel, P. A.: Verapamil competitively inhibits α_1-adrenergic and muscarinic but not β-adrenergic receptors in rat myocardium. *J. Cardiovasc. Pharmacol.* **4:**515–520, 1982.

Konishi, M. and Su, C.: Role of endothelium in dilator responses of spontaneously hypertensive rat arteries. *Hypertension* **5:**881–886, 1983.

Lamping, K. G., Marcus, M. L., and Dole, W. P.: Removal of the endothelium potentiates canine large coronary artery constrictor responses to 5-hydroxytryptamine in vivo. *Circulation Res.* **57:**46–54, 1985.

Leitjen, P., Saida, K., and van Breemen, C.: Norepinephrine-induced intracellular Ca^{++} release from vascular smooth muscle. *J. Cardiovasc. Pharmacol.* **7** (suppl. 6)**:**s38–s42,, 1985.

Lincoln, T. M.: Effects of nitroprusside and 8-bromo-cyclic GMP on the contractile activity of the rat aorta. *J. Pharmacol. Exp. Ther.* **224:**100–107, 1983.

Long, C. J. and Stone, T. W.: The release of endothelium-derived relaxant factor is calcium dependent. *Blood Vess.* **22:**205–208, 1985.

Lues, I. and Schumann, H.-J.: Effect of removing the endothelial cells on the reactivity of rat aortic segments to different alpha-adrenoceptor agonists. *Naunyn Schmiedebergs Arch. Pharmacol.* **328:**160–163, 1984.
Malta, E., Schini, V., and Miller, R. C.: The role of efficacy in the assessment of the actions of α-adrenoceptor agonists in rat aorta with endothelium. *J. Pharm. Pharmacol.* **38:**209–213, 1986a.
Malta, E., Schini, V., and Miller, R. C.: Effect of endothelium on basal and α-adrenoceptor stimulated calcium fluxes in rat aorta. *Naunyn-Schmiedeberg's Arch. Pharmacol.* **334:**63–70, 1986b.
Matsuda, H., Kuon, E., Holtz, J., and Busse, R.: Endothelium-mediated dilations contribute to the polarity of the arterial wall in vasomotion induced by α-adrenergic agonists. *J. Cardiovasc. Pharmacol.* **7:**680–688, 1985.
Miller, R. C.; and Stoclet, J. C.: Modulation by endothelium of contractile responses in rat aorta in absence and presence of flunarizine. *Br. J. Pharmacol.* **86:**655–661, 1985.
Miller, V. M. and Vanhoutte, P.M.: Endothelial α-adrenoceptors in canine pulmonary and systemic blood vessels. *Eur. J. Pharmacol.* **118:**123–129, 1985.
Miller, R. C., Mony, M. C., Schini, V., Schoeffter, P., and Stoclet, J. C.: Endothelial mediated inhibition of contraction and increase in cyclic GMP levels evoked by the alpha-adrenoceptor agonist B-HT 920 in rat isolated aorta. *Br. J. Pharmacol.* **83:**903–908, 1984.
Miller, R. C., Schoeffter, P., and Stoclet, J. C.: Insensitivity of calcium-dependent endothelial stimulation in rat isolated aorta to the calcium entry blocker, flunarizine. *Br. J. Pharmacol.* **85:**481–487, 1985.
Murakami, K., Karaki, H., and Urakawa, N.: Role of endothelium in the contractions induced by norepinephrine and clonidine in rat aorta. *Jap. J. Pharmacol.* **39:**357–364, 1985.
Nghiem, C., Swamy, V. C., and Triggle, D. J.: Inhibition by D 600 of norepinephrine- and clonidine-induced responses of the aortae from normotensive (WKY) and spontaneously hypertensive rats (SHR) *Life Sci.* **30:**45–49, 1982.
Nilius, B., Hess, P., Lansman, J. B., and Tsien, R. W.: A novel type of cardiac channel in ventricular cells. *Nature* **316:**443–446, 1985.
Nowycky, M. C., Fox, A. P., and Tsien, R. W.: Three types of neuronal calcium channel with different calcium agonist sensitivity. *Nature* **316:**440–443, 1985.
Oshiro, M. E. M., Paiva, A. C. M., and Paiva, T. B.: Endothelium-dependent inhibition of the use of extracellular calcium for the arterial response to vasoconstrictor agents. *Gen. Pharmacol.* **16:**567–572, 1985.
Rapoport, R. M.: Cyclic guanosine monophosphate inhibition of contraction may be mediated through inhibition of phosphalidylinositol hydrolysis in rat aorta. *Circ. Res.* **58:**407–410, 1986.
Rapoport, R. M. and Murad, F.: Agonist-induced endothelium-dependent relaxation in rat thoracic aorta may be mediated through cGMP. *Circulation Res.* **52:**352–357, 1983.

Rapoport, R. M., Schwartz, K., and Murad, F.: Effects of Na^+, K^+-pump inhibitors and membrane depolarizing agents on acetylcholine-induced endothelium-dependent relaxation and cyclic GMP accumulation in rat aorta. *Eur. J. Pharmacol.* **110:**203–209, 1985a.

Rapoport, R. M., Schwartz, K., and Murad, F.: Effects of sodium-potassium pump inhibitors and membrane-depolarizing agents on sodium nitroprusside-induced relaxation and cyclic guanosine monophosphate accumulation in rat aorta. *Circ. Res.* **57:**164–170, 1985b.

Rimele, T. J. and Vanhoutte, P. M.: Effects of inhibitors of arachidonic acid metabolism and calcium entry on responses to acetylcholine, potassium and norepinephrine in the isolated canine saphenous vein. *J. Pharmacol. Exp. Ther.* **225:**720–726, 1983.

Rubanyi, G. M., Schwartz, A., and Vanhoutte, P. M.: The calcium agonists Bay K 8644 and (+)202,791 stimulate the release of endothelial relaxing factor from canine femoral arteries. *Eur. J. Pharmacol.* **117:**143–144, 1985.

Schini, V., Malta, E., and Miller, R. C.: Effect of endothelium and carbachol on α-adrenoceptor agonist-stimulated uptake and efflux of ^{45}Ca in rat isolated aorta. *Naunyn Schmiedebergs Arch. Pharmacol.*, in press, 1987.

Schoeffter, P. and Miller, R. C.: Role of sodium-calcium exchange and effects of calcium entry blockers on endothelial-mediated responses in rat isolated aorta. *Mol. Pharmacol.* **30:**53–57, 1986.

Schramm, M., Thomas, G., Towart, R., and Franckowiak, G.: Novel dihydropyridine with positive inotropic action through activation of Ca^{++} channels. *Nature* **303:**535–537, 1983.

Sercombe, R., Verrecchia, C., Oudart, N., Dimitriadou, V., and Seylaz, J.: Pial artery responses to norepinephrine potentiated by endothelium removal. *J. Cerebr. Blood Flow Metabol.* **5:**312–317, 1985.

Singer, H. A. and Peach, M. J.: Calcium- and endothelium-mediated vascular smooth muscle relaxation in rabbit aorta. *Hypertension* **4** (suppl. II): 19–25, 1982.

Singer, H. A. and Peach, M. J.: Endothelium-dependent relaxations of rabbit aorta. 1. Relaxation stimulated by arachidonic acid. *J. Pharmacol. Exp. Ther.* **226:**790–795, 1983a.

Singer, H. A. and Peach, M. J.: Endothelium-dependent relaxation of rabbit aorta. 2. Inhibition of relaxation stimulated by methacholine and A 23187 with antagonists of arachidonic acid metabolism. *J. Pharmacol. Exp. Ther.* **226:**796–801, 1983b.

Spedding, M.: Direct inhibitory effects of some 'calcium-antagonists' and trifluoperazine on the contractile proteins in smooth muscle. *Br. J. Pharmacol.* **79:**225–231, 1983a.

Spedding, M.: Functional interactions of calcium-antagonists in K^+-depolarized smooth muscle. *Br. J. Pharmacol.* **80:**485–488, 1983b.

Spedding, M., Schini, V., Schoeffter, P., and Miller, R. C.: Calcium channel activation does not increase release of endothelial-derived relaxant factors (EDRF) in rat aorta although tonic release of EDRF may modulate calcium channel activity in smooth muscle. *J. Cardiovasc. Pharmacol.* **8:**113–137, 1986.

Spokas, E. G., Folco, G., Quilley, J., Chander, P., and McGiff, J. C.: Endothelial mechanism in the vascular action of hydralazine. *Hypertension* **5** (suppl. 1):107–111, 1983.
van Breemen, C., Aaronson, P., Loutzenhiser, R., and Meisheri, K.: Ca^{++} movements in smooth muscle. *Chest* **78** (suppl):157–165, 1980.
Verrecchia, C., Sercombe, R., and Seylaz, J.: Influence of endothelium on noradrenaline-induced vasoconstriction in rabbit central ear artery. *Clin. Exp. Pharmacol. Physiol.* **12**:169–179, 1985.
Waelbroeck, M., Robberecht, P., de Neef, P., and Christophe, J.: Effects of verapamil on the binding of rat heart muscarinic receptors: Evidence for an allosteric site. *Biochem. Biophys. Res. Commun.* **121**:340–345, 1984.
Weber, A.: The mechanism of the action of caffeine on sarcoplasmic reticulum. *J. Gen. Physiol.* **52**:760–772, 1968.
White, R. E. and Carrier, G. O.: $Alpha_1$- and $alpha_2$-adrenoceptor agonist-induced contraction in rat mesenteric artery upon removal of endothelium. *Eur. J. Pharmacol.* **122**:349–352, 1986.
Winquist, R. J., Bunting, P. B., and Schofield, T. L.: Blockade of endothelium-dependent relaxation by the amiloride analog dichlorobenzamil: Possible role of Na^+/Ca^{++} exchange in the release of endothelium-derived relaxant factor. *J. Pharmacol. Exp. Ther.* **235**:644–650, 1985.

Chapter 13

Endothelium-Derived Relaxing Factor Relaxes Vascular Smooth Muscle by Cyclic GMP-Mediated Effects on Calcium Movements

P. Collins, M. J. Lewis, and A. H. Henderson

1. Introduction

The relaxant effect both of endothelium-derived relaxing factor (EDRF) and of the nitrovasodilators on vascular smooth muscle is associated with activation of soluble guanylate cyclase (Rapoport et al., 1983b; Busse et al., 1985) and increased intracellular levels of cyclic guanosine monophosphate (cyclic GMP) (Katsuki et al., 1977; Ignarro et al., 1981; Rapoport et al., 1983a; Rapoport and Murad, 1983; Griffith et al., 1985a). How cyclic GMP regulates contraction of smooth muscle has only recently begun to emerge.

Rapoport and colleagues (1983a) showed an association between endothelium-dependent relaxations, raised tissue levels of cGMP, and reduced phosphorylation of the myosin-light chains in vascular smooth muscle. The extent to which this dephosphorylation of the myosin light chains might mediate mechanical relaxation was not known, nor was it clear whether reduction of this calcium-calmodulin-dependent process (Aksoy and Murphy, 1983) was a direct consequence of the raised cyclic GMP levels or secondary to a reduction of cytosolic calcium. Recently Hathaway and colleagues (1985) have shown that cyclic GMP-dependent protein kinase in fact does not lead to altered activity of myosin light chain kinase nor alter its requirement for calmodulin. A more likely explanation for the relaxant effect of EDRF and nitrovasodilators (and of cyclic GMP itself) might therefore be that they reduce the level of activator Ca^{2+} within the cell.

There are a number of mechanisms by which, alone or in combination, cyclic GMP may be able to mediate a reduction of free cytosolic Ca^{2+}. It may reduce the influx of Ca^{2+} through one or other of its membrane "channels," it may enhance extrusion of Ca^{2+}, or it may reduce intracellular calcium stores or the release of Ca^{2+} from these stores.

2. Calcium Influx

2.1. EDRF

There are few published studies on the effects of EDRF on the control of Ca^{2+} in vascular smooth muscle. Godfraind and colleagues (1985) showed lower ^{45}Ca content in rat aortas with intact endothelium than in preparations denuded of endothelium: They pointed out that this could be caused by either decreased Ca^{2+} entry or increased Ca^{2+} efflux. Ratz and colleagues (1985) showed that norepinephrine-stimulated ^{45}Ca influx was reduced by acetylcholine in endothelium-containing preparations, and suggested that this effect was EDRF-mediated.

A series of studies have recently been conducted using rabbit aortic preparations to measure and compare ^{45}Ca influx, ^{45}Ca efflux, and mechanical responses to EDRF, to a nitrovasodilator, and to a lipid-soluble analog of cyclic GMP (Collins et al., 1984, 1985a,b, 1986a,b). Ca^{2+} influx was measured using ^{45}Ca (Meisheri et al., 1980). The method has been validated by a number of workers (Meisheri et al., 1980; Gleason et al., 1985; Loutzenhiser and Epstein, 1985), and the short period of ^{45}Ca accumulation is considered to provide a valid measure of Ca^{2+} influx in vascular smooth muscle (Meisheri et al., 1980, 1981). Previous measurements of ^{45}Ca flux across smooth muscle cell membranes have been confounded by extracellular calcium (bound and free); this problem is overcome by removing extracellular calcium with EGTA in the cold (van Breemen et al., 1982). Measurements were made under a variety of conditions—in preparations with and without endothelium; with stimulation of EDRF activity by acetylcholine and testing of the specificity of EDRF-mediated responses by dithiothreitol, borohydride, phenidone, and phenylhydrazine—known inhibitors of EDRF activity (Griffith et al., 1984); in resting preparations and in preparations constricted either by norepinephrine or KCl; and with flurbiprofen where appropriate to exclude additional cyclooxygenase-dependent influences. Parallel studies on mechanical responses were conducted in isometrically mounted, opened-ring preparations; in which Ca^{2+} influx has been shown to be similar to that in the unstretched preparations used for the ^{45}Ca flux studies (Gleason et al., 1985).

The results of these experiments clearly demonstrate that EDRF-mediated vascular relaxation is associated with a reduction of Ca^{2+} influx (Figs. 1 and 2) (Collins et al., 1984, 1985a,b, 1986a,b). The increase in Ca^{2+} influx induced by norepinephrine is reduced by acetylcholine in preparations with intact endothelium (Fig. 1). This acetylcholine-stimulated endothelium-dependent effect is blocked by the EDRF inhibitors—dithiothreitol, borohydride, or phenidone. It was incidentally observed that Ca^{2+} influx is slightly greater in the presence of endothelium than in its absence, both in resting and in norepinephrine-constricted strips of the rabbit aorta (though not in other preparations such as rat aorta or the canine coronary artery) (Collins et al., 1986b); this however is not caused by EDRF since it is unaffected by EDRF inhibitors and it appears to be caused by a cyclooxygenase product since it is abolished by flurbiprofen (Fig. 1). In the presence of flurbiprofen, the reduction in Ca^{2+} influx induced by acetylcholine in norepinephrine-treated preparations with intact endothelium was greater after blocking of this opposing endothelium-mediated effect. Thus the reduction of calcium influx coincides precisely with all conditions in which EDRF-mediated relaxation occurs. This was shown to be true also in other preparations such as aortic strips of the rat and coronary artery strips of the dog (Collins et al., 1986b). Thus, EDRF activity is associated with a reduction of Ca^{2+} influx.

2.2. Nitrovasodilators, 8-Bromo-cyclic GMP

The influence of nitrovasodilators on Ca^{2+} influx in vascular smooth muscle has been the subject of a number of studies with what at first sight appear to be conflicting results. As part of the investigation described above into the effects of EDRF on Ca^{2+} influx and mechanical relaxation, the effects of sodium nitroprusside and of the lipid-soluble analog of cGMP, 8-bromo-cyclic GMP, were studied in the rabbit aorta (Collins et al., 1984, 1985a,b, 1986a). Both agents similarly reduced Ca^{2+} influx in norepinephrine-treated preparations, under conditions in which they caused relaxation (lower panel of Fig. 1, middle and right panels of Fig. 2). Their effects were independent of the presence of the endothelium.

2.3. Cyclic GMP and Calcium Influx

The vascular relaxing action of EDRF, nitrovasodilators, and 8-bromo-cyclic GMP is associated in each case with a reduction of Ca^{2+} influx.

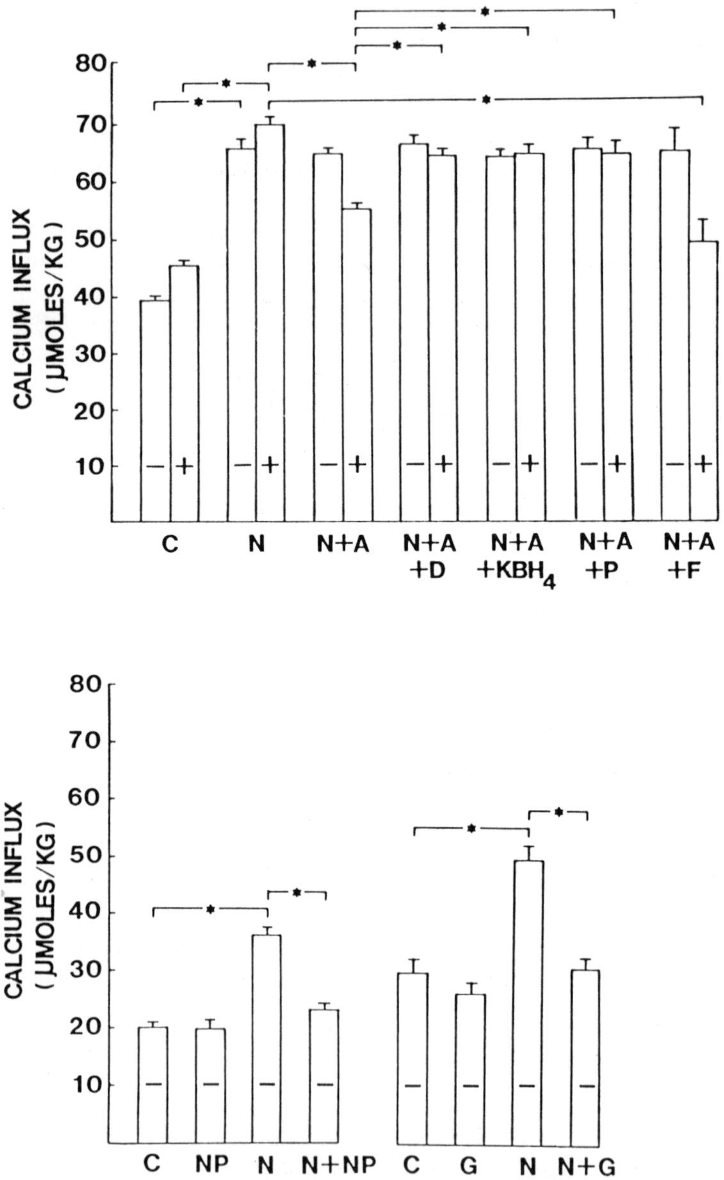

Fig. 1. (Upper panel) Calcium influx into rabbit aortas with (+) and without (−) endothelium. Preparations were incubated with ^{45}Ca for 1.5 min and with different agents as shown (for times indicated below), each agent then remaining present during incubation with ^{45}Ca. (C) Control, no added agent. (N) Norepinephrine (10^{-5}*M*) 4.5 min. (A) Acetylcholine (10^{-6}*M*) 1.5 min. (D) Dithiothreitol

Since these three agents have in common the property of elevating cellular levels of cyclic-GMP, the implication is that it is the cyclic nucleotide that mediates this reduction of Ca^{2+} influx. Thus cyclic GMP appears to cause mechanical relaxation at least in part by reducing calcium influx.

2.4. Specificity for "Receptor-Operated Channels"?

Both mechanical relaxation (Fig. 2, Table 1) and the inhibition of Ca^{2+} influx (Fig. 3) by EDRF, sodium nitroprusside, and 8-bromo-cGMP were greater when the rabbit aortas were contracted with norepinephrine than when they were constricted to a similar extent by high concentrations of potassium chloride (KCl). A review of published studies on the effects of nitrovasodilators on Ca^{2+} influx reveals that in almost all those studies in which influx was stimulated by the agonists norepinephrine or histamine, nitrovasodilators reduced Ca^{2+} influx (Haeusler and Thorens, 1975; Karaki et al., 1980, 1984; Hester, 1985; Matlib et al., 1985); when Ca^{2+} influx was stimulated with high extracellular potassium, however, nitrovasodilators generally had no effect (Haeusler and Thorens, 1975; Zsoter et al., 1977; Karaki et al., 1980; Ozaki et al., 1981; Karaki et al., 1984; Hester, 1985) except in one report (Kreye et al., 195), in which the findings were later refuted (Kreye, 1980). There is some evidence that nitrovasodilators cause also greater mechanical relaxation in agonist-constricted than in

← Fig. 1 (*Continued*)

($5 \times 10^{-4} M$) 1 min. (KBH_4) Potassium borohydride ($5 \times 10^{-4} M$) 1 min. (P) Phenidone ($5 \times 10^{-4} M$) 1 min. (F) Flurbiprofen ($10^{-5} M$) 90 min. $n > 12$ in each case; bars indicate mean ± SEM; $^{*}p < 0.05$; NS, not significant. Note that norepinephrine increases ^{45}Ca influx, and that this is reduced in an endothelium-dependent manner by the additional presence of acetylcholine. The effect of acetylcholine is inhibited by the three EDRF inhibitors—dithiothreitol, borohydride, and phenidone. Acetylcholine had no measurable effect in resting preparations not stimulated by norepinephrine. Note incidentally that ^{45}Ca influx was slightly *higher* in the presence of endothelium than in denuded preparations, both in resting and in norepinephrine-stimulated preparations; this difference was not abolished by flurbiprofen, implying that it was cyclooxygenase-dependent. (Lower panel) Influence of nitroprusside and 8-bromo-cyclic GMP on Ca^{2+} influx into resting or norepinephrine-stimulated endothelium-free (−) preparations of rabbit aorta. ^{45}Ca incubations for 1.5 min. Experimental protocol and conventions as for upper panel. Norepinephrine (N) was added 3 min before the addition of sodium nitroprusside (NP) ($10^{-5} M$) or 8-bromo-cyclic GMP (G) ($10^{-3} M$); NP was added 5 min before ^{45}Ca; and G was added 10 min before ^{45}Ca. $n > 18$ in each case. Note that the norepinephrine-stimulated increase of ^{45}Ca influx was inhibited by sodium nitroprusside and by 8-bromo-cyclic GMP, neither of which had measurable effects in resting preparations.

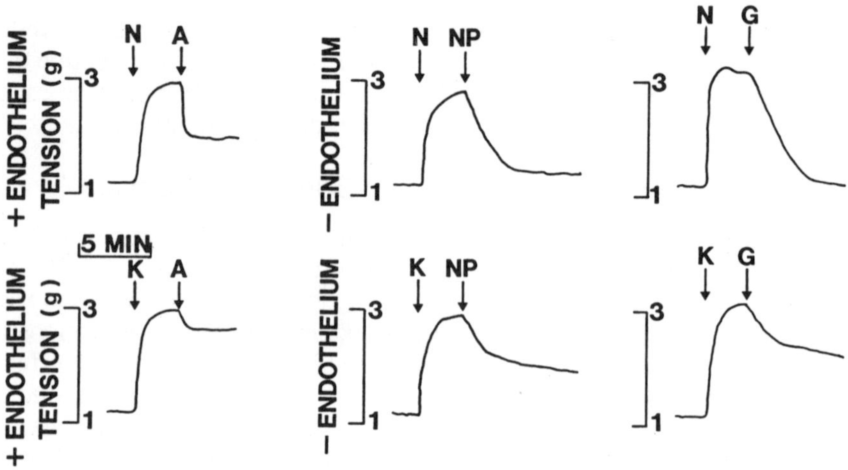

Fig. 2. Representative traces comparing relaxations in rabbit aortas contracted to similar extent with norepinephrine (N) ($10^{-5}M$) (upper traces) or with KCl (K) (120 m*M*) (lower traces). The left panel shows endothelium-dependent responses to acetylcholine (A) ($10^{-6}M$). The middle panel shows endothelium-independent responses to sodium nitroprusside (NP) ($10^{-5}M$). The right panel shows endothelium-independent responses to 8-bromo-cyclic GMP (G) ($10^{-3}M$). Note the greater relaxations to all agents in N-constricted preparations than in K-constricted preparations. The differences are significant when compared between groups, $n > 12$.

potassium-constricted arteries (Lincoln, 1983; Karaki et al., 1984, 1985). 8-Bromo-cyclic GMP likewise has been reported to have a greater mechanical relaxant effect in arteries constricted by norepinephrine than in arteries constricted by potassium (Lincoln, 1983). Rapoport and colleagues (1985) have reported a greater effect of sodium nitroprusside on cyclic GMP levels as well as on mechanical relaxation in norepinephrine-constricted than in potassium-constricted preparations of rat aorta.

Table 1
Maximum Relaxant Responses of Rabbit Aortic Preparations Constricted to a Similar Extent by Norepinephrine (N) or KCl, to EDRF (Endothelium-Dependent, 10^{-6} *M* Acetylcholine-Induced), and (in the Absence of Endothelium) to Sodium Nitroprusside (NP) and 8-Bromo-cyclic GMP (G)[a]

	N, $10^{-5}M$	KCl, 120 m*M*	*n*	*p*
EDRF	54 ± 4%	24 ± 1%	9	< 0.001
NP, $10^{-5}M$	106 ± 11%	47 ± 4%	6	< 0.001
G, $10^{-3}M$	106 ± 5%	41 ± 7%	6	< 0.001

[a]Relaxant responses (mean ± SEM) are expressed as % of contraction. Statistical comparison (Student's *t*-test) between relaxant responses in N-constricted and in KCl-constricted preparations.

In aortic preparations of rabbit and rat in the resting state, acetylcholine, sodium nitroprusside, and 8-bromo-cyclic GMP likewise caused no relaxation (unpublished observations). In line with this lack of mechanical effect, none of these interventions reduced Ca^{2+} influx in either of these preparations in the resting state (Collins et al., 1984, 1985a, 1986a,b). Cyclic GMP levels, however, are increased in noncontracted preparations of the rat aorta both by acetylcholine (in the presence of endothelium) and by nitrovasodilators (Rapoport et al., 1985).

In the rabbit aorta, norepinephrine is thought to activate only what in model terms are considered to be receptor-operated Ca^{2+} channels [norepinephrine-induced constriction is associated with a slight depolarization only at higher concentrations (Mekata, 1974; Leijten et al., 1985)], and KCl is thought to activate only voltage-operated channels (Meisheri et al., 1981). The greater effect of these cyclic GMP-mediated interventions in agonist-constricted preparations than in resting or potassium-constricted preparations suggests a relatively selective effect of EDRF on receptor-operated channels. The finding that cyclic GMP levels (Rapoport et al., 1985), as well as the mechanical and Ca^{2+} influx response to 8-bromo-cyclic GMP, were differentially affected suggests that the selectivity of the response may lie at more than one level. The selectivity would appear to be less than complete. The precise mechanisms underlying the concept of receptor-operated channels remain at present unknown.

3. Calcium Efflux

3.1. EDRF

As part of the study on Ca^{2+} fluxes and mechanical relaxation in the rabbit aorta described above, calcium efflux was also measured, using the method of Aaronson and van Breemen (1981). This method includes the achievement of a steady state and the removal of extracellular calcium by EGTA before studying the effects of interventions, so that it allows for meaningful measurements of ^{45}Ca efflux (Aaronson and van Breemen 1981; Godfraind et al., 1982).

EDRF was shown to inhibit the transient norepinephrine-induced increase of ^{45}Ca efflux (Fig. 4; Collins et al., 1985a, 1986a). Norepinephrine caused a transient increase in Ca^{2+} efflux rate. In preparations with endothelium, this is reduced by acetylcholine. The presence of endothelium slightly reduced the transient norepinephrine-evoked increase of efflux; this effect of the endothelial cells is markedly augmented by stimulation with acetylcholine. The endothelium-dependent effect of acetylcholine was blocked by the EDRF inhibitors, but not by flurbiprofen, demonstrating that it is mediated by EDRF (Fig. 4).

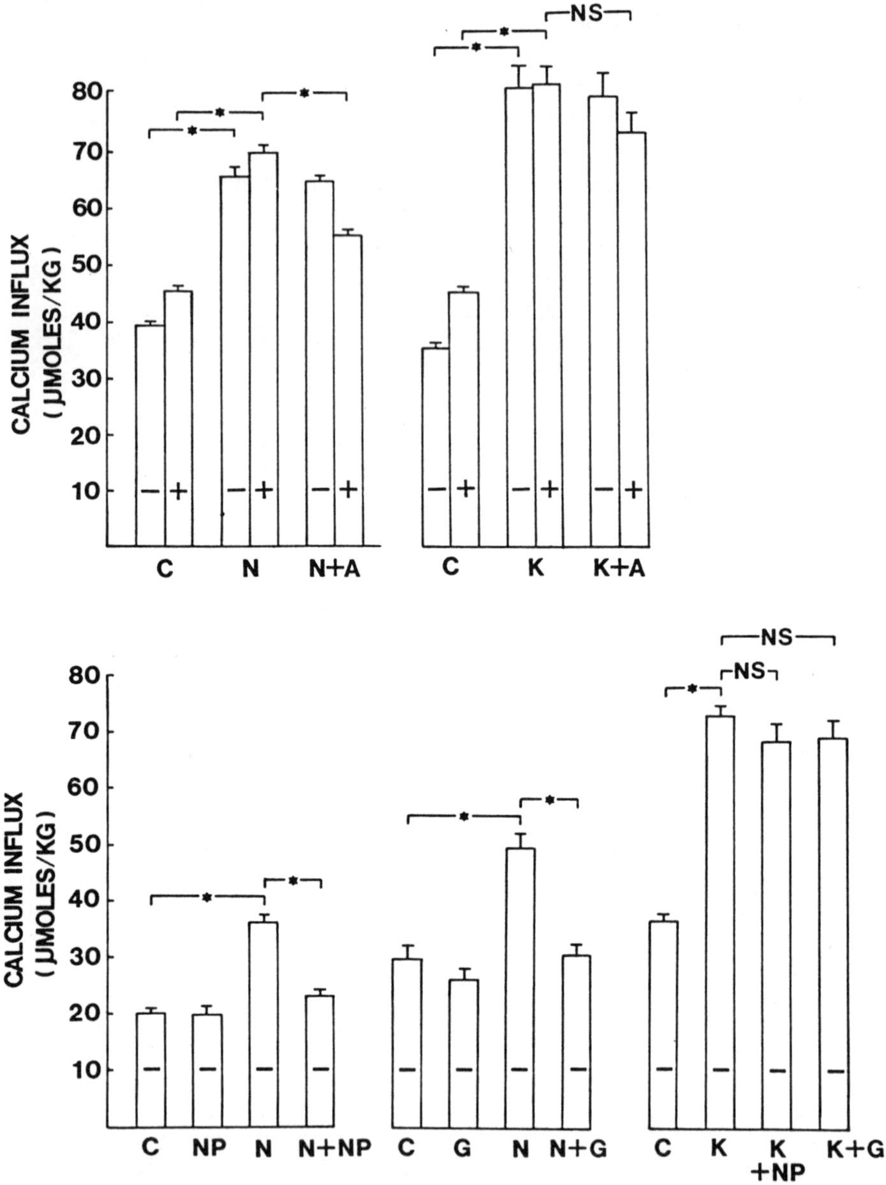

Fig. 3. Influence of EDRF, sodium nitroprusside, and 8-bromo-cyclic GMP on calcium influx, compared in norepinephrine- and K-stimulated preparations of rabbit aorta with (+) and without (−) endothelium. ^{45}Ca incubations for 1.5 min. Experimental protocol and conventions as in Fig. 1. Norepinephrine (N) (10^{-5}*M*) and KCl (120 m*M*) were added 3 min before the addition of acetylcholine

3.2. Nitrovasodilators, 8-Bromo-cyclic GMP

The effect of nitrovasodilators on Ca^{2+} efflux in vascular smooth muscle has been studied by several groups (Hester et al., 1979; Nakazawa and Imai, 1981; Itoh et al., 1983; Hester, 1985; Matlib et al., 1985). The findings are generally more conflicting than reports of the effects of nitrovasodilators on calcium influx, probably because of differences in the methods used. For example, in only one study with an intact vascular smooth muscle preparation was an agonist (histamine) used to stimulate calcium efflux (Matlib et al., 1985). This study, with isolated canine coronary artery preparation, employed a chelating agent (EGTA) to remove extracellular Ca^{2+} (whereas most other studies have not) and demonstrated that isosorbide dinitrate inhibited ^{45}Ca efflux in association with an increase in the tissue content of cyclic GMP. In aortic preparations of the rabbit, sodium nitroprusside and 8-bromo-cyclic GMP, like EDRF, both reduce the transient norepinephrine-induced increase of ^{45}Ca efflux (Collins et al., 1985b, 1986a) (Fig. 4). The persistence of this effect in the absence of extracellular Ca^{2+} confirms that it is caused by a reduction of efflux itself and not attributable to an effect on net efflux secondary to increased influx. The effect of EDRF could not be studied in the absence of extracellular calcium, because EDRF production is dependent on the presence of extracellular Ca^{2+} (Long and Stone 1985a,b; Griffith et al., 1985b; 1986).

3.3. Implications of Efflux Data

The agents studied were added after the ^{45}Ca labeling of the preparations, so precluding indirect effects caused by reduced loading of intracellular stores. The decrease in ^{45}Ca efflux might be caused by increased calcium sequestration, since sodium nitroprusside depressed phasic contractions in norepinephrine-stimulated rat aortas in the absence of extracellular calcium (Lincoln, 1983). There are, however, alternative explana-

---

← Fig. 3 (*Continued*)

(A) ($10^{-6}M$), nitroprusside (NP) ($10^{-5}M$), or 8-bromo-cyclic GMP (G) ($10^{-3}M$). A was added 1.5, NP 5, and G 10 min before ^{45}Ca. Bars indicate mean ± SEM; $n > 18$; $^*p < 0.05$; NS, not significant. Note that the endothelium-dependent acetylcholine-induced reduction of ^{45}Ca influx was 67 ± 8% in N-stimulated preparations, which is significantly more ($p < 0.01$) than the 25 ± 9% in K-stimulated preparations; likewise NP- and G-induced reductions of ^{45}Ca influx were 83 ± 2% and 90 ± 13%, respectively, in N-stimulated preparations, compared with 16 ± 8% and 11 ± 10% ($p < 0.01$ in each case) in K-stimulated preparations ($n > 12$).

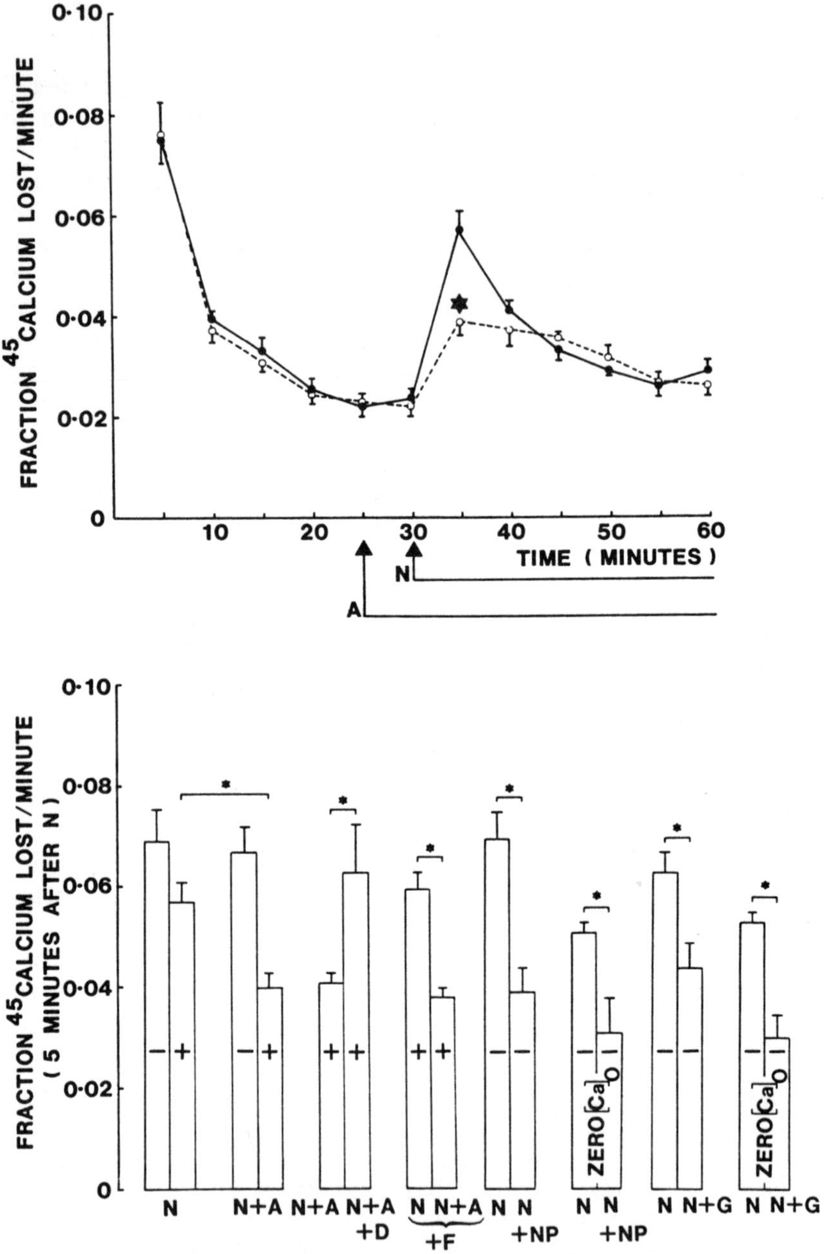

Fig. 4. (Upper panel) Efflux rates from preparations with endothelium, as influenced by the addition of norepinephrine (N) ($10^{-5}M$) at near steady efflux rate, and as additionally influenced by acetylcholine (A) ($10^{-6}M$), added at time shown and present in subsequent aliquots of buffer. Solid line, only N added;

tions for this observation. The transience of the effects on ^{45}Ca efflux observed in the experiments described above argues against a sustained increase in calcium sequestration. It also seems unlikely on theoretical grounds, insofar as it would lead to intracellular accumulation of Ca^{2+}. Another possible explanation for the findings is that these agents directly reduce Ca^{2+} efflux. This too seems unlikely in view of recent evidence that cyclic GMP, by activating cyclic GMP-dependent protein kinase, stimulated ATP-dependent calcium extrusion across the sarcolemma of vascular smooth muscle (Popescu et al., 1985). The most likely explanation for the efflux data is a reduction in the amount of calcium released from intracellular stores. We therefore may conclude that raised levels of cyclic GMP—as induced by EDRF, sodium nitroprusside, or 8-bromo-cyclic GMP—impair the intracellular release of calcium induced by agonists such as norepinephrine.

4. Biochemical Mechanisms

Breakdown of phospholipids of the cell membrane by agonist agents such as norepinephrine gives rise to the formation of a number of second messengers within the cell (Berridge and Irvine, 1984). Two of these second messengers, inositol 1,4,5-trisphosphate (IP_3) and diacylglycerol (DAG), are derived from phosphatidylinositol and probably regulate cal-

← Fig. 4. (*Continued*)

broken line, N+A added. Mean ± SEM, $n = 5$, $*p < 0.05$ cf efflux rate during period 0–5 min after adding only N, i.e., point of maximal divergence of traces. (Lower panel) Effect of various agents on N-stimulated ^{45}Ca efflux rates from preparations with (+) or without (−) endothelium. Agents were first added at time before addition of N, as indicated below. Protocol is illustrated in upper panel; presented are rates during the period from 0 to 5 min after adding N (N.B.: No agent influenced ^{45}Ca efflux rate before or more than 10 min after addition of N). (N) Norepinephrine ($10^{-5}M$). (A) Acetylcholine ($10^{-6}M$) 5 min. (D) Dithiothreitol ($5 \times 10^{-4}M$) 5 min. (F) Flurbiprofen ($10^{-5}M$) 255 min. (NP) Sodium nitroprusside ($10^{-5}M$) 10 min. (G) 8-Bromo-cyclic GMP ($10^{-3}M$) 10 min. Buffer contained 1.5 mM $CaCl_2$, except where shown. Bars indicate mean ± SEM, $n = 6$, $*p < 0.05$. Note that acetylcholine caused endothelium-dependent reduction of transient, norepinephrine-induced increase in ^{45}Ca efflux, and that this effect of acetylcholine was inhibited by the EDRF inhibitor, dithiothreitol (and other EDRF inhibitors, data not shown), but not by the cyclooxygenase inhibitor flurbiprofen. Sodium nitroprusside and 8-bromo-cyclic GMP likewise reduced norepinephrine-stimulated ^{45}Ca efflux, in denuded preparations and in the presence or absence of extracellular Ca^{2+}.

cium movements. IP_3 releases Ca^{2+} from internal stores in skinned smooth muscle cells of the rat and rabbit (Yamamoto and van Breemen, 1985; Hashimoto et al., 1986). The role of DAG is less clear, but Campbell and colleagues (1985) suggest that DAG or its product, phosphatidic acid, is responsible for receptor-induced calcium influx. Since cyclic GMP inhibits phosphatidylinositol breakdown in vascular smooth muscle (Rapoport, 1986) as well as in platelets (Takai et al., 1981; McNicol and MacIntyre, 1986), this could explain the reductions both of Ca^{2+} influx and of intracellular Ca^{2+} release observed in the studies summarized in this chapter.

5. Concluding Remarks

It is clear that EDRF-mediated relaxation of vascular smooth muscle is associated with a reduction of Ca^{2+} influx and also a reduction of norepinephrine-stimulated Ca^{2+} efflux. Sodium nitroprusside, whose relaxant action is similarly associated with a rise in cyclic GMP, and 8-bromo-cyclic GMP, a lipid-soluble analog of cyclic GMP itself, exert the same effects. The similar findings with these three interventions imply that they are mediated by cyclic GMP.

The reduction of Ca^{2+} influx and the mechanical relaxation induced by these three agents is greater when the preparations are constricted by norepinephrine, whose action is considered to be through receptor-operated Ca^{2+} channels, than when they are resting or when they are constricted to a similar mechanical degree by high concentrations of potassium, which is considered to cause constriction through voltage-operated channels. This suggests that the reduction of calcium influx that these agents cause may be effected predominantly through modulating the receptor-operated channels. The similar findings with all three agents indicate that this is a cyclic GMP-mediated effect, though a report that cyclic GMP levels are also differentially responsive to sodium nitroprusside (Rapoport et al., 1985) in norepinephrine and potassium-stimulated arteries suggests that the selectivity may operate at more than one level.

Reduction of the transient norepinephrine-induced rise in Ca^{2+} efflux is interpreted as reflecting a reduction in the release of intracellular Ca^{2+}. Thus cyclic GMP appears also to influence the amount of calcium that is released by constrictor agonists such as norephinephrine from intracelllar stores.

There is evidence too that cyclic GMP leads to stimulation of active Ca^{2+} extrusion across the vascular smooth muscle cell membrane (Popescu et al., 1985).

A reduction of Ca^{2+} influx and of intracellular release, together with stimulation of active Ca^{2+} efflux, will each act toward reducing the calcium content of vascular smooth muscle, as has been reported in the presence of endothelium (Godfraind et al., 1985), and toward reducing free cytosolic calcium, as has been described with sodium nitroprusside in potassium-depolarised ferret portal vein (Morgan and Morgan, 1984). A reduction in free cytosolic Ca^{2+} will reduce the level of contraction. The previously described dephosphorylation of myosin light chains is likely to be mediated by a reduction of cytosolic Ca^{2+} concentration, as suggested by Johnson and Lincoln (1985), since this phosphorylation is now known to be a calcium-calmodulin-dependent process that is not directly influenced by cyclic GMP-dependent kinase (Aksoy and Murphy, 1983; Hathaway et al., 1985).

EDRF, like the nitrovasodilators, appears to relax vascular smooth muscle by increasing the level of cyclic GMP. It appears that cyclic GMP mediates mechanical relaxation by reducing the concentration of free cytosolic Ca^{2+} as a consequence of modulating calcium movements at a number of control sites. In this respect, the role of cyclic GMP in vascular smooth muscle may be regarded as analogous, but opposite, to that of cyclic AMP in heart muscle.

References

Aaronson, P. and van Breemen, C.: Effects of sodium gradient manipulation upon cellular calcium, ^{45}Ca fluxes and cellular sodium in the guinea pig taenia coli. *J. Physiol.* **319:**443–461, 1981.

Aksoy, M. O. and Murphy, R. A.: Regulation of the Dynamic Properties of Smooth Muscle: Ca^{++}-Stimulated Cross-Bridge Phosphorylation, in *Biochemistry of Smooth Muscle* vol. 1 (Stephens, N. L., ed.) CRC, Florida, 1983.

Berridge, M. J. and Irvine, R. F.: Inositol triphosphate, a novel second messenger in cellular signal transduction. *Nature* **312:**315–321, 1984.

Busse, R., Trogisch, G., and Bassenge, E.: The role of endothelium in the control of vascular tone. *Basic Res. Cardiol.* **80:**475–490, 1985.

Campbell, M. D., Deth, R. C., Payne, R. A., and Honeyman, T. W.: Phosphoinositide hydrolysis is correlated with agonist-induced calcium influx and contraction in the rabbit aorta. *Eur. J. Pharmacol.* **116:**129–136, 1985.

Collins, P., Griffith, T. M., Henderson, A. H., and Lewis, M. J.: Endothelium and calcium flux in rabbit aortic preparations. *J. Physiol.* (Lond.) **360:**63P, 1984.

Collins, P., Griffith, T. M., Henderson, A. H., and Lewis, M. J.: Endothelium and calcium flux in rabbit aortic preparations. *Br. J. Pharmacol.* **85:**342P, 1985a.

Collins, P., Griffith, T. M., Henderson, A. H., and Lewis, M. J.: 8-Bromo-cGMP and calcium flux in arterial smooth muscle. *Br. J. Pharmacol.* **85:**279P, 1985b.

Collins, P., Griffith, T. M., Henderson, A. H., and Lewis, M. J.: Endothelium-derived relaxing factor alters calcium fluxes in rabbit aorta: A cyclic guanosine monophosphate-mediated effect. *J. Physiol.* (Lond.) *381:*427–437, 1986.

Collins, P., Chappell, S. P., Griffith, T. M., Lewis, M. J., and Henderson, A. H.: Differences in basal endothelium-derived relaxing factor activity in different artery types. *J. Cardiovasc. Pharmacol.* **8:**1158–1162, 1986b

Gleason, M. M., Ratz, P. H., and Flaim, S. F.: Measurement of calcium influx in tethered rings of rabbit aorta under tension. *Am. J. Physiol.* **249:**H470–H476, 1985.

Godfraind, T., Miller, R. C., and Socrates Lima, J.: Selective α_1- and α_2-adrenoceptor agonist-induced contraction and ^{45}Ca fluxes in the rat isolated aorta. *Br. J. Pharmacol.* **77:597**–604, 1982.

Godfraind, T., Egleme, C., and Osachie, J. A.: Role of endothelium in the contractile response of rat aorta to α-adrenoceptor agonists. *Clin. Sci.* **68** (suppl. 10):65S–71S, 1985.

Griffith, T. M., Edwards, D. H., Lewis, M. J., Newby, A. C., and Henderson, A. H.: The nature of endothelium-derived relaxant factor. *Nature* **308:**645–647, 1984.

Griffith, T. M., Edwards, D. H., Lewis, M. J., and Henderson, A. H.: Evidence that cyclic guanosine monophosphate (cGMP) mediates endothelium-dependent relaxation. *Eur. J. Pharmacol.* **112:**195–202, 1985a.

Griffith, T. M., Edwards, D. H., Collins, P., Lewis, M. J., and Henderson, A. H.: Endothelium-derived relaxant factor. *J. Roy. Coll. Phys. Lond.* **19:**74–79, 1985b.

Griffith, T. M., Edwards, D. H., Newby, A. C., Lewis, M. J., and Henderson, A. H.: Production of endothelium-derived relaxant factor (EDRF) is dependent on oxidative phosphorylation and extracellular calcium. *Cardiovasc. Res.* **20:**7–12, 1986.

Haeusler, G. and Thorens, S.: Pharmacology of vasoactive antihypertensives. *Blood Vess.* **12:**339, 1975.

Hashimoto, T., Hirata, M., Itoh, T., Kanmura, R., and Kuriyama, H.: Inositol 1,4,5-trisphosphate activates pharmacomechanical coupling in smooth muscle of the rabbit mesenteric artery. *J. Physiol.* **370:**605–618, 1986.

Hathaway, D. R., Konicki, M. V., and Coolican, S. A.: Phosphorylation of myosin light chain kinase from vascular smooth muscle by cAMP- and cGMP-dependent protein kinases. *J. Mol. Cell. Cardiol.* **17:**841–850, 1985.

Hester, R. K.: Effects of 2-nicotinamidoethyl nitrate on agonist-sensitive Ca^{++} release and Ca^{++} entry in rabbit aorta. *J. Pharmacol. Exp. Ther.* **233:**100–111, 1985.

Hester, R. K., Weiss, G. B., and Fry, W. J.: Differing actions of nitroprusside and D-600 on tension and ^{45}Ca fluxes in canine and renal arteries. *J. Pharmacol. Exp. Ther.* **208:**155–60, 1979.

Ignarro, L. J., Lippton, H., Edwards, J. C., Baricos, W. H., Hyman, A. L., Kadowitz, P. J., and Gruetter, C. A.: Mechanism of vascular smooth muscle relaxation by organic nitrates, nitrites, nitroprusside and nitric oxide: Evidence for the involvement of S-nitrothiols as active intermediates. *J. Pharmacol. Exp. Ther.* **218:**739–749, 1981.

Itoh, J., Kuriyama, H., and Ueno, H.: Mechanism of the nitroglycerin-induced vasodilatation in vascular smooth muscles of rabbit and pig. *J. Physiol.* (Lond.) **343:**233–252, 1983.

Johnson, R. M. and Lincoln, T. M.: Effects of nitroprusside, glyceryl trinitrate, and 8-bromo-cyclic GMP on phosphorylase a formation and myosin light chain phosphorylation in rat aorta. *Mol. Pharmacol.* **27:**3333–3342, 1985.

Karaki, H., Hester, R. K., and Weiss, G. B.: Cellular basis of nitroprusside-induced relaxation of graded responses to norepinephrine and potassium in canine renal arteries. *Arch. Int. Pharmacodyn. Ther.* **245:198**–210, 1980.

Karaki, H., Nakagawa, H., and Urakawa, N.: Comparative effects of verapamil and sodium nitroprusside on contraction and ^{45}Ca uptake in smooth muscle of rabbit aorta, rat aorta and guinea pig taenia coli. *Br. J. Pharmacol.* **81:**393–400, 1984.

Karaki, H., Nakagawa, H., and Urakawa, N.: Age-related changes in the sensitivity to verapamil and nitroprusside of vascular smooth muscle of rabbit aorta. *Br. J.Pharmacol.* **85:**223–228, 1985.

Katsuki, S., Arnold, W., Mittal, C., and Murad, F.: Stimulation of guanylate cyclase by sodium nitroprusside, nitroglycerin and nitric oxide in various tissue preparations and comparison to the effects of sodium azide and hydroxylamine. *J. Cyclic Nucleotide Res.* **3:**23–25, 1977.

Kreye, V. A. W.: Sodium nitroprusside; approaches towards the elucidation of the mode in action. *Trends Pharmacol. Sci.* **1:**384–388, 1980.

Kreye, V. A. W., Baron, G. D., Luth, J. B., and Schimdt-Gayk, H.: Mode of action of sodium nitroprusside on vascular smooth muscle. *Naunyn Schmiedebergs Arch. Pharmacol.* **288:**381–402, 1975.

Leitjen, P., Cauvin, C., Lodge, N., Saida, K., and van Breeman, C.: Ca^{++} sources mobilised by α_1-receptor activation in vascular smooth muscle. *Clin. Sci.* **68:**(suppl. 10)**:**47s–50s, 1985.

Lincoln, T. M.: Effects of nitroprusside and 8-bromo-cyclic GMP on the contractile activity of the rat aorta. *J. Pharmacol. Exp. Ther.* **224:**100–107, 1983.

Long, C. J. and Stone, T. W.: The release of endothelium-dependent relaxant factor is calcium dependent. *Br. J. Pharmacol.* **84:**152P, 1985a.

Long, C. J. and Stone, T. W.: The release of endothelium-derived relaxant factor is calcium dependent. *Blood Vess.* **22:**205–208, 1985b.

Loutzenhiser, R. and Epstein, M.: Activation mechanism of human renal artery: Effects of KCl, norepinephrine and nitrendipine upon tension development and ^{45}Ca influx. *Eur. J. Pharmacol.* **106:**47–52, 1985.

Matlib, M. A., Dube, G. P., Millard, R. W., Lathrop, D. A., Baik, Y. H., Sakai, K., DiSalvo, J., and Schwartz, A.: Studies on the mode of action of isosorbide dinitrate: A physiological and biochemical approach. *Am. Heart J.* **110**:204–212, 1985.

McNicol, A. and MacIntyre, D. E.: Reversal of agonist-induced phosphoinositide hydrolysis in human platelets by cAMP, cGMP and 1,2-diacylglycerol. *Br. J. Pharmacol.* **86**:546P, 1986.

Meisheri, L. D., Palmer, L. F., and van Breemen, C.: The effects of amrinone on contractility, calcium intake and cAMP in smooth muscle. *Eur. J. Pharmacol.* **61**:159–165, 1980.

Meisheri, K. D., Hwang, O., and van Breemen, C.: Evidence for two separate Ca^{++} pathways in smooth muscle plasma lemma. *J. Membr. Biol.* **59**:19–25, 1981.

Mekata, F.: Current spread in the smooth muscle of the rabbit aorta. *J. Physiol.* (Lond.) **242**:143–155, 1974.

Morgan, J. P. and Morgan, K. G.: Alteration of cytoplasmic ionised calcium levels in smooth muscle by vasodilators in the ferret. *J. Physiol.* (Lond.) **357**:539–551, 1984.

Nakazawa, M. and Imai, S.: Effect of nitroglycerin on Ca efflux in the coronary artery. *Jpn. J. Pharmacol.* **31**:621–625, 1981.

Ozaki, H., Shibata, S., Kitano, H., Masumoto, P., and Ishida, Y.: A comparative study of the relaxing effects of nitroprusside and verapamil on human umbilical vessels. *Blood Vess.* **18**:321–329, 1981.

Popescu, L. M., Panoiu, C., Hinescu, M., and Nutu, O.: The mechanism of cGMP-induced relaxation in vascular smooth muscle. *Eur. J. Pharmacol.* **107**:393–394, 1985.

Rapoport, R. M.: Cyclic guanosine monophosphate inhibition of contraction may be mediated through inhibition of phosphatidylinositol hydrolysis in rat aorta. *Circ. Res.* **58**:407–410, 1986.

Rapoport, R. M. and Murad, F.: Agonist-induced endothelium-dependent relaxation in rat thoracic aorta may be mediated through cGMP. *Circ. Res.* **52**:351–357, 1983.

Rapoport, R. M., Draznin, M. B., and Murad, F.: Endothelium-dependent relaxation in rat aorta may be mediated through cyclic GMP-dependent protein phosphorylation. *Nature* **306**:174–176, 1983a.

Rapoport, R. M., Draznin, M. B., and Murad, F.: Endothelium-dependent vasodilator- and nitrovasodilator-induced relaxation may be mediated through cyclic GMP formation and cyclic GMP-dependent protein phosphorylation. *Trans. Assoc. Am. Physicians* **96**:19–30, 1983b.

Rapoport, R. M., Schwartz, K., and Murad, F.: Effect of sodium-potassium pump inhibitors and membrane-depolarising agents on sodium nitroprusside-induced relaxation and cyclic guanosine monophosphate accumulation in rat aorta. *Circ. Res.* **57**:164–170, 1985.

Ratz, P. H., Gleason, M. M., and Flaim, S. F.: Acetylcholine relaxes norepinephrine-contracted rabbit thoracic aorta by inhibiting calcium influx and release. *Fed. Proc.* **43**:837, 1985.

Takai, Y., Karbuchi, K., Matsubara, T., and Mishiyuka, Y.: Inhibitory action of guanosine 3′, 5′-monophosphate on thrombin-induced phosphatidylinositol turnover and protein phosphorylation in human platelets. *Biochem. Biophys. Res. Comm.* **101:**61–67, 1981.

van Breemen, C., Aaronson, P. I., Cauvin, C. A., Loutzenhiser, R. D., Mangel, A. W., and Saida, K.: The Calcium Cycle in Arterial Smooth Muscle, in *Calcium Blockers* (Flaim, S. F. and Zelis, R., eds.) Urban and Schwarzenberg, Baltimore, Munich, 1982.

Yamamoto, H. and van Breemen, C.: Inositol-1, 4, 5-triphosphate releases calcium from skinned cultured smooth muscle cells. *Biochem. Biophys. Res. Comm.* **130:**270–274, 1985.

Zsoter, T. T., Henein, N. F., and Wolchinsky, C.: The effect of sodium nitroprusside on the uptake of ^{45}Ca from rabbit and rat vessels. *Eur. J. Pharmacol.* **45:**7–12, 1977.

Chapter 14

Heterogeneity in Endothelium-Dependent Relaxations

Acute, Chronic, and Evolutionary Modulations

Virginia M. Miller and Paul M. Vanhoutte

1. Introduction

Since the identification by Furchgott and Zawadzki (1980) of the role of the endothelium as a modulator of vascular smooth muscle, attention has focused on the characterization of endothelium-dependent responses and on the identification of the substance(s) that mediates such responses in mammalian blood vessels. It soon became obvious that the amplitude and direction of endothelium-dependent responses to a given agonist varied among vessels derived from different vascular beds (De Mey and Vanhoutte, 1982). For example, acetylcholine initiates endothelium-dependent relaxations in systemic arteries of the dog, whereas in the comparable veins such relaxations were absent or blunted. Since the endothelia of veins and arteries are chronically exposed to quite different physical and chemical conditions, one explanation for the variations in responsiveness among them is that exposure to a given environment modulates the function of the endothelial cells and therefore affects the expression of endothelium-dependent responses.

Evidence for the modulation of endothelial cell function by alterations in the chemical environment (i.e., hormonal status, oxygen tension) or physical environments (i.e., shear stress to which the cells are exposed) is derived primarily from experiments utilizing endothelial cells in culture. Experiments that manipulate these factors in vivo are often difficult to interpret, in that changes in the environment of the blood vessels invariably affect both the vascular smooth muscle and the endothelial cells. Models that alter endothelial cell environment in vivo with subsequent study of the endothelium-dependent responses of the blood vessels in vitro address this issue. This chapter will focus on such studies.

2. Hormones

2.1. Acute Effects

2.1.1. Catecholamines

2.1.1.1. Alpha-Adrenoceptors. Alpha-adrenoceptors have been identified in endothelial cells in culture in which their activation is associated with changes in adenylate and guanylate cyclase activity (Wagner et al., 1972; Venter et al., 1975; Bounassisi and Venter, 1976; Karnushina et al., 1982). Cocks and Angus (1983) first demonstrated endothelium-dependent relaxations to norepinephrine in canine and porcine coronary arteries contracted with a thromboxane analog. Norepinephrine initiated endothelium-dependent relaxations in the presence of both $alpha_1$ and beta-adrenergic blockade. The partial $alpha_2$-adrenoceptor agonist, clonidine, mimicked the relaxations induced by norepinephrine; they are prevented by selective $alpha_2$-adrenergic antagonists (Fig. 1). Thus, the relaxations can be attributed to activation of $alpha_2$-adrenoceptors on the endothelium. Endothelial $alpha_2$-adrenergic receptors have also been identified in femoral arteries and pulmonary veins of the dog (Fig. 2; Miller and Vanhoutte, 1985). Endothelium-dependent relaxations to $alpha_2$-adrenergic activation as evidenced by relaxations during contractions with either prostaglandin $F_{2\alpha}$ or 5-hydroxytryptamine are difficult to demonstrate in femoral veins or pulmonary arteries of the same species (Miller and Vanhoutte, 1985), in which a relaxation may be masked in part by direct $alpha_2$-adrenergic activation of the smooth muscle. $Alpha_2$-adrenergic, endothelium-dependent responses are less pronounced in mesenteric and renal arteries compared to femoral, carotid, and coronary arteries of the dog, even though $alpha_2$-adrenergic activation of the smooth muscle is weak (Angus et al., 1986). These regional variations in $alpha_2$-adrenergically mediated endothelium-dependent responses could be explained either by differences in density of endothelial $alpha_2$-adrenergic receptors among blood vessels or by nonuniform release or different efficacy of endothelium-derived relaxing factor(s) in individual vascular beds.

The removal of the endothelium enhances norepinephrine-induced contractions of isolated rings of systemic, pulmonary, and cerebral arteries

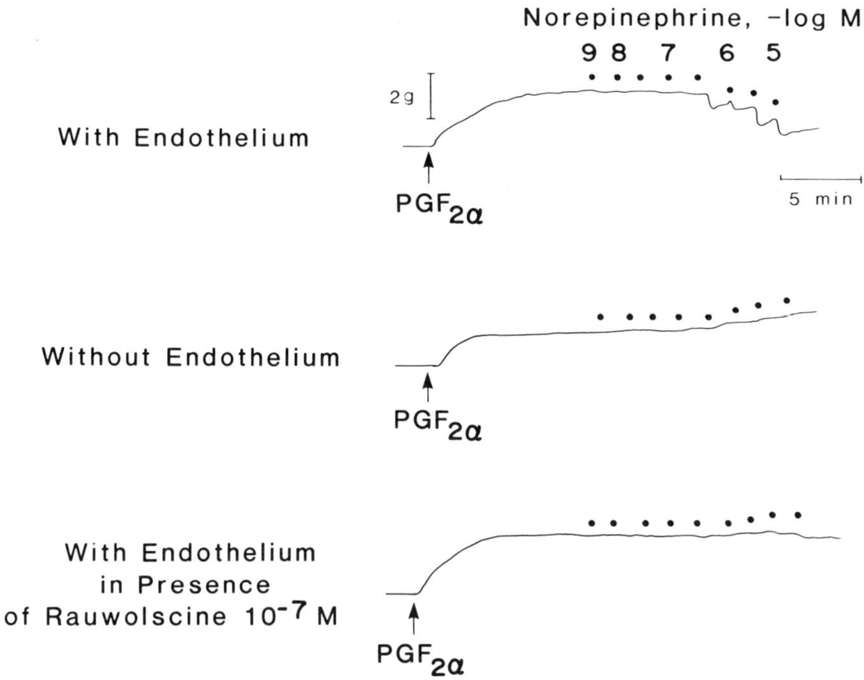

Fig. 1. Isometric tension recordings in three isolated rings prepared from the same canine coronary artery and contracted with prostaglandin $F_{2\alpha}$ ($2 \times 10^{-6}M$; $PGF_{2\alpha}$). In a ring with endothelium (upper) norepinephrine (10^{-9}, 10^{-8}, 5×10^{-8}, 10^{-7}, 5×10^{-7}, 10^{-6}, 10^{-5}, and $5 \times 10^{-5}M$) at the consecutive dots (•), caused relaxation. This was not seen in the ring without endothelium (middle) or if the preparation was treated with the selective $alpha_2$-adrenergic antagonist rauwolscine. The experiments were performed in the presence of prazosin ($10^{-7}M$) and propranolol ($5 \times 10^{-6}M$) throughout to inhibit $alpha_1$- and beta-adrenoceptors, respectively. They confirm the original observations of Cocks and Angus, 1983 (from Vanhoutte, 1986, with permission).

of cats, dogs, rabbits, and rats (Egleme et al., 1984; Lues and Schumann, 1984; Carrier and White, 1985; Godfraind et al., 1985; Matsuda et al., 1985; Miller and Vanhoutte, 1985; Sercombe et al., 1985) and pulmonary and systemic veins of the dog (Miller and Vanhoutte, 1985). In aortic preparations taken from rats, removal of the endothelium also enhances contractions evoked by selective $alpha_1$-adrenergic agonists (methoxamine, phenylephrine) (Lues and Schumann, 1984; Carrier and White, 1985; Martin et al., 1986). Enhanced contraction to epinephrine, norepinephrine, and phenylephrine was also reported during *in situ* infusion of the drugs in iliac arteries of conscious dogs when the endothelium was removed by

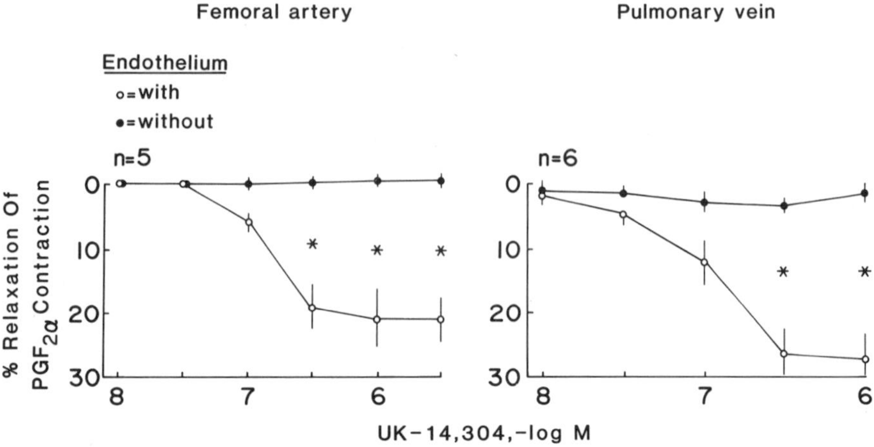

Fig. 2. Relaxations induced by UK 14,304 during contractions evoked by prostaglandin $F_{2\alpha}$ ($2 \times 10^{-6}M$) in paired rings of canine femoral arteries (n = 5, left) and pulmonary veins (n = 6, right) with (○) and without (●) endothelium. The experiments were performed in the presence of prazosin ($10^{-7}M$) and propranolol ($5 \times 10^{-6}M$). Data are shown as means ± SEM, and are expressed as percent of the initial contraction to prostaglandin $F_{2\alpha}$, which did not differ between rings with and without endothelium (100% = 6.2 ± 1.9 and 4.0 ± 0.8 g for femoral arteries and pulmonary veins, respectively). *Denotes that the difference between rings with and without endothelium is statistically significant ($p < 0.05$) (from Miller and Vanhoutte, 1985, with permission).

balloon catheter (Young and Vatner, 1986). Likewise, removal of the endothelium shifts the concentration–response curve to norepinephrine to the left in pulmonary arteries of the dog and that to norepinephrine and selective $alpha_2$-adrenergic agonists in femoral veins of the same species (Miller and Vanhoutte, 1985). These results are explained best by the removal of tonic or "spontaneous" release of endothelium-derived relaxing factor (e.g., Martin et al., 1986; Rubanyi et al., 1985). If endothelium-derived relaxing factor(s) acts as a functional antagonist of the contractile activation of the vascular smooth muscle, then tonic release of the factor(s) would be more effective in modulating responses initiated by partial agonists or agonists of low efficacy than of responses initiated by full agonists of high efficacy (Martin et al., 1986).

2.1.1.2. Beta-Adrenoceptors. Norepinephrine increases the formation of actin and decreases the motility of aorta-derived endothelial cells in culture; in both cases a beta-adrenergic mechanism seems to be involved (Bottaro et al., 1985; Welles et al., 1985). In isolated canine coronary arteries, removal of the endothelium reduces relaxations induced by isoproterenol (Rubanyi and Vanhoutte, 1985a) during contractions evoked by prostaglandin $F_{2\alpha}$, but not potassium chloride. Although these results suggest modulation of beta-adrenergically mediated responses by the endothelium, they are also consistent with the notion of tonic release of endothelium-derived relaxing factor(s) antagonizing partial (prostaglandin $F_{2\alpha}$), but not full (KCl), activation of the vascular smooth muscle.

2.1.2. Vasopressin and Oxytocin

In the canine basilar artery, both vasopressin and oxytocin in a physiologically relevant range of concentrations (10^{-12} to $10^{-8}M$) initiate endothelium-dependent relaxations by stimulation of V_1 vasopressinergic receptors (Katusic et al., 1984, 1986; *see* chapter by Katusic and Shepherd, in this volume). In the femoral artery of the same species, only contractions caused by direct activation of the smooth muscle are observed with vasopressin (Katusic et al., 1984). These results suggest that in physiological situations in which release of vasopressin may be high (i.e., hemorrhagic or septic shock), the hormone's activation of the endothelium would enhance blood flow to the brain while reducing flow to the limbs (Vanhoutte et al., 1984).

2.1.3. Glucagon

Glucagon added to isolated helical strips of cerebral, coronary, femoral, mesenteric, and renal arteries and veins of the dog initiated relaxation and increased the intracellular levels of cyclic adenosine monophosphate. These relaxations were not modified, however, by the presence or absence of endothelium (Okamura et al., 1986).

2.2. Chronic Effects

2.2.1. Steroid Hormones

2.2.1.1. Estrogens. The demonstration of binding sites for estrogen in the cytosol of endothelial cells (Colburn and Buonassisi, 1978) suggests that estrogens (and to a lesser extent progesterone and testosterone that may inhibit or compete with estrogen binding) could effect endothelial cell function and possibly the production or release of endothelium-derived vasoactive substances. Generally, estrogen treatment does not affect growth or DNA synthesis of endothelial cells in culture (Taggart and Stout, 1980; Covazier et al., 1984; Maca et al., 1978; Seillan et al., 1983). The number of actin-myosin stress fibers within endothelial cells may be influenced by the hormones, however (White et al., 1983), with greater number of stress fibers being present in endothelial cells from male compared to female rats. Also, the effect of estrogen treatment on the metabolism of arachidonic acid by endothelial cell cultures and therefore the production of vasoactive prostaglandins seems to depend upon the origin of the endothelial cell line and the length of exposure to the hormone. Thus, the release of prostacyclin by the calcium ionophore A23187 was not affected by estrogen treatment in endothelial cells derived from human umbilical vein either grown in the presence of the hormone or after 30 min of exposure to it (Covazier et al., 1984; Witter and DiBlasi, 1984). In cells derived from aortas of female piglets, however, the addition of estrogen to the growth media initiated two- to fivefold increases in the production of prostaglandins (including prostacyclin, prostaglandin $F_{2\alpha}$, and prostaglandin E_2) only after 48 h of exposure to the hormone. By contrast, such increases were not observed in endothelial cells derived from aortas of male piglets (Seillan et al., 1983).

Indirect evidence for chronic modulation of endothelium-mediated vascular responses by estrogens comes from the observation that treatment with the hormone augmented the relaxation to acetylcholine of the uterine artery of the guinea pig (Bell and Coffey, 1982). These relaxations were not eliminated by the section of the cholinergic vasodilator nerves to the uterus, but were reduced by inhibitors of protein synthesis. However, since acetylcholine initiates relaxations in most mammalian arteries by its action on the endothelium (*see* Furchgott, 1984; Vanhoutte et al., 1986), it is tempting to speculate that estrogens modulate this effect. This interpretation is confirmed by studies in which the endothelium-dependent responses to acetylcholine were compared in the femoral arteries of castrated female rabbits with or without treatment with 17 β-estradiol (for 4 d; Gisclard and Vanhoutte, 1985). The endothelium-dependent relaxations to 10^{-9} to $10^{-7}M$ acetylcholine were greater in estrogen-treated animals, an effect that was not eliminated by prior treatment of the blood vessels with inhibitors of prostaglandin production. These inhibitors also do not abolish the increased uterine blood flow in estrogen-primed rabbits (Mueller et al., 1978). Since the treatment with estrogens failed to alter endothelium-dependent relaxations initiated by the calcium ionophore A23187 (which are not mediated by receptors on the cell membrane of the endothelial cells), the hormone probably modulates vascular responses by alteration of endothelial receptors not associated with the production of prostaglandins.

2.2.1.2. Glucocorticoids. Although glucocorticoid hormones do not affect endothelial cell replication (Maca et al., 1978; Jarvelainen et al., 1985; Longnecker et al., 1982), they may have a stabilizing effect on endothelial cellular membranes (Jarvelainen et al., 1985). Chronic exposure of endothelial cell cultures to either dexamethasone or hydrocortisone increases protein synthesis (Maca et al., 1978; Nichols et al., 1983; Mendelsohn et al., 1982). In other cellular systems (rat peritoneal leukocytes), hydrocortisone-treatment is associated with inhibition of phospholipase A_2, an enzyme that releases arachidonic acid from membrane phospholipids (Blackwell et al., 1980). Treatment of bovine aortic endothelial cells with hydrocortisone results in an inhibition of the bradykinin-induced production of prostacyclin, whereas that induced by arachidonic acid is unaffected (Crutchley et al., 1985). Although these different observations all suggest that chronic exposure to glucocorticoids can affect endothelial function, it is unknown whether or not they also modulate endothelium-dependent relaxations.

2.2.2. Insulin

Insulin binds to cloned and primary endothelial cells in culture (Kaiser et al., 1982, 1983; Jialal et al., 1984; King and Johnson, 1985) and is transported from the luminal to the smooth muscle surface of endothelial cells with minimal degradation (King and Johnson, 1985). Thus the vascular endothelium represents a primary regulatory compartment of the level of insulin in the blood and the tissues.

Evidence for direct effects of insulin on endothelial cell function is sparse. Insulin increases the formation of micropinocyotic vesicles in the capillary endothelium of rats made diabetic with streptozotocin (Osterby et al., 1978). The hormone also stimulates DNA synthesis, as measured by tritiated thymidine incorporation, in subconfluent (Kaiser et al., 1983), but not in confluent (Taggart and Stout, 1980), endothelial cell cultures. Thus insulin may be an important serum factor affecting endothelial cell growth and repair following intimal damage.

Vascular responsiveness is impaired in the absence of insulin in diabetic humans and animal models of the disease (*see* Altura et al., 1979). Two responses with an endothelium-dependent component, histamine and reactive hyperemia (*see* Vanhoutte et al., 1986), are reduced in diabetic compared to nondiabetic humans (Ajjam et al., 1985). Direct evaluation of endothelium-dependent responses has been made in mesenteric arteries of rats with alloxan-induced diabetes (Fortes et al., 1983, 1984). The endothelium-dependent relaxations to acetylcholine were similar in blood vessels of control and diabetic rats, but those to histamine and bradykinin were reduced in the alloxan-treated animals. Topical application of insulin (in concentrations that did not alter serum levels of the hormone) restored normal relaxations to both agonists in the diabetic rat. In control animals the responses to histamine and bradykinin were unaffected by inhibitors of cyclooxygenase and phospholipase A_2; therefore, the metabolism of arachidonic acid probably does not play a role in the response to these agents in the blood vessels studied. Since the relaxations to acetylcholine were unaffected by diabetes, the production, release, or activity on vascular smooth muscle of endothelium-derived relaxing factor(s) may not be impaired by the absence of insulin.

3. Innervation

Innervation of vertebrate blood vessels appears to follow a progression from primarily cholinergic innervation in fish to primarily adrenergic innervation in mammals (Fig. 3; Burnstock, 1969). In the ventral aorta of a teleost, the rainbow trout, an endothelium-dependent relaxation to the calcium-ionophore is observed; only endothelium-independent contractile responses are seen with muscarinic activation, however (Miller and Vanhoutte, 1986a). Endothelium-dependent relaxations to acetylcholine are present in aortas of amphibia (frog) and reptiles (cayman) in which adrenergic stimulation of the smooth muscle by exogenous norepinephrine is prominent. These phylogenetic comparisons suggest that the degree of adrenergic innervation may well help determine endothelium-dependent responsiveness to muscarinic activation. This interpretation is confirmed by experiments performed following chronic sympathetic denervation. Rings of the ear artery of the rabbit were studied 2, 4, 6, and 8 wk after removal of the superior cervical ganglia with interruption of the greater and anterior auricular nerves. The maximal relaxation induced by methacholine was less in denervated compared to unoperated control arteries. The effect of denervation was maximal after 4 wk; the relaxations evoked by sodium nitroprusside were comparable in innervated and denervated arteries (Fig. 4; Mangiarua and Bevan, 1986). Since sodium nitroprusside and the endothelium-derived relaxing factor(s) released by muscarinic activation elicit relaxation of smooth muscle by a similar biochemical process (activation of guanylate cyclase; e.g., Rapoport and Murad, 1983), it is unlikely that denervation *per se* affected the ability of the smooth muscle to relax. Hence, these findings suggest that chronic denervation of mammalian arteries alters the production and/or release of endothelium-derived relaxing factor(s).

4. Blood Flow

The shearing force of the blood along the intimal surface with the subsequent denudation of endothelial cells may promote the development of atherosclerotic plaques (*see* Schwartz et al., 1985). This notion has focused attention toward the examination of the impact of shear stress and blood flow velocity on the morphology and function of endothelial cells.

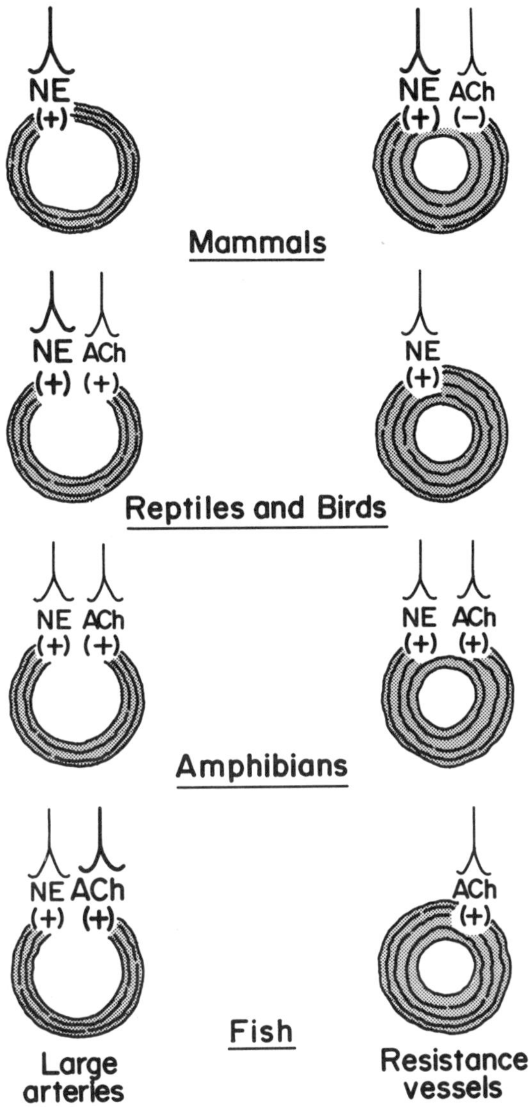

Fig. 3. Schematic of the phylogenetic distribution of innervation in arteries from vertebrates. Plus (+) indicates contraction of the smooth muscle; minus (−) indicates inhibition of the smooth muscle. Thickness of lines indicates prominence of adrenergic (NE; norepinephrine) or cholinergic (ACh; acetylcholine) innervation (redrawn from Burnstock, 1969).

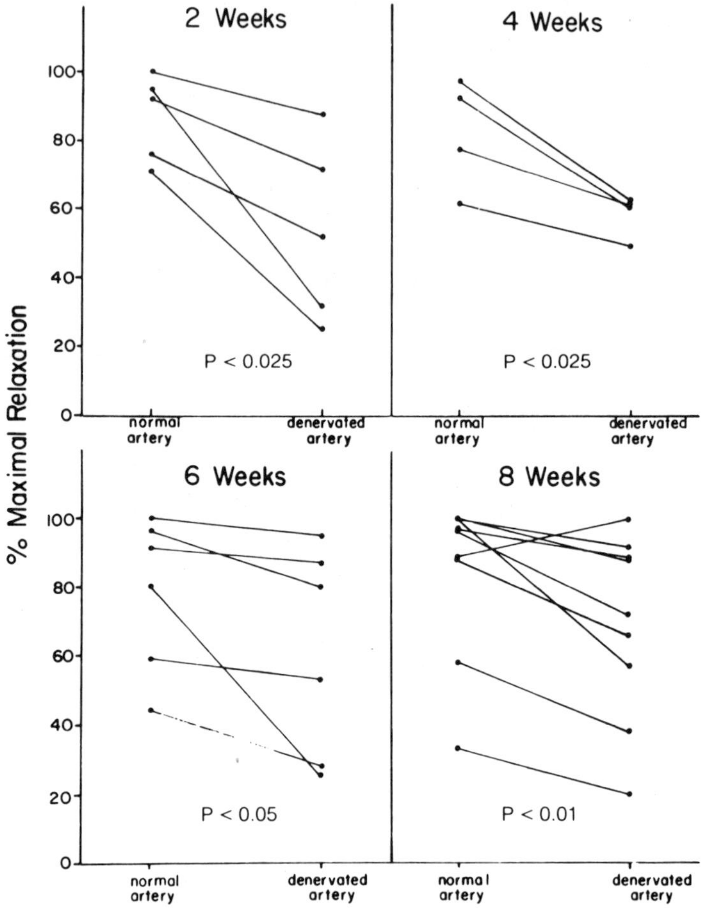

Fig. 4. Maximal relaxations to methacholine ($10^{-6}M$) of normal and denervated ear arteries in growing rabbits 2, 4, 6, and 8 wk following surgical denervation. Arteries were contracted with a concentration of norepinephrine to obtain 50% of the maximum contraction to the drug; relaxations are expressed as a percentage of the maximal relaxation to sodium nitroprusside ($10^{-5}M$) (reprinted from Mangiarua and Bevan, 1986, with permission).

After acute (1 h) coarctation of the canine aorta, sloughing of endothelial cells occurred at areas of high shear stress or at areas of low shear stress accompanied by high turbulance (Fry, 1968). After chronic (3–12 mo) coarctation in animal (Zarins et al., 1981) and in human autopsy material (Zarins et al., 1983), the formation of atherosclerotic plaques was prevalent in areas of unstable flow and low or absent in areas of high flow velocities. Areas proximal to aortic coarctation (high flow velocities) in rabbits are also associated with increased replication of endothelial cells (Langille et al., 1986). In these studies, the effects of shear stress and turbulance cannot be separated. To do so, a measured fluid shear stress can be generated by a rotary cylinder or plate and applied to monolayers of cultured endothelial cells or to the intimal surface of isolated vascular segments. Such studies show that shear stress alone (as high as 2000 dyne/cm^2) does not produce denudation of the endothelium from the canine aorta (Langille, 1984). Cultured endothelial cells respond to shear stress by losing their polygonal pattern in monolayer; they become elongated in the direction of the applied force (Remuzzi et al., 1984; Dewey et al., 1981; Eskin et al., 1984). A similar change in orientation of endothelial cells is also observed in vivo (Reidy and Langille, 1980). Other aspects of the morphology of endothelial cells affected by flow are the orientation of centrioles (Rogers et al., 1985) and the orientation and density of actin-myosin fibers (White et al., 1983; Wong et al., 1983; Franke et al., 1984). The synthesis of prostacyclin and the rate of pinocytosis also are increased when shear stress is applied to endothelial cells in culture (Dewey et al., 1981; Davies et al., 1984; Grabowski et al., 1985; Frangos et al., 1985; van Grondelle et al., 1984). These observations imply that changes in shear stress can probably affect several functions of the endothelial cells. These include endothelium-dependent relaxations.

4.1. Acute Conditions

Evidence is emerging that suggests that endothelial cells may act as sensors and effectors to modulate the tone of underlying smooth muscle in response to local changes in blood flow. Thus, in the intact dog, the acute flow-induced dilatation of coronary and femoral arteries is reduced when the endothelium is removed (Holtz et al., 1983; Kaiser and Sparks, 1986; Pohl et al., 1986; Smieski et al., 1985). The inhibitor of cyclooxygenase, indomethacin, does not alter the endothelium-dependent component of flow-induced dilatation of canine epicardial coronary arteries (Holtz et al., 1984), suggesting that an endothelium-derived factor other than prostacyclin is involved in mediating the dilation. This conclusion has been confirmed in vitro by the biochemical analysis and the bioassay of the perfusate flowing through a canine femoral artery during acute increases in perfusion rate (Rubanyi et al., 1986). The perfusate through femoral arteries with endothelium initiated relaxation of coronary artery without endothelium (bioassay ring) contracted with prostaglandin $F_{2\alpha}$ (Fig. 5). If flow was increased acutely, the relaxations of the bioassay ring were enhanced, demonstrating that a diffusible substance capable of relaxing vascular smooth muscle was released in response to flow. Although the produc-

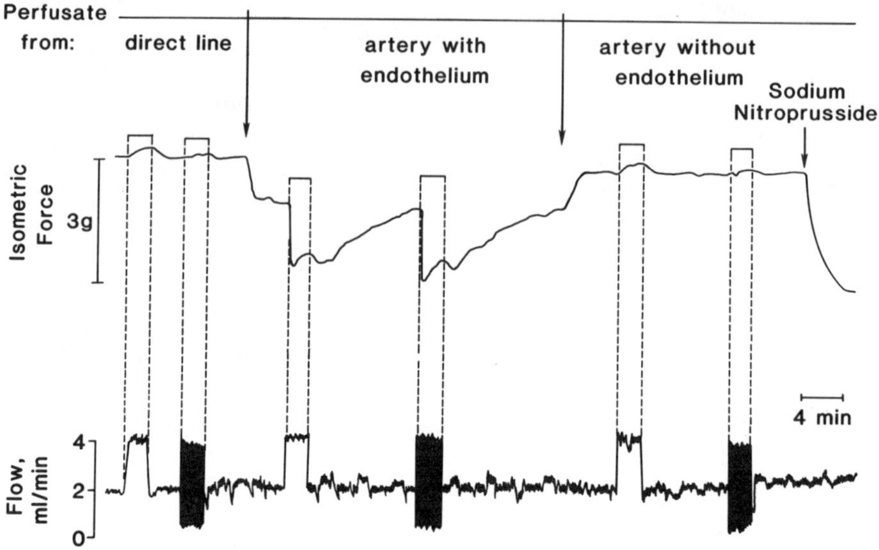

Fig. 5. Changes in isometric force of a ring of canine circumflex coronary artery without endothelium in response to sudden increases in steady flow rate (from 2 to 4 mL/min) or to pulsatile flow during superfusion with buffered salt solution passed through a stainless steel cannula (direct line, left) or through a segment of femoral artery with (middle) or without (right) endothelium. Tone of the coronary arterial ring was estimated by relaxation induced by sodium nitroprusside ($10^{-5}M$) (from Rubanyi et al., 1986, with permission).

tion of prostacyclin by the arteries with endothelium was enhanced during the periods of increased flow, inhibition of prostacyclin production by indomethacin does not reduce the flow-induced relaxation. Since superoxide dismutase prolonged the half-life of the endothelium-derived substance released in response to increases in flow, the substance shares chemical characteristics with the endothelium-derived relaxing factor released by acetylcholine (Rubanyi and Vanhoutte, 1986). Another similarity between the endothelium-derived relaxing factor(s) released in response to flow and acetylcholine is that the responses to both are diminished by methylene blue, an inhibitor of guanylate cyclase (Kaiser et al., 1986).

4.2. Chronic Conditions

In the rabbit, the ligation of the external carotid artery resulted in a 70% reduction in blood flow through the common carotid artery; after a 2-wk exposure to the reduced flow, the lumen of the common carotid artery was decreased by 21% (Langille and O'Donnell, 1986). If prior to the ligation the endothelium of the common carotid artery is removed either by rubbing with a balloon catheter or by infusion of Triton X, no reduction of luminal dimensions was observed when the blood flow was decreased. Although structural alterations of the blood vessel wall may cause this response, alteration in the production and release of endothelium-derived vasoactive factors in response to the decreases in flow may contribute.

Blood flow was chronically reduced in the femoral artery of the dog by application of a clamp, which abruptly reduces blood flow to 75% of the unrestricted value. The clamp remained in place for 6 wk, at which time the vessels proximal to the clamp were removed and their endothelium-dependent responses studied in vitro. Neither the magnitude nor the threshold for the endothelium-dependent relaxations initiated by acetylcholine differed between clamp- and sham-operated arteries (Fig. 6; Miller et al., 1986). Since only blood flow, and not pulse pressure, mean pressure,

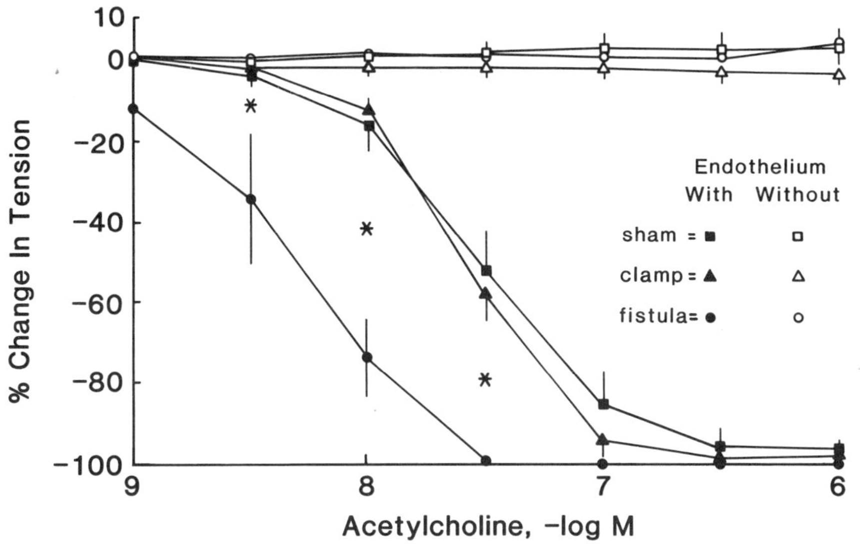

Fig. 6. Concentration–response curves to acetylcholine in arteries with (closed symbols) and without (open symbols) endothelium from sham- (□), fistula- (○, n =5), and clamp- (△, n = 5) operated vessels. Responses of sham-operated vessels were not different between clamp- and fistula-operated groups and were combined (n = 10) for clarity. Values are presented as percent of changes in tension (means ± SEM) from contraction to norepinephrine. Responses of vessels with endothelium were significantly different from those without endothelium. *Denotes statistical significance ($p < 0.05$) between fistula- and sham-operated vessels by Student's t test for paired observations (from Miller et al., 1986, with permission).

or oxygen tension, differed between clamp- and sham-operated arteries, it was concluded that chronic reductions in blood flow did not diminish the endothelium-dependent relaxations to acetylcholine. In contrast, the endothelium-dependent relaxations to arachidonic acid (which are mediated mainly by products of cyclooxygenase; De Mey et al., 1982) were greater in clamp- than in sham-operated arteries; these results suggest that chronic decreases in flow affect endothelial cell function selectively.

Chronic increases in blood flow by the creation of a fistula between the femoral artery and vein of the hind limb of the dog resulted in augmented endothelium-dependent relaxations to muscarinic (Fig. 7) and alpha$_2$-adrenergic agonists (Miller et al., 1986). Indomethacin did not diminish the relaxations, demonstrating that the modulation of prostacyclin is not responsible for the augmented responses. Under bioassay conditions, the relaxation of rings of unoperated femoral arteries (without endothelium) to acetylcholine is greater when they are superfused with effluent from a fistula-operated artery than from a sham-operated artery. Thus the production and release of endothelium-derived relaxing factor(s) released in response to acetylcholine must be enhanced in arteries that have been exposed to chronic increases in blood flow (Miller and Vanhoutte, 1986b). Endothelium-dependent relaxations to acetylcholine and alpha$_2$-adrenergic stimulation are also enhanced in fistula-operated veins (Fig. 7; Miller et al., 1986). Since both flow and oxygen-concentration are increased to arterial levels in these veins, however, it is not possible to identify the modulating parameter, and both probably contribute.

It is unknown how changes in shear stress modify endothelial responsiveness. One possibility is that flow (shear stress) may acutely or chronically alter the glycocalyx of the endothelial cells and unmask receptors that mediate the release of endothelium-derived vasoactive substances. The glycocalyx of the endothelial cells can be disarranged by certain serum proteins, which results in exposure of binding sites and altered permeability of the cells (*see* Ryan, 1986; Baldwin and Winlove, 1984; Schneeberger and Hamelin, 1984).

5. Oxygen

5.1. Acute Effects

Acute exposure of rings of canine cerebral, coronary, femoral, and pulmonary arteries to a gas mixture composed of 95% N_2 and 5% CO_2 produces a contraction that is significantly greater in rings with endothelium (De Mey and Vanhoutte, 1982, 1983; Rubanyi and Vanhoutte, 1985b;

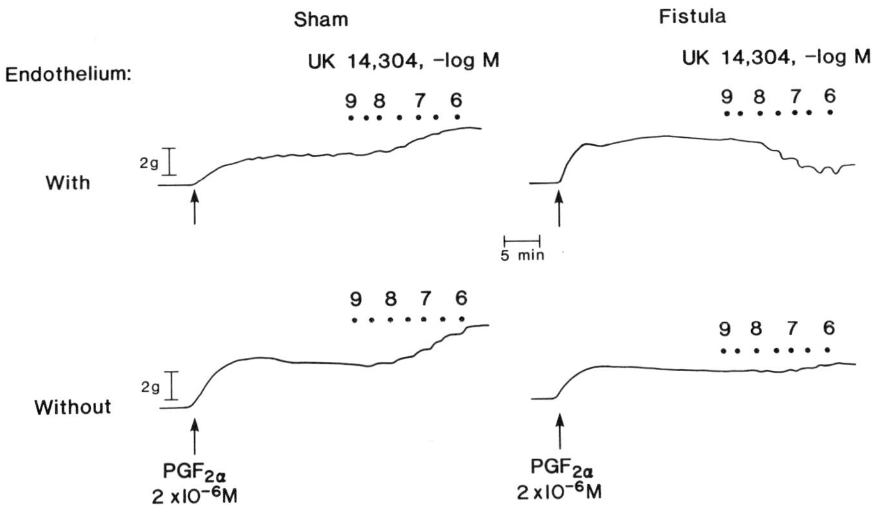

Fig. 7. Example of response to α_2-adrenergic agonist, UK 14,304, in sham- and fistula-operated canine femoral veins with and without endothelium contracted with prostaglandin $F_{2\alpha}$ ($PGF_{2\alpha}$). A prominent endothelium-dependent relaxation is observed in fistula-operated femoral vein. Contractile response to UK 14,304 of vascular smooth muscle (without endothelium) from fistula-operated vein is reduced compared with that from sham-operated vein (from Miller et al., 1986, with permission).

Katusic and Vanhoutte, 1986). An endothelium-dependent contraction to anoxia is present in pulmonary veins, but not femoral veins of the same species (De Mey and Vanhoutte, 1982). Since a contraction can be initiated in strips of coronary artery without endothelium layered with a strip of femoral vein containing endothelium, hypoxia must cause the release of a diffusible endothelium-derived contracting substance. The contractile response is not diminished by inhibition of cyclooxygenase, lipoxygenase, or phospholipase A_2, thus eliminating the metabolism of arachidonic acid in the production of this contractile factor (Rubanyi and Vanhoutte, 1985b). Acute hypoxia also prevents endothelium-dependent relaxations to acetylcholine, thrombin, and vasopressin (De Mey and Vanhoutte, 1980, 1983; Katusic and Vanhoutte, 1986).

5.2. Chronic Effects

Several functions of cultured endothelial cells can be affected by changes in partial pressure of oxygen. The effects depend upon the origin of the cells. For example, in endothelial cells derived from the pulmonary artery and the aorta of cattle exposed to the same partial pressure of oxygen tensions, the activity of the enzymes, pyruvate kinase, phosphofructokinase, and cytochrome aa_3 were similar. The activity of pyruvate kinase, however, was greater in freshly isolated endothelial cells of pulmonary origin (normal PO_2 = 40 mm Hg) compared to cells prepared from the aorta (normal PO_2 = 90 mm Hg) (Cummiskey et al., 1981). Hypoxia (3% O_2) increased the activity of angiotensin-converting enzyme in pulmonary arterial endothelial cells (Krulewitz and Fanburg, 1984) and decreased synthesis of collagen in aortic endothelial cells (Levene et al., 1982).

Endothelium-dependent relaxations to $alpha_2$-adrenergic activation are prominent in blood vessels chronically exposed to high flow and high partial pressures of oxygen: coronary, renal, mesenteric (Cocks and Angus, 1983; Angus et al., 1986), femoral (Miller and Vanhoutte, 1985; Matsuda et al., 1985) arteries and pulmonary veins (Miller and Vanhoutte, 1985). They are absent in systemic veins (Miller and Vanhoutte, 1985). The creation of a fistula, however, between the femoral artery and vein of the dog elevates chronically both blood flow and oxygen tension in the vein (Miller et al., 1986). In rings of vein proximal to the fistula, the contractile responses to norepinephrine are less than those of sham-operated veins. The contractile responses to selective $alpha_2$-adrenoceptor agonists is also reduced. These observations suggest that chronic exposure of the veins to increased flow and a high partial pressure of oxygen decreases the $alpha_2$-adrenergic responsiveness of the venous smooth muscle. Modulation of endothelial $alpha_2$-adrenoceptors is also evident in that fistula-operated veins with endothelium display a markedly augmented inhibitory response to selective $alpha_2$-adrenergic agonists (Fig. 7). These observations imply that the vascular response to neuronally released and circulating catecholamines could be altered by flow and oxygen tension.

6. Concluding Remarks

The ability of the endothelium to modulate the tone of the underlying vascular smooth muscle in several classes of vertebrates identifies these cells as an important regulator of cardiovascular function that appeared early in the evolution of the phylum (Miller and Vanhoutte, 1986a). The physiological importance of this regulatory role of the endothelial cells is that these cells can act as sensors of local, acute changes in the properties of the blood circulating (content of catecholamines and other hormones), partial pressure of oxygen, and shear stress. As for other regulatory mechanisms within the cardiovascular system, the local function of the endothelium adapts to chronic changes. This adaptation of the function of endothelial cells to chronic exposure to various physical and chemical factors may explain in part why the endothelium-dependent responses of arteries differ from those of veins. It could also contribute to pathological changes in vascular reactivity associated with atherosclerosis and vascular grafting.

References

Ajjam, Z. S.. Barton, S., Corbett, M., Owens, D., and Marks, R.: Quantitative evaluation of the dermal vasculature of diabetics. *Q. J. Med.* **54:**229–239, 1985.

Altura, B. M., Halevy, S., and Turlapaty, P. D. M. V.: Vascular smooth muscle in diabetes and its influence on the reactivity of blood vessels. *Adv. Microcirc.* **8:**118–150, 1979.

Angus, J. A., Cocks, T. M., and Satoh, K.: α_2-Adrenoceptors and endothelium-dependent relaxation in canine large arteries. *Br. J. Pharmacol.* **88:**767–777, 1986.

Baldwin, A. L. and Winlove, C.P.: Effects of perfusate composition on the binding of ruthemium red and gold colloid to glycocalyx of rabbit aortic endothelium. *J. Histochm. Cytochem.* **32:**259–266, 1984.

Bell, C. and Coffey, C.: Factors influencing oestrogen-induced sensitization to acetylcholine of guinea-pig uterine artery. *J. Reprod. Fertil.* **66:**133–137, 1982.

Blackwell, G. J., Carnuccio, R., Di Rosa, M., Flower, R. J., Parente, L., and Persico, P.: Macrocortin: A polypeptide causing the antiphospholipase effect of glucocorticoids. *Nature* **287:**147–149, 1980.

Bottaro, D., Shepro, D., Peterson, S., and Hechtman, H. B.: Serotonin, histamine and norepinephrine mediation of endothelial and vascular smooth muscle cell movement. *Am. J. Physiol.* **248:**C252–C257, 1985.

Buonassisi, V. and Venter, J. C.: Hormone and neurotransmitter receptors in an established vascular endothelial cell line. *Proc. Natl. Acad. Sci. USA* **73:**1612–1616, 1976.

Burnstock, G.: Evolution of the autonomic innervation of visceral and cardiovascular systems in vertebrates. *Pharmacol. Rev.* **21:**247–324, 1969.

Carrier, G. O. and White, R. E.: Enhancement of alpha-1 and alpha-2 adrenergic agonists induced vasoconstriction by removal of endothelium in rat aorta. *J. Pharmacol. Exp. Ther.* **232:**682–687, 1985.

Cocks, T. M. and Angus, J. A.: Endothelium-dependent relaxation of coronary arteries by noradrenaline and serotonin. *Nature* **305:**627–630, 1983.

Colburn, P. and Buonassisi, V.: Estrogen-binding sites in endothelial cell cultures. *Science* **201:**817–819, 1978.

Covazier, E., Dupuy, E., Dosne, A. M., and Maclouf, J.: Minimal effect of estrogens on endothelial cell growth and production of prostacyclin. *Thromb. Res.* **34:**303–310, 1984.

Crutchley, D. J., Ryan, U. S., and Ryan, J. W.: Glucocorticoid modulation of prostacyclin production in cultured bovine endothelial cells. *J. Pharmacol. Exp. Ther.* **233:**650–655, 1985.

Cummiskey, J. M., Simon, L. M., Theodore, J., Ryan, U. S., and Robin, E. D.: Bioenergetic alterations in cultivated pulmonary artery and aortic endothelial cells exposed to normoxia and hypoxia. *Exp. Lung Res.* **2:**155–163, 1981.

Davies, P. F., Dewey, C. F., Jr., Bussolari, S. R., Gordon, E. J., and Gimbrone, M. A., Jr.: Influence of hemodynamic forces on vascular endothelial function. *J. Clin. Invest.* **73:**1121–1129, 1984.

De Mey, J. G. and Vanhoutte, P. M.: Anoxia and endothelium-dependent reactivity of the canine femoral artery. *J. Physiol.* **335:**65–74, 1983.

De Mey, J. G. and Vanhoutte, P. M.: Heterogeneous behavior of the canine arterial and venous wall. *Circ. Res.* **51:**439–447, 1982.

De Mey, J. G. and Vanhoutte, P. M.: Interaction between Na^+,K^+ exchanges and the direct inhibitory effect of acetylcholine on canine femoral arteries. *Circ. Res.* **46:**826–836, 1980.

De Mey, J. G., Claeys, M., and Vanhoutte, P. M.: Endothelium-dependent inhibitory effects of acetylcholine, adenosine triphosphate, thrombin and arachidonic acid in the canine femoral artery. *J. Pharmacol. Exp. Ther.* **222:**166–173, 1982.

Dewey, C. F., Jr., Bussolari, S. R., Gimbrone, M. A., Jr., and Davis, P. F.: The dynamic response of vascular endothelial cells to fluid shear stress. *J. Biomech. Eng.* **103:**177–185, 1981.

Egleme, C., Godfraind, T., and Miller, R. C.: Enhanced responsiveness of rat isolated aorta to clonidine after removal of endothelial cells. *Br. J. Pharmacol.* **81:**16–18, 1984.

Eskin, S. G., Ives, C. L., McIntire, L. V., and Navarro, L. T.: Response of cultured endothelial cells to steady flow. *Microvasc. Res.* **28:**87–94, 1984.

Fortes, Z. B., Garcia-Leme, J., and Scivoletto, R.: Vascular reactivity in diabetes mellitus: Possible role of insulin on the endothelial cell. *Br. J. Pharmacol.* **83:**635–643, 1984.

Fortes, A. B., Garcia Leme, J., and Scivoletto, R.: Vascular reactivity in diabetes mellitus: Role of the endothelial cell. *Br. J. Pharmacol.* **79:**771–781, 1983.

Frangos, J. A., Eskin, S. G., McIntire, L. V., and Ives, C. L.: Flow effects on prostacyclin production by cultured human endothelial cells. *Science* **227:**1477–1479, 1985.

Franke, R.-P., Grafe, M., Schmittler, H., Seiffge, D., Mittermayer, C., and Drenckhahn, D.: Induction of human vascular endothelial stress fibres by fluid shear stress. *Nature* **307:**648–649, 1984.

Fry, D. L.: Acute vascular endothelial changes associated with increased blood velocity gradients. *Circ. Res.* **22:**165–197, 1968.

Furchgott, R. F.: Endothelium-Dependent Responses: The Last Three Years, in *Mechanisms of Vasodilatation* Vanhoutte, P. M. and Vatner, S. F., eds.) Biblthca. Cardiol. no. 38, Karger, Basel, 1984.

Furchgott, R. F. and Zawadzki, J. V.: The obligatory role of endothelial cells in the relaxation of arterial smooth muscle by acetylcholine. *Nature* **288:** 373–376, 1980.

Gisclard, V. and Vanhoutte, P. M.: Estrogens and endothelium-dependent responses in the femoral artery of the rabbit. *Physiologist* **28:**324, 1985.

Godfraind, T., Egleme, C., and Osachie, I. A.: Role of endothelium in the contractile response of rat aorta to alpha-adrenoceptor agonists. *Clin. Sci.* **68** (suppl. 10)**:**65s–71s, 1985.

Grabowski, E. F., Jaffe, E. A., and Weskler, B. B.: Prostacyclin production by cultured endothelial cell monologues exposed to step increases in shear stress. *J. Lab. Clin. Med.* **105:**36–43, 1985.

Holtz, J., Busse, R., and Giesler, M.: Flow-dependent dilatation of canine epicardial coronary arteries in vivo and in vitro: Mediated by the endothelium *Naunyn Schmiedebergs Arch. Pharmacol.* **322:**R44, 1983.

Holtz, J., Forstermann, U., Pohl, U., Giesler, M., and Bassenge, E.: Flow-dependent, endothelium-mediated dilation of epicardial coronary arteries. *J. Cardiovasc. Pharmacol.* **6:**1161–1169, 1984.

Jarvelainen, H., Pellinieni, T. T., and Ronnemaa, T.: The stabilizing effect of glucocorticoids on human endothelial cells in culture. *Scand. J. Clin. Lab. Invest.* **45:**223–228, 1985.

Jialal, I., King, G L., Buchwald, S., Kahn, C. R., and Crettaz, M.: Processing of insulin by bovine endothelial cells in culture. *Diabetes* **33:**794–800, 1984.

Kaiser, L. and Sparks, H. V.: Mediation of flow dependent arterial dilation by endothelial cells. *Circ. Shock* **18:**109–114, 1986.

Kaiser, L., Hull, S. S., and Sparks, H. V.: Methylene blue and ETYA block flow dependent dilation in canine femoral artery. *Am. J. Physiol.* **25:**H974–H981, 1986.

Kaiser, N., Vlodavsky, I., Tur-Sinai, A., Fuks, Z., and Cerasi, E.: Binding internalization and degradation of insulin in vascular endothelial cells. *Diabetes* **31:**1077–1083, 1982.

Kaiser, N., Vlodavsky, I., Tur-Sinai, A., Fuks, Z., and Cerasi, E.: Insulin binding and degradation in vascular endothelial-cells: Modulation by cell growth and culture organization. *Endocrinology* **113:**228–234, 1983.

Karnushina, I. L., Spatz, M., and Bembry, J.: Cerebral endothelial cell culture. I. The presence of β_2- and α_2-adrenergic receptors linked to adenylate cyclase activity. *Life Sci.* **30:**849–858, 1982.

Katusic, Z. S. and Vanhoutte, P. M.: Anoxic contractions in isolated canine cerebral arteries. Contribution of endothelium-derived factors, metabolites of arachidonic acid and calcium entry. *J. Cardiovasc. Pharmacol.* **8:**S97–S101, 1986.

Katusic, Z. S., Shepherd, J. T., and Vanhoutte, P. M.: Vasopressin causes endothelium-dependent relaxations of the canine basilar artery. *Circ. Res.* **55:**575–579, 1984.

Katusic, Z. S., Shephard, J. T., and Vanhoutte, P. M.: Oxytocin causes endothelium-dependent relaxations of canine basilar arteries by activating V_1-vasopressinergic receptors. *J. Pharmacol. Exp. Ther.* **236:**166–170, 1986.

King, G. L. and Johnson, S. M.: Receptor-mediated transport of insulin across endothelial cells. *Science* **227:**1583–1586, 1985.

Krulewitz, A. H. and Fanburg, B. L.: The effect of oxygen tension on the *in vitro* production and release of angiotensin-converting enzyme by bovine pulmonary artery endothelial cells. *Am. Rev. Respir. Dis.* **130:**866–869, 1984.

Langille, B. L.: Integrity of arterial endothelium following acute exposure to high shear stress. *Biorheology* **21:**333–346, 1984.

Langille, B. L. and O'Donnell, F.: Reductions in arterial diameter produced by chronic decreases in blood flow are endothelium-dependent. *Science* **231:** 405–407, 1986.

Langille, B. L., Reidy, M. A., and Kline, R. L.: Injury and repair of endothelium at sites of flow disturbances near abdominal aortic coarctations in rabbits. *Atherosclerosis* **6:**146–154, 1986.

Levene, C. I., Kapoor, R., and Heale, G.: The effect of hypoxia on synthesis of collagen and glycosaminoglycans by cultured pig aortic endothelium. *Atherosclerosis* **44:**327–337, 1982.

Longnecker, J. P., Kilty, L. A., and Johnson, L. K.: Glucocorticoid influence on growth of vascular wall cells in culture. *J. Cell. Physiol.* **113:**197–202, 1982.

Lues, I. and Schumann, H. J.: Effect of removing the endothelial cells on the reactivity of rat aortic segments to different α-adrenoceptor agonist. *Naunyn Schmiedebergs Arch. Pharmacol.* **328:**160–163, 1984.

Maca, R. D., Fry, G. L., and Hoak. J. C.: The effects of glucocorticoids on cultured human endothelial cells. *Br. J. Haematol.* **38:**501–509, 1978.

Mangiarua, E. I. and Bevan, R. D.: Altered endothelium-mediated relaxation after denervation of growing rabbit ear artery. *Eur. J. Pharmacol.* **122:**149–152, 1986.

Martin, W., Furchgott, R. F., Villani, G. M., and Jothianandan, B.: Depression of contractile responses in rat aorta by spontaneously released endothelium-derived relaxing factor. *J. Pharm. Exp. Ther.* **237:**529–538, 1986.

Matsuda, H., Kuon, E., Holtz, J., and Busse, R.: Endothelium-mediated dilations contribute to the polarity of the arterial wall in vasomotion induced by alpha$_2$-adrenergic agonists. *J. Cardiovasc. Pharmacol.* **7:**680–688, 1985.

Mendelsohn, F. A., Lloyd, C. J., Kachel, C., and Funder, J. W.: Glucocorticoid induction of angiotensin converting enzyme production from bovine endothelial cells in culture and rat lung in vivo. *Clin. Exp. Pharmacol. Physiol.* **7** (suppl).:57–62, 1982.

Miller, V. M. and Vanhoutte, P. M.: Endothelial alpha$_2$ adrenoceptors in canine pulmonary and systemic blood vessels. *Eur. J. Pharmacol.* **118:**123–129, 1985.

Miller, V. M. and Vanhoutte, P.M.: Endothelium-dependent responses in lower vertebrates. *Blood Vessels* **23:**225–235, 1986a.

Miller, V. M. and Vanhoutte, P. M.: Mechanism of augmented endothelium-dependent relaxations in fistula-operated arteries of the dog. *Physiologist* **29:**158, 1986b.

Miller, V. M., Aarhus, L. L., and Vanhoutte, P. M.: Modulation of endothelium-dependent responses by chronic alterations of blood flow. *Am. J. Physiol.* **251**:H520–H527, 1986.

Mueller, D., Stoehr, B., Phernetton, T., and Rankin, J. H. G.: The effects of indomethacin and meclofenamate on estrogen induced vasodilation in the rabbit uterus. *Proc. Soc. Exp. Biol. Med.* **159**:25–29, 1978.

Nichols, N. R., Lloyd, C. J., Mendelsohn, F. A., and Funder, J. W.: Glucocorticoid induced proteins in bovine endothelial cells. *Mol. Cell Endocrinol.* **32**:245–254, 1983.

Okamura, T., Miyazaki, M., and Toda, N.: Responses of isolated dog blood vessels to glucagon. *Eur. J. Pharmacol.* **125**:395–401, 1986.

Osterby, R., Gundersen, H. J., and Christensen, N. J.: The acute effect of insulin on capillary endothelial cells. *Diabetes* **27**:745–749, 1978.

Pohl, U., Holtz, J., Busse, R., and Bassenge, E.: Crucial role of endothelium in the vasodilator response to increased flow in vivo. *Hypertension* **8**:37–44, 1986.

Rapoport, R. M. and Murad, F.: Agonist-induced endotheium dependent relaxation in rat thoracic aorta may be mediated through cGMP. *Circ. Res.* **52**:352–357, 1983.

Reidy, M. A. and Langille, B. L.: The effect of local blood flow patterns on endothelial cell morphology. *Exp. Mol. Pathol.* **32**:276–289, 1980.

Remuzzi, A., Dewey, C. F., Jr., Davies, P. F., and Gimbrone, M. A., Jr.: Orientation of endothelial cells in shear fields in vitro. *Biorheology* **21**:617–630, 1984.

Rogers, K. A., McKee, N. H., and Kalnins, V. I.: Preferential orientation of centrioles toward the heart in endothelial cells of major blood vessels is reestablished after reversal of a segment. *Proc. Natl. Acad. Sci. USA* **82**:3272–3276, 1985.

Rubanyi, G. M. and Vanhoutte, P. M.: Endothelium-removal decreases relaxations of canine coronary arteries caused by beta-adrenergic agonists and adenosine. *J. Cardiovasc. Pharmacol.* **7**:139–144, 1985a.

Rubanyi, G. M. and Vanhoutte, P. M.: Hypoxia relaxes a vasoconstrictor substance from the canine vascular endothelium. *J. Physiol.* **364**:45–56, 1985b.

Rubanyi, G. M. and Vanhoutte, P. M.: Superoxide anions and hyperoxia inactivate endothelium-derived relaxing factor(s). *Am. J. Physiol.* **250**:H822–H827, 1986.

Rubanyi, G. M., Lorenz, R. R., and Vanhoutte, P. M.: Bioassay of endothelium-derived relaxing factor(s). Inactivation by catecholamines. *Am. J. Physiol.* **249**:H95–H101, 1985.

Rubanyi, G. M., Romero, J. C., and Vanhoutte, P. M.: Flow-induced release of endothelium-derived relaxing factor. *Am. J. Physiol.* **205**:H1145–H1149, 1986.

Ryan, U. S.: The endothelial surface and response to injury. *Fed. Proc.* **45**:101–108, 1986.

Schneeberger, E. E. and Hamelin, M.: Interaction of serum proteins with lung endothelial glycocalyx: Its effect on endothelial permeability. *Am. J. Physiol.* **247**:H206–H217, 1984.

Schwartz, S. M., Reidy, M. R., and Clowes, A.: Kinetics of atherosclerosis: A stem cell model. *Ann. NY Acad. Sci.* **454**:292–304, 1985.

Seillan, C., Ody, C., Russo-Marie, F., and Dual, D.: Differential effects of sex steroids on prostaglandin secretion by male and female cultured piglet endothelial cells. *Prostaglandins* **26**:3–21, 1983.

Sercombe, R., Verrecchia, C., Oudart, N., Dimitriadou, V., and Seylaz. J.: Pial artery responses to norepinephrine potentiated by endothelium removal. *J. Cereb. Blood Flow Metab.* **5**:312–317, 1985.

Smieski, V., Kozik, J., and Dolezel, S.: Role of endothelium in the control of arterial diameter by blood flow. *Blood Vessels* **22**:247–251, 1985.

Taggart, H. and Stout, R. W.: Control of DNA synthesis in cultured vascular endothelial and smooth muscle cells. *Atherosclerosis* **37**:549–557, 1980.

van Grondelle, A., Worthen, G. S., Ellis, D., Mathias, M. M., Murphy, R. C., Striefe, R. J., Reeves, J. T., and Voelkel, N. F.: Altering hydrodynamic variables influences PGI_2 production by isolated lungs and endothelial cells. *J. Appl. Physiol.* **57**:388–395, 1984.

Vanhoutte, P. M.: Could the absence or malfunction of vascular endothelium precipitate the occurrence of vasospasm? *J. Mol. Cell. Cardiol.* **18**:679–689, 1986.

Vanhoutte, P. M., Katusic, Z. S., and Shepherd, J. T.: Vasopressin induces endothelium-dependent relaxations of cerebral and coronary, but not of systemic arteries. *J. Hyperten.* **2** (suppl. 3):421–422, 1984.

Vanhoutte, P. M., Rubanyi, G. M., Miller, V. M., and Houston, D. S.: Modulation of vascular smooth muscle contraction by the endothelium. *Ann. Rev. Physiol.* **48:** 307–320, 1986.

Venter, J. C., Buonassisi, V., Bevan, R., Heinemann, S., and Bevan, J. A.: Hormone and neurotransmitter receptors on vascular endothelium. *Blood Vessels* **12**:381–382, 1975.

Wagner, R. C., Kreiner, P., Barrnett, R. J., and Bitensky, M. W.: Biochemical characterization and cytochemical localization of a catecholamine-sensitive adenylate cyclase in isolated capillary endothelium. *Proc. Natl. Acad. Sci. USA* **69**:3175–3179, 1972.

Welles, S. L., Shepro, D., and Hechtman, H. B.: Vasoactive amines modulate actin cables (stress fibers) and surface area in cultured bovine endothelium. *J. Cell Physiol.* **123**:337–342, 1985.

White, G. E., Gimbrone, M. A., Jr., and Fujiwara, K.: Factors influencing the expression of stress fibers in vascular endothelial cells in situ. *J. Cell Biol.* **97**:416–424, 1983.

Witter, F. R. and DiBlasi, M. C.: Effect of steroid hormones on arachidonic acid metabolites of endothelial cells. *Obstet. Gynecol.* **63**:747–751, 1984.

Wong, A. J., Pollard, T. D., and Herman, I. M.: Actin filament stress fibers in vascular cells *in vivo*. *Science* **219**:867–869, 1983.

Young, M. A. and Vatner, S. F.: Enhanced adrenergic constriction of iliac artery with removal of endothelium in conscious dogs. *Am. J. Physiol.* **250**:H892–897, 1986.

Zarins, C. K., Giddens, D. P., Bharadvaj, B. K., Sottiurai, V. S., Malon, R. F., and Glagov, S.: Quantitative correlation of plaque localization with flow velocity profiles and wall shear stress. *Circ. Res.* **53**:502–514, 1983.

Zarins, C. K., Bomberger, R. A., and Glagov, S.: Local effects of stenosis: Increased flow velocity inhibits atherogenesis. *Circulation* **64** (suppl. II): 221–227, 1981.

Chapter 15

Endothelium-Dependent Regulation of Resting Levels of Cyclic GMP and Cyclic AMP and Tension in Pulmonary Arteries and Veins

Louis J. Ignarro, Russell E. Byrns, and Keith S. Wood

1. Introduction

Relaxation of vascular smooth muscle elicited by numerous endothelium-independent and -dependent vasodilators appears to be mediated and/or modulated by intracellular cyclic GMP. A variety of "nitrogen oxide-containing vasodilators" such as organic nitrate and nitrite esters, inorganic and organic nitroso compounds, *S*-nitrosothiols, and nitric oxide (NO) all activate the soluble heme-containing enzyme guanylate cyclase, elevate levels of cyclic GMP in vascular smooth muscle, and cause vasodilatation (*see* reviews by Ignarro et al., 1984a; Ignarro and Kadowitz, 1985). All of these vasodilators generate or release NO, which then reacts with heme-containing guanylate cyclase to generate the nitrosyl-heme adduct, which represents the activated state of guanylate cyclase (Ignarro et al., 1982a, 1984b; Wolin et al., 1982). This activated form of guanylate cyclase produced by NO-heme is kinetically identical to that produced by protoporphyrin IX, the immediate precursor to heme (Wolin et al., 1982). Methylene blue, a lipophilic vital biological stain, inhibits the NO-heme-elicited activated state of guanylate cyclase without appreciably affecting basal activity or protoporphyrin IX-activated enzyme (Ignarro et al., 1982a; Wolin et al., 1982). Methylene blue, at concentrations that do not alter resting tension in isolated vascular strips or rings, abolishes both cyclic GMP accumulations and relaxations elicited by the nitrogen oxide-containing vasodilators (Gruetter et al., 1979, 1980, 1981a,b; Ignarro et al., 1981; Kukovetz et al., 1981; Keith et al., 1982; Edwards et al., 1984).

Many endothelium-dependent vasodilators also promote the accumulation of cyclic GMP and relaxation of isolated arterial preparations (Holzmann, 1982; Diamond and Chu, 1982; Rapoport and Murad, 1983; Rapoport et al., 1984; Ignarro et al., 1984c, 1985; Furchgott et al., 1984; Martin et al., 1985). Veins are far less responsive in this manner, and studies with bovine pulmonary veins revealed that only bradykinin causes endothelium-dependent accumulation of cyclic GMP and relaxation (Harbison et al., 1985). None of the endothelium-dependent vasodilators directly activates guanylate cyclase. Yet these agents stimulate formation of cyclic GMP in vascular smooth muscle, and this response, as well as relaxation, is antagonized by methylene blue and related inhibitors of the activated state of soluble guanylate cyclase (Ignarro et al., 1984c, 1985, 1986a; Holzmann, 1982; Martin et al., 1985). Therefore, the endothelium-dependent vasodilators must indirectly activate soluble guanylate cyclase, and the activated enzyme form generated must possess properties that are remarkably similar to those of the NO-heme-activated form of guanylate cyclase. Indeed, recent studies from this laboratory indicate that both arterial and venous endothelium release substances that directly activate soluble guanylate cyclase, purified from bovine lung, in a methylene blue-sensitive manner (Ignarro et al., 1986b).

Some endothelium-dependent vasodilators stimulate the accumulation of cyclic GMP, but also that of cyclic AMP in arteries. These include arachidonic acid, bradykinin, and vasoactive intestinal peptide (Ignarro et al., 1985; Ignarro and Kadowitz, 1985; Harbison et al., 1985). Methylene blue abolishes the accumulation of cyclic GMP and partially inhibits the relaxation. Indomethacin abolishes the accumulation of cyclic GMP and partially inhibits relaxation. A combination of the two antagonists abolishes relaxation. These studies indicate clearly that not only cyclic GMP, but also cyclic AMP, may be involved in mediating endothelium-dependent relaxant responses. This dual mechanism does not appear to be involved in veins, however. Neither arachidonic acid nor vasoactive intestinal peptide relaxes pulmonary veins, and bradykinin causes accumulation of cyclic GMP, but not cyclic AMP in veins (Harbison et al., 1985).

The objective of the present chapter is to provide new information on: (1) the role of the endothelium in regulating resting levels of cyclic nucleotides and tone; (2) novel properties of endothelium-derived relaxing factor(s) (EDRF); and (3) differences in responsiveness between arteries and veins. The studies reported here were conducted with isolated rings of bovine intrapulmonary arteries and veins.

2. Resting Levels of Cyclic GMP and Cyclic AMP

To determine the resting or basal levels of both cyclic GMP and cyclic AMP, isolated rings of pulmonary arteries and veins were equilibrated under tension, as is done in all experiments designed to study the actions of relaxant drugs. Many early experiments revealed that phenylephrine did not alter the levels of cyclic GMP or cyclic AMP in rings that had been equilibrated under tension. Full equilibration of rings in Kreb's bicarbonate was necessary, however, since failure to do so resulted in highly variable and significantly lower levels of both cyclic nucleotides. Furthermore, experimentation disclosed that no detectable quantities of either cyclic GMP or cyclic AMP were present in fresh bovine pulmonary arterial or venous endothelium. Likewise, no appreciable activity of soluble guanylate cyclase was found in the endothelium, whereas both arterial and venous medial layers (smooth muscle) possessed significant quantities of NO- and porphyrin-activatable enzyme. Therefore, the endothelium alters enzymatic activities and cyclic nucleotide levels by influencing biochemical pathways in the smooth muscle cells. Figure 1 illustrates the influence of the endothelium on arterial levels of cyclic GMP and cyclic AMP. Resting cyclic GMP levels were three- to fourfold higher in endothelium-intact than in denuded artery. Methylene blue and hemoglobin markedly lowered cyclic GMP levels in arteries with than in those without endothelium. The mechanism of action of methylene blue is attributed to inhibition of the activated state of soluble guanylate cyclase. This interpretation implies that arterial endothelial cells release a factor such as EDRF that activates guanylate cyclase by mechanisms related to those manifested by NO. This view is supported by the observed negligible effect of methylene blue in denuded artery. Hemoglobin, which cannot permeate cells, probably lowers cyclic GMP levels via a direct interaction with released EDRF. It is noteworthy in this regard that hemoproteins have an extremely high binding affinity for NO and are capable of abolishing NO-elicited relaxations of vascular smooth muscle (Gruetter et al., 1979, 1980). Preliminary experiments indicate that oxyhemoglobin and oxymyoglobin abolish both endothelium-dependent and NO-elicited relaxation. These inhibitory effects are dramatically reversed by cyanide, which reacts with hemoproteins to form the corresponding cyanomethemoprotein.

Resting levels of cyclic AMP were twofold higher in arteries with than in those without endothelium (Fig. 1). Whereas methylene blue failed to alter these levels, indomethacin lowered levels of cyclic AMP only in arteries with endothelium. Preliminary data indicate that indomethacin elicits its effects by inhibiting cyclooxygenase and the formation of prostacyclin, which is the endothelium-derived substance responsible for the stimulation of the accumulation of cyclic AMP. Arachidonic acid relaxes

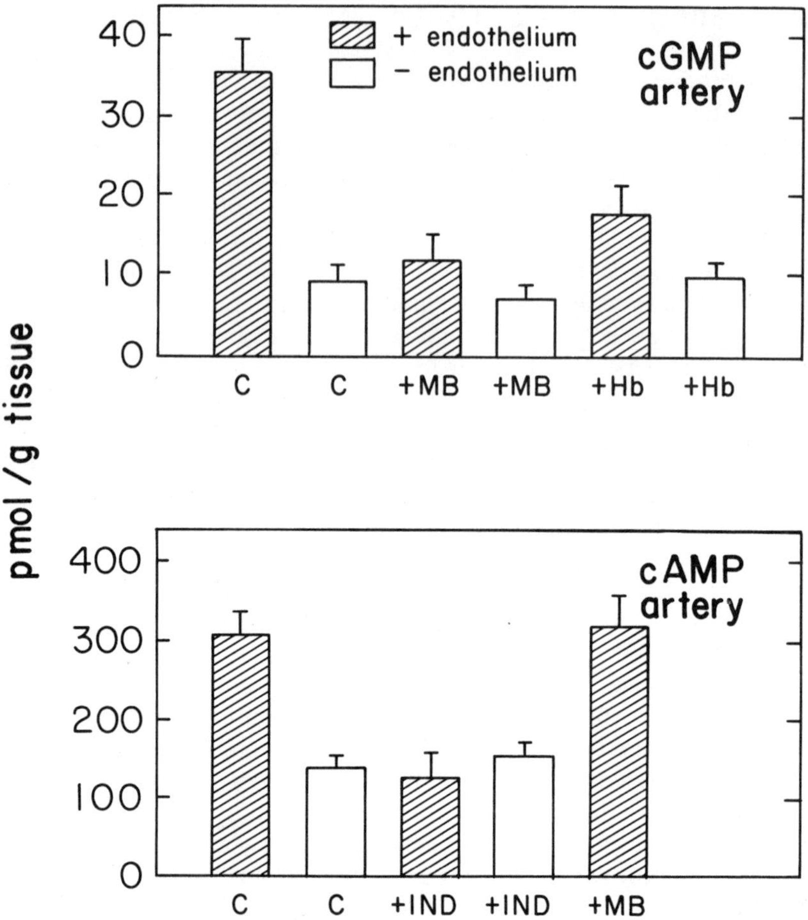

Fig. 1. Influence of endothelium on resting levels of cyclic GMP and cyclic AMP in the bovine pulmonary artery. Rings were prepared from the second branch of the main pulmonary artery and equilibrated under 6*g* tension for 2 h. Methylene blue (MB; $10^{-5}M$), hemoglobin (Hb; $10^{-6}M$), or indomethacin (IND; $10^{-5}M$) were added to the organ chambers for 15 min. Control (C) and drug-treated rings were freeze-clamped and assayed for cyclic nucleotides (Ignarro et al., 1984c, 1985). Data represented the mean ± SEM using 6–8 rings from 3–4 animals.

bovine pulmonary artery with but not those without endothelium by elevating both cyclic GMP and cyclic AMP levels, and prostacyclin is responsible for the stimulated accumulation of cyclic AMP (Ignarro et al., 1985).

Veins with endothelium also had three- to fourfold higher resting levels of cyclic GMP than did veins without endothelium (Fig. 2). Methylene

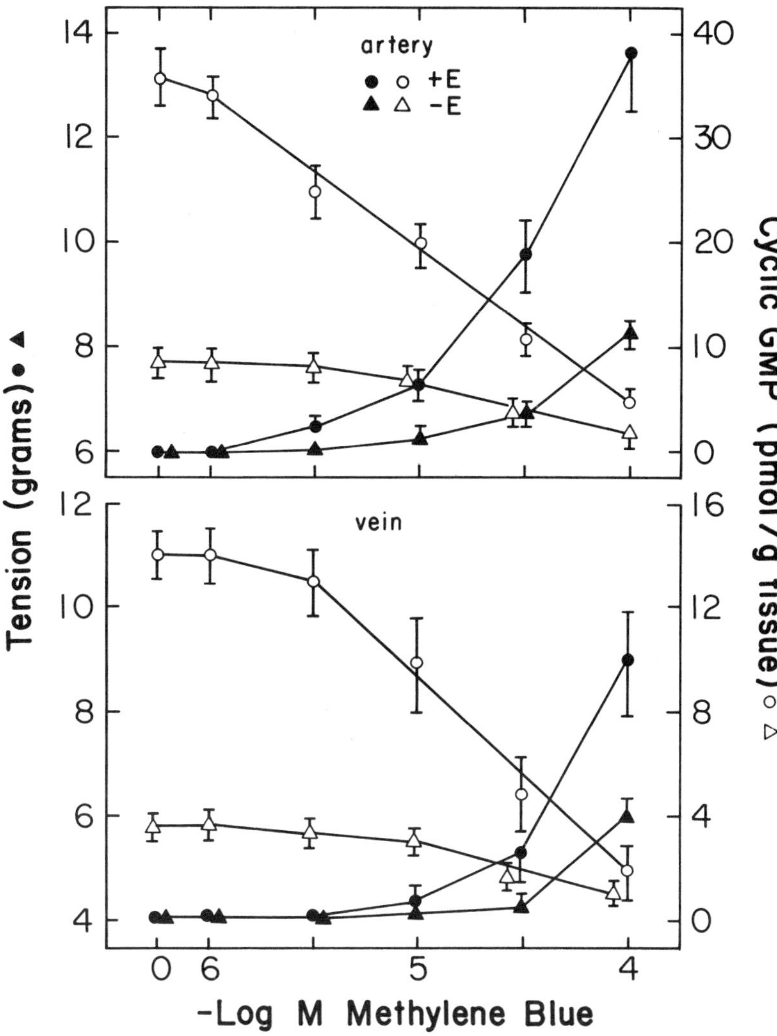

Fig. 2. Influence of methylene blue on resting levels of cyclic GMP and tension in isolated bovine pulmonary arteries and veins with or without endothelium. Rings were prepared from the second branch, equilibrated for 2 h under 6*g* (arteries) or 4*g* (veins) of tension, freeze-clamped 15 min after addition of methylene blue, and assayed for cyclic GMP (Ignarro et al., 1984c, 1985). Data represent the mean ± SEM using 8–10 rings from 4–5 animals.

blue markedly lowered levels of cyclic GMP in veins with endothelium only. Cyclic AMP levels were comparable with or without endothelium, and indomethacin failed to influence venous levels of cyclic AMP. These observations, together with the findings that bradykinin causes endothelium-dependent relaxation and accumulation of cyclic GMP, but not cyclic AMP, in veins, indicate clearly that the endothelium of bovine pulmonary vein can generate EDRF.

3. Endothelium-Dependent Regulation of Vascular Cyclic GMP Levels

3.1. Methylene Blue

Resting levels of cyclic GMP in arteries with endothelium were very sensitive to the lowering effect of methylene blue (Fig. 2). Declining levels of cyclic GMP were closely associated with increased tension development. The reason for this is that a progressive decline in smooth muscle concentrations of the relaxant cyclic GMP (Schultz et al., 1979; Napoli et al., 1980) results in increased tone or contraction. Arteries without endothelium were much less responsive to methylene blue, and similar effects occurred only at very high concentrations of the agent (Fig. 2). Therefore, the arterial endothelium generates a factor(s) capable of elevating resting levels of cyclic GMP in smooth muscle, and this process is highly sensitive to the inhibitory action of methylene blue.

Veins were qualitatively similar to arteries in that resting levels of cyclic GMP were higher in rings with than without endothelium, and methylene blue caused a decline in cyclic GMP levels accompanied by an increase in tension (Fig. 2). Venous levels of cyclic GMP were significantly lower than arterial levels both in the presence and absence of endothelium.

Thus, both arterial and venous endothelium possess factors that stimulate accumulation of cyclic GMP in the underlying smooth muscle. It is conceivable that this factor, which likely is EDRF, permeates smooth muscle cells to activate soluble guanylate cyclase, a biochemical event that is easily inhibited by methylene blue.

3.2. MB 22948

MB 22948 has been reported to be a selective inhibitor of cyclic GMP-specific phosphodiesterase and has been employed to examine the consequences of tissue accumulation of cyclic GMP (Kukovetz et al., 1981, 1982; Holzmann, 1982). Figure 3 illustrates the effects of MB 22948 on resting levels of cyclic GMP and tension. Similar to the observations made for methylene blue rings with endothelium, were more sensitive than those without to the effects of MB 22948. The action of MB 22948 was opposite in direction, however, to that of methylene blue. Thus, resting levels

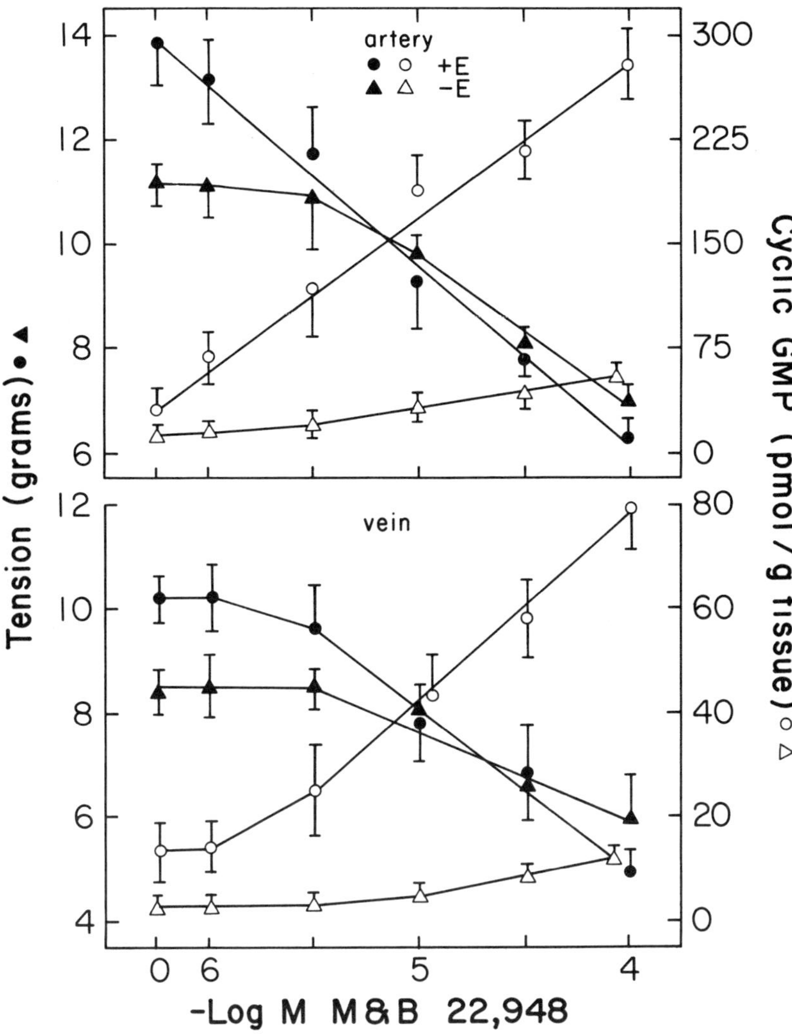

Fig. 3. Influence of MB 22948 on resting cyclic GMP levels and tension in bovine pulmonary arteries and veins with or without endothelium. Rings were prepared and handled as described for Fig. 2, except that submaximal precontractions with phenylephrine were elicited and MB 22948 was added at peak contractile response for 15 min. Data represent the mean ± SEM using 6–8 rings isolated from 3–4 animals.

of cyclic GMP increased, and this was accompanied by decreases in tension (relaxation). Clearly, although denuded rings were less sensitive than rings with endothelium, they did respond well to higher concentrations of MB 22948. The reason for this, of course, is that MB 22948 inhibits smooth muscle phosphodiesterase (independently of the endothelium), thereby allowing cyclic GMP to accumulate. Rings with endothelium are more sensitive than those without it, merely because the turnover of cyclic GMP is much higher in the former. Thus, the action of MB 22948 is not endothelium-dependent (Kukovetz et al., 1981, 1982).

4. Dependence of Activity of EDRF on the Diameter of the Blood Vessels

An excellent inverse correlation was found between the diameter of arteries or veins and the apparent activity of EDRF. The word apparent is used since EDRF has not yet been isolated or identified, and, therefore, no direct measurement of it is possible. The observations presented below may have physiological implications in the regulation of regional blood flow and pathophysiological relevance in the vasodilator actions of certain autacoids. Arteries and veins of various branches off the main intrapulmonary vessels from small lungs (2-yr-old cows) and large lungs (5–8-yr-old cows) were studied.

4.1. Artery

Arterial rings with endothelium prepared from larger arterial branches maintained tone to phenylephrine very well. Rings of smaller diameter, however, could not maintain contractile tone, and this condition worsened as the diameter became smaller (Fig. 4). This remarkable impairment in the ability to maintain tone was directly dependent on endothelial integrity. All rings without endothelium maintained tone very well. Rings with endothelium prepared from the second arterial branches were more sensitive or responsive to phenylephrine than were corresponding rings without intima (Fig. 4). In smaller branches, however, it appeared that the denuded rings were the more responsive ones. It is possible, however, that the rapid loss of contractile tone masked the true or inherent responsiveness of the smaller rings to phenylephrine. Another interesting observation was that during the initial equilibrium of vascular rings under tension, the rings with endothelium required more frequent adjustments of length to maintain applied tension (assigned load) than did the corresponding denuded rings prepared from the same vessel segment.

Methylene blue markedly enhanced contractile responsiveness to phenylephrine in rings with, but not those without, endothelium (Fig. 4). The greater responsiveness of the former to methylene blue is attributed to a higher cyclic GMP turnover.

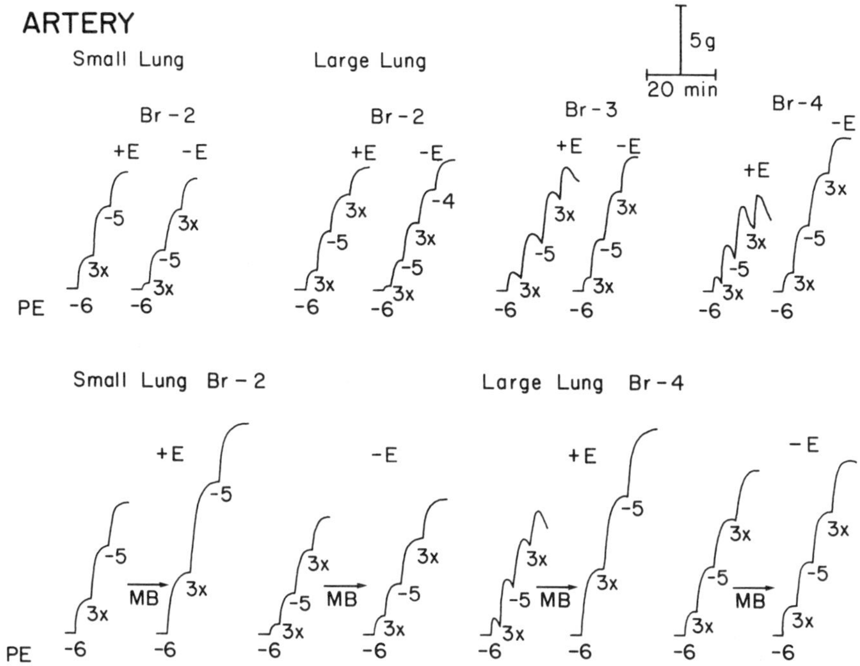

Fig. 4. Influence of the diameter of the bovine pulmonary artery and endothelium on phenylephrine-elicited contractions and responsiveness to methylene blue. Representative tracings are illustrated. Where indicated, methylene blue (10^{-6}*M*) was added for 15 min. Abbreviations: Br, branch; E, endothelium; PE, phenylephrine; MB, methylene blue. Diameters (mm) were; 3, Br-2, small lung; 4, Br-2, large lung; 2.5, Br-3, large lung; 1.5, Br-4, large lung. Initial resting tensions (*g*) were: 6, Br-2; 4, Br-3; 3, Br-4. Rings were equilibrated under tension for 2 h before eliciting contractions.

Hemoglobin produced effects that were similar to those of methylene blue (Fig. 5). Ferricyanide, an oxidant that works like methylene blue in inhibiting guanylate cyclase activation by NO, does not antagonize relaxation by vascular smooth muscle by NO because the highly charged ferricyanide (divalent anion) cannot permeate cells to reach the guanylate cyclase (Gruetter et al., 1979, 1980). Thus, ferricyanide failed to alter arterial responsiveness to phenylephrine (Fig. 5). Through its capacity to elevate levels of cyclic GMP, MB 22948 caused an impairment in the ability of arteries with endothelium to maintain contractions to phenylephrine (Fig. 5). In the presence of MB 22948, the larger rings with endothelium behaved more like rings of small diameter. Working by inhibition of cyclic GMP formation, methylene blue counteracted the effects of MB 22948.

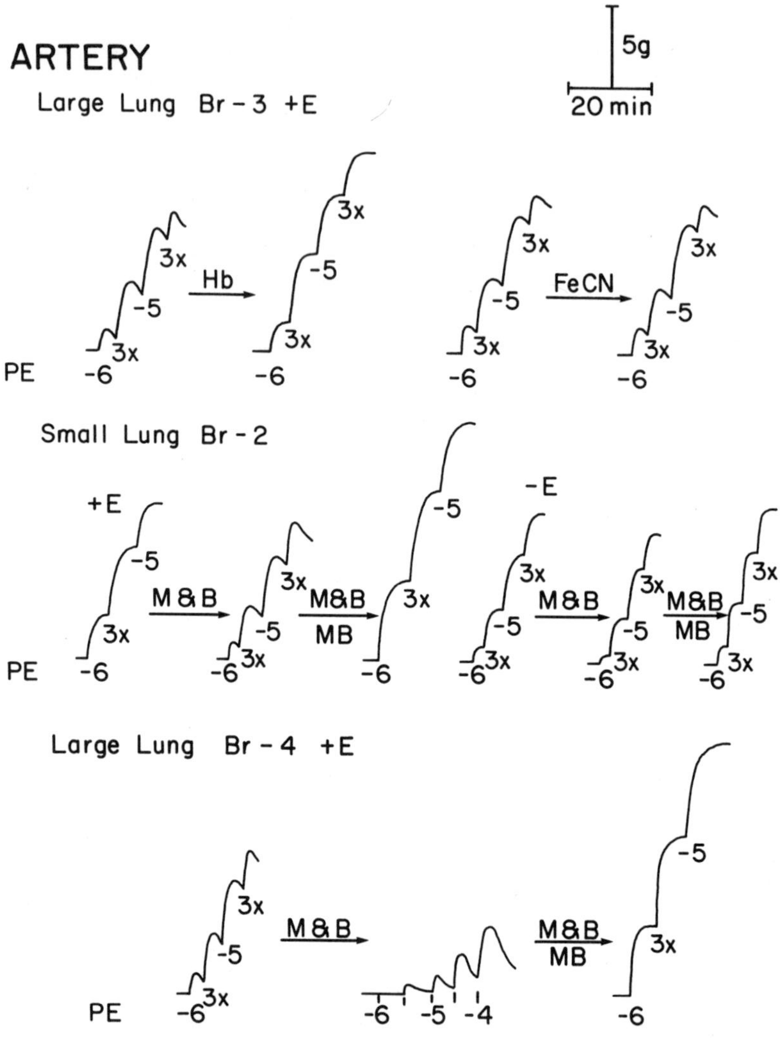

Fig. 5. Influence of various agents on contractile responsiveness of phenylephrine in the bovine pulmonary artery. Representative tracings are illustrated. Abbreviations: Br, branch; E, endothelium; PE, phenylephrine; Hb, hemoglobin ($10^{-6}M$); FeCN, ferricyanide ($10^{-5}M$); MB, methylene blue ($10^{-6}M$). Test agents were added for 15 min. Diameters (mm) were: 3, Br-2, small lung; 2.5, Br-3, large lung; 1.5, Br-4, large lung. Initial resting tensions (*g*) were: 6, Br-2; 4, Br-3; 3, Br-4. Rings were equilibrated under tension for 2 h before eliciting contractions.

These observations suggest that the endothelium is responsible for the progressive impairment of arteries of decreasing diameters to maintain contractions to phenylephrine. Moreover, the data suggest that cyclic GMP may be an important determinant of the level of tension or contractile tone. In fact, resting levels of cyclic GMP increase as the arterial diameter becomes smaller (*see* section 5). Hemoglobin antagonizes endothelium-dependent relaxations (Martin et al., 1985; Griffith et al., 1985), accumulation of cyclic GMP (Martin et al., 1985), and activation of soluble guanylate cyclase in the presence of acetylcholine (Ignarro et al., 1986b). In the present study, hemoglobin produced similar effects to those of methylene blue, which may be attributed to the inactivation of EDRF in the intercellular compartment, since hemoglobin cannot permeate cells.

Additional evidence for greater activity of EDRF in the smaller resistance vessels is presented in Fig. 6. Rings of smaller diameter were clearly more sensitive to endothelium-dependent relaxation elicited by acetylcholine, but not to endothelium-independent relaxation produced by NO. Relaxations to both acetylcholine and NO were abolished, however, by methylene blue. Similarly, hemoglobin inhibited both relaxations. Neither methylene blue nor hemoglobin affected responses to isoproterenol, which causes endothelium-independent accumulation of cyclic AMP. These observations again demonstrate the remarkably similar properties of EDRF and NO.

4.2. Vein

In general, veins behaved qualitatively similarly to arteries with respect to the dependence of apparent activity of EDRF on the diameter (Fig. 7). The principal difference between the arteries and the veins was quantitative. That is, venous rings with endothelium were slightly less responsive than arterial rings to phenylephrine and methylene blue. Whereas acetylcholine contracted veins with endothelium, bradykinin elicited relaxation that was abolished by methylene blue and inhibited by hemoglobin (Fig. 7). Hemoglobin also abolished NO-elicited relaxation of venous rings.

Venous endothelial cells therefore possess EDRF. The failure of acetylcholine, arachidonic acid, vasoactive intestinal peptide, histamine, and thrombin to relax the bovine pulmonary vein (Ignarro et al., 1984c, 1985; Harbison et al., 1985) can be attributed to a lack of specific surface receptors on the endothelial cells for these agonists. The general divalent cation ionophore, A23187, does elicit endothelium-dependent relaxation of the bovine pulmonary vein (Carl A. Gruetter, personal communication).

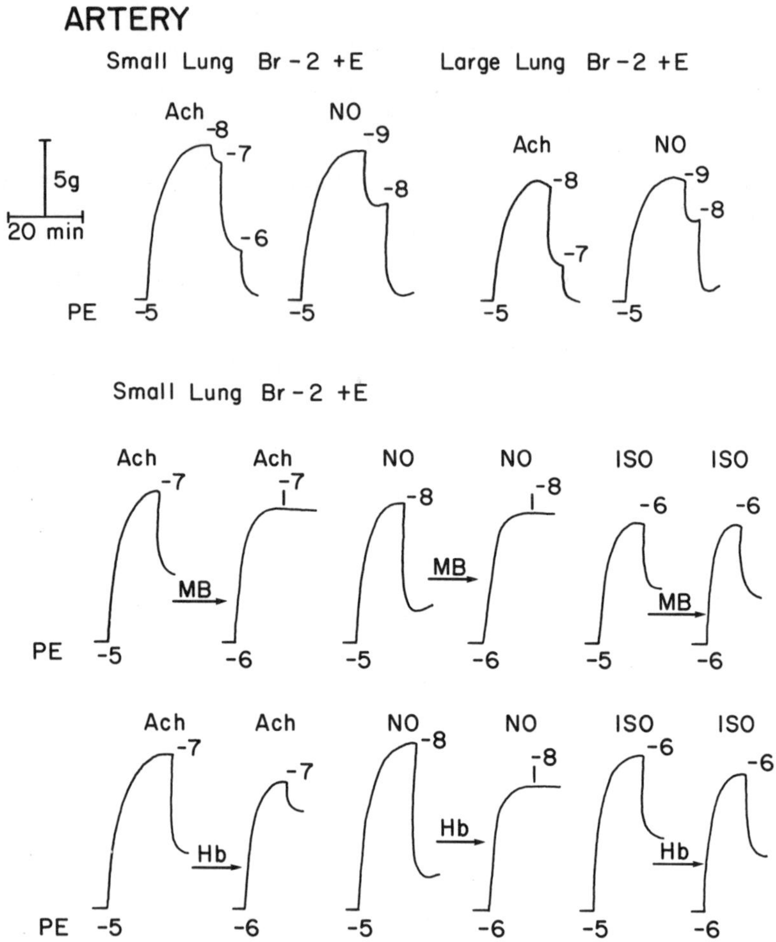

Fig. 6. Effects of methylene blue and hemoglobin on endothelium-dependent and -independent relaxations in the bovine pulmonary artery. Representative tracings are shown. Abbreviations: Br, branch; E, endothelium; PE, phenylephrine; Ach, acetylcholine; NO, nitric oxide; ISO, isoproterenol; MB, methylene blue ($10^{-6}M$); Hb, hemoglobin ($10^{-6}M$). Test agents were added for 15 min. Diameters (mm) were: 3, Br-2, small lung; 4, Br-2, large lung. Rings were equilibrated under 6*g* tension for 2 h before eliciting contractions.

5. Relationship Between Resting Cyclic Nucleotide Levels and Endothelial Integrity in Arteries and Veins

As discussed earlier, resting levels of cyclic GMP were three- to four-fold higher in the presence than in the absence of endothelium. Resting

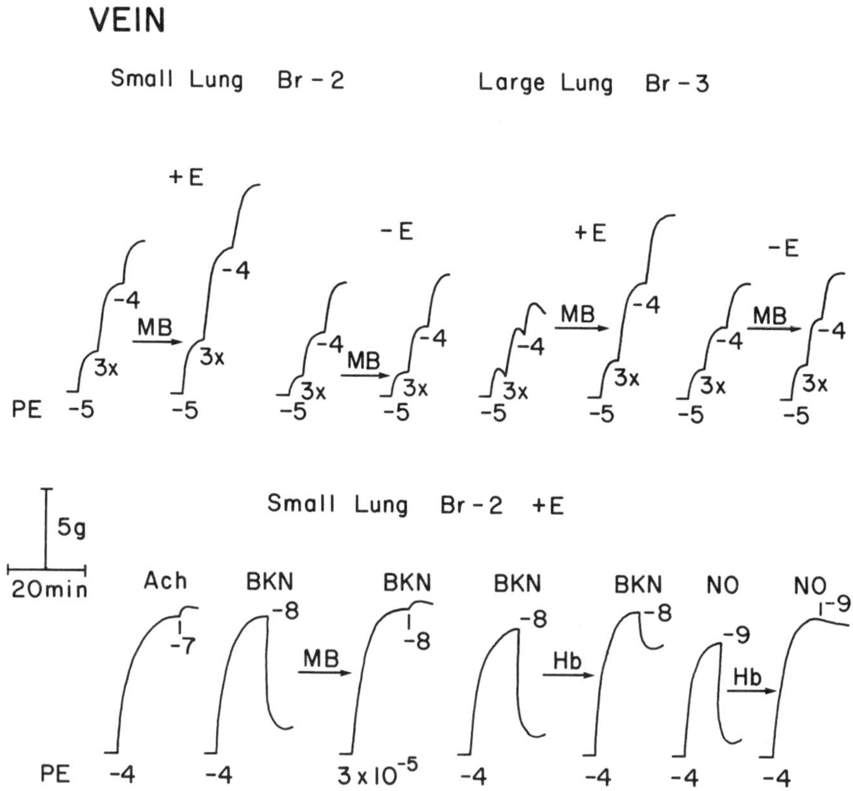

Fig. 7. Influence of endothelium, vessel diameter, and methylene blue on contractile responsiveness and effects of relaxants in the bovine pulmonary vein. Representative tracings are shown. Abbreviations: Br, branch; E, endothelium; PE, phenylephrine; Ach, acetylcholine; BKN, bradykinin; NO, nitric oxide; MB, methylene blue (10^{-6}*M*); Hb, hemoglobin (10^{-6}*M*). Test agents were added for 15 min. Diameters (mm) were: 5, Br-2, small lung; 3.5, Br-3, large lung. Rings were equilibrated under 4*g* tension for 2 h before eliciting contractions.

cyclic AMP levels were twofold higher in the arteries with than in those without endothelium, but no such differences were apparent in veins. As expected, resting levels of cyclic GMP correlated inversely with the diameter of blood vessels with, but not of those without, intima (Table 1). Thus, a close correlation existed between cyclic GMP levels and apparent activity of EDRF. The higher the resting levels of cyclic GMP, the greater the impairment of the blood vessel to maintain tone or contractile tension to phenylephrine. Methylene blue lowered the levels of cyclic GMP back toward levels characteristic of vessels without endothelium. Thus, the progressive impairment of the maintenance of tone can be attributed to elevated concentrations of cyclic GMP in smooth muscle.

Levels of cyclic AMP arteries with endothelium were independent of the diameter of the blood vessels. These observations suggest that cyclic GMP, rather than cyclic AMP, is the principle determinant of the impairment of tension maintenance in endothelium-containing pulmonary arteries.

6. Some Unique Properties of Putative EDRF

6.1. Direct Activation of Soluble Guanylate Cyclase

Arterial and venous rings with but not those without endothelium directly activate the soluble, heme-containing form of guanylate cyclase purified to homogeneity from bovine lung (Ignarro et al., 1986b). Figure 8 highlights some of the more important findings. The general experimental procedure involved the rapid transfer of the rings from organ chambers (after equilibration and testing relaxant responses to acetylcholine and bradykinin) to warmed and oxygenated mixtures for the assay of guanylate cyclase, some of which contained additional test agents (*see* Ignarro et al., 1986b, for details). Enzyme activation by rings with endothelium was markedly enhanced by the addition of acetylcholine to the enzyme reaction mixtures. Methylene blue and hemoglobin inhibited enzyme activation by endothelium-containing rings in the absence or presence of acetylcholine. Atropinized arterial rings also activated guanylate cyclase, but acetylcholine failed to increase enzyme activation (Ignarro et al., 1986b).

In the veins with endothelium, acetylcholine failed to influence activation of the enzyme (Fig. 8). Bradykinin, however, increased the activation, and this was blocked by methylene blue. These observations are strikingly similar to the actions of acetylcholine (no relaxation) and bradykinin (marked relaxation) on contracted, endothelium-containing venous rings in organ chambers.

These data suggest strongly that EDRF was responsible for the observed direct activation of soluble guanylate cyclase. Again, the properties of EDRF in this regard were similar to those of NO. Guanylate cyclase activation by NO requires the presence of heme attached to the enzyme (Craven and DeRubertis, 1978, 1983; Craven et al., 1979; Ignarro et al., 1982, 1986c; Ohlstein et al., 1982). Similarly, enzyme activation by EDRF is also heme-dependent (preliminary findings). The only guanylate cyclase activators known to have an absolute requirement for heme are NO and phenyl radical (Ignarro et al., 1984b,d). EDRF behaves similarly, therefore, to NO and phenyl radicals.

6.2. Stability of EDRF

The half-life of the putative EDRF was reported to be 6–7 s (Griffith et al., 1984a,b; Rubanyi et al., 1985). In the experiments described in

Table 1
Influence of Vessel Diameter and Endothelium on Resting Levels of Cyclic GMP and Cyclic AMP in Bovine Pulmonary Artery and Vein[a]

Vessel and branch		±E	Cyclic GMP		Cyclic AMP
			Control	Methylene blue, $10^{-5}M$	Control
			pmol/g tissue		
Artery					
Small lung,	2nd	+	19 ± 2.1[b]	7 ± 0.8[c]	316 ± 22[b]
		−	6 ± 0.5	5 ± 0.3	152 ± 10
Large lung,	2nd	+	28 ± 1.9[b]	12 ± 1.1[c]	307 ± 19[b]
		−	7 ± 0.8	7 ± 0.9	168 ± 11
	3rd	+	39 ± 2.6[b]	17 ± 1.1[b,c]	336 ± 20[c]
		−	9 ± 0.7	7 ± 0.6	180 ± 14
	4th	+	55 ± 4.5[b]	18 ± 2.3[b,c]	304 ± 18[b]
		−	10 ± 0.9	8 ± 0.9	141 ± 15
Vein					
Small lung,	2nd	+	9 ± 0.7[b]	3 ± 0.1[c]	110 ± 8
		−	3 ± 0.2	3 ± 0.2	124 ± 7
Large lung,	2nd	+	16 ± 0.9[b]	4 ± 0.2[c]	126 ± 9
		−	4 ± 0.3	3 ± 0.2	117 ± 6
	3rd	+	30 ± 2.4[b]	9 ± 0.9[c]	98 ± 8
		−	6 ± 0.6	5 ± 0.8	112 ± 9
	4th	+	43 ± 3.1[b]	12 ± 0.8[b,c]	131 ± 7
		−	8 ± 0.7	9 ± 0.7	114 ± 9

[a]All rings were mounted under optimal tension, equilibrated, and depolarized with KCl. Following washing and further equilibration, methylene blue was added for 15 min where indicated. Phenylephrine was then added to all rings to elicit submaximal contractions. Rings were freeze-clamped at peak contractile responses, extracted, and assayed for cyclic nucleotides. Small and large lungs were from cows that were 2 and 5 yr old, respectively. Values represent the mean ± SEM using 8–10 rings from 4-5 animals.

[b]Significantly different ($p < 0.05$) from values corresponding to minus (−) endothelium.

[c]All values corresponding to plus (+) endothelium are significantly different ($p < 0.05$) from cyclic GMP control values plus (+) endothelium.

the preceding section, a half-life of 40–45 s for EDRF was found (Ignarro et al., 1986b). Differences in oxygen tension and temperature are some factors that can alter the stability of EDRF. A recent report indicated that lowering the oxygen tension increased the half-life of EDRF to 24–49 s (Fostermann et al., 1984). Other factors may also alter EDRF stability such as the presence of sulfhydryls, reducing agents, and antioxidants (Griffith et al., 1984a,b; 1985). It is of interest, and perhaps importance, to note that the stability of NO in air or water is about 8–15 s, and the decomposition of NO (oxidation to NO_2 and higher oxides of nitrogen) is markedly accelerated by oxygen and heat.

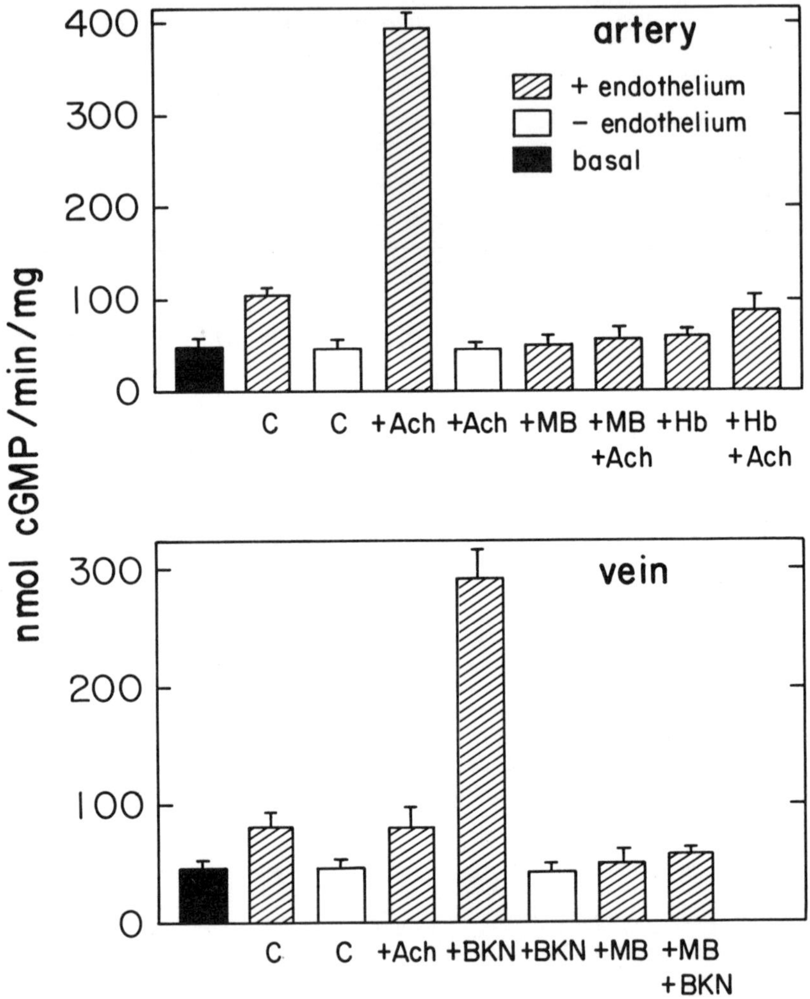

Fig. 8. Direct activation of purified soluble guanylate cyclase by endothelium-containing arteries and veins and enhancement by endothelium-dependent vasodilators. Rings were first equilibrated under tension for 2 h and tested for endothelial integrity by assessing responsiveness of phenylephrine-contracted rings to acetylcholine (arteries) or bradykinin (veins). Following washing and further equilibration, rings were rapidly transferred to guanylate cyclase reaction mixtures that had been oxygenated and warmed to 37 °C (Ignarro et al., 1986b). Reaction mixtures were incubated at 37 °C in air for 10 min and then assayed for cyclic

6.3. Chemical Reactivity of EDRF

Reports indicate that EDRF is inactivated by sulfhydryl agents, certain reductants or antioxidants, phenylhydrazine, and hemoglobin (Griffith et al., 1984a, 1985; Martin et al., 1985). Experiments in this laboratory have been only partially consistent with these reports. Whereas dithiothreitol appears to inactivate EDRF, no such effect could be found with glutathione or cysteine, ascorbate, bisulfite, borohydride, α-tocopherol, or phenylhydrazine. In fact, phenyl radical, which is generated from phenylhydrazine in the presence of trace amounts of hemoglobin, relaxes vascular smooth muscle (Ignarro et al., 1984d) and is additive with acetylcholine in the bovine pulmonary artery. Hemoglobin has been suggested to inhibit EDRF activity by binding with it (Martin et al., 1985; Griffith et al., 1985). Methemoglobin, myoglobin, and metmyoglobin also inhibit EDRF activity (unpublished observations), and these findings are expected if EDRF possesses hemoprotein-binding properties similar to those of NO or phenyl radical. Caution must be exercised in the interpretation of observations made with many antioxidants such as dithiothreitol and nordihydroguaiaretic acid because such agents nonselectively depress contractions as well as relaxations (Ignarro and Kadowitz, 1985; Furchgott, 1984).

6.4. Sites of Action of Agents Affecting EDRF Activity

The experimental procedures employed to test the effects of EDRF on guanylate cyclase permitted an analysis of potential sites of action of the tested agents (Ignarro et al., 1986b). For example, if vascular rings are pretreated with agents added only to the chambers and then transferred to reaction mixtures containing acetylcholine, then the tested agent will have no effect on the purified enzyme, and one can deduce whether or not generation or release of EDRF has been impaired. Atropinized rings do not release EDRF upon exposure to acetylcholine in enzyme reaction mixtures. On the other hand, rings treated with methylene blue respond

← Fig. 8 (*Continued*)

GMP. The ordinate signifies rate of cyclic GMP formation. Control rings (C) received no further additions. Where indicated acetylcholine (Ach, 10^{-6} *M*), bradykinin (BKN, 10^{-7}*M*), methylene blue (MB, 10^{-5}*M*), or hemoglobin (Hb, 10^{-5}*M*) was added to reaction mixtures just prior to addition of rings. Data represent the mean ± SEM using 6–8 rings from 3–4 animals. Values for controls with (+) endothelium, but not those without (−) endothelium, are significantly different ($p < 0.05$) from those for basal activity. Methylene blue and hemoglobin significantly ($p < 0.05$) decreased cyclic GMP formation when compared with controls with endothelium.

fully to acetylcholine added to the reaction mixtures, because although guanylate cyclase of smooth muscle is inhibited by methylene blue, the purified enzyme contained in reaction mixtures is not. Therefore, one may conclude that methylene blue does not impair the generation or release of EDRF. If methylene blue is added to reaction mixtures, however, then the purified enzyme will be inhibited and acetylcholine-elicited release of EDRF will not result in activation of the enzyme. Another procedure involves adding acetylcholine, a tested agent, and rings to enzyme-free reaction mixtures and delaying addition of the enzyme to reaction mixtures for 20–30 s. This method allows any interaction between released EDRF and the tested agent to occur prior to interaction of EDRF with subsequently added purified enzyme. Using this method, it was found that hemoproteins, dithiothreitol, and certain other agents either sequestered or inactivated the released EDRF so that enzyme activation could not be observed. Experiments are in progress to classify the sites of action of numerous reductants, oxidants, spin-trap reagents, hemoproteins, NO-trapping reagents, phenyl radical ligands, fatty acid binding ligands, and other classes of tested agents.

7. Conclusions and Future Direction

The following fundamental concepts have been derived from our studies on the properties of the putative EDRF:

1. Arterial and venous EDRF directly activate soluble guanylate cyclase by heme-dependent mechanisms, and activation is inhibited by methylene blue, brilliant cresyl blue, and riboflavin via mechanisms entailing inhibition of the activated state of guanylate cyclase.
2. Certain agents such as hemoproteins and very select reductants sequester and/or inactivate EDRF. Many analogous substances do not share these inhibitory actions, however.
3. Blood vessels of smaller diameter possess greater activities or amounts of EDRF than larger vessels, and this appears to account for the impaired ability of the smaller vessels to maintain steady state contractions to alpha-adrenergic agonists. Potassium-elicited contractions remain unimpaired, however, and this may be closely related to the markedly reduced capacity of endothelium-dependent vasodilators to relax potassium-contracted arterial rings.
4. Resting levels of cyclic GMP appear to be intimately coupled to activity of EDRF, and endothelial cells may therefore play an important physiological and/or pathophysiological role in the regula-

tion of vascular levels of cyclic GMP and associated responses of the smooth muscle. In this regard, endothelium-dependent vasodilatation may not be as important for acetylcholine as it might be for numerous autacoids that dilate very small resistance vessels. It is of interest, therefore, that smaller arteries appear to be more reactive than larger ones to endothelium-dependent vasodilators.

5. Not only arteries but also veins possess the capacity to generate EDRF. The failure of most endothelium-dependent arterial relaxants to relax veins is attributed to the lack of specific receptors on venous endothelial cells and not to the lack of venous endothelial EDRF.
6. EDRF is not very stable, but its half-life can be extended from 6–7 to 40–45 s by controlling certain environmental conditions. Additional experimentation to further stabilize EDRF is a prerequisite to its successful isolation and characterization.
7. Preliminary experiments suggest that generation and/or release of EDRF is dependent on mitochondrial production of ATP. Among important activities within mitochondria are fatty acid oxidation and ammonia production. Thus, EDRF may be chemically related to oxidation products of fatty acids or ammonia, either of which can be very unstable.

Studies are in progress to define systems in endothelial cells capable of catalyzing two-electron oxidation of ammonia to hydroxylamine and four-electron oxidation of ammonia to NO. Hydroxylamine, once formed, may be rapidly converted to NO through the catalytic properties of intracellular catalase resulting in the formation of unstable NO-catalase. NO-catalase would rapidly and steadily release the lipophilic radical species, NO, which could then readily penetrate smooth muscle cells to cause guanylate cyclase activation, cyclic GMP accumulation, and smooth muscle relaxation. Preliminary data indicate that hydroxylamine and azide are potent relaxants of phenylephrine-, but not potassium-, contracted endothelium-containing areterial rings. Oxyhemoproteins and methylene blue are potent inhibitors of these relaxant effects, as well as of the relaxing action of NO and acetylcholine.

A schematic is presented in Fig. 9 that highlights certain aspects of the above conclusions. Endothelium-dependent vasodilators interact with specific endothelial surface receptors and trigger EDRF formation via mechanisms involving calcium and oxygen. One or more precursors of EDRF are likely to exist. EDRF, which possesses properties similar to NO, is lipophilic and exits the endothelial cell by penetrating cell membranes on either side. Lipophilic organic nitrate and nitrite esters permeate smooth muscle cells and release NO intracellularly. NO itself can rapidly

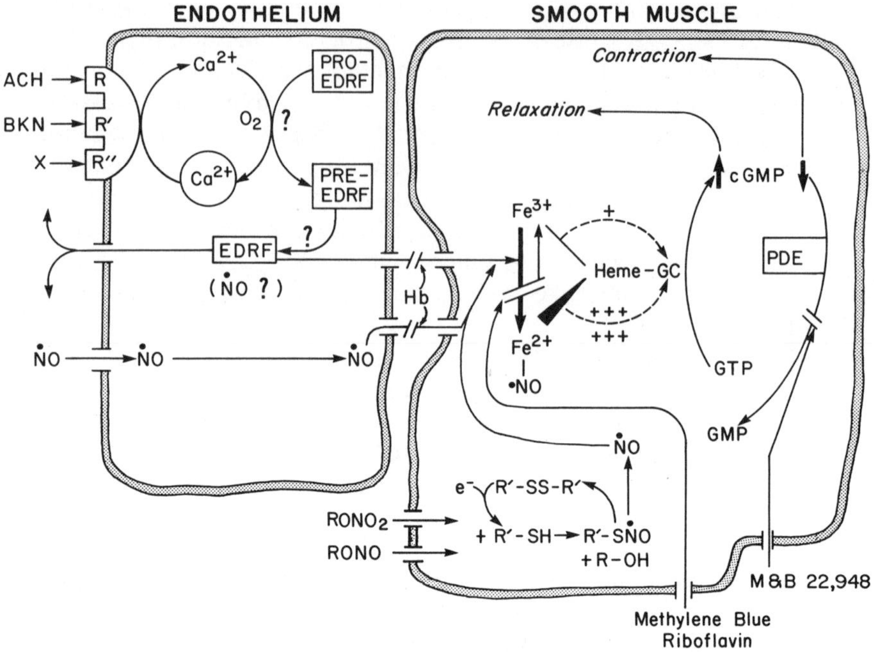

Fig. 9. Schematic illustration of possible mechanism of generation and action of EDRF. Abbreviations: ACH, acetylcholine; BKN, bradykinin; X, other endothelium-dependent vasodilators; R, R′, R″, specific receptors; NO, nitric oxide; $RONO_2$, organic nitrate ester; RONO, organic nitrite ester; R′-SH, thiol; R′-SNO, *S*-nitrosothiol; EDRF, endothelium-derived relaxing factor(s); PRO- and PRE-EDRF, precursors of EDRF; GC, soluble guanylate cyclase; Hb, hemoglobin; PDE, phosphodiesterase. Endothelium-dependent vasodilators interact with surface receptors of the endothelial cells, whereas nitrogen oxide-containing vasodilators permeate the smooth muscle cell. Methylene blue, riboflavin, and MB 22948 permeate cells, whereas hemoglobin does not. EDRF is lipophilic and permeates endothelial and smooth muscle cell membranes. EDRF possesses properties that are very similar to those of NO. EDRF and NO require enzyme-bound heme to activate guanylate cyclase, and the latter takes the form of the ternary complex: activator-Fe^{2+} heme-enzyme. Increases and decreases in resting cyclic GMP levels in smooth muscle equilibrated under tension result in relaxation and contraction, respectively.

permeate all cells. NO reacts with the heme moiety of soluble guanylate cyclase and, through reductive mechanisms, generates the nitrosyl-heme (NO-Fe^{2+}-heme) adduct of guanylate cyclase, which represents the activated state of the enzyme. EDRF permeates the smooth muscle cell and also generates an activated state of heme-containing guanylate cyclase. What this activated enzyme form is, and whether or not EDRF is related to NO, are unknown at present. The extracellular inhibitory action of hemoglobin on EDRF and NO activity suggests that EDRF and NO may be chemically similar. Lipophilic methylene blue and riboflavin inhibit the activated, but not basal, state of guanylate cyclase. Inhibition of cyclic GMP phosphodiesterase activity would cause cyclic GMP to accumulate within the cell, especially if cyclic GMP turnover is high as when guanylate cyclase is in its activated state. The consistent observation that resting or basal levels of cyclic GMP are elevated in endothelium-containing vessels and that these are more sensitive than blood vessels without endothelium to both methylene blue and MB 22948 suggest that basal release of EDRF maintains smooth muscle guanylate cyclase in a partially activated state. Further increases in cyclic GMP accumulation would trigger relaxation, whereas decreases would trigger contraction or an increase in tension.

Future advancements in this area of research will come from studies designed to stabilize, trap, isolate, identify, and characterize the properties of endothelium-derived relaxing substances. As more sophisticated techniques are developed for working with small blood vessels and perhaps cultured endothelial cells, more information should evolve that will enable investigators to achieve the ultimate goal of identifying vascular EDRF(s).

References

Craven, P. A. and DeRubertis, F. R.: Restoration of the responsiveness of purified guanylate cyclase to nitrosoguanidine, nitric oxide, and related activators by heme and heme proteins: Evidence for the involvement of the paramagnetic nitrosyl-heme complex in enzyme activation. *J. Biol. Chem.* **253:**8433–8443, 1978.

Craven, P. A. and DeRubertis, F. R.: Requirement for heme in the activation of purified guanylate cyclase by nitric oxide. *Biochim. Biophys. Acta* **745:**310–321, 1983.

Craven, P. A., DeRubertis, F. R., and Pratt, D. W.: Electron spin resonance study of the role of NO-catalase in the activation of guanylate cyclase by NaN_3 and NH_2OH: Modulation of enzyme responses by heme proteins and their nitrosyl derivatives. *J. Biol. Chem.* **245:**8213–8222, 1979.

Diamond, J. and Chu, E. B.: Possible role for cyclic GMP in endothelium-dependent relaxation of rabbit aorta by acetylcholine. Comparison with nitroglycerin. *Res. Commun. Chem. Pathol. Pharmacol.* **41:**369–381, 1983.

Edwards, J. C., Ignarro, L. J., Hyman, A. L., and Kadowitz, P. J.: Relaxation of intrapulmonary artery and vein by nitrogen oxide-containing vasodilators and cyclic GMP. *J. Pharmacol. Exp. Ther.* **228:**33–42, 1984.

Forstermann, U., Trogisch, G., and Busse, R.: Species-dependent differences in the nature of endothelium-derived vascular relaxing factor. *Eur. J. Pharmacol.* **106:**639–643, 1984.

Furchgott, R. F.: The role of endothelium in the responses of vascular smooth muscle to drugs. *Ann. Rev. Pharmacol.* **24:**175–197, 1984.

Furchgott, R. F., Cherry, P. D., Zawadzki, J. V., and Jothianandan, D.: Endothelial cells as mediators of vasodilation of arteries. *J. Cardiovasc. Pharmacol.* **6:**S336–S343, 1984.

Griffith, T. M., Edwards, D. H., Lewis, M. J., and Henderson, A. H.: Evidence that cyclic guanosine monophosphate (cGMP) mediates endothelium-dependent relaxation. *Eur. J. Pharmacol.* **112:**195–202, 1985.

Griffith, T. M., Edwards, D. H., Lewis, M. J., Newby, A. C., and Henderson, A. H.: The nature of endothelium-derived vascular relaxant factor. *Nature* (Lond.) **308:**645–647, 1984a.

Griffith, T. M., Henderson, A. H., Edwards, D. H., and Lewis, M. J.: Isolated perfused rabbit coronary artery and aortic strip preparations: The role of endothelium-derived relaxant factor. *J. Physiol.* (Lond.) **351:**13–24, 1984b.

Gruetter, C. A., Barry, B. K., McNamara, D. B., Gruetter, D. Y., Kadowitz, P. J., and Ignarro, L. J.: Relaxation of bovine coronary artery and activation of coronary arterial guanylate cyclase by nitric oxide, nitroprusside and a carcinogenic nitrosoamine. *J. Cyclic Nucleotide Res.* **5:**211–224, 1979.

Gruetter, C. A., Barry, B. K., McNamara, D. B., Kadowitz, P. J., and Ignarro, L. J.: Coronary arterial relaxation and guanylate cyclase activation by cigarette smoke, N^1-nitrosonornicotine and nitric oxide. *J. Pharmacol. Exp. Ther.* **214:**9–15, 1980.

Gruetter, C. A., Gruetter, D. Y., Lyon, J. E., Kadowitz, P. J., and Ignarro, L. J.: Relationship between cyclic GMP formation and relaxation of coronary arterial smooth muscle by glyceryl trinitrate, nitroprusside, nitrite and nitric oxide: Effects of methylene blue and methemoglobin. *J. Pharmacol. Exp. Ther.* **219:**181–186, 1981a.

Gruetter, C. A., Kadowitz, P. J., and Ignarro, L. J.: Methylene blue inhibits coronary arterial relaxation and guanylate cyclase activation by nitroglycerin, sodium nitrite and amyl nitrite. *Can. J. Physiol. Pharmacol.* **59:**150–156, 1981b.

Harbison, R. G., Wood, K. S., Hyman, A. L., Kadowtiz, P. J., and Ignarro, L. J.: Endothelium-dependent and independent regulation of cyclic GMP formation and relaxation of intrapulmonary artery by peptides (abstract). *Fed. Proc.* **44:**1235, 1985.

Holzmann, S.: Endothelium-induced relaxation by acetylcholine associated with larger rises in cyclic GMP in coronary arterial strips. *J. Cyclic Nucleotide Res.* **8:**409–419, 1982.

Ignarro, L. J. and Kadowitz, P. J.: Pharmacological and physiological role of cyclic GMP in vascular smooth muscle relaxation. *Ann. Rev. Pharmacol.* **25:**171–191, 1985.

Ignarro, L. J., Harbison, R. G., Wood, K. S., and Kadowitz, P. J.: Dissimilarities between methylene blue and cyanide on relaxation and cyclic GMP formation in endothelium-intact intrapulmonary artery caused by nitrogen oxide-

containing vasodilators and acetylcholine. *J. Pharmacol. Exp. Ther.* **236:**30–36, 1986a.
Ignarro, L. J., Harbison, R. G., Wood, K. S., and Kadowitz, P. J.: Activation of purified soluble guanylate cyclase by endothelium-derived relaxing factor from intrapulmonary artery and vein: Stimulation by acetylcholine, bradykinin and arachidonic acid. *J. Pharmacol. Exp. Ther.* **237:**893–900, 1986b.
Ignarro, L. J., Adams, J. B., Horwitz, P. M., and Wood, K. S.: Activation of soluble guanylate cyclase by NO-hemoproteins involves NO-heme exchange: Comparison of heme-containing and heme-deficient enzyme forms. *J. Biol. Chem.* **261:**4997–5002, 1986c.
Ignarro, L. J., Harbison, R. G., Wood, K.S., Wolin, M. S., McNamara, D. B., Hyman, A. L., and Kadowitz, P. J.: Differences in responsiveness of intrapulmonary artery and vein to arachidonic acid: Mechanism of arterial relaxation involves cyclic GMP and cyclic AMP. *J. Pharmacol. Exp. Ther.* **233:**560–569, 1985.
Ignarro, L. J., Gruetter, C. A., Hyman, A. L., and Kadowitz, P. J.: Molecular Mechanisms of Vasodilatation, in *Dopamine Receptor Agonists,*Poste, G. and Crooke, S. T. eds., Plenum, New York, 1984a.
Ignarro, L. J., Wood, K. S., and Wolin, M. S.: Regulation of purified soluble guanylate cyclase by porphyrins and metalloporphyrins: A unifying concept. *Adv. Cyclic Nucleotide Res.* **17:**267–274, 1984b.
Ignarro, L. J., Burke, T. M., Wood, K. S., Wolin, M. S., and Kadowitz, P. J.: Association between cyclic GMP accumulation and acetylcholine-elicited relaxation of bovine intrapulmonary artery. *J. Pharmacol. Exp. Ther.* **228:**682–690, 1984c.
Ignarro, L. J., Wood, K. S., Ballot, B., and Wolin, M. S.: Guanylate cyclase from bovine lung: Evidence that enzyme activation by phenylhydrazine is mediated by iron-phenyl hemoprotein complexes. *J. Biol. Chem.* **259:**5923–5931, 1984d.
Ignarro, L. J., Wood, K. S., and Wolin, M. S.: Activation of purified soluble guanylate cyclase by protoporphyrin IX. *Proc. Natl. Acad. Sci. USA* **79:**2870–2873, 1982a.
Ignarro, L. J., Degnan, J. N., Baricos, W. H., Kadowitz, P. J., and Wolin, M. S.: Activation of purified guanylate cyclase by nitric oxide requires heme: Comparison of heme-deficient, heme-reconstituted and heme-containing forms of soluble enzyme from bovine lung. *Biochim. Biophys. Acta* **718:**49–59, 1982b.
Ignarro, L. J., Lipton, H., Edwards, J. C., Baricos, W. H., Hyman, A. L., Kadowitz, P. J., and Gruetter, C. A.: Mechanism of vascular smooth muscle relaxation by organic nitrates, nitrites, nitroprusside and nitric oxide: Evidence for the involvement of S-nitrosothiols as active intermediates. *J. Pharmacol. Exp. Ther.* **218:**739–749, 1981.
Keith, R. A., Burkman, A. M., Sokoloski, T. D., and Fertel, R. H.: Vascular tolerance to nitroglycerin and cyclic GMP generation in rat aortic smooth muscle. *J. Pharmacol. Exp. Ther.* **221:**525–531, 1982.

Kukovetz, W. R., Holzmann, S., and Poch, G.: Function of cyclic GMP in acetylcholine-induced contraction of coronary smooth muscle. *Naunyn Schmiedebergs Arch. Pharmacol.* **319:**29–33, 1982.

Kukovetz, W. R., Poch, G., and Holzmann, S.: Cyclic Nucleotides and Relaxation of Vascular Smooth Muscle, in *Vasodilatation* Vanhoutte, P. M. and Leusen, I. eds., Raven, New York, 1981.

Martin, W., Villani, G. M., Jothianandan, D., and Furchgott, R. F.: Selective blockade of endothelium-dependent and glyceryl trinitrate-induced relaxation by hemoglobin and by methylene blue in the rabbit aorta. *J. Pharmacol. Exp. Ther.* **232:**708–716, 1985.

Napoli, S. A., Gruetter, C. A., Ignarro, L. J., and Kadowitz, P. J.: Relaxation of bovine coronary arterial smooth muscle by cyclic GMP, cyclic AMP and analogs. *J. Pharmacol. Exp. Ther.* **212:**469–473, 1980.

Ohlstein, E. H., Wood, K. S., and Ignarro, L. J.: Purification and properties of heme-deficient hepatic soluble guanylate cyclase: Effects of heme and other factors on enzyme activation by NO, NO-heme, and protoporphyrin IX. *Arch. Biochem. Biophys.* **218:**187–98, 1982.

Rapoport, R. M. and Murad, F.: Agonist-induced endothelium-dependent relaxation in rat thoracic aorta may be mediated through cGMP. *Circ. Res.* **52:**352–357, 1983.

Rapoport, R. M., Draznin, M. B., and Murad, F.: Mechanisms of adenosine triphosphate-, thrombin-, and trypsin-induced relaxation of rat thoracic aorta. *Circ. Res.* **55:**468–479, 1984.

Rubanyi, G. M., Lorenz, R. R., and Vanhoutte, P. M.: Bioassay of endothelium-derived relaxing factor(s): Inactivation by catecholamines. *Am. J. Physiol.* **249:**H95–H101, 1985.

Schultz, K. D., Bohme, E., Kreye, V. A. W., and Schultz, G.: Relaxation of hormonally stimulated smooth muscular tissues by the 8-bromo derivative of cyclic GMP. *Naunyn Schmiedebergs Arch. Pharmacol.* **306:**1–9, 1979.

Wolin, M. S., Wood, K. S., and Ignarro, L. J.: Guanylate cyclase from bovine lung: A kinetic analysis of the regulation of the purified soluble enzyme by protoporphyrin IX, heme and nitrosyl-heme. *J. Biol. Chem.* **257:**13312–13320, 1982.

Chapter 16

Endothelium-Dependent Reponses of Cerebral Arteries

Z. S. Katusic and J. T. Shepherd

1. Introduction

Studies on isolated canine cerebral arteries provide evidence that the responses of these vessels to a variety of vasoactive agents depends on the presence of their endothelial cells (Vanhoutte and Katusic, 1986). Large cerebral arteries contribute to the total vascular resistance in the brain (Heistad and Kontos, 1983); thus endothelium-dependent phenomena may actively participate in regulation of cerebral blood flow. This chapter will review briefly the studies characterizing endothelium-dependent responses in cerebral arteries of different species.

2. Endothelium-Dependent Relaxations

2.1. Acetylcholine

Endothelium-dependent relaxations in response to acetylcholine have been demonstrated in isolated cerebral arteries of the cat. This inhibitory response can be blocked by atropine, indicating that activation of muscarinic receptors is responsible (Lee, 1982). In the dog, no endothelium-dependent relaxations are evident in the basilar artery.

In the intact organism, the only sources of acetylcholine probably are the cholinergic nerve endings. Cholinesterases degrade the liberated acetylcholine avidly, however, so that the cholinergic transmitter probably never reaches the intima in concentrations sufficient to activate endothelial muscarinic receptors. Thus, the data obtained in the posterior auricular artery of the cat and in large canine coronary arteries indicate that acetylcholine released by cholinergic nerve endings does not affect the endothelial cells (Cohen et al., 1984; Brayden and Bevan, 1985). Therefore, in large arteries it is unlikely that endothelium-dependent relaxations to acetylcholine participate in the regulation of the smooth muscle tone in vivo. This may be different at the microcirculatory level, where the contacts between autonomic nerves and the intima are much closer. Parnavelas et al. (1985) have demonstrated the presence of the enzyme choline acetyltransferase (which plays a key role in the synthesis of the cholinergic transmitter in neurons) in certain endothelial cells of cerebral blood vessels. This raises the possibility that endothelial synthesis of acetylcholine may contribute to the control of the function of the underlying smooth muscle.

2.2. Bradykinin

Bradykinin causes endothelium-dependent relaxations in a number of peripheral arteries (Chand and Altura, 1981; Cherry et al., 1982; Regoli, 1984). It also causes endothelium-dependent relaxations in the canine basilar artery (Fig. 1; Katusic et al., 1986). It is unknown whether or not the endothelium-dependent response to bradykinin contributes to the local regulation of cerebral blood flow under normal or pathologic conditions.

2.3. Thrombin

Thrombin causes endothelium-dependent relaxations in coronary, femoral, saphenous, and pulmonary arteries of the dog (De Mey et al., 1982; Vanhoutte and Miller, 1985). It also causes endothelium-dependent relaxations in the canine basilar artery of the dog (Fig. 2; White et al., 1984; Katusic et al., 1984). The inhibitory response to thrombin in cerebral arteries is transient. It is followed by a contraction caused by direct activation of the vascular smooth muscle. These observations on isolated cerebral arteries, when extrapolated to the intact organism, suggest that production of this coagulation factor could play a role in pathogenesis of cerebral vasospasm only following damage of endothelial cells or during extravasation of blood as seen with subarachnoid hemorrhage.

2.4. Vasopressin and Oxytocin

Vasopressin, at concentrations corresponding to circulating levels of the hormone in vivo, causes endothelium-dependent relaxations of the canine basilar artery (Fig. 3; Katusic et al., 1984). Oxytocin produces a

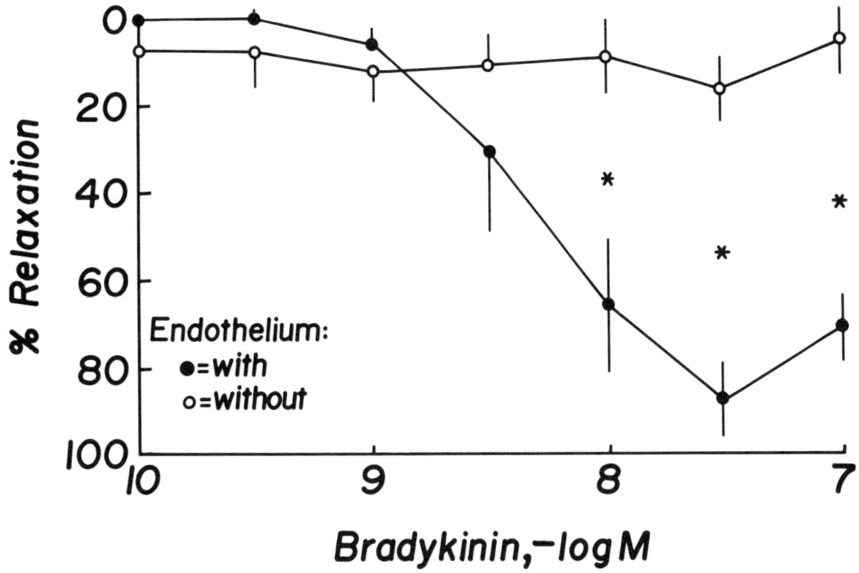

Fig. 1. Cumulative concentration–response curves to bradykinin in canine basilar arteries with and without endothelium. The relaxations were obtained in rings contracted with prostaglandin $F_{2\alpha}$ ($2 \times 10^{-6} M$). The data are shown as means ± SEM and are expressed as percent of the maximal relaxations induced by papaverine. The asterisk denotes that the difference between the two types of preparations is statistically significant (modified from Katusic et al., 1986).

similar effect, but it is less potent than vasopressin. The inhibitory responses to both neuropeptides is antagonized in a competitive manner by the V_1-vasopressinergic antagonist, $d(CH_2)_5$Tyr(Me)AVP, suggesting that effects are mediated by activation of V_1-vasopressinergic receptors (Katusic et al., 1986). It is unlikely that metabolites of arachidonic acid are involved in these responses. Indeed, inhibitors of cyclooxygenase and lipoxygenase do not affect the inhibitory effect of vasopressin (Vanhoutte et al., 1984). The endothelium-dependent relaxation to vasopressin is significantly reduced by inhibition of Na^+-K^+-ATPase with ouabain or incubation of the artery in potassium-free solution (unpublished observations). It is uncertain whether the inhibition of Na^+-K^+-ATPase involves the endothelium, the smooth muscle, or both. Inhibition of Na^+-K^+-ATPase does not, however, affect relaxations induced by papaverine, indicating that the effects of ouabain and potassium-free solution on the inhibitory responses to vasopressin are not caused by nonselective antagonism.

Perivascular application of vasopressin and oxytocin do not alter the diameter of cerebral pial arterioles (Lassof and Altura, 1980). Cerebral

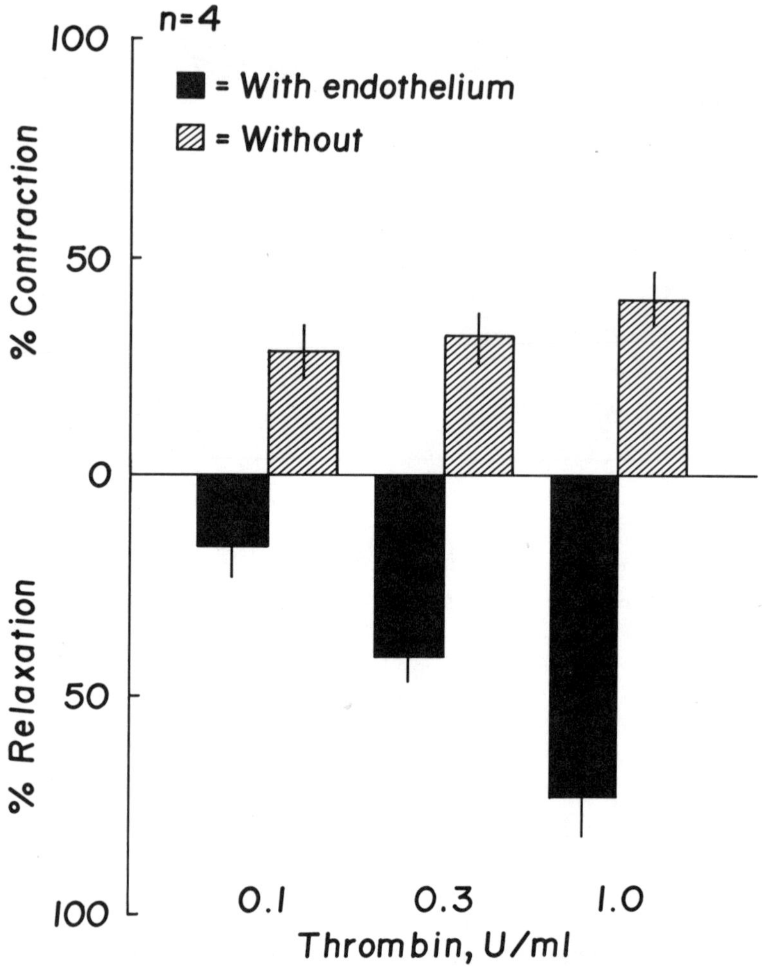

Fig. 2. Effects of increasing doses of bovine thrombin on contractions of canine basilar arteries (with and without endothelium), induced by prostaglandin $F_{2\alpha}$ ($2 \times 10^{-6} M$). Thrombin caused transient relaxations (shown), followed by contraction, in arteries with endothelium; in denuded rings, thrombin caused only contraction. The changes in tension are expressed as percent of the response to prostaglandin $F_{2\alpha}$ and shown as means ± SEM (from Katusic et al., 1984, by permission of the American Heart Association).

blood flow increases, however, during intracarotid infusion of vasopressin (Kozniewska and Skolasinska, 1982). By contrast, in the systemic circulation, vasopressin is a potent vasoconstrictor and, in physiological concentrations, may help to regulate total peripheral vascular resistance (Cowley

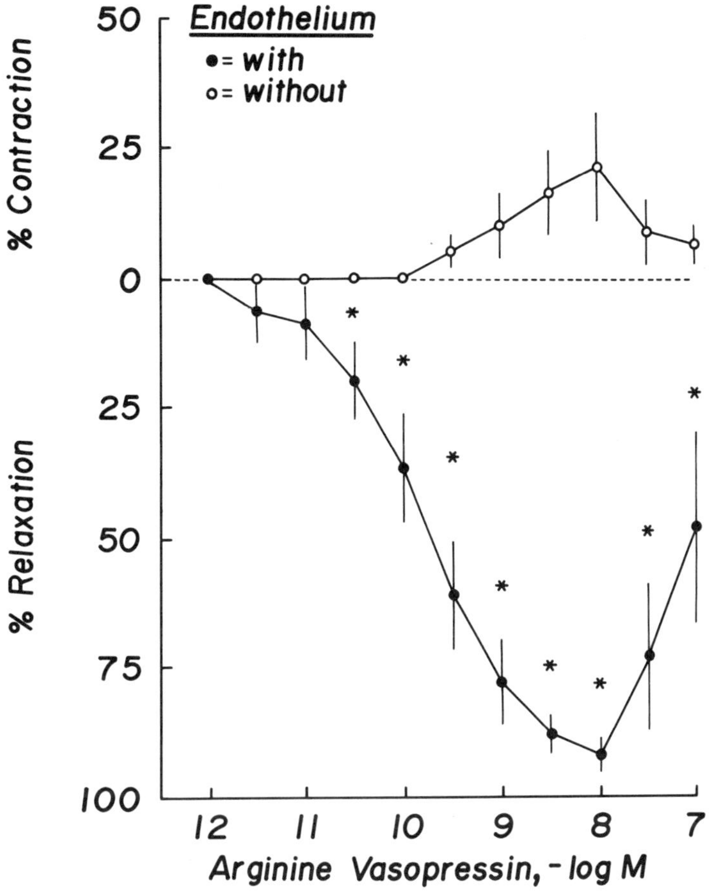

Fig. 3. Cumulative concentration–response curves to arginine-vasopressin in rings of canine basilar arteries with and without endothelium. The relaxations were obtained in rings that developed myogenic tone. The data are shown as means ± SEM and are expressed as percent of the maximal relaxations induced by papaverine. The asterisk denotes that the difference between the two types of preparations is statistically significant (from Katusic et al., 1984, by permission of the American Heart Association).

et al., 1983). Thus, the hormone may affect peripheral and cerebral vascular resistance in opposite directions. An increase in systemic combined with a decrease in cerebral vascular resistance would tend to redistribute blood flow from the peripheral to the cerebral circulation (Fig. 4; Vanhoutte et al., 1984). The decrease in cerebral resistance could be explained best by the action of vasopressin on the endothelial cells since the hormone,

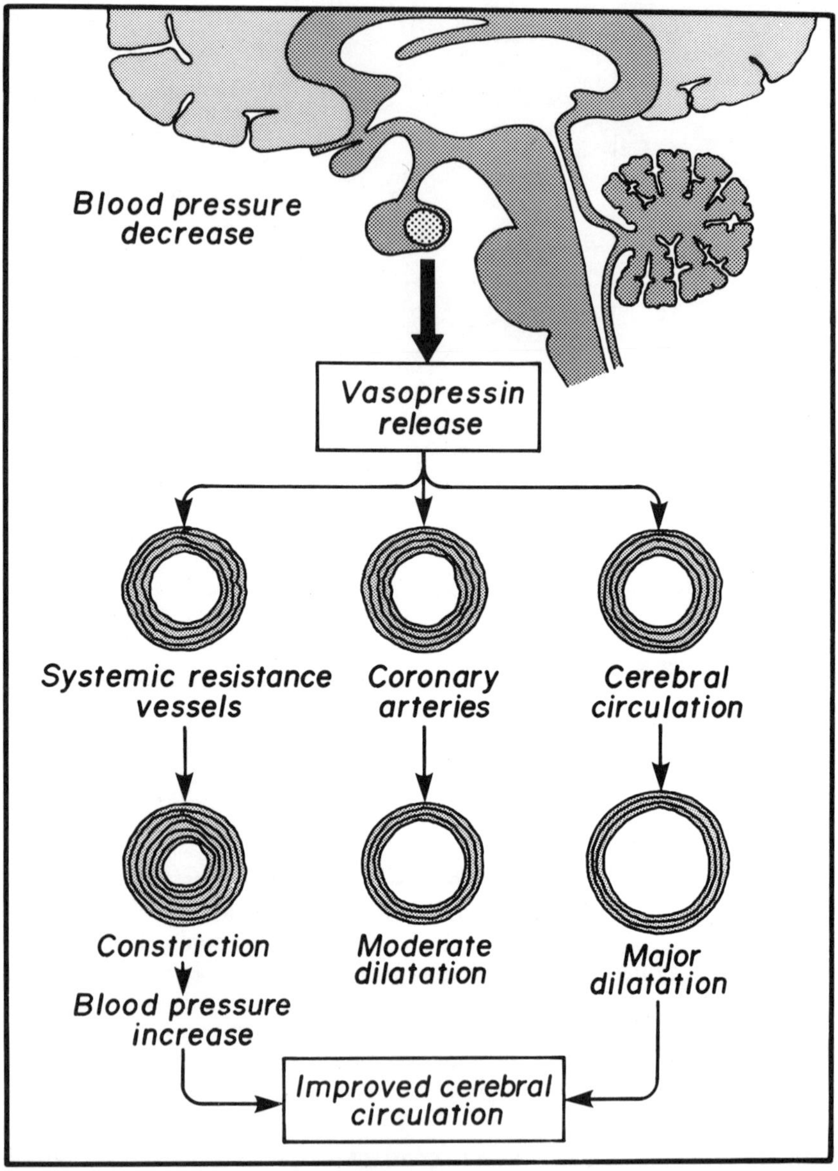

Fig. 4. The selective, endothelium-dependent relaxing effect of vasopressin on cerebral (and to a lesser extent coronary) arteries, combined with its constrictor effect on systemic resistance vessels, favors the redistribution of blood flow to the cerebral circulation (from Vanhoutte et al., 1984, by permission).

at least in the dog, does not cross the blood–brain barrier (Wang et al., 1981). The plasma levels of vasopressin are increased during hemorrhage, septic shock, acute intracranial hypertension, and congestive heart failure. Under these conditions, redistribution of the available supply of blood from the periphery to the brain may become particularly important and could significantly contribute to the maintenance of cerebral blood flow.

3. Endothelium-Dependent Contractions

3.1. Arachidonic Acid, Calcium Ionophore A23187, and Acetylcholine

Arachidonic acid, the calcium ionophore A23187, and acetylcholine cause endothelium-dependent contractions in quiescent isolated canine basilar arteries (unpublished observations). The excitatory effects of all three substances are prevented by inhibitors of cyclooxygenase. Endothelium-dependent contractions to arachidonic acid also occur in systemic and pulmonary veins (De Mey et al., 1982; Vanhoutte and Miller, 1985). Why these substances produce endothelium-dependent responses in cerebral arteries that are opposite to those observed in systemic arteries (Furchgott, 1984; Vanhoutte et al., 1986) remains to be established.

3.2. Anoxia

In the canine basilar artery, anoxia causes contractions of quiescent preparations only if the endothelium is present (Fig. 5; Katusic and Vanhoutte, 1986). Anoxia also augments contractions of these arteries evoked by prostaglandin $F_{2\alpha}$, 5-hydroxytryptamine, KCl, and uridine 5′-triphosphate; removal of the endothelium considerably reduces the anoxic contractions. The role of the metabolites of arachidonic acid in the responses to anoxia is complex. Inhibitors of cyclooxygenase do not prevent endothelium-dependent contractions to anoxia. The blockers of cyclooxygenase, however, unmask anoxic contractions in arteries without endothelium. These findings would be explained if inhibition of the enzyme diverts arachidonic acid to the lipoxygenase pathway with resulting formation of metabolites that accelerate the entry of calcium (Rimele and Vanhoutte, 1983). Endothelium-dependent relaxations to vasopressin and thrombin are abolished under anoxic conditions (Katusic and Vanhoutte, 1986). These observations suggest that both the release of endothelium-derived constrictor factor(s) and the inhibition of the production of endothelium-derived relaxing factor(s) may contribute to anoxic cerebral vasospasm (Fig. 6).

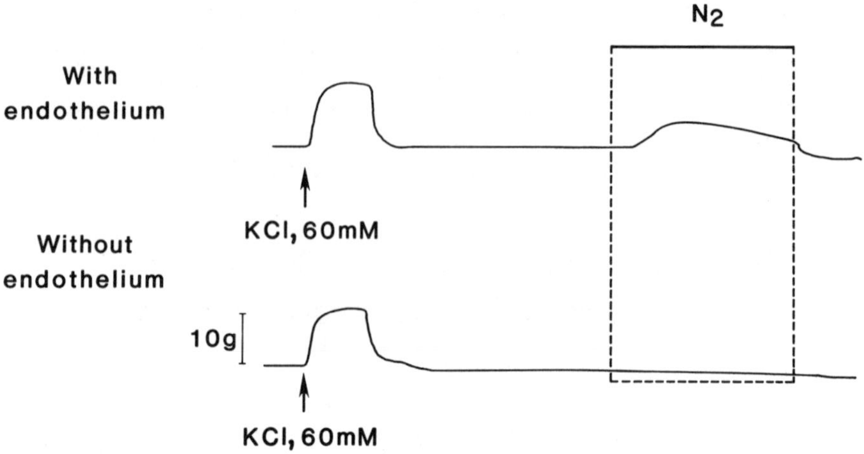

Fig. 5. Isometric tension recordings in rings of the same canine basilar artery with (upper) and without (lower) endothelium. N_2, anoxia, was obtained by switching the mixture gassing the organ chamber from 95% O_2–5% CO_2 to 95% N_2–5% CO_2. The period of anoxia lasted 10 min. For comparison, the contractions to potassium chloride (60 m*M*) are shown (from Katusic and Vanhoutte, 1986, by permission).

3.3. Potassium

Increased concentration of extracellular potassium (3–20 m*M*) causes rhythmic changes in tension in the basilar artery of the dog (Katusic et al., 1985). The rhythmic activity induced by potassium is not affected by tetrodotoxin, phentolamine, propranolol, bretylium tosylate, atropine, methiothepin, or cimetidine, indicating that the release of norepinephrine or the activation of muscarinic, S_1-serotonergic, and H_2-histaminergic receptors are not involved. Removal of the endothelium, exposure to inhibitors of cyclooxygenase, inhibition of Na^+-K^+-ATPase with ouabain and potassium-free solution, or blockade of calcium entry prevented the rhythmic activity. Exogenous prostaglandin $F_{2\alpha}$ and E_2 increased the amplitude and the frequency of the rhythmic activity and reversed the effect of the cyclooxygenase inhibitors. By contrast, prostacyclin causes transient inhibition of the rhythmic activity. These observations could be explained if the increase in extracellular K^+ concentration caused cyclic alterations in the endothelial production of prostacyclin and other prostaglandins. Prostacyclin would activate Na^+-K^+-ATPase and thus initiate the relaxation phase of the rhythmic activity, whereas the production of other prostaglandins might be responsible for the contraction phase.

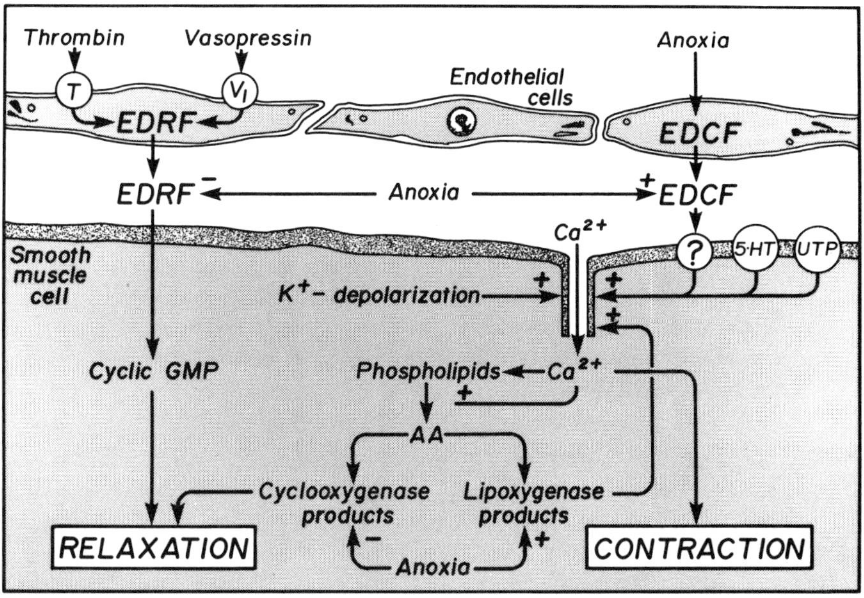

Fig. 6. Possible effects of anoxia on events in endothelial and smooth muscle cells that may explain anoxic contractions of canine cerebral arteries. Certain interactions are derived from work on systemic or coronary smooth muscle. AA, arachidonic acid; EDCF, endothelium-derived contracting factor(s); EDRF, endothelium-derived relaxing factor(s); GMP, guanylate 3′,5′-monophosphate; 5-HT, serotonergic receptor; UTP, receptor for uridine 5′-triphosphate; T, thrombin receptor; V_1, vasopressinergic receptor; ?, unknown action; +, facilitation; −, inhibition (from Katusic and Vanhoutte, 1986, by permission).

3.4. Stretch

Stretch applied to isolated canine basilar arteries with endothelium causes the development of active tension that is abolished in preparations without endothelium (Katusic et al., 1987). The endothelium-dependent contractions of the basilar artery induced by stretch are prevented by calcium entry blockade and by inhibition of cyclooxygenase. Thus the contractile effect of stretch appears to be mediated by an increased production and/or augmented release of prostaglandins from the endothelium, followed by the influx of extracellular Ca^{2+} (Fig. 7). Leukotriene-synthetase inhibitors do not affect the response to stretch, indicating that leukotrienes probably are not involved.

There is substantial evidence that elevating intravascular pressure causes resistance vessels to contract to less than their previous diameter

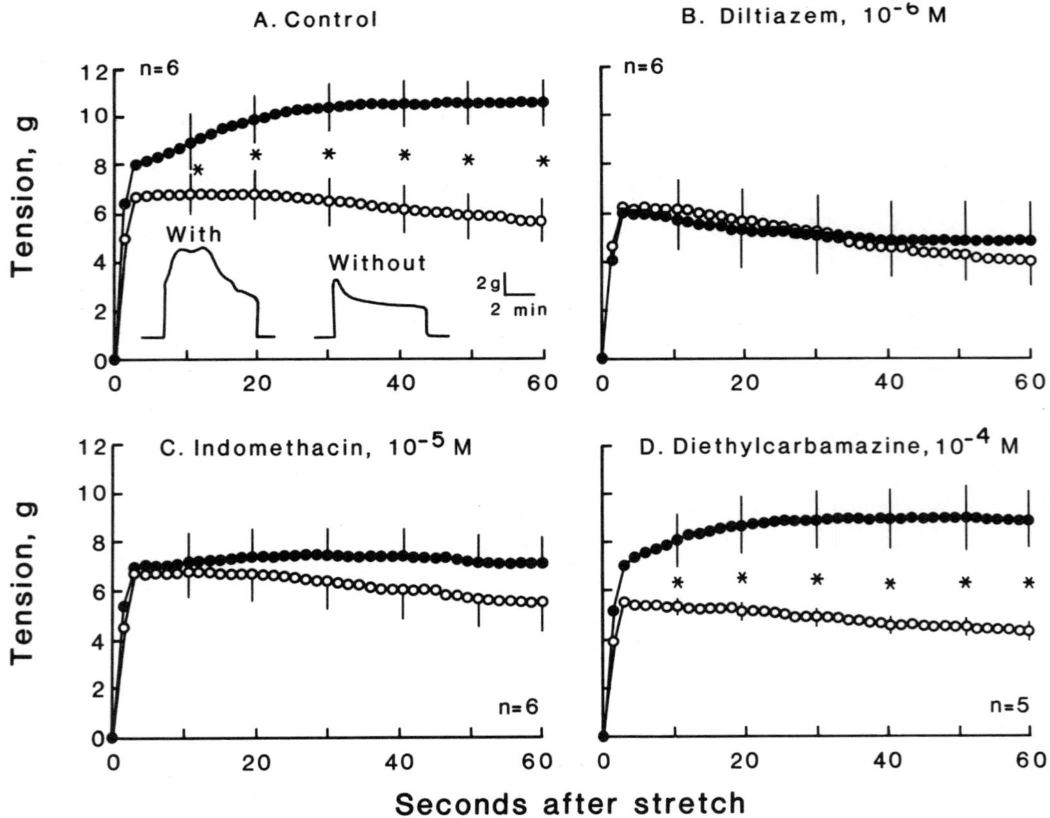

Fig. 7. Responses of canine basilar arteries with (●) and without (○) endothelium to stretch. Following the rapid rise in passive tension, active tension is developed in response to stretch in rings with endothelium (original tracing in inset and panel A). This response is not seen in rings without endothelium nor in rings with endothelium treated with either diltiazem (10^{-6}*M*) or indomethacin (10^{-5}*M*; inset, panels A, B, and C). Diethylcarbamazine (10^{-4}*M*) does not affect response to stretch (panel D). The data are shown as means + SEM and expressed in absolute values. *The difference between the rings with

(autoregulation; e.g., Johnson, 1980). The exact mechanism underlying autoregulation is unknown. Endothelial cells may play a role in this phenomenon at the level of the cerebral blood vessels. Indeed, acute changes in transmural pressure stimulate the synthesis of cyclooxygenase products by endothelial cells, which in turn may trigger contraction of underlying cerebral vascular smooth muscle (Kontos et al., 1980). The endothelium-dependent increases in active tension observed in the basilar artery induced upon stretching develop within a few seconds, a time course that is very similar to that of the autoregulatory response of the cerebral circulation to increased transmural pressure in vivo (Stroudgaard and Paulson, 1984). Since large arteries in the brain contribute significantly to cerebral vascular resistance, it seems logical to conclude that products of cyclooxygenase released from the arterial endothelium may contribute to the autoregulatory response (Fig. 7; *see* Chapter 2).

References

Brayden, J. E. and Bevan, J. A.: Neurogenic muscarinic vasodilatation in cat. An example of endothelial cell-independent cholinergic relaxation. *Circ. Res.* **56:**205-211, 1985.

Chand, N. and Altura, B. M.: Acetylcholine and bradykinin relax intrapulmonary arteries by acting on endothelial cells: Role in lung vascular disease. *Science* **213:**1376–1379, 1981.

Cherry, P. D., Furchgott, R. F., and Zawadzki, J. V.: Role of endothelial cells in relaxation of isolated arteries by bradykinin. *Proc. Natl. Acad. Sci. USA* **79:**2106–2110, 1982.

Cohen, R. A., Shepherd, J. T., and Vanhoutte, P. M.: Neurogenic cholinergic prejunctional inhibition of sympathetic beta-adrenergic relaxation in the canine coronary artery. *J. Pharmacol. Exp. Ther.* **229:**417–421, 1984.

Cowley, R. A., Quillen, E. W., and Skelton, M. M.: Role of vasopressin in cardiovascular regulation. *Fed. Proc.* **43:**3170–3176, 1983.

De Mey, J. G., Claeys, M., and Vanhoutte, P. M.: Endothelium-dependent inhibitory effects of acetylcholine, adenosine triphosphate, thrombin and arachidonic acid in the canine femoral artery. *J. Pharmacol. Exp. Ther.* **222:**166–173, 1982.

Furchgott, R. F.: The role of endothelium in the responses of vascular smooth muscle to drugs. *Ann. Rev. Pharamcol. Toxicol.* **24:**175–197, 1984.

Heistad, D. D. and Kontos, H. A.: Cerebral Circulation, in *Handbook of Physiology* section 2 *The Cardiovascular System* vol. III, *Peripheral Circulation and Organ Blood Flow* part 1 (Shephard, J. T. and Abboud, F. M., eds.) American Physiological Society, Washington DC, 1983.

Johnson, P. C.: The Myogenic Response, in *Handbook of Physiology* Section 2 *The Cardiovascular System* vol. II *Vascular Smooth Muscle* (Bohr, D. F.,

Somlyo, A. P., Sparks, H. V., Geiger, T. R., and Geiger, S. R., eds.) American Physiological Society, Washington DC, 1980.

Katusic, Z. S. and Vanhoutte, P. M.: Anoxic contractions in isolated canine cerebral arteries. Contribution of endothelium-derived factors, metabolites of arachidonic acid and calcium entry. *J. Cardiovasc. Pharmacol.* **8:**S97–S101, 1986.

Katusic, Z. S., Shepherd, J. T., and Vanhoutte, P. M.: Vasopressin causes endothelium-dependent relaxations of the canine basilar artery. *Circ. Res.* **55:**575–579, 1984.

Katusic, Z. S., Shepherd, J. T., and Vanhoutte, P. M.: Potassium causes endothelium-dependent rhythmic activity in canine basilar artery. *Pharmacologist* **27:**233, 1985.

Katusic, Z. S., Shepherd, J. T., and Vanhoutte, P. M.: Oxytocin causes endothelium-dependent relaxations of canine basilar arteries by activating V_1-vasopressinergic receptors. *J. Pharmacol. Exp. Ther.* **236:**166–170, 1986.

Katusic, Z. S., Shepherd, J. T., and Vanhoutte, P. M.: Endothelium-dependent contractions to stretch in canine basilar arteries. *Am. J. Physiol.* H671–H673, 1987.

Kontos, H. A., Wei, E. P., Povlishock, J. T., Deitrich, W. D., and Magieras, C. J.: Cerebral arteriolar damage by arachidonic acid and prostaglandin G_2. *Science* **209:**1242–1245, 1980.

Kozniewska, E. and Skolasinska, K.: The Effect of Vasopressin on Cerebral Blood Flow and Its Modulation by Central Adrenergic System, in *Effects of Nerves and Neurotransmitters* (Heistad, D. D. and Marcus, M. L., eds.) Elsevier, Amsterdam, 1982.

Lassof, S. and Altura, B. M.: Do pial terminal arterioles respond to local perivascular application of the neurohypophyseal peptide hormones, vasopressin and oxytocin? *Brain Res.* **136:**266–273, 1980.

Lee, T. J. F.: Cholinergic mechanism in the large cat cerebral artery. *Circ. Res.* **50:**870–879, 1982.

Parnavelas, J. G., Kelley, W., and Burnstock, G.: Ultrastructural localization of choline acetyltransferase in vascular endothelial cells in rat brain. *Nature* **316:**724–725, 1985.

Regoli, D.: Regulation of precapillary vessels: The kallikrenin-kinin system. *J. Cardiovasc. Pharmacol.* **6** (suppl. 2):S401–S413, 1984.

Rimele, T. J. and Vanhoutte, P. M.: Effects of inhibitors of arachidonic acid metabolism and calcium entry on responses to acetylcholine, potassium, and norepinephrine in the isolated canine saphenous vein. *J. Pharamcol. Exp. Ther.* **225:**720–728, 1983.

Stroudgaard, S. and Paulson, O. B.: Cerebral autoregulation. *Stroke* **15:**413–416, 1984.

Vanhoutte, P. M. and Katusic, Z. S.: Endothelium-Dependent Responses in Cerebral Blood Vessels, in *Neural Regulation of Brain Circulation* (Owman, C. and Hardebo, J. E., eds.) Elsevier, Amsterdam, 1986.

Vanhoutte, P. M. and Miller, V. M.: Heterogeneity of endothelium-dependent responses in mammalian blood vessels. *J. Cardiovasc. Pharmacol.* **7** (suppl. 3):S12–S23, 1985.

Vanhoutte, P. M., Rubanyi, G. M., Miller, V. M., and Houston, D. S.: Modulation of vascular smooth muscle contraction by the endothelium. *Ann. Rev. Physiol.* **48:**307–320, 1986.

Vanhoutte, P. M., Katusic, Z. S., and Shepherd, J. T.: Vasopressin induces endothelium-dependent relaxations of cerebral and coronary, but not of systemic arteries. *J. Hypertension* **2** (suppl. 3):421–422, 1984.

Wang, B. C., Share, L., Guyton, J. T., and Kimura, T.: Changes in vasopressin concentration in pressure and cerebrospinal fluid in response to hemorrhage in anesthetized dogs. *Neuroendocrinology* **33:**61–66, 1981.

White, R. P., Shiresawa, Y., and Robertson, J. T.: Comparison of responses elicited by alpha-thrombin in isolated canine basilar, coronary mesenteric and renal arteries. *Blood Vess.* **21:**12–22, 1984.

Chapter 17

Endothelium, Blood Flow, and Vascular Responses in Large Coronary and Iliac Arteries of the Conscious Dog

Mark A. Young and Stephen F. Vatner

1. Introduction

The majority of insight into mechanisms of endothelium-mediated vascular responses has been gained from experiments in vitro (Furchgott and Zawadzki, 1980; De Mey and Vanhoutte, 1981; De Mey et al., 1982; Furchgott, 1983). It is also important to understand whether these mechanisms apply in the intact conscious animal. This is especially important since isolated vessels in vitro often do not respond in a manner identical to the same vessels examined in vivo (Feigl, 1983; Young and Vatner, 1986b). In this regard our laboratory (Young and Vatner, 1986a,b; 1987) and several others (Lamping et al., 1985a,b; Pohl et al., 1986; Kaiser et al., 1986; Kaiser and Sparks, 1986) have examined the question of endothelium-mediated responses in vivo, studying large arteries to the hindlimb and to the heart. Use of the hindlimb vessels offers the primary advantage of the ability to control or restrict blood flow without severely compromising oxygen delivery to the skeletal muscle of the hindlimb. Furthermore the metabolic requirements of the hindlimb do not dramatically affect the responsiveness of the large vessel, as is the case in the coronary circulation (Macho and Vatner, 1981). Nevertheless, studies in both the hindlimb and coronary circulations have elucidated four mechanisms controlling large arteries in the conscious animal: (1) direct, endothelium-independent vasodilatation; (2) endothelium-mediated, blood flow-independent vasodilatation; (3) endothelium-mediated, blood flow-dependent vasodilatation; and (4) endothelium-mediated protection against vasoconstriction.

These mechanisms of vascular control were studied in conscious dogs, which were previously instrumented with Doppler ultrasonic flow transducers for measurement of blood flow, miniature ultrasonic dimension transducers for measurement of external vascular diameter, and hydraulic occluders for restriction or occlusion of blood flow in either the iliac (Fig. 1) (Young and Vatner, 1986a) or coronary circulation (Hintze and Vatner, 1983). Typically the changes in iliac or coronary blood flow are representative of the changes in the small resistance vessels, whereas the responses of the large vessels are reflected directly by changes in transit time of ultrasound transmitted between the dimension transducers (Vatner et al., 1980). In this way, direct and continuous measurements of the reactivity of large and small vessels can be obtained in the resting conscious dog.

2. Direct, Endothelium-Independent Vasodilatation

Since the experiments of Winbury et al. (1969) and Fam and McGregor (1964, 1968), vasodilatation of the large coronary arteries has been thought to be an important mechanism mediating the therapeutic benefit of the nitrates. This preferential effect on large arteries was confirmed in the conscious dog (Macho and Vatner, 1981; Hintze and Vatner, 1984) in which intravenous administration of nitroglycerin caused profound and prolonged vasodilatation of the large coronary artery, whereas the resistance vessels dilated only transiently and actually constricted slightly at a time when the large coronary arterial diameter was still maximally elevated. The direct, endothelium-independent nature of the vasodilatation induced by the nitrate has been demonstrated many times in isolated vessels that dilate upon administration of nitroglycerin regardless of whether the endothelium is intact or removed (Gruetter et al., 1981; Rapoport and Murad, 1983). The nitrate-induced vasodilatation is believed to depend upon an increase in cyclic GMP, which is unaltered by removal of the endothelium. We have recently confirmed the endothelium independence of the nitroglycerin-induced vasodilatation of the iliac artery in conscious dogs (Young and Vatner, 1986a). In such dogs resting quietly, nitroglycerin was infused intraarterially (1 μg/kg/min for 5 min), which resulted in an immediate rise in the diameter of the iliac artery without concomitant systemic effects as indicated by the lack of changes in arterial pressure or heart rate. The dilatation of the iliac artery induced by nitroglycerin was equivalent whether iliac blood flow rose slightly during the infusion of nitroglycerin

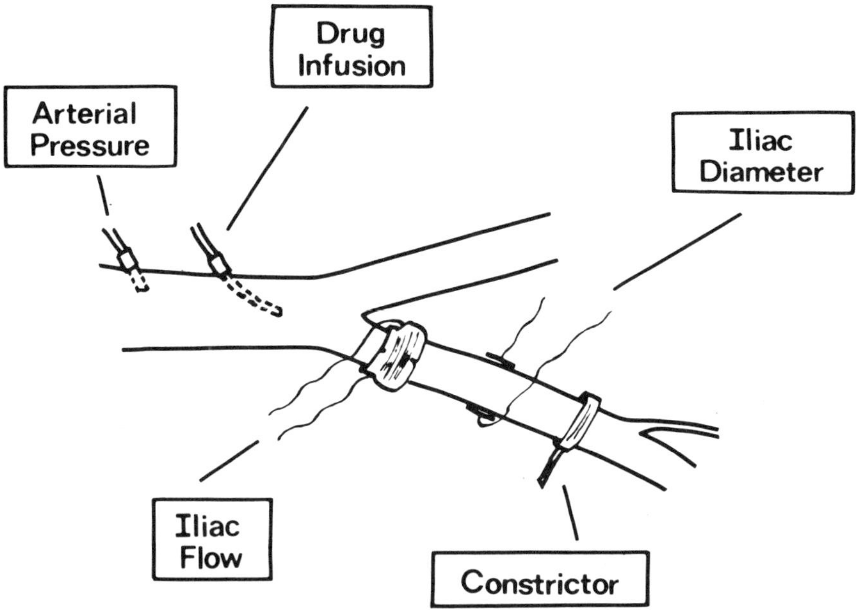

Fig. 1. Diagram of the method used to assess the responsiveness of the iliac artery. Dogs were studied in the conscious state, beginning 2 wk following surgery.

or whether it was held constant (Fig. 2). Following removal of the endothelium, the dilatation of the iliac artery in response to nitroglycerin was also unchanged. As with studies in vitro, this unaltered responsiveness to nitroglycerin indicates the endothelium independence of the phenomenon, and furthermore, confirms that the response of the vascular smooth muscle was not adversely affected by the mechanical abrasion of the endothelial removal. The endothelium independence of vasodilatation to nitroglycerin has also been confirmed in the femoral artery of anesthetized dogs (Pohl et al., 1986), as well as in the coronary circulation of conscious dogs (Chu et al., 1985).

3. Endothelium-Mediated, Blood Flow-Independent Vasodilatation

Acetylcholine is the prototype of agents that induce the endothelium to release a vasodilating substance, which in turn relaxes vascular smooth muscle. The abolition of acetylcholine-induced vasodilatation has been used as a functional criterion to establish adequate removal of endothelial cells. Yet in the conscious or anesthetized animal, intraarterial administration

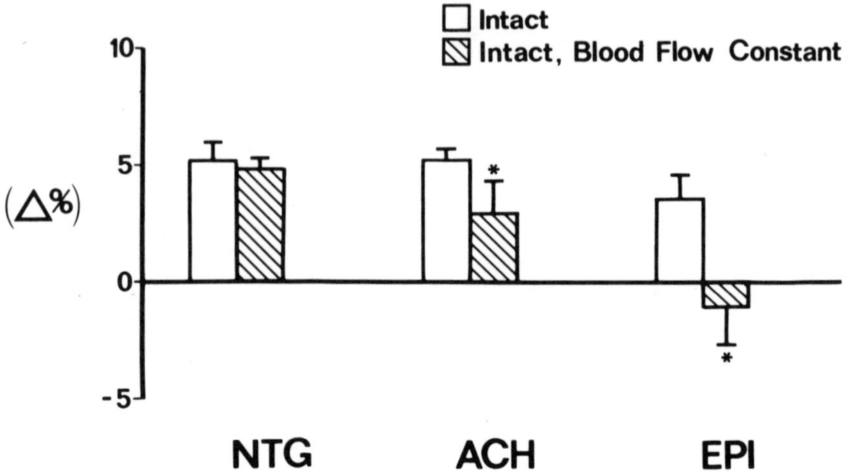

Fig. 2. The effects of controlling iliac blood flow on the dilatation produced by nitroglycerin (NTG, 1.0 μg/kg/min), acetylcholine (ACh, 0.1 μg/kg/min), and epinephrine (EPI, 0.01 μg/kg/min). The dilation with ACh was attenuated, and that with EPI abolished, by holding blood flow constant. Asterisks indicate responses significantly different from intact, $p < 0.05$.

of acetylcholine also results in profound increases in blood flow, an effect that is obviously not demonstrable in isolated vessels. Since it has been demonstrated that increases in blood flow alone can cause substantial dilatations of large arteries (Hilton, 1959; Lie et al., 1970; Ingebrigtsen and Leraand, 1970; Gerova et al., 1981; Hintze and Vatner, 1984), it is important to demonstrate that the acetylcholine-induced vasodilatation is caused by a direct effect on the endothelial cells, rather than an indirect one mediated by the increases in blood flow. In studies of the iliac circulation, intraarterial infusion of acetylcholine (0.1 μg/kg/min) elevated iliac blood flow by 140% and increased iliac artery diameter by approximately 4.5%. When the increase in iliac blood flow was prevented by partial or complete inflation of the hydraulic constrictor, the increase in iliac arterial diameter was only slightly reduced, indicating that the effect of acetylcholine on the large artery was not mediated by changes in blood flow (Figs. 2 and 3) (Young and Vatner, 1986a, 1987). Such a dilatation mediated by the direct action of acetylcholine (i.e., in the absence of increased blood flow) has also been demonstrated in the coronary circulation of conscious dogs (Cox et al., 1983). Also, in the hindlimb of anaesthetized dogs (Kaiser and Sparks, 1986), topical application of acetylcholine produced increases in the diameter of the femoral artery, though the hindlimb was pump-perfused and blood flow did not change with the application of the drug.

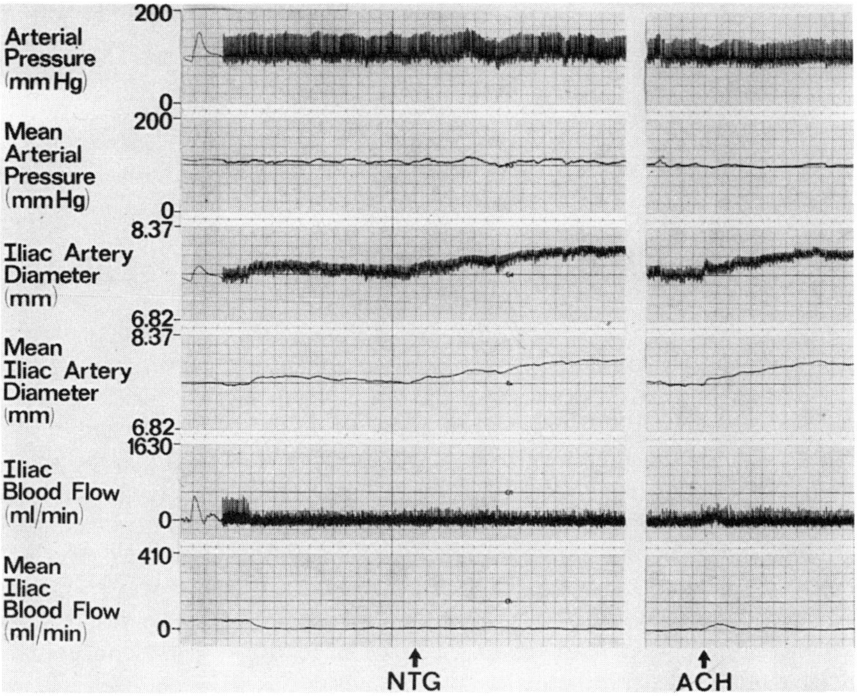

Fig. 3. Effects of nitroglycerin (NTG; 1.0 μg/kg) and acetylcholine (ACh; 0.1 μg/kg) injected as a bolus when the iliac artery was occluded and blood flow was held to zero. Nitroglycerin and acetylcholine each increased the diameter of the iliac artery in the absence of blood flow, illustrating the blood flow independence of the response that they evoke.

The endothelium-dependent nature of the acetylcholine-induced vasodilatation is demonstrated in studies of the iliac circulation in which removal of the endothelium completely abolished the acetylcholine induced vasodilatation (Young and Vatner, 1986a). In contrast to many studies conducted in vitro (Furchgott, 1983), however, infusion of acetylcholine following removal of the endothelium in the conscious dog did not result in vasoconstriction of the large vessels. In the coronary circulation of the conscious dog, abolition of the acetylcholine-induced vasodilatation by removal of the endothelium is more difficult to demonstrate than in the hindlimb. In one preliminary report, Chu et al. (1985) described that removal of the endothelium from the left circumflex coronary artery of conscious dogs, eliminating the dilatation in response to left atrial injection of acetylcholine 1 d following removal of the endothelium. Yet the response had returned to normal within 4 d, despite the persistent absence of endothelial cells. We have also demonstrated that removal of the endothelium from conscious dogs studied 1–2 d following the removal of the endothelium abolished dilatation of the left circumflex coronary artery to acetylcholine, yet the time course of the reappearance of the dilatation with acetylcholine is still questionable, as is the extent to which the return of physiological responsiveness is caused by regeneration of the endothelium.

4. Endothelium-Dependent, Blood Flow-Mediated Vasodilatation

It has long been recognized that acute or chronic increases in blood flow or shear stress can increase the diameter of large arteries (Hilton, 1959; Lie et al., 1970; Ingebrigtsen and Leraand, 1970; Gerova et al., 1981). The increased diameter of large vessels is believed thereby to effectively reduce large artery resistance, accommodating the elevated demand for blood flow. In fact chronic increases in blood flow or shear stress are believed to be responsible for poststenotic dilation and eventual development of vascular aneurism (Rodbard, 1975). Although it was proposed in the 1970s (Rodbard, 1975) that the endothelium may be responsible for mediating the increases in diameter of large arteries, this concept has been confirmed experimentally only recently.

A striking example of a blood flow-mediated dilatation of large arteries is the reactive increase in diameter of the left circumflex coronary artery in response to the release of a temporary coronary occlusion (Hintze and Vatner, 1984). Typically the release of the occlusion is followed by an increase in coronary blood flow above control levels with the peak response occurring almost immediately (Fig. 4). In contrast, the response of the large vessel is slower to develop, and the peak increase in diameter occurs at a time when the blood flow has returned almost to control levels. The reactive dilatation of the large vessels was diminished only slightly by treatment with aminophylline or indomethacin, but was abolished by restricting blood flow to control levels upon release of the occlusion (Fig. 5), thus demonstrating the blood flow-dependent nature of the response.

The involvement of the endothelium in reactive dilatation has not yet been conclusively demonstrated, yet a preliminary report by Inoue et al. (1985) indicates that reactive dilation of large coronary arteries in anesthetized dogs is abolished following removal of the endothelium. Another study in anesthetized dogs has demonstrated that increases in blood flow can attenuate the constriction of the left anterior descending coronary artery during intracoronary infusions of serotonin (Lamping et al., 1985a).

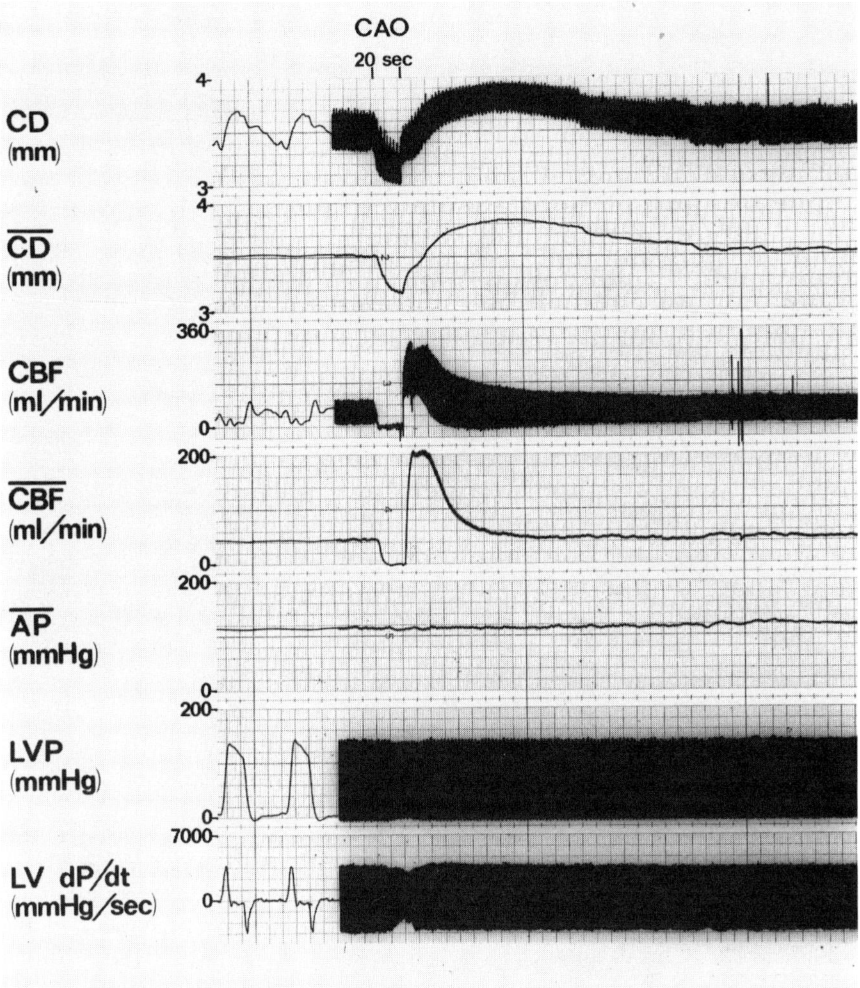

Fig. 4. The effects of a 20-s coronary occlusion (CAO) of the left circumflex coronary artery proximal to the measurement of left circumflex coronary arterial diameter are shown on recordings of phasic and mean coronary dimensions (CD), phasic and mean coronary blood flow (CBF), mean arterial pressure (AP), left ventricular (LV) pressure (P), and LV dP/dt. After release of the coronary occlusion, CBF increased dramatically, resulting in a typical reactive hyperemia. Coronary dimensions increased slowly after release of the occlusion, reaching a peak approximately 60 s later, at a time when coronary blood flow had already returned to control (Hintze and Vatner, 1984, by permission of the American Heart Association).

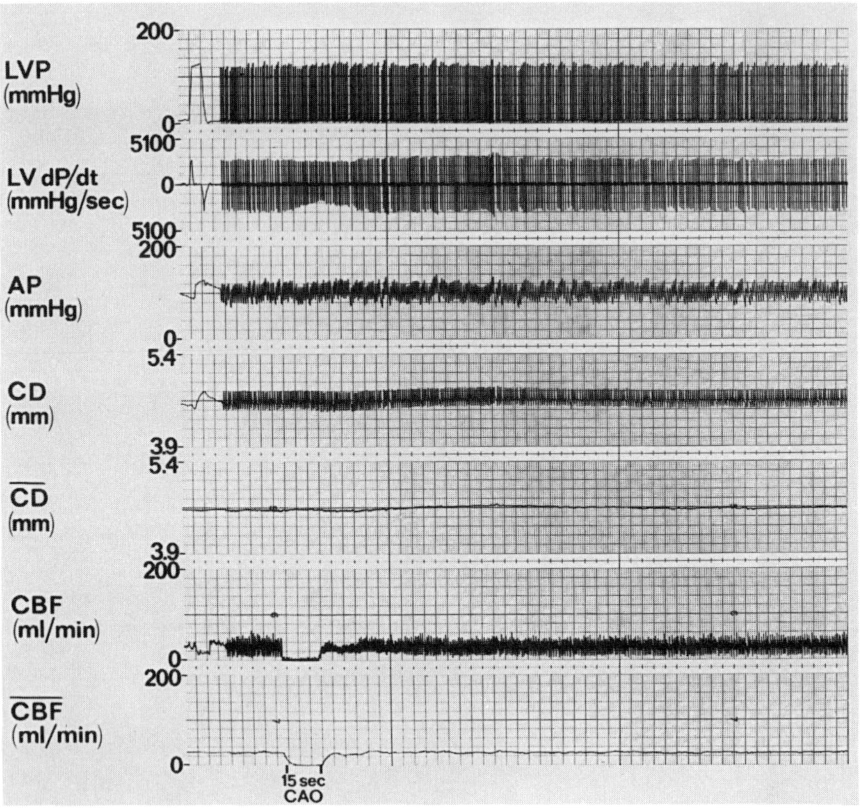

Fig. 5. An example of the effects of a brief coronary artery occlusion (CAO) with flow held constant upon release of occlusion in a conscious dog. After release of the coronary occlusion, reactive dilation was not observed. Abbreviations are the same as in Fig. 4. (from Hintze and Vatner, 1984, by permission of the American Heart Association).

Blood flow-dependent dilatation of large arteries has been repeatedly demonstrated using a variety of other techniques. In anesthetized dogs with iliac (Lie et al., 1970; Ingebrigtsen and Leraand, 1970), femoral (Kaiser and Sparks, 1986), or coronary (Gerova et al., 1981) arteriovenous shunts, increases in shunt blood flow are accompanied by proportionate increases in diameter of the large artery. Increases in blood flow to the hindlimb and diameter of large arteries have also been observed by infusion of acetylcholine 6–8 cm distal to the point of measurement of the dimension (Pohl et al., 1986). The authors' laboratory has demonstrated blood flow-dependent vasodilatation with the infusion of epinephrine in the iliac circulation of conscious dogs (Fig. 2) (Young and Vatner, 1987). Intraarterial infusion of epinephrine increased iliac blood flow by approximately 100% and produced a rise in iliac diameter of 3.5%. When blood flow was restricted to baseline levels during the infusion of epinephrine, the rise of iliac arterial diameter was prevented, illustrating a blood flow-dependent mechanism. Removal of the endothelium from the iliac artery also prevented the rise in iliac diameter despite the fact that blood flow still rose during the infusion of epinephrine. This observation further suggests that the blood flow-dependent increases in diameter of large arteries are mediated via the endothelium. Similar results demonstrating the endothelium dependence of the blood flow-mediated response have been demonstrated in the hindlimb circulation by Pohl et al. (1986) and Kaiser and Sparks (1986).

5. Endothelium-Mediated Protection Against Vasoconstriction

In addition to the elimination of vasodilatation, removal of the endothelium unmasked a vasoconstriction of the iliac artery in response to the infusion of epinephrine despite the fact that the iliac blood flow still rose (Fig. 6). This observation indicates that blood flow-mediated dilatation in response to epinephrine competes with a vasoconstrictor effect; in the absence of the endothelium the vasodilator component is lost and the catecholamine causes constriction of the large artery. Yet epinephrine did not constrict the large artery when the endothelium was intact and blood flow was held constant (Fig. 2) (Young and Vatner, 1987), indicating that the endothelial cell layer may prevent epinephrine-induced vasoconstriction even in the absence of an increase of blood flow. The possibility that the endothelium limits adrenergic vasoconstriction is supported by the observation that removal of the endothelium enhances the vasoconstriction of the iliac artery in response to intraarterial infusion of either norepinephrine or phenylephrine (Young and Vatner, 1986a) (Fig. 7). The mechanisms of the potentiated vasoconstriction to epinephrine, norepinephrine, or phenylephrine are uncertain at this point. It may, however, be caused by the absence of a diffusion barrier normally presented by the intact endothelium, the absence of significant amounts of monoamine oxidase normally present in the intact endothelial cells, or the potentiation of the alpha-adrenergic constrictor effects of epinephrine, norepinephrine, and phenylephrine by locally adhered platelets and the serotonin that they release. The dramatic enhancement of serotonin-induced vasoconstriction by the removal of the endothelium has been demonstrated in both the femoral (Blackshear et al., 1985) and coronary (Lamping et al., 1985b) circulation of anesthetized dogs.

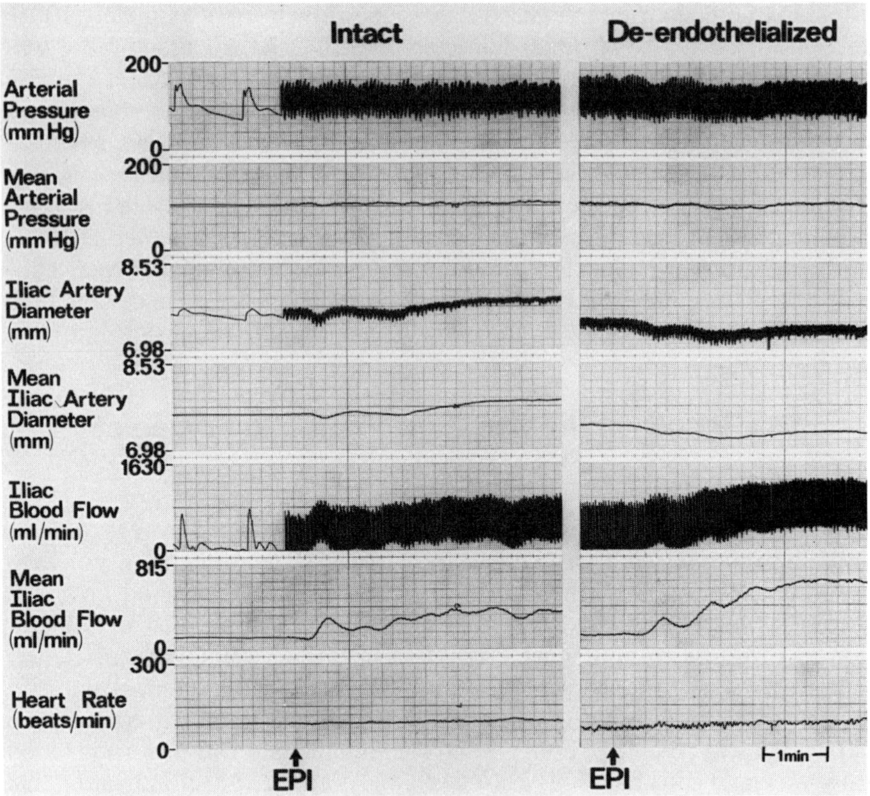

Fig. 6. Example of hemodynamic responses of phasic and mean arterial pressure, iliac artery diameter, iliac blood flow, and heart rate to infusion of epinephrine (EPI) in a dog with intact iliac arteries (intact) and in same dog after removal of endothelium (de-endothelialized). Removal of the endothelium reversed the dilatation observed with EPI to vasoconstriction (from Young and Vatner, 1986a, by permission).

6. Concluding Remarks

Experiments in intact animals have demonstrated an important role of the endothelial cell layer in mediating vascular responses of large arteries. The endothelium mediates vasodilatation in response to drugs activating endothelial receptors (i.e., acetylcholine) and in response to increases in blood flow (as reactive hyperemia or infusion of vasodilator agents). Finally the endothelium serves to limit the extent of vasoconstriction in response to the vasoactive amines, epinephrine, phenylephrine, norepinephrine, and serotonin. Whether the endothelium modulates the vascular response to physiologic maneuvers such as exercise or autonomic neural activation

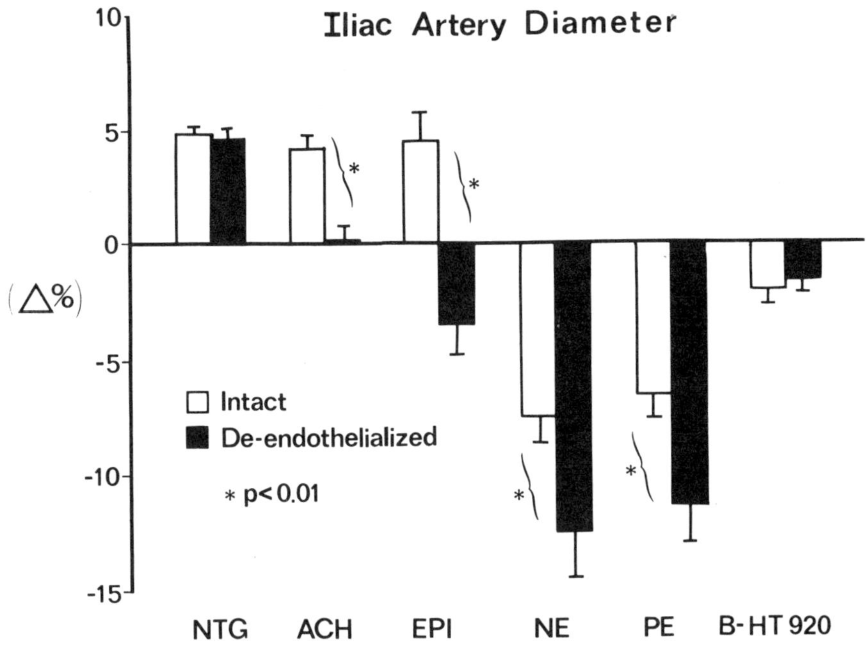

Fig. 7. Average ± SE effects for nitroglycerin (NTG), acetylcholine (ACh), epinephrine (EPI), norepinephrine (NE), phenylephrine (PE), and B-HT 920 are compared on iliac artery diameter in dogs with endothelium intact and in same dogs after removal of the endothelium. Values represent peak changes in iliac arterial diameter at the end of 5 min intraaortic infusion (from Young and Vatner, 1986a, by permission).

remains to be demonstrated. In a pathologic context, recent information indicates that endothelium-dependent vasodilatation may be inhibited in large arteries isolated from monkeys with atherosclerosis (Harrison et al., 1986). This involvement has been extended to the human coronary circulation in a study by Ludmer et al. (1986), who demonstrated that infusion of acetylcholine into the coronary artery of patients with atherosclerosis of the artery results in vasoconstriction, whereas nonatherosclerotic patients, acetylcholine produced only vasodilatation. At this point the data are inconclusive, but an intriguing possibility is that the evidence of impaired endothelium-mediated vasodilatation might well presage disease of the large arteries as occurs in coronary atherosclerosis.

Acknowledgments

Supported in part by US Public Health Service Grants, HL33065, HL33743, and HL33107, and NIRA 34372.

References

Blackshear, J. L., Orland, C., Garnic, J. D., and Hollenberg, N. K.: Differential large and small vessel responses to serotonin in the dog hindlimb *in vivo*: Role of the $5HT_2$ receptor. *J. Cardiovasc. Pharmacol.* **7:**42–49, 1985.

Chu, A., Murdock, R. H., Stakely, A., and Cobb, F. R.: Return of basal vasomotor tone and endothelial mediated vasomotor function in absence of endothelium. (abstract) *Circulation* **72:**III–264, 1985.

Cox, D. A., Hintze, T. H., and Vatner, S. F.: Effects of actylcholine on large and small arteries in conscious dogs. *J. Pharmacol. Exp. Ther.* **225:**764–769, 1983.

De Mey, J. G. and Vanhoutte, P. M.: Role of the intima in cholinergic and purinergic relaxation of isolated canine femoral arteries. *J. Physiol.* **316:** 347–355, 1981.

De Mey, J. G., Claeys, M., and Vanhoutte, P. M.: Endothelium-dependent inhibitory effects of acetylcholine, adenosine triphosphate, thrombin and arachidonic acid in the canine femoral artery. *J. Pharmacol. Exp. Ther.* **222:**166–173, 1982.

Fam, W. M. and McGregor, M.: Effect of coronary vasodilator drugs on retrograde flow in areas of chronic myocardial ischemia. *Circ. Res.* **15:**355–365, 1964.

Fam, W. M. and McGregor, M.: Effect of nitroglycerin and dipyridamole on regional coronary resistance. *Circ. Res.* **22:**649–659, 1968.

Feigl, E. O.: Coronary physiology. *Physiol Rev.* **63:**1–205, 1983.

Furchgott, R. F.: Role of endothelium in responses of vascular smooth muscle. *Circ. Res.* **53:**557–573, 1983.

Furchgott, R. F. and Zawadzki, J. V.: The obligatory role of endothelial cells in the relaxation of arterial smooth muscle by acetylcholine. *Nature* **288:**373–376, 1980.

Gerova, M., Gero, J., Barta, E., Dolezel, S., Smiesko, V., and Levicky, V.: Neurogenic and myogenic control of conduit coronary a: A possible interference. *Basic Res. Cardiol.* **76:**503–507, 1981.

Gruetter, C. A., Gruetter, D. Y., Lyon, J. E., Kadowitz, P. J., and Ignarro, L. J.: Relationship between cyclic guanosine 3′:5′-monophosphate formation and relaxation of coronary arterial smooth muscle by glyceryl trinitrate, nitroprusside, nitrate and nitric oxide: Effects of methylene blue and methemoglobin. *J. Pharmacol. Exp. Ther.* **219:**181–186, 1981.

Harrison, D. G., Freiman, P. C., and Mitchell, G. G.: Atherosclerosis significantly impairs vasoactive responses to a variety of humoral agents. *Blood Vess.* **23:**73, 1986.

Hilton, S. M.: A peripheral arterial conducting mechanism underlying dilation of the femoral artery and concerned in functional vasodilation in skeletal muscle. *J. Physiol.* **149:**93–111, 1959.

Hintze, T. H. and Vatner, S. F.: Comparison of effects of nifedipine and nitroglycerin on large and small coronary arteries and cardiac function in conscious dogs. *Circ. Res.* **53** (suppl. I):139–146, 1983.

Hintze, T. H. and Vatner, S. F.: Reactive dilation of large coronary arteries in conscious dogs. *Circ. Res.* **54:**50–57, 1984.

Ingebrigtsen, R. and Leraand, S. : Dilation of a medium-sized artery immediately after local changes of blood pressure and flow as measured by ultrasonic technique. *Acta Physiol. Scand.* **79:**552–558, 1970.

Inoue, T., Tomoike, H., Hisano, K., and Mohri, M.: Endothelium functions as a flow sensor in reactive dilation of epicardial coronary artery in conscious dogs. *Circulation* **72:**III–81, 1985.

Kaiser, L. and Sparks, H. V.: Mediation of flow dependent arterial dilation by endothelial cells. *Circ. Shock* **18:**109–114, 1986.

Kaiser, L., Hull, S. S., and Sparks, H. V.: Methylene blue and ETYA block flow-dependent dilation in canine femoral artery. *Am. J. Physiol.* **250:**H974–H981, 1986.

Lamping, K. G., Marcus, M. L., and Dole, W. P.: Flow dependent inhibition of 5-hydroxytrypamine (5HT) constriction of large coronary arteries *in vivo*. *Fed. Proc.* **44:**1561, 1985a.

Lamping, K. G., Marcus, M. L., and Dole, S. P.: Removal of the endothelium potentiates canine large coronary artery constrictor responses to 5-hydroxytryptamine *in vivo*. *Circ. Res.* **57:**46–54, 1985b.

Lie, M., Sejersted, O. M., and Kiil, F.: Local regulation of vascular cross section during changes in femoral arterial blood flow in dogs. *Circ. Res.* **27:** 727–737, 1970.

Ludmer, P. L., Selwyn, A. P., Shook, T. L., Wayne, R. R., Mudge, G. H., Alexander, R. W., and Ganz, P.: Paradoxical vaso constriction induced by acetylcholine in atherosclerotic coronary arteries. *N. Engl. J. Med.* **315:**1046–1051, 1986.

Macho, P. and Vatner, S. F.: Effects of nitroglycerin and nitroprusside on large and small coronary vessels in conscious dogs. *Circulation* **64:**1101–1107, 1981.

Pohl, U., Holtz, J., Brusse, R., and Bassenge, E.: Crucial role of endothelium in the vasodilator responses to increased flow *in vivo*. *Hypertension* **8:**37–44, 1986.

Rapoport, R. M. and Murad, F.: Agonist-induced endothelium-dependent relaxation in rat thoracic aorta may be mediated through cGMP. *Circ. Res.* **52:**352–357, 1983.

Rodbard, S.: Vascular caliber. *Cardiology* **60:**4–49, 1975.

Vatner, S.F., Pagani, M., Manders, W. T., and Pasipoularides, A. D.: Alpha adrenergic vasoconstriction and nitroglycerin vasodilation of large coronary arteries in the conscious dog. *J. Clin. Invest.* **65:**5–14, 1980.

Winbury, M. M., Howe, B. B., Hefner, M. A.: Effect of nitrates and other coronary dilators on large and small coronary vessels: An hypothesis for the mechanism of action of nitrates. *J. Pharmacol. Exp. Ther.* **168:**70–95, 1969.

Young, M. A. and Vatner, S. F.: Enhanced adrenergic vasoconstriction of the iliac artery with removal of endothelium in the conscious dog. *Am. J. Physiol.* **250:**H892–H897, 1986a.

Young, M. A. and Vatner, S. F.: Blood flow and endothelium mediated vasomotion of iliac arteries in conscious dogs. *Circ. Res.*, in press, 1987.
Young, M. A. and Vatner, S. F.: Brief review: Regulation of large coronary arteries. *Circ. Res.* **59:**579–596, 1986b.

Chapter 18

Endothelium-Dependent Responses in Large Arteries and in the Microcirculation

J. A. Angus, T. M. Cocks, C. E. Wright, K. Satoh, and G. R. Campbell

1. Introduction

Studies focusing on specific components of the blood vessel wall have yielded an important secret of how the "microenvironment" can control large and small artery diameter. As a result of the now classic discovery of endothelium-derived relaxing factor (EDRF) by Furchgott and Zawadzki (1980), any discussion of the physiology or pathology of vascular reactivity must consider the integrity and function of the cells lining the intimal surface.

This chapter reviews our experiments concerned with the reactivity of large and small blood vessels. It deals with the identification of specific receptors for serotonin and norepinephrine on the endothelium. The reactivity of large arteries to endothelium-dependent dilator agents is compared over five arteries in the dog and in the coronary arteries of cow and pig. The reactivity of large arteries to EDRF is assessed in vitro and in vivo after the removal of the endothelium and the subsequent growth of a neointima. At the level of the microcirculation, we assessed whether resistance vessel responsiveness to endothelium-dependent vasodilator agents and to reactive hyperemia were altered by hypertension or hypercholesterolemia in the hindquarter vasculature of conscious rabbits.

These studies emphasize the regional and species-dependent differences in population of receptors on endothelial cells and smooth muscle and demonstrate that marked alterations in reactivity of larger arteries can occur when the architecture of the blood vessel wall is altered.

2. EDRF and Reactivity of Large Arteries In Vivo

2.1. Sonomicrometry in Femoral and Coronary Arteries under Conditions of Controlled Flow and Pressure

To test whether endothelial cells were obligatory for the relaxation caused by acetylcholine in blood-perfused arteries, femoral arteries and left anterior descending coronary arteries of the greyhound were perfused with heparinized blood under conditions of controlled flow and perfusion pressure (by means of a distal Starling resistor). These experimental conditions were considered necessary to allow for precise measurement of the changes in diameter of the large arteries caused by acetylcholine-induced release of EDRF and not by changes in blood flow or confounded by the fall in perfusion pressure (Fig. 1). These latter events occur as a result of acetylcholine acting in the distal resistance vessels.

The external diameter of the arteries was measured at two points 1–2 cm apart by pairs of sonomicrometer crystals attached to the upper and lower surfaces of the artery (Angus et al., 1983; Angus and Cocks, 1984). Acetylcholine (0.1–1 μM) infused directly into the arteries proximal to the crystals caused increase in diameter (5–15%) at both sites that was sustained as long as the drug infusion was maintained (Fig. 2). Removal of the endothelium by intraarterial balloon inflation at the distal crystal site abolished the relaxation in response to acetylcholine, substance P, bradykinin, or adenosine triphosphate, but not to nitroglycerin. Thus the proximal endothelial cells could not apparently transfer the EDRF that they produce via the blood to relax the artery just 1 cm downstream. This suggests that (1) EDRF was diluted in the blood to below threshold concentration, or (2) EDRF was released only from the abluminal surface onto the underlying smooth muscle cells, or (3) hemoglobin or other proteins in the blood had reduced the concentration of free EDRF. In any event these experiments suggest that it is unlikely that EDRF is a circulating hormone. Sodium nitroprusside also dilated the distal vasculature, but appeared to have little effect on the diameter of large arteries in either the uncontrolled (Fig. 1) or controlled preparations, regardless of the presence of the endothelium (Fig. 3). The concentration of sodium nitroprusside infused intraarterially was very high, as shown by the delayed depressor

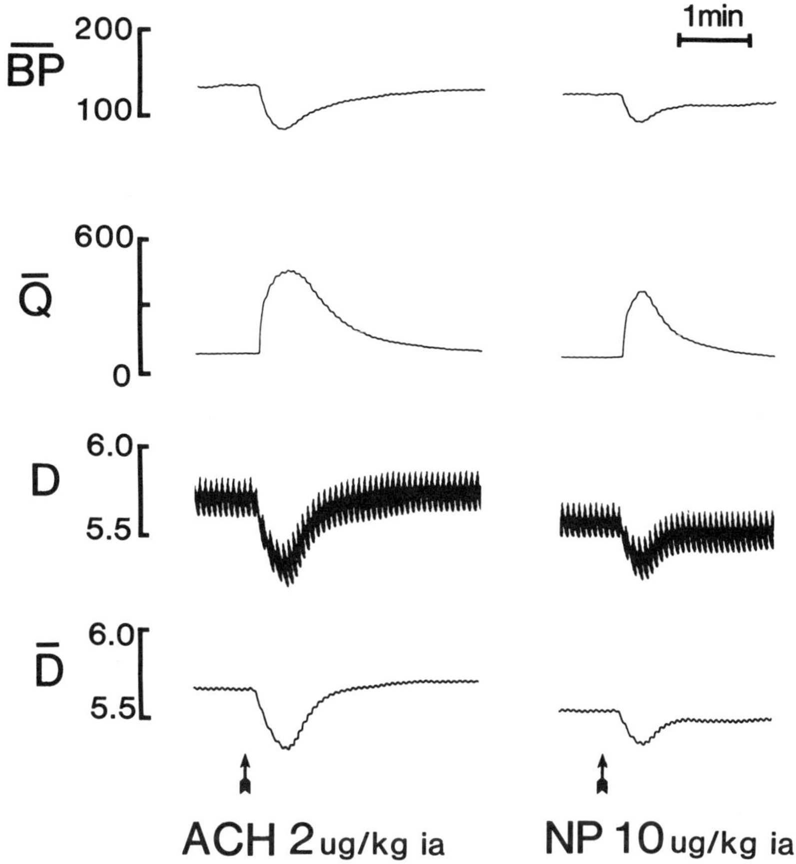

Fig. 1. Recording of femoral hemodynamics in an anesthetized dog. Acetylcholine (Ach, left) and sodium nitroprusside (NP, right) were given as bolus intraarterial (ia) injections into the femoral artery. Traces from top are BP, systemic blood pressure, mm Hg; Q, femoral blood flow (electromagnetic flowmeter, mL/min); D, $\overline{D}$ phasic and mean femoral diameter, mm. (Angus J. A., unpublished data).

action as the drug reached the systemic circulation in the controlled preparation (Fig. 3). In contrast, acetylcholine had no effect on the systemic circulation presumably because it was destroyed rapidly in the blood (Fig. 2).

Acetylcholine and substance P applied to the adventitial surface of the coronary and femoral arteries also increased the diameter if the endothelium was intact. The concentrations required for topical application of both agonists were 50–100 times higher than those required for the intraarterial route (Angus et al., 1983). This illustrates that substances released from nerve endings within the vessel wall could play a role in vasodilatation by causing the release of EDRF.

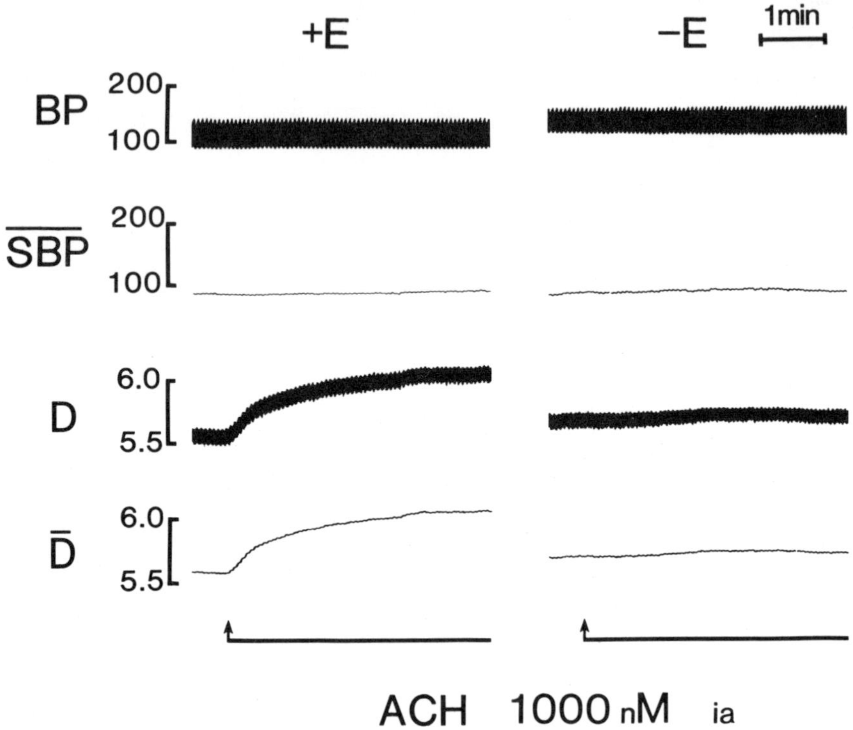

Fig. 2. Records from the femoral artery from the same dog as in Fig. 1 after controlling blood flow and distal resistance. Traces are: BP, systemic blood pressure, mm Hg; SBP, side branch pressure in the blood perfused artery, mm Hg; D and $\overline{D}$, phasic and mean femoral diameter (mm). Acetylcholine (1000 n*M*) was infused into the artery before (+E, left) and after (−E, right) endothelium removal (Angus J. A., unpublished data).

2.2. Chronic Measurement of the Diameter of the Carotid Artery after Removal of the Endothelium

2.2.1. Effect of Acetylcholine

To examine the long-term effects of the removal of the endothelium and subsequent intimal repair on vascular reactivity, greyhound dogs were surgically prepared with flow transducers (Doppler), sonomicrometer crystals, and intraarterial catheters (Herd and Barger, 1964) on both common carotid arteries. During the surgery, the endothelium was removed from one carotid artery with a balloon catheter (Fogarty). Three days later, as the dogs rested quietly unanesthetized on a padded table, acetylcholine was infused into the carotid artery with intact endothelium. This infusion caused a small increase in carotid diameter, after an initial transient fall that coincided with the fall in carotid pressure and the cranial vasodilatation (Fig. 4).

By contrast, on the same day the left carotid artery in which the endothelium had been removed at surgery only showed the passive fall in diameter as the pressure fell and the flow increased. Over the next 10 d, the responses in the artery without endothelium gradually returned to be similar to those in the control artery indicative of a recovery of the intima by regrowth of the endothelial cells from adjoining areas. At autopsy 4 wk later, the arteries from 10 dogs showed intimal thickening of 2–20 smooth muscle cells (whose cytoplasm contained many synthetic organelles) ar-

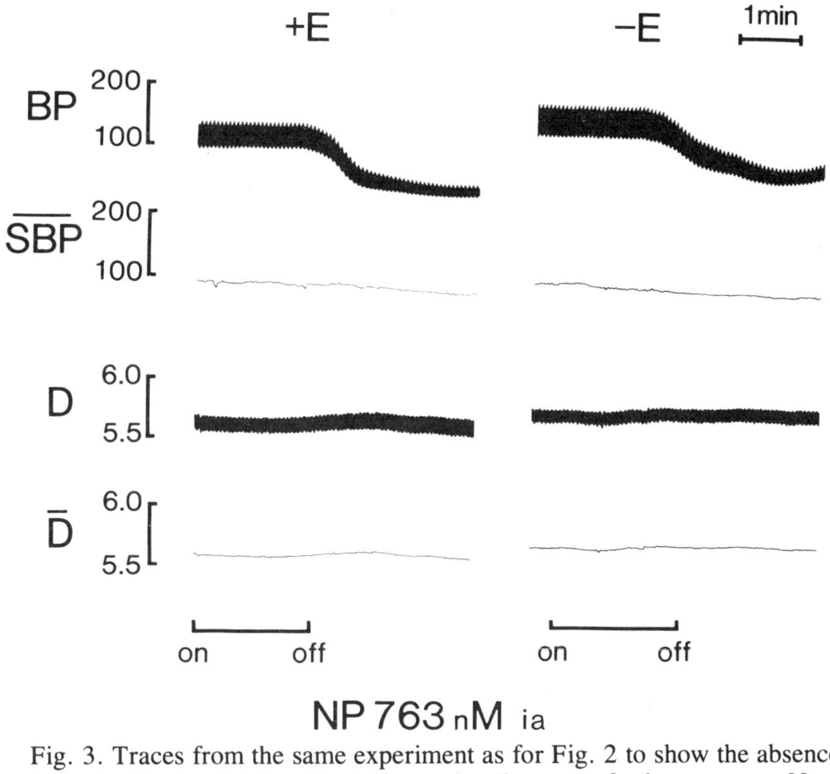

Fig. 3. Traces from the same experiment as for Fig. 2 to show the absence of effect of sodium nitroprusside (NP) on the diameter of a large artery. Note the dramatic depressor response in the systemic circulation after the drug left the controlled femoral artery circuit (Angus J. A., unpublished data). Symbols same as in Fig. 2.

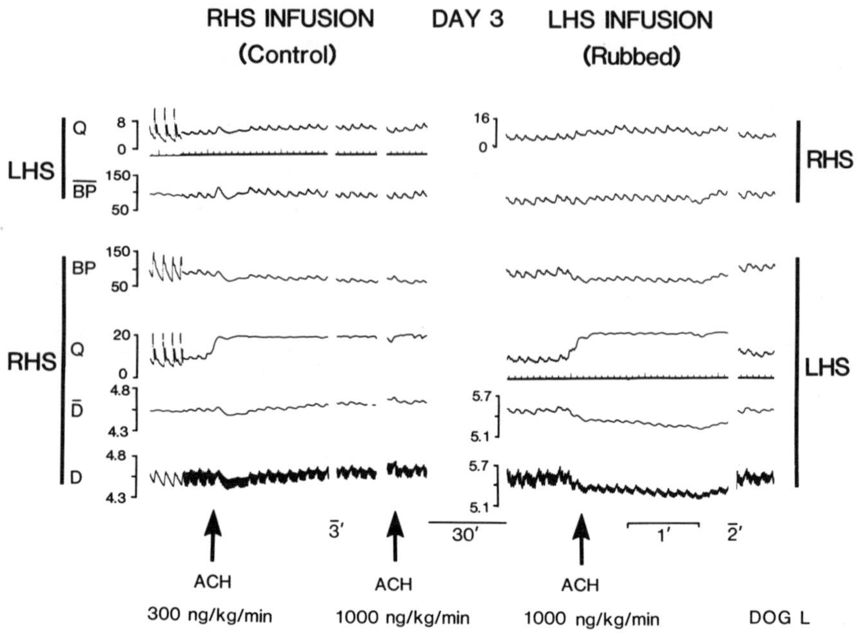

Fig. 4. Recording of intracarotid infusion of acetylcholine in a conscious greyhound 3 d after surgery at which the endothelium on the common carotid was left intact (control, right carotid, RHS) or removed by balloon (rubbed, left carotid, LHS). Measurements are from top, left panels: left carotid flow (Q, kHz Doppler shift, phasic and mean) mean left carotid blood pressure ($\overline{BP}$, mm Hg); right carotid blood pressure (BP), flow (Q) mean, and phasic carotid diameter (D, mm). Acetylcholine was infused into the right carotid (left panel) and 30 min later into the left carotid (right panel) (Angus J. A., unpublished data).

ranged in eccentric or concentric patterns, but generally covered by endothelium (Angus et al., 1986c). These arteries with thickened intima release EDRF in response to acetylcholine (*see* section 3.5).

2.2.2. Effect of Change in Blood Flow

In the coronary circulation, a change in blood flow *per se* is sufficient to cause an endothelium-dependent relaxation of large arteries (Holtz et al., 1984). This phenomenon was confirmed in the chronically instrumented carotid artery. At 4 wk, dogs were anesthetized, and a shunt was placed between the distal common carotid artery and the external jugular vein. Opening of the shunt increased the blood flow velocity (measured by Doppler flowmeter) in the carotid artery and always resulted in an increase in the diameter (Fig. 5). The time course and pattern of the diameter changes

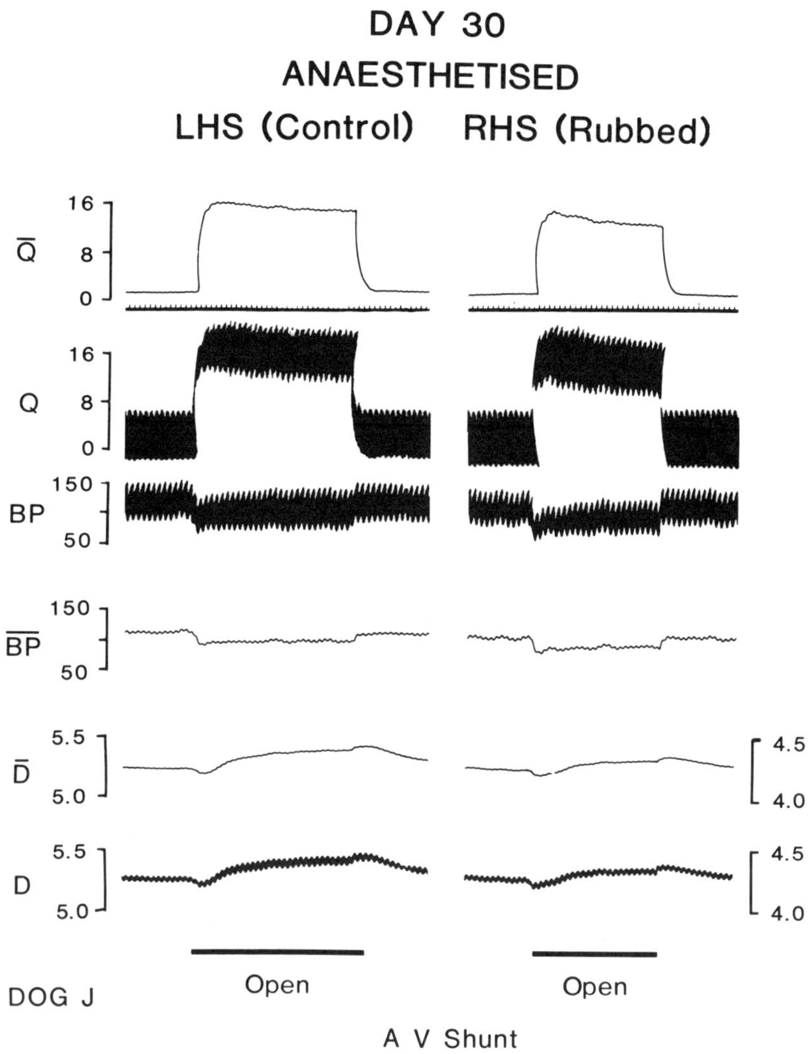

Fig. 5. Polygraph records of carotid artery blood flow (Q, kHz Doppler shift), carotid artery pressure (BP, $\overline{BP}$ mm Hg), and carotid artery diameter (D, $\overline{D}$, mm) in an anesthetized dog 4 wk after instrumentation and removal of endothelium from the right carotid artery (RHS, rubbed) or left undisturbed (LHS, control). Opening the distal carotid to jugular vein shunt for about 2 min (A-V shunt) caused a marked increase in blood flow velocity and increase in diameter, even in the presence of reduced pressure (Angus J. A., unpublished data).

did not follow the pressure trace, except when the flow was abruptly increased or decreased. The artery with previously damaged intima, responded in a similar pattern as the control artery, indicative of the presence of endothelial cells at this time.

These chronic experiments illustrate that the diameter of large arteries can be altered in vivo (1) by passive changes in transmural pressure, (2) by the release of EDRF evoked by agonists such as acetylcholine, or (3) by changes in velocity of blood flow if the endothelium is present.

3. Reactivity of Large Arteries In Vitro

3.1. EDRF Is Released by Norepinephrine and Serotonin

Loss of endothelial cells may have at least two important effects on the reactivity of large arteries. First, a denuded intimal surface would present an attractive site for platelet adhesion and release reaction leading to high local concentrations of autacoids such as serotonin. Second, if vasoconstrictor substances such as serotonin also stimulate the release of EDRF in normal arteries, then removal of endothelial cells would result in a greater direct constrictor action. This may occur in coronary artery spasm when focal loss of endothelial cells has occurred. With this in mind, the effect of the removal of endothelial cells was examined on cumulative concentration–response curves to four coronary constrictor substances in isolated rings of large coronary arteries (Cocks and Angus, 1983). In coronary arteries of the greyhound, mongrel dogs, and the pig, the maximal contractile response to serotonin was more than doubled in rings denuded of endothelium, and the EC_{50} concentration moved to the left compared to that obtained in arteries with intact endothelium (Fig. 6). This has been confirmed in canine coronary arteries in vivo and in vitro (Cohen et al., 1983; Lamping et al., 1985). Similarly, in the presence of propranolol, the concentration–response curves to norepinephrine were amplified and slightly shifted to the left by the removal of the endothelium (Fig. 6). This did not occur with two other coronary constrictor agents, K^+, or the thromboxane mimetic U46619, ruling out a nonspecific effect of intimal damage.

The amplified response for norepinephrine and serotonin are consistent with at least two hypotheses. First, the loss of endothelium could remove a barrier and a site of uptake or metabolism for these two amines across the artery wall allowing for higher concentrations of the amines at the level of the medial smooth muscle. Second, when the endothelium is intact, the amines stimulate the endothelial cells to release EDRF that coincidentally attenuates the direct constrictor response of the medial smooth muscle to the amine. The loss of the endothelium thus uncovers the full

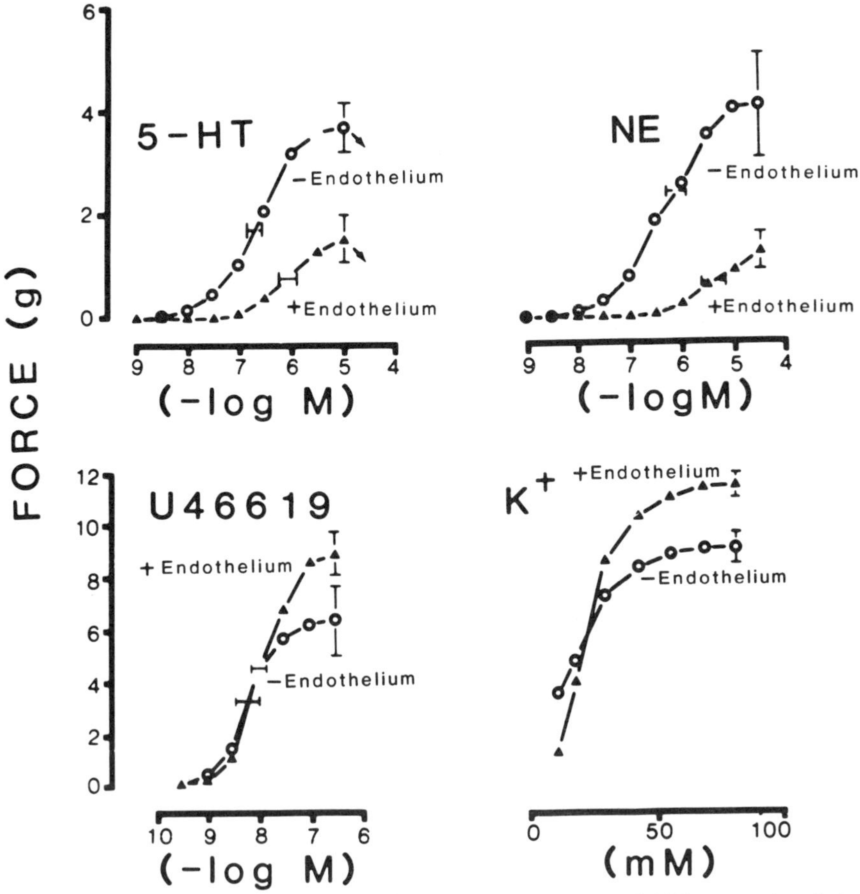

Fig. 6. Concentration–contraction curves to serotonin (5-HT), norepinephrine (NE) (in the presence of propranolol 1 *μM*), thromboxane mimetic (U46619), and K^+ in ring segments of pig coronary artery in the absence or presence of endothelium (data redrawn from Cocks and Angus, 1983).

direct constrictor response. More direct evidence for this latter hypothesis was obtained by contracting porcine arteries and observing a concentration-dependent relaxation to norepinephrine in the presence of high concentrations of propranolol (Cocks and Angus, 1983). In the coronary artery of the greyhound, the alpha$_1$-adrenoceptor blocker, prazosin was required to directly observe the relaxation to norepinephrine, whereas serotonin only relaxed coronary arteries of the dogs and pigs when the 5-HT$_2$-serotonergic blocker ketanserin was present. Thus a unifying hypothesis is that the endothelial cells on dog and pig large coronary arteries have receptors for norepinephrine and serotonin that cause the release of EDRF.

3.2. Classification of Endothelial Alpha$_2$-Adrenoceptors

In contracted coronary arteries of dogs and pigs the endothelium-dependent relaxation response to norepinephrine was mimicked by epinephrine in the presence of propranolol and prazosin. The alpha$_1$-adrenoceptor agonist methoxamine was without effect, and the preferential alpha$_2$-adrenoceptor agonist clonidine partially relaxed the rings when the endothelium was present. The nonselective alpha-adrenoceptor blocker phentolamine caused a parallel shift to the right of the relaxation–response curve to norepinephrine (Cocks and Angus, 1983). Direct evidence that the alpha-adrenoceptor on endothelial cells was of the alpha$_2$-subtype was that the selective alpha$_2$-adrenergic agonist UK 14,304 could relax contracted canine coronary rings (in the presence of propranolol and prazosin) provided the endothelium was present (Angus et al., 1986a). The concentration–relaxation curves to UK 14,304 were displaced in an apparently competitive fashion by idazoxan, the relatively selective alpha$_2$-adrenoceptor antagonist (Fig. 7).

3.3. Distribution of Alpha$_2$-Adrenoceptors on Endothelium

The distribution of alpha$_2$-adrenoceptors on endothelial cells was examined in five large arteries of the dog. To ensure that the most favorable conditions were present to observe relaxations mediated by alpha$_2$-adrenoceptors, all arteries were contracted with U46619 at a concentration that caused about 80% of its maximal effect. These conditions would avoid relaxation by EDRF being masked by an excessive contraction to U46619. Each ring segment of the coronary, carotid, renal, mesenteric, and femoral artery was stretched passively (before applying U46619) to a level of resting force, determined from the Laplace relationship to be equivalent to a transmural pressure of approximately 70 mm Hg (Angus et al., 1986a). The alpha$_2$-mediated maximal relaxation to norepinephrine in the presence of propranolol and prazosin was measured as a percentage of the maximal relaxation to the endothelium-independent vasodilator drug, nitroglycerin, and were in the order of 70% in the coronary; 34% in the carotid, 19% in the femoral; 7% in the renal; and 2% in the mesenteric arteries (Fig. 8). To determine whether the same rank order applied for another endothelium-dependent agonist, the maximal response to substance P was compared across the five blood vessels. Substance P showed a higher efficacy than norepinephrine, but a similar order in the degree of maximum relaxation (i.e., coronary, 100%; carotid, 80%; femoral, 71%; renal, 49%; and mesenteric, 41%). The same rank order for two endothelium-dependent vasodilator agents would suggest that the release of EDRF is site-dependent,

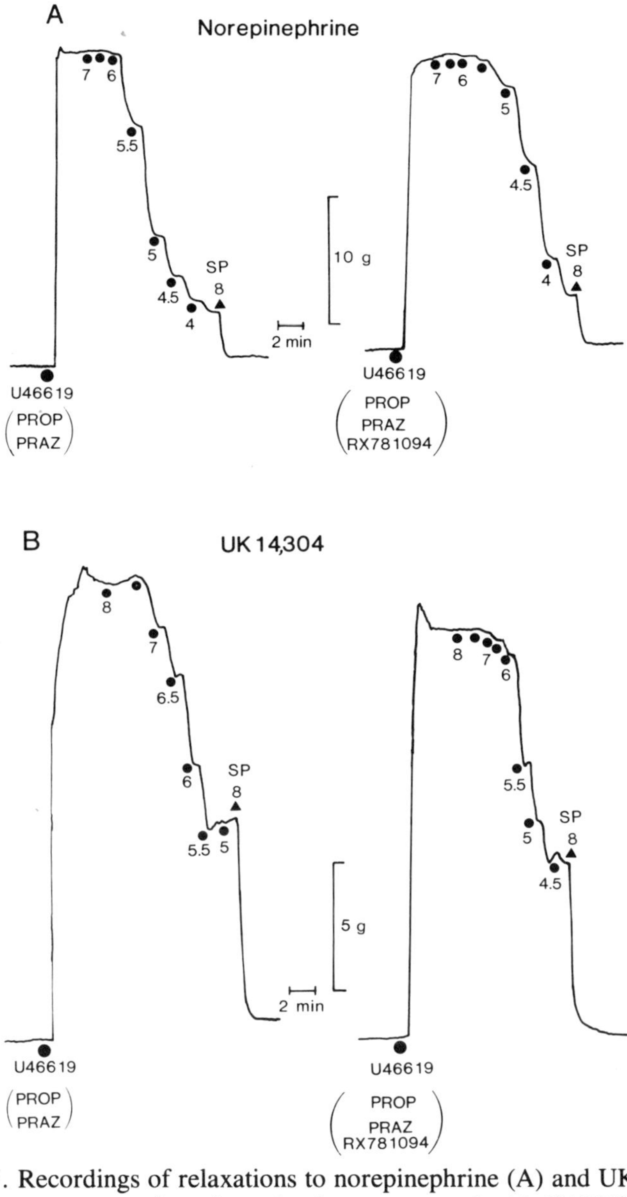

Fig. 7. Recordings of relaxations to norepinephrine (A) and UK14304. (B) in four coronary artery rings from the dog, contracted with U46619 (30 n*M*) in the presence of propranolol (prop, 3 μM) and prazosin (praz, 1 μM). Right traces were in the presence of idazoxan (RX781094, 1 μM). Substance P (SP) was applied where indicated (reproduced from Angus et al., 1986a, with permission of Macmillan Publishers Ltd.).

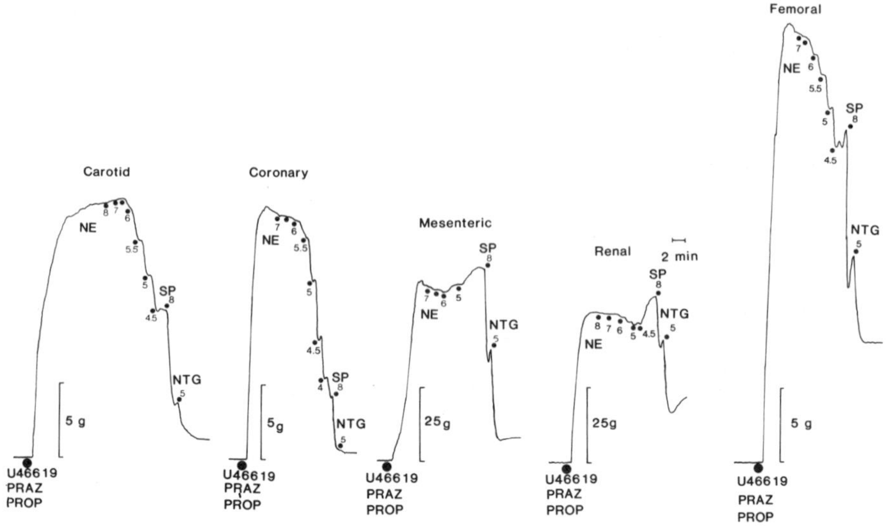

Fig. 8. Traces of relaxations to norepinephrine (NE) in five large arterial rings of the greyhound. Each ring was contracted with U46619 (EC_{80}) in the presence of prazosin (praz, 1 μM) and propranolol (prop, 1 μM). Substance P (SP) and nitroglycerin (NTG) were given where indicated. Numbers refer to $-\log M$ concentration (Angus J. A., Cocks T. M., and Satoh K., unpublished data).

or that the efficacy of EDRF on smooth muscle varies, being greatest in the coronary and weakest in the renal and mesenteric arteries in the dog.

This correlation does not appear to hold for other species. In the bovine coronary artery, no alpha$_2$-adrenoceptor mediated relaxation was observed, but this artery was readily relaxed by bradykinin, a response that was endothelium-dependent (Angus et al., 1986b). In the pig, the coronary and mesenteric arteries readily responded to norepinephrine in the presence of prazosin and propranolol (Table 1). In small branches (200–400 μm in diameter) of the coronary artery from the dog and pig contracted with K^+, norepinephrine could also relax the vessels in the presence of prazosin and propranolol (Angus et al., 1986b). This suggested that endothelial cells in the microvasculature (at least in that of the heart of these species) also possess alpha$_2$-adrenoceptors.

The resistance of some arteries with intact and functional endothelium to relax to norepinephrine in the presence of prazosin clearly could be explained by (1) the absence of alpha$_2$-adrenoceptors on endothelial cells, or (2) the poor efficacy of norepinephrine at alpha$_2$-adrenoceptors releasing EDRF, or (3) the presence of alpha$_2$-adrenoceptors on the medial smooth muscle cells that cause contraction and thus counter the relaxation to EDRF. Evidence that this latter event may be operative is that renal

Table 1
Comparison of Relaxation Responses to Norepinephrine in Large Arteries Contracted In Vitro by U46619 in the Presence of Propranolol and Prazosin[a]

	Artery				
Species	Coronary	Carotid	Mesenteric	Renal	Femoral
Pig	+++++[b]	++	+++++	+++	++
Cattle	0	—	—	—	—
Dog					
Mongrel	+++	++	+	+	++
Greyhound	++++	+++	+	+	++

[a]Relaxation scale: +++++, 70–100%; ++++, 40–70%; +++, 20–40%; ++, 10–20%; +, 0–10%; —, no experiment.
[b]No prazosin.

and femoral arteries of the pig contract further to norepinephrine in the absence of endothelium and the presence of prazosin (Angus et al., 1986b).

These data suggest that the integrated response of the artery wall to norepinephrine will depend upon at least four populations of adrenoceptors; alpha$_2$- on the endothelium; and alpha$_1$-, alpha$_2$- and beta- on the smooth muscle (Fig. 9). Responses to norepinephrine should be interpreted with caution when the integrity of the endothelium has not been considered.

3.4. Comparison of Endothelium-Dependent Agonists in Five Large Arteries

Full concentration–response curves were obtained for acetylcholine, substance P, bradykinin, ATP, substance P, and the endothelium-indeendent dilator nitroglycerin in rings of five large arteries contracted with U46619 (EC_{80}). The logistic-fitted EC_{50} values for relaxation to acetylholine ranged from $-7.09 + 0.18$ (log $M \pm 1$ SEM) in the carotid to -7.59 ± 0.18 in the mesenteric artery, a 3.2-fold difference (Fig. 10). Similarly, the EC_{50} values for substance P differed by less than twofold and that for ATP by fivefold over the five arteries; the carotid artery was the least sensitive. In contrast the sensitivity to bradykinin differed by 20-fold (from the EC_{50} of -8.64 ± 0.09 in the renal to -7.33 ± 0.16 in the carotid artery). The arterial rank order for the maximal relaxation as a percentage of the contraction to U46619 did not correlate with the sensitivity (EC_{50}). For example, substance P relaxed the coronary to a maximum of $88.6 \pm 2.9\%$, but the mesenteric artery, only to a maximum of $32.3 \pm 4.1\%$ (Table 2). In general the coronary and carotid arteries relaxed to a greater extent than the other blood vessels. This may reflect the favorable functional antagonism between EDRF (or nitroglycerin) and

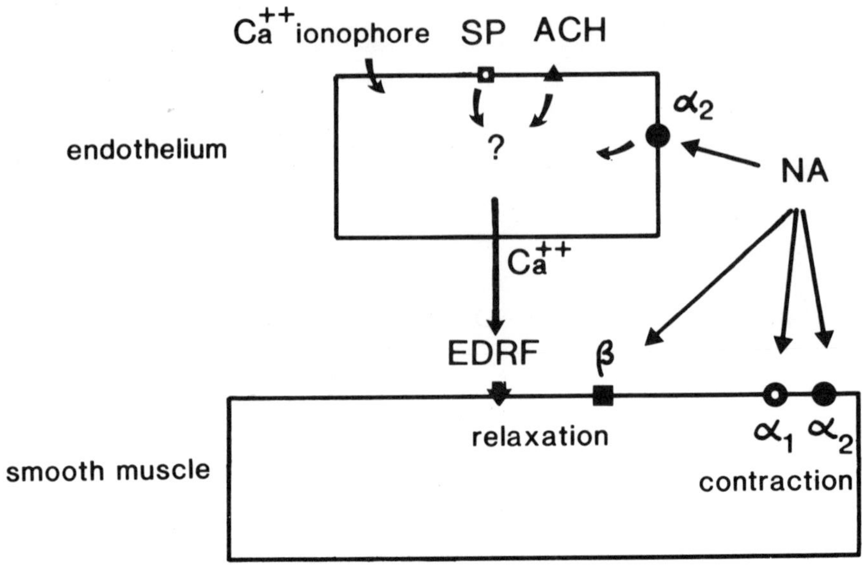

Fig. 9. Proposed receptors for norepinephrine (NA) and other vasoactive agonists on endothelial cells and smooth muscle cells. Abbreviations: α, alpha-adrenoceptor; β, beta-adrenoceptor; ACH, acetylcholine; SP, substance P; EDRF, endothelium-dependent relaxing factor.

the tone generated by U46619 in these particular arteries. Importantly, the sensitivity to substance P and acetylcholine was quite similar in the different blood vessels, indicating similar densities of receptors and/or similar efficacy of these agonists at the different locations on endothelial cells. This was not the case with bradykinin and, to a lesser extent, ATP.

The sensitivity to nitroglycerin varied by 15-fold from the most sensitive (coronary; $EC_{50} = -7.28$) to the least sensitive (mesenteric, -6.11) artery. This may be the consequence of a difference in activity of guanylate cyclase in the smooth muscle or to a difference in distribution of the nitroglycerin receptors. EDRF and nitroglycerin probably relax smooth muscle through the generation of cyclic GMP (Rapoport and Murad, 1983). Clearly if there was a variation in the production of cyclic GMP among arteries, the pattern of curves for the different blood vessels should have been similar for substance P and nitroglycerin.

3.5. Carotid Artery Reactivity 4 wk after Endothelial Denudation

Removal of endothelial cells experimentally by balloon rubbing of the intimal surface initiates a repair reaction from migrating medial smooth

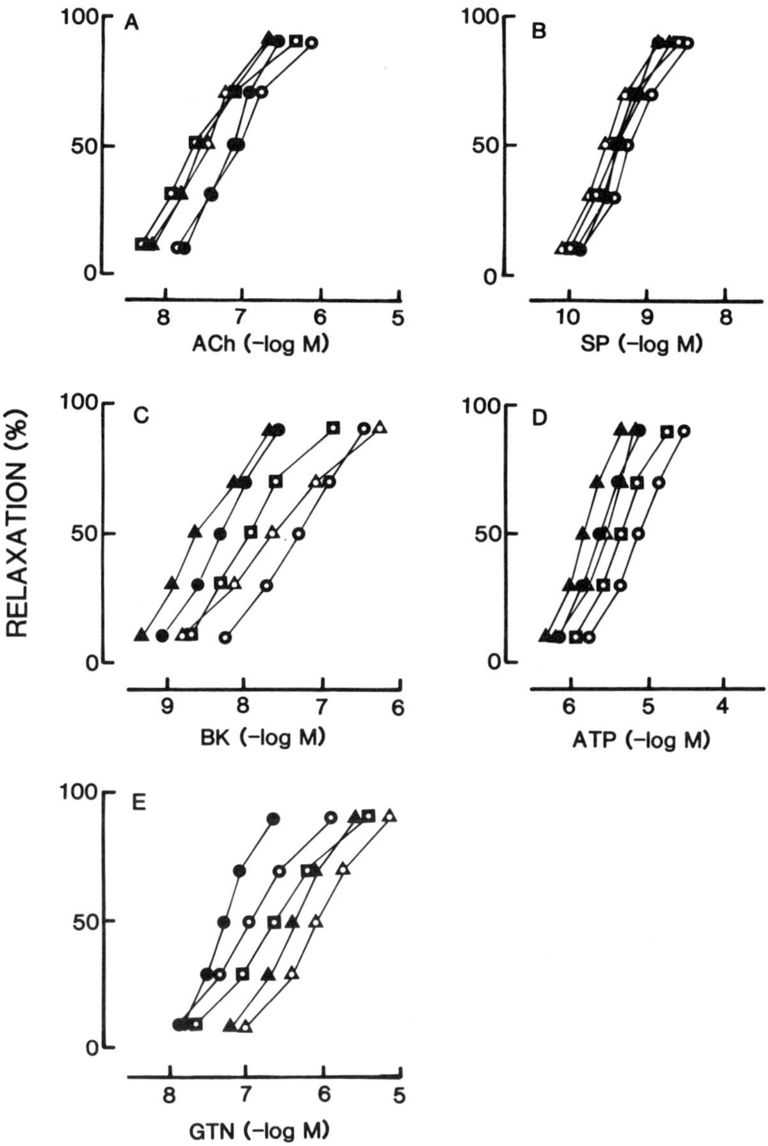

Fig. 10. Concentration–relaxation curves for acetylcholine (ACh), substance P (SP), bradykinin (BK), ATP, and nitroglycerin (GTN) in the coronary (●), carotid (○), femoral (□), renal (▲), and mesenteric arteries (△) of the dog. There were 6–8 arteries in each group (error bars at EC_{50} are within the symbols). Relaxations are expressed as % of maximal responses to each vasodilator agent (Satoh K. and Angus J., unpublished data).

Table 2
Comparison by Rank Order of Maximum Relaxations to Five Vasodilator Agents in Five Large Arteries of the Dog In Vitro[a]

Agent	Endothelium-dependent	Order of relaxation
Acetylcholine	Yes	Co > Ca = F > R = M
Subtance P	Yes	Co > Ca = F > R > M
Bradykinin	Yes	Co > R > F > M > Ca
ATP	Yes	Co > Ca > R = M = F
Nitroglycerin	No	Co > Ca > M = R > F

[a]Abbreviations: Co, coronary; Ca, carotid; F, femoral; R, renal; M, mesenteric (from Satoh K. and Angus J., unpublished data).

muscle cells. By 2 wk, a neointima consisting of smooth muscle cells with few myofilaments, but large numbers of synthetic organelles surrounded by extracellular matrix forms (Clowes et al., 1983). This thickened intima is often covered by endothelium. This increased intimal thickness could offer a significant barrier to EDRF and thus prevent relaxations to acetylcholine. To investigate this possibility, greyhound dogs were anesthetized and the endothelium from a 10-cm length of one common carotid artery was removed under sterile conditions by means of a balloon catheter; the other artery served as control. Four weeks later, rings of both carotid arteries were taken and suspended in organ baths. A length–tension curve was obtained in each ring to determine the precise resting force to apply to reach a similar passive stretch (equivalent to a transmural pressure of 70 mm Hg) in arteries of different geometry (Mulvany and Halpern, 1977; Angus et al., 1986a). Cumulative contraction curves (one per ring) to norepinephrine, serotonin, U46619, and K^+ were not different as regards maximal force between control or rings with previous intimal damage (Fig. 11), but were significantly more sensitive (lower EC_{50} values) by 3.4-, 2.1.-, and 1.4-fold for norepinephrine, serotonin, and K^+, respectively. The histological examination of each ring with the light microscope [thick sections (2–4 μm) stained with methylene blue] indicated that control rings had either one or no smooth muscle cells in the intima, but more than 90% covering with endothelial cells. Arterial rings previously subjected to intimal damage showed mostly eccentric intimal thickening of 2–20 smooth muscle cells, largely covered by endothelial cells. To test the integrity of these endothelial cells, ring segments were submaximally contracted with either serotonin or with U46619. There was no difference in the sensitivity (EC_{50}) or in the maximal relaxation to acetylcholine or nitroglycerin when comparing responses from control and damaged rings

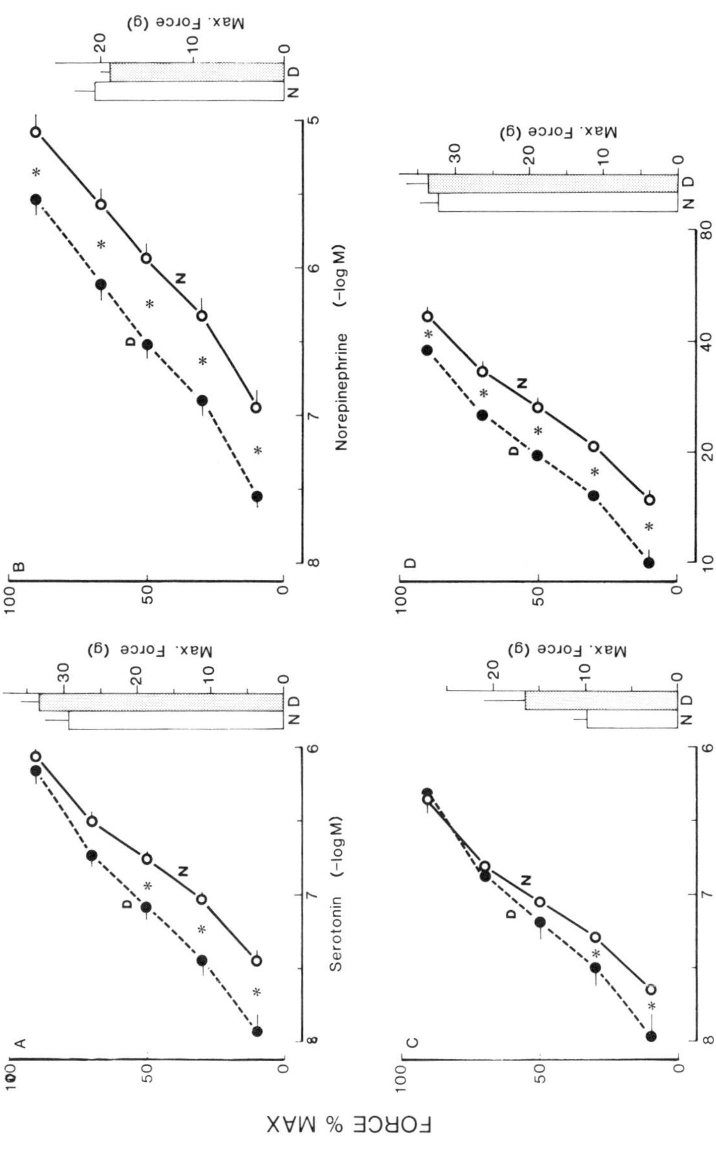

Fig. 11. Comparison of concentration–contraction curves in rings of carotid artery of the dog, 4 wk after removal of the endothelium (damaged, D, solid circles) or with endothelium left intact at surgery (normal, N, open circles). Lines are logistic fitted data to compare sensitivity with SEM bars. Asterisks indicate significant differences. Histograms are maximal responses developed by the rings in the two groups (Satoh K., Angus J., Cocks T., and Campbell G., unpublished data).

(Fig. 12). There was a small enhancement in the relaxation to adenosine in the rings contracted with U46619, but not in those contracted with serotonin.

These findings suggest that there is a small (less than fourfold) increase in sensitivity to norepinephrine and to serotonin in carotid arteries with neointima. In these studies the neointima was well covered with endothelial cells that responded normally to acetylcholine, suggesting that the neointima did not constitute an effective barrier to EDRF. The relaxation to the endothelium-independent vasodilators (nitroglycerin and adenosine) was essentially unchanged.

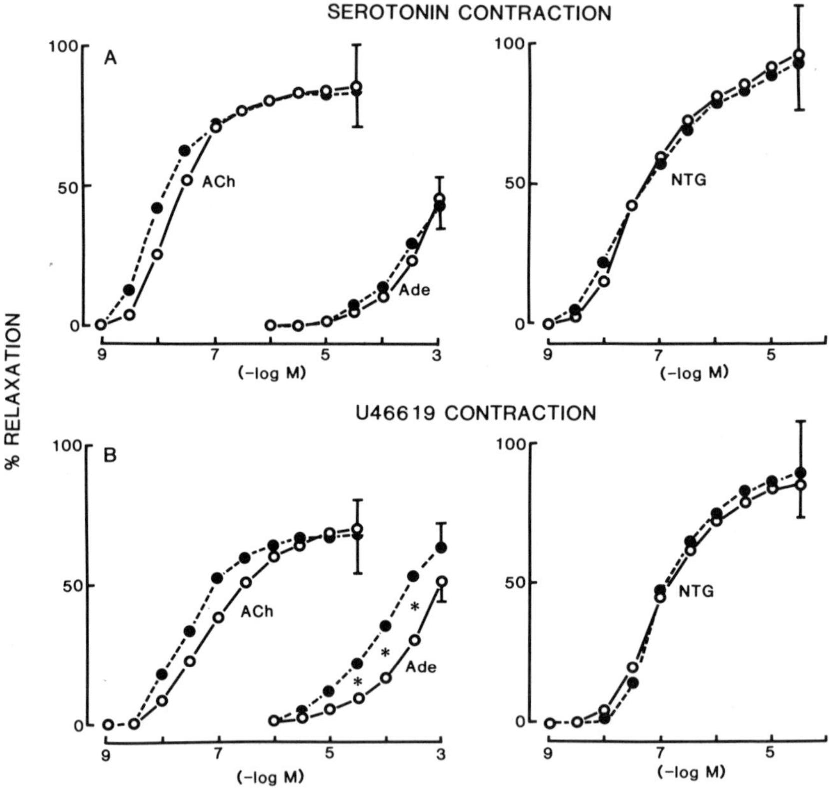

Fig. 12. Concentration–relaxation curves to acetylcholine (ACh), adenosine (Ade), and nitroglycerin (NTG) in normal (open circles) and damaged (closed circles) carotid artery rings (endothelium removed 4 wk previously). The relaxations are expressed as percentage of the resting contraction to serotonin (0.3 μM, top) or to U46619 (0.3 μM, bottom).

4. Reactivity of Microcirculation to EDRF

4.1. Effect of Hypertension on Reactivity of the Hindquarter of the Rabbit

A reduction in the production of a powerful local vasodilator substance is an attractive hypothesis to at least partly explain the development of hypertension. The identification of EDRF as a powerful dilator of large and small arteries prompted the evaluation of the reactivity of an intact vascular bed in conscious normotensive and hypertensive rabbits. First, the animals were anesthetized, and pulsed Doppler-flow transducers were implanted on the lower abdominal aorta just proximal to the iliac bifurcation. A catheter was placed in the aorta proximal to the flow probe (Herd and Barger, 1964) to allow for local intraarterial infusion of agonists. The kidneys were left undisturbed (sham) or wrapped in cellophane (wrap) to cause perinephritic hypertension over the next 5 wk. On the 35th day after surgery, the rabbits were treated with mecamylamine (10 mg/kg) and propranolol 0.5 mg/kg to prevent autonomic reflexes. Blood pressures (mm Hg measured in the ear artery) before and after the ganglion blockade were 79.4 and 67.4 for sham and 124.8 and 103.6 for kidney-wrapped rabbits, respectively. The changes in hindquarter vascular resistance (pressure in ear artery/blood flow in the hindquarter) obtained with intraarterial infusions of acetylcholine, adenosine, and serotonin were characterized by (1) the range from resting to maximal dilatation, (2) the sensitivity (ED_{50}), and (3) the average slope about ED_{50} (Wright et al., 1987). There were no significant differences in ED_{50} values for each of the three dilator drugs between normotensive and hypertensive animals. In the hypertensive rabbits, however, the range and slope of the curves for all agents were about twice that of the normotensive rabbits. These results suggest that hypertrophy of the muscle in the precapillary vessels make them a nonspecific amplifier of changes in vascular resistance induced by a variety of endothelium-dependent (acetylcholine) and endothelium-independent dilator drugs (such as adenosine). Since responses to acetylcholine mimicked those of adenosine in the two groups, an alteration in the response to EDRF in the resistance vessels seems unlikely in this experimental model of hypertension (5 wk of duration).

4.2. Effect of Cholesterol on EDRF

An increase in plasma cholesterol levels from dietary intake could affect production and/or release of EDRF by the endothelial cells. Experiments were designed to test whether this could occur in the resistance vessels of the hindquarters of normotensive and hypertensive rabbits—areas of the vascular bed well distal from the classical atheromatous lesions seen in the aorta or other major arteries (*see* section 5).

Rabbits were fed 1% cholesterol chow or normal chow for 4 wk and had their kidneys wrapped in cellophane or left undisturbed. On the day of the experiment (5 wk postsurgery) autonomic reflexes were blocked pharmacologically and vascular resistance in the hindquarter was measured during intravenous infusions of increasing doses of acetylcholine, adenosine, or serotonin (Wright and Angus, 1986). The sensitivity (ED_{50}) to the three dilator agonists was altered less than twofold for any treatment group compared with the normotensive group on normal chow. The resting vascular resistance increased (over that of the sham-operated rabbits on normal chow, 100%) in sham-operated animals receiving cholesterol (121%), hypertensive rabbits on normal chow (223%), and hypertensive rabbits receiving cholesterol (229%). These values are consistent with the increase in viscosity of plasma during the cholesterol diet (plasma cholesterol approximated 1000 mg%) and with the presence of medial hypertrophy in the resistance vessels of hypertensive animals. The maximal vasodilator response to acetylcholine, but not that for adenosine or serotonin, was significantly reduced by hypercholesterolemia, hypertension, or a combination of both. This study suggested that the function of the intima of resistance vessels of the hindquarters of the rabbit was not altered greatly by raised plasma cholesterol or by hypercholesterolemia combined with hypertension as far as the sensitivity to the endothelium dependent agonist acetylcholine was concerned. In addition, the cholesterol diet and/or hypertension had no effect on reactive hyperemia induced by balloon inflation of the abdominal aorta for 5–80 s in the hindquarters (Wright and Angus, 1986).

5. EDRF and Coronary Atheroma

To study the effects of atheroma on responses to EDRF in coronary arteries, rings were prepared from arteries with concentric atheromatous lesions, near the origin of the left anterior descending coronary artery in rabbits fed a 1% cholesterol diet for 5 mo. A coronary resistance vessel that was obviously clear of atheroma was prepared from a distal side branch from the same heart. Both vessels were suspended on 40-μm diameter wires in the same myograph (Mulvany and Halpern, 1977). Potassium contracted the vessels to a steady level of force. Addition of acetylcholine caused the small vessel to relax, but caused the large atheromatous artery to contract (Fig. 13). At the highest concentrations, the small vessel began to

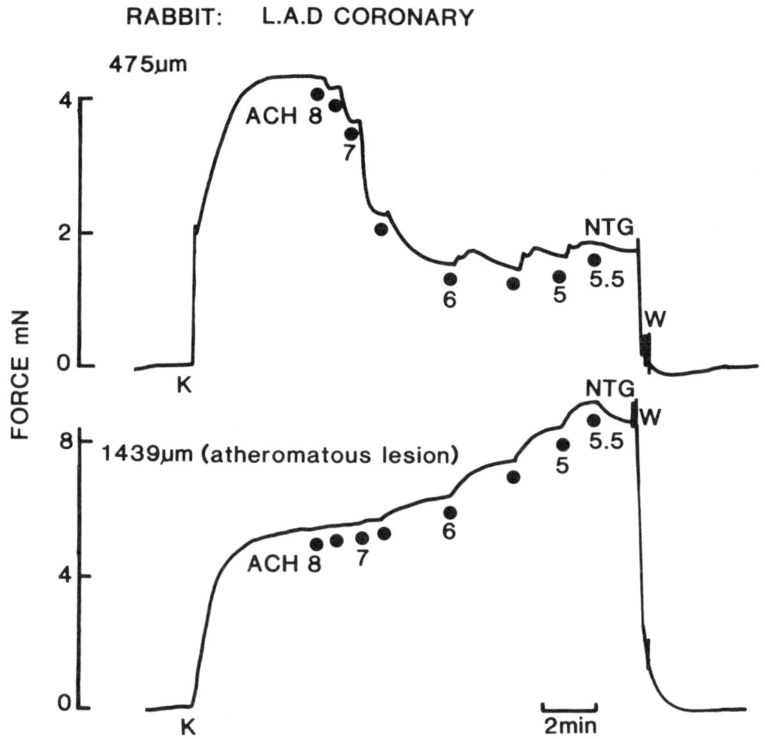

Fig. 13. Isometric force recording in a segment of microvessel (top) and a segment of large coronary artery with concentric atheromatous plaque (bottom) studied in the same myograph. Cumulative concentrations of acetylcholine (−log *M*) were applied during steady contraction obtained with K^+ 30 m*M* (histology of vessels, *see* Figs. 14 and 15).

contract weakly; only further contractions were observed in the large artery. These observations are consistent with a dual effect of acetylcholine, to cause the release of EDRF and to stimulate the smooth muscle cells to contract. If the EDRF cannot penetrate the neointima (as in the atheromatous vessel) the contraction component of the response would proceed unchecked and be amplified. Substance P relaxed the small vessel, but had no effect on the large artery (data not shown). The vessels were fixed for light microscopy, and both showed generally intact endothelium, except at the sites where the suspending wires had damaged the wall (Figs. 14 and 15). The concentric neointima that had formed in the large artery was filled with extracellular cholesterol and lipid-engorged smooth muscle cells and macrophage-derived foam cells (Fig. 15). In three other large coronary arteries taken from rabbits on a normal diet, only relaxation, similar to that shown here for the small artery, was observed.

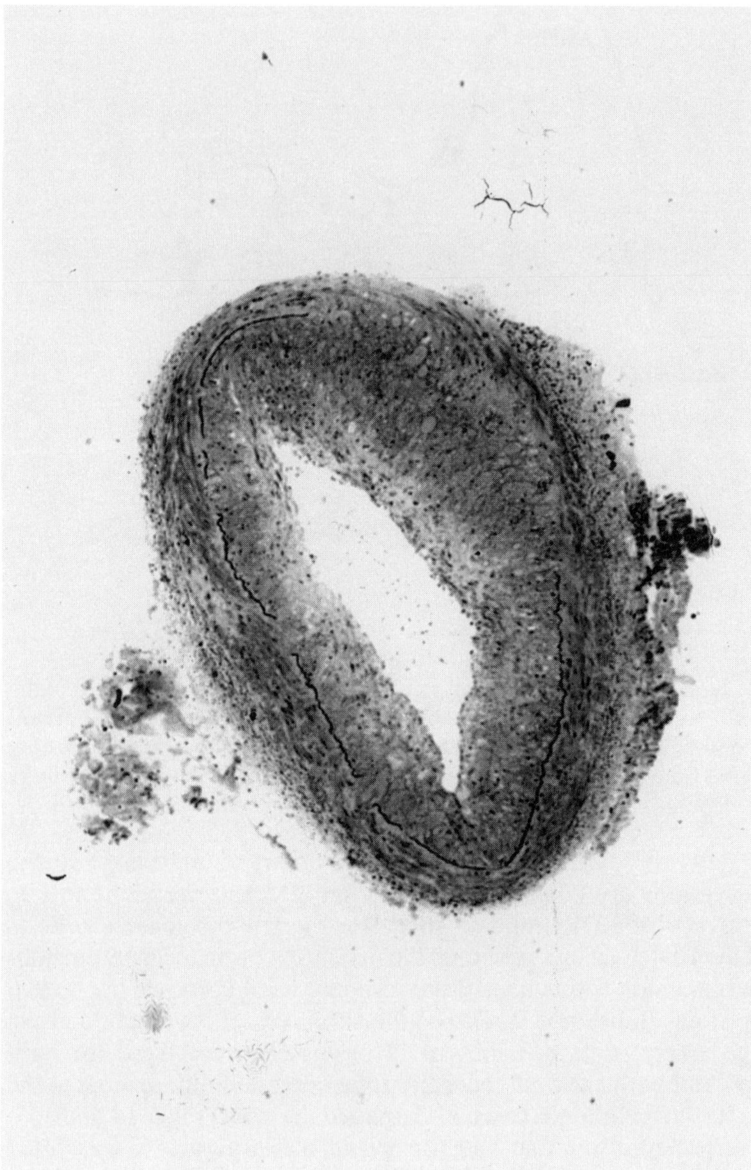

Fig. 14A. Transverse section through the left coronary artery of a rabbit fed at 1% cholesterol supplemented diet for 5 mo. Note the atheromatous plaque completely lining the lumen of the vessel (×63).

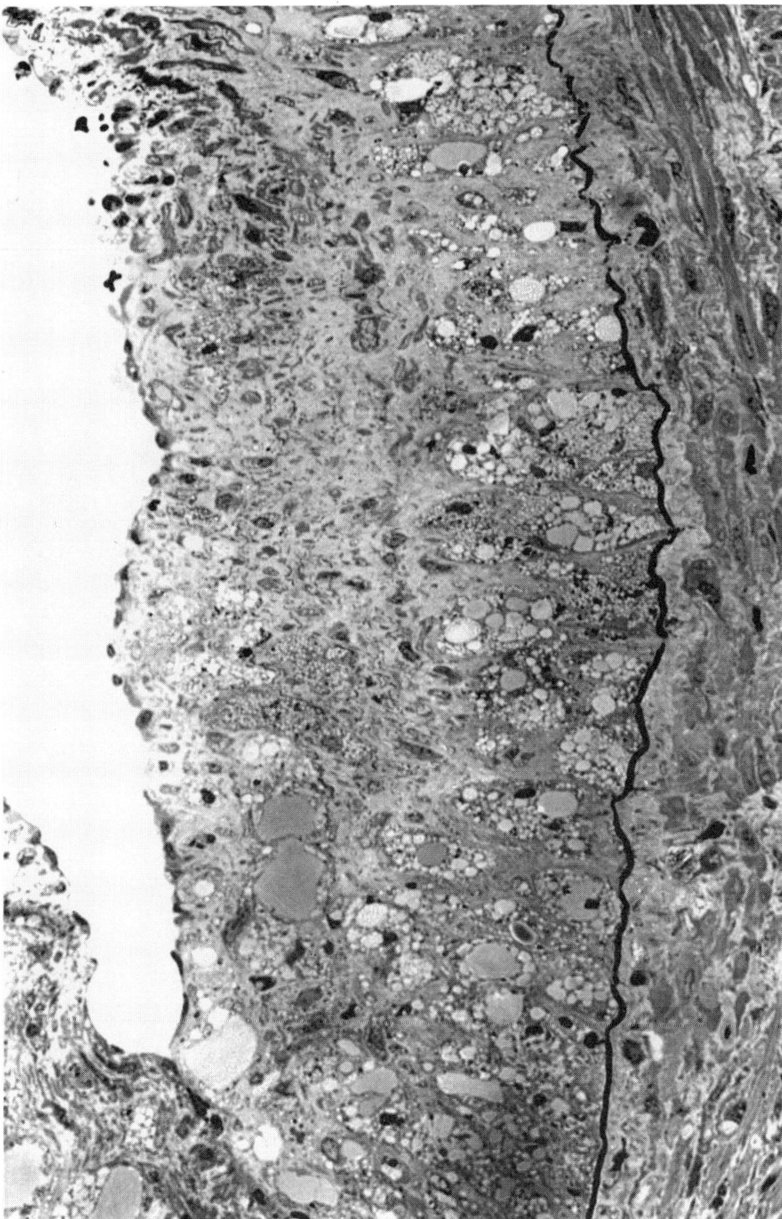

Fig. 14B. Enlargement of intima from Fig. 14A. Note that the vessel is lined by an endothelium and contains extracellular cholesterol and a large number of lipid-engorged cells (×350).

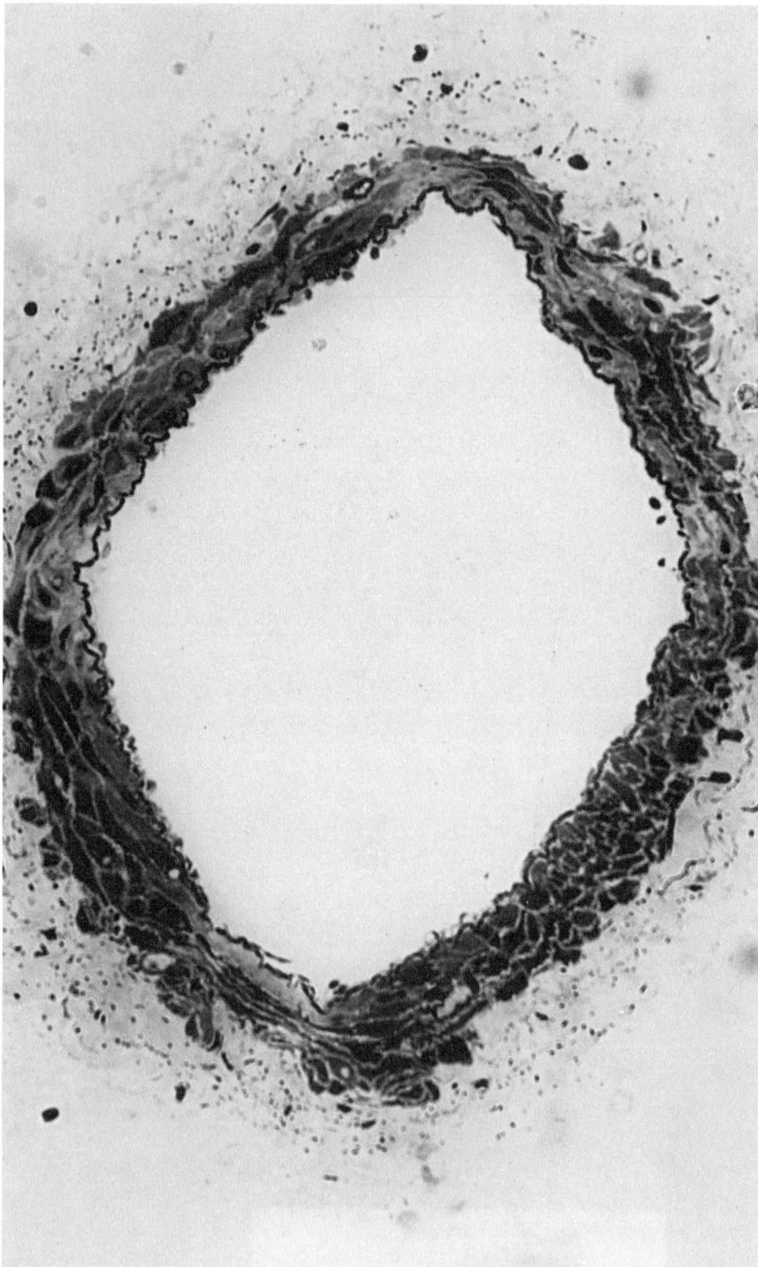

Fig. 15. Transverse section through a coronary microvessel from the same heart as in Fig. 14. Note the vessel is lined by an endothelium, and there is no evidence of atheromatous plaque formation or intimal thickening (×430).

We concluded that an increase in plasma cholesterol *per se* does not interfere with the release of and/or the response to EDRF, but that atheromatous lesions of large coronary arteries may be a sufficient barrier to trap or destroy EDRF released from the obviously intact endothelium. Similar conclusions have been reported for the aorta from rabbits fed 0.3% cholesterol for 16 wk (Coene et al., 1985).

6. Conclusions

EDRF is released from large arteries in vivo to relax the underlying artery wall. It does not appear to circulate in blood even for short distances. When blood flow and perfusion pressure are controlled, the diameter of the artery increases in response to EDRF released by acetylcholine, but not to nitroglycerin or sodium nitroprusside. In the carotid artery of chronically instrumented greyhounds, acetylcholine can increase diameter only if endothelium is present.

Coronary endothelial cells release EDRF in response to serotonin and norepinephrine. The response to serotonin is not caused by activation of 5-HT_2-serotonergic receptors; that to norepinephrine is mediated by alpha$_2$-adrenoceptors. Endothelial alpha$_2$-adrenoceptors are present in other large arteries of the dog and coronary resistance vessels of dog and pig. The distribution of bradykinin receptors on endothelial cells appears to vary widely across five large arteries of the dog, a phenomenon not shared with substance P or acetylcholine. Hypertension and hypercholesterolemia do not appear to alter the release of or the response to EDRF in the resistance vessels of the hindquarter of the conscious rabbit. Neo-intimal thickening in large arteries *per se* does not appear to constitute an important barrier to EDRF. When neo-intimal thickening involves foam cell formation and the deposition of extracellular cholesterol, however, the response to EDRF is absent, suggesting that the lipid involvement has trapped the factor preventing it from reaching the medial smooth muscle.

Taken together, these studies suggest that the environment of an artery in which there is neo-intimal thickening, with lipid involvement, would reflect (1) some enhanced sensitivity of the smooth muscle to constrictor stimuli; (2) prevention of EDRF from reaching the media; and (3) encroachment of the atheroma in the lumen, which perhaps results in a hyperresponse, especially if the perfusion pressure is low. Is this the scenario for large artery vasospasm?

Acknowledgments

This work was supported by an Institute Grant from the National Health and Medical Research Council of Australia. We thank Matthew Le Duc, Peter Coles, Eileen Wright, David Prior, and Juliana Wong for excellent technical assistance.

References

Angus, J. A. and Cocks, T. M.: Role of endothelium in vascular responses to norepinephrine, serotonin and acetylcholine. *Bibl. Cardiol.* **38:**43–52, 1984.

Angus,. J. A., Campbell, G. R., Cocks, T. M., and Manderson, J. A.: Vasodilatation by acetylcholine is endothelium-dependent: A study by sonomicrometry in canine femoral artery in vivo. *J. Physiol.* (Lond.) **344:**209–222, 1983.

Angus, J. A., Cocks, T. M., and Satoh, K.: Alpha-adrenoceptors and endothelium-dependent relaxation in canine large arteries. *Br. J. Pharmacol.* **88:**767–777, 1986a.

Angus, J. A., Cocks, T. M., and Satoh, K. Alpha-adrenoceptors on endothelial cells. *Fed. Proc.* **45:**2355–2359, 1986b.

Angus, J. A., Wright, C. E., Cocks, T. M., Satoh, K., and Campbell, G.: Endothelium-dependent relaxation is unaltered by hypertension, cholesterol or intimal thickening. *Clin. Exp. Pharmacol. Physiol.* **13:**289–293, 1986c.

Clowes, A. W., Reidy, M. A., and Clowes, M. M.: Kinetics of cellular proliferation after arterial injury. 1. Smooth muscle cell growth in the absence of endothelium. *Lab. Invest.* **49:**327–333, 1983.

Cocks, T. M. and Angus, J. A.: Endothelium-dependent relaxation of coronary arteries by noradrenaline and serotonin. *Nature* **305:**627–630, 1983.

Coene, M.-C., Herman, A. G., Jordaens, F., Van Hove, C., Verbeuren, T. J., and Zonnekeyn, L.: Endothelium-dependent relaxations in isolated arteries of control and hypercholesterolemic rabbits. *Br. J. Pharmacol.* **85:**267P, 1985.

Cohen, R. A., Shepherd, J. T., and Vanhoutte, P. M.: 5-Hydroxytryptamine can mediate endothelium-dependent relaxation of coronary arteries. *Am. J. Physiol.* **245:**H1077–H1080, 1983.

Furchgott, R. F. and Zawadzki, J. V.: The obligatory role of endothelial cells in the relaxation of arterial smooth muscle by acetycholine. *Nature* (Lond.) **288:**373–376, 1980.

Herd, J. A. and Barger, A. C.: Simplified technique for chronic catheterization of blood vessels. *J. Appl. Physiol.* **19:**791–792, 1964.

Holtz, J., Forsterman, U., Pohl, U., Giesler, M., and Bassenge, E.: Flow-dependent, endothelium-mediated dilation of epicardial coronary arteries in conscious dogs: Effects of cyclooxygenase inhibition. *J. Cardiovas. Pharmacol.* **6:**1161–1169, 1984.

Lamping, K., Marcus, M. L., and Dole, W. P.: Removal of the endothelium potentiates canine large coronary artery constrictor responses to 5-hydroxytryptamine in vivo. *Circ. Res.* **57**:46–54, 1985.

Mulvany, M. J. and Halpern, W.: Contractile properties of small arterial resistance vessels in spontaneously hypertensive and normotensive rats. *Circ. Res.* **41**:19–26, 1977.

Rapoport, R. M. and Murad, F.: Agonist-induced endothelium-dependent relaxation in rat thoracic aorta may be mediated through cGMP. *Circ. Res.* **52**:352–357, 1983.

Wright, C. E. and Angus, J. A.: Effects of hypertension and hypercholesterolemia on vasodilatation in the rabbit. *Hypertension* **8**:361–371, 1986.

Wright, C. E., Angus, J. A., and Korner, P. I.: Vascular amplifier properties in renovascular hypertension in conscious rabbits: Hindquarter responses to constrictor and dilator stimuli. *Hypertension* **9**:122–131, 1987.

Chapter 19

Endothelium-Dependent Responses in the Peripheral Microcirculation

T. M. Griffith, D. H. Edwards, R. L. Davies, T. J. Harrison, and K. T. Evans

1. Introduction

Experimental studies of endothelium-dependent relaxation have to date largely focused on the properties of pharmacologically and physiologically vasoactive agents in conduit vessels. The phenomenon has been demonstrated in vitro (Furchgott, 1983) and in vivo (Angus et al., 1983), although it has become apparent that species and vessel differences exist (Furchgott, 1983; Kalsner, 1985; Katusic et al., 1984). There is evidence that injury to the arterial wall, as occurs in experimental models of atheroma (Habib et al., 1984) or hypertension (Konishi and Su, 1983), may functionally impair endothelium-dependent responses. These observations may have clinical relevance. Relatively little is known, however, about endothelium-dependent relaxation in the microvasculature, where it may have an important physiological role, since small vessels regulate peripheral resistance and hence govern the distribution of blood flow in the intact circulation.

The role of the endothelium in the microvasculature can be interpreted only in the context of the many other factors that influence vessel tone and caliber. Of primary importance among these are the intrinsic properties of smooth muscle. Bayliss in 1902 observed enhanced constriction in response to elevation of intravascular pressure, the so called myogenic response, and this phenomenon has since been observed in blood vessels in a variety of vascular beds, although it is more easily elicited in small vessels (Johnson, 1980). Such behavior is an intrinsic property of vascular smooth muscle and does not depend on adventitial nerves or circulating humoral factors. In principle, the myogenic response constitutes a positive feedback mechanism that could theoretically lead to instability in the circulation (Folkow, 1964), and recent studies in the isolated central artery of the rabbit ear suggest the existence of a more sophisticated pressure-sensitive, negative feedback loop that is set so as to minimize changes in vessel diameter following either increases or decreases in intraluminal pressure (Speden, 1984). Such a feedback mechanism would predispose to stability by making vessel diameters insensitive to fluctuations in blood pressure. The influence of the endothelium on such phenomena, however, is unknown. The situation is also complicated by the fact that the pharmacological responsiveness of small arteries and arterioles is highly dependent on circumferential wall stress (Gore, 1972). Furthermore, as an artery constricts, less active stress development is required for further constriction to occur (Speden, 1985). Small blood vessels are also capable of exhibiting spontaneous rhythmic contraction, and it appears that circumferential wall stress is the critical determinant of the frequency of this contraction (Burger et al., 1984), so that such phasic activity is consequently affected by any physical or pharmacological intervention that alters wall stress. Intravascular pressure, therefore, can clearly exert an important influence on the function of vascular smooth muscle in a variety of ways.

Through the release of endothelium-derived relaxing factor (EDRF), either basally (Griffith et al., 1984a,b) or when stimulated by a variety of pharmacological agents (Furchgott and Zawadzki, 1980; Furchgott, 1983), the microvascular endothelium may have the potential to modulate the behavior of the vessel wall in an important way. A number of peptides, including vasoactive intestinal polypeptide (VIP), substance P, and calcitonin gene-related peptide are present in nerve terminals in the adventitia of both conduit and microvessels (Weihe et al., 1984; Zaidi et al., 1985). These agents appear to dilate conduit vessels mainly through an endothelium-dependent mechanism (Furchgott, 1983; Davies and Williams, 1984; D'Orleans-Juste et al., 1985; Brain et al., 1985), and their circulating plasma levels are probably derived from nerve endings (Andersson et al., 1983, 1984; Zaidi et al., 1985), so it is likely that their diffusion through the vessel wall can result in the release of EDRF. On physical grounds alone, such a process is more likely to be important in small than in large blood vessels. There are, however, qualitative differences between the endothelium and smooth muscle of small vessels and their large vessel counterparts. For example, the enzyme xanthine oxidase is present in microvascular endothelium (Jarasch et al., 1981), and small vessels are more dependent on sources of extracellular calcium for contraction (Sutter et al., 1977; Cauvin et al., 1982). Consequently, it is possible that regional differences in the nature of endothelial cells and smooth muscle responsiveness to EDRF exist. It is also of interest that myoendothelial junctions are found in resistance vessels, although the functional significance of this observation is unknown (Hammerson and Hammerson, 1985).

A further mechanism exists through which endothelium may regulate the behavior of resistance vessels; namely, flow-dependent dilatation (Schretzenmayr, 1933; Hilton, 1959; Gerova et al., 1981). Dilatation of conduit arteries secondary to increases in flow has been convincingly shown to be endothelium-dependent (Holtz et al., 1983a,b, 1984), and probably occurs through release of EDRF (Rubanyi et al., 1986). Flow-dependent release of EDRF in resistance vessels may have the potential to inhibit the myogenic-type responses already discussed and provide an important regulatory link between intravascular pressure, flow, and local vessel tone. Indeed, such a mechanism might act as a "brake" to the theoretical instability that results from the classical Bayliss response, by superimposing an opposing positive feedback loop, which results in enhanced dilatation in response to increased flow, and thus, by inference, perfusion pressure.

Apart from such primarily mechanical and hydrodynamic determinants of microvascular tone are, of course, important metabolic control mechanisms mediated through pH, extracellular potassium concentration, and substances such as adenosine (Sparks, 1980). In this context, it is also of considerable interest that the production of EDRF itself is linked to metabolic processes; namely, the generation of adenosine triphosphate (ATP) by mitochondria within endothelial cells (Griffith et al., 1986). The complexity of the interactions possible thus provides for great subtlety, and probably also redundancy, in the homeostatic mechanisms that control microvascular tone.

In order to investigate the influence of EDRF in resistance vessels, novel X-ray microscopial techniques that permit simultaneous imaging of vessels with sizes in the range 25–800 μm were developed. Such techniques obviate the necessity for traumatic dissection of blood vessels that can alter their responsiveness and confer the particular advantage of measuring internal vessel caliber, so that from a fluid mechanical point of view, it is not necessary to determine vessel wall thickness. An isolated, buffer-perfused rabbit ear preparation was chosen as the experimental model. This is two-dimensional, which greatly simplifies analysis and, at the flow rates used, is probably free of complicating secondary metabolic effects. The work concentrates specifically on the role played by endothelium-derived relaxing factor (EDRF) in modulating the vessel diameter changes that result from incremental alterations in flow and constriction by pharmacological agents that do not themselves stimulate EDRF release. An attempt has been made to relate the data to the branching topology of the vascular network.

2. Methods

2.1. Preparation

Male New Zealand white rabbits (2.5 kg) were killed, the ears removed, and the central artery cannulated and perfused by a peristaltic pump with oxygenated Holman's buffer with continuous monitoring of perfusion pressure (Griffith et al., 1984b). Dextran (5%, mw, 80,000) was dissolved in the buffer to increase its viscosity and thus eliminate artifactual rises in perfusion pressure on addition of viscous radiographic contrast medium to the perfusion system. The final viscosity of the buffer was 2.23 mPas (milliPascal seconds) and its behavior Newtonian, at shear rates of 30–300/s. Perfusate temperature was maintained at 35 $\pm$ 9 °C with free outflow at atmospheric pressure from draining veins. Stabilization was allowed for 90 min before the experiments were started. A 10-mL air-filled compliance chamber, connected to the perfusion circuit, damped phasic pressure variations to less than 5% of mean pressure at any given flowrate. Zero perfusion pressure corresponded to zero flow.

2.2. X-Ray Techniques. Contrast Medium

The X-ray microscope used in this study (Ely, 1980; Davies, 1983) consisted of a triode electron gun with electromagnetic focusing of the electron beam by an extended pole lens (Mulvey and Newman, 1973) onto a copper target. The system was operated in reflection rather than transmission mode to allow for the use of high beam currents (Bellman, 1953). Final X-ray source size was approximately 4 μm at a beam current of 500 μA. An electrostatically self-focused microchannel plate introduced a gain factor of 70,000 into the intensification system, which was based on a gadolinium oxysulfide phosphor suitable for low-kilovoltage techniques. Static imaging on conventional X-ray film and real-time imaging on a television monitor were possible. Technical details have been given elsewhere (Davies et al., 1986).

The perfusion circuit was designed so that the introduction of low volumes (0.5 mL) of contrast medium (iohexol, Nyegard) into the system did not produce pressure artifacts, and Holman's solution, free of contrast medium, immediately perfused the bed after completion of the X-ray exposure. In view of the vasodilator properties of iodinated contrast medium, a low concentration of 100 mg iodine/mL (iohexol 300 mg iodine/mL diluted in Holman's buffer) was chosen. To obtain satisfactory images with such low concentrations of contrast medium, exposures of 12–15 s and low-kilovoltage (10–25 kV) techniques were necessary. Preliminary real-time experiments were performed with the image-intensification system to synchronize X-ray exposures with the rapid transit of contrast medium through the vascular bed. The maximal pressure fall induced by the contrast medium itself occurred toward the end of, or after, termination of the radiographic exposure. This was less than 2 mm Hg in unstimulated preparations (perfusion pressure approximately 30–40 mm Hg) and less than 8 mm Hg in constricted preparations (100–150 mm Hg). Calculations of diameter changes corresponding to these small pressure falls based on the data of Fig. 1 indicated that systematic errors would be most significant in highly activated beds, and then be of the order of 20–25 μm (i.e., approximately 5%). The influence of the contrast medium itself on the accuracy of the data was therefore ignored.

2.3. Quantitation

An IBAS Kontron Semi-interactive Image Analysis System (Kontron Electronics, Munich, FRG) was used to measure vascular diameters from conventional radiographic film using software specifically developed for this purpose. The geometric edges of a vessel under study were first identified visually on a display monitor, and the analysis system then computed the absolute diameter. Magnification factors were determined by placing a small calibrated gold grid on the rabbit ear within the field of view. Diameter measurements of each vessel selected for study were made six times and then meaned. The accuracy of this protocol, as assessed by fine tungsten wires calibrated by electron microscopy, was within 5% for vessels of approximately 70 μm in size and considerably higher for larger vessels such as the central ear artery or small veins. Experiments were performed at geometric magnifications of approximately 4, further magnification being provided electronically by the image-analysis system. Although vessels with a diameter of larger than 25 μm could be imaged, for radiographic reasons, accurate quantitation was possible only in vessels of over 70–80 μm; that is, in the central ear artery (G0) and its first three generations of branch vessels (G1–G3). To obtain spatially "averaged" diameter estimates, measurements were made on five representative non-tapering vessels in each of these generations. Data from each generation and from other preparations were then pooled and meaned. The diameters of small veins and venules were obtained by the same method. Standard errors were always less than 5% of mean diameter and have not been included on the figures for the sake of clarity. Up to 1000 separate diameter

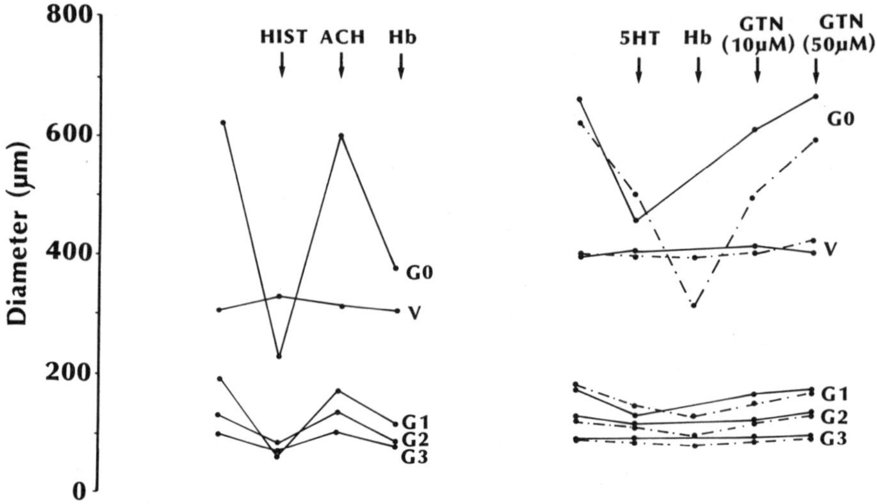

Fig. 1. (Left) Data from a representative experiment showing constriction by histamine (Hist, 1 μM) and inhibition of acetylcholine-induced dilatation (ACh, 1 μM) by hemoglobin (Hb, 1 μM). Diameter changes are plotted for the central artery (G0) and the first three generation of branches (G1–G3). Responses were qualitatively similar in vessels down to about 25 μm in diameter. The constrictor stimulus of this concentration of histamine was greatest in the central artery and diminished progressively in successive branch generations. In contrast, the relative dilator effects of acetylcholine were approximately the same in all vessels. In this example, constrictor responses were completely reversed in all generations by 1 μM acetylcholine. Venules and small veins (v) behaved passively. (Right) Pooled data comparing the dilator effects of glyceryl trinitrate (GTN, 10 and 50 μM) in preparations constricted by 1 μM serotonin (5-HT) before (——) and after (—·—·) inhibition of basal EDRF activity by hemoglobin (n = 9). As with 1 μM histamine, constrictor responses to this relatively high pharmacological stimulus were progressively smaller in successive branch generations and were enhanced, particularly in G0, by inhibition of basal EDRF activity. GTN itself was equipotent in all vessels in G0 to G3, however, since diameter decreases resulting from the combination of serotonin and hemoglobin were reversed by about 60% by 10 μM GTN and almost completely reversed by 50 μM GTN. Formal comparison of the sensitivities of the vessels to GTN before and after inhibition of basal EDRF activity is not possible, since hemoglobin decreases the sensitivity of vascular smooth muscle to nitrovasodilators (Martin et al., 1985). Note again the lack of active responses in small veins (v).

measurements were made in each experiment. If experiments were repeated after an interval of 60 min, measurements were always within 10%, and in many cases agreement was within 5%. Intermittent exposure to iohexol did not therefore significantly impair endothelium-dependent dilatation or constrictor responses.

2.4. Agents Used To Induce Tone. Hemoglobin

Tone was induced in the bed by serotonin or histamine, which do not stimulate the release of EDRF from rabbit aortic endothelial cells (Griffith et al., 1984a,b; Griffith, 1985). To provide evidence that this is also the case in the rabbit ear, combinations of these agents were studied. When preparations were first constricted by serotonin (1 μM), low concentrations (0.001–0.1 μM) of histamine introduced into the perfusate produced further constriction only. This was also the case when the protocol was reversed. Furthermore, if the bed was constricted with histamine in the presence of ketanserin (3 μM), vasodilatation was not induced by serotonin, indicating that endothelium-dependent relaxation is not unmasked by S_2-serotonergic antagonism, as in the canine coronary artery (Cocks and Angus, 1983). Hemoglobin (1 μM) was chosen as an inhibitor of endothelium-dependent dilatation since its action is specific against EDRF and it has no direct constrictor action on vascular smooth muscle at this concentration (Cocks and Angus, 1985; Martin et al., 1985). Hemoglobin was prepared either by lysis of washed human red blood cells and purification by Sephadex G200 gel chromatography, as described previously (Edwards et al., 1986), or from commercially available human hemoglobin reduced by the method of Martin et al. (1985). If responses of an isolated segment of the central ear artery were studied after reaming with a fine piece of thread to remove its endothelium, acetylcholine, substance P, and hemoglobin (all up to 1 μM) were devoid of vasomotor activity. Their activity was also reduced or absent in intact beds that had been perfused with air, which is known to damage endothelium, confirming that their action was caused by stimulation and inhibition of EDRF-dependent effects.

3. Network Analysis

From microradiographs of the complete network of the rabbit ear preparation obtained at different magnifications, it was possible to determine the average number of branches arising from vessels in each generation of vessels. The central artery (generation G0) was found to have about 20 side branches (generation G1), and to simplify analysis, the bed was treated as a series-coupled network of branching tubes, ignoring the hydrodynamic effects of any anastomoses between vessels in G2 and G3. Nonparabolic velocity profiles at branch points were neglected since entry lengths (the distance along each vessel before parabolic flow is reestablished), even at the highest flow-rate studied, were less than 5% of vessel length in each generation. For high Reynolds flow, this entry length is classically given by $0.03 \times \text{Re} \times d$, and for low values (approximately

1 or less) is approximately $0.6 \times d$ (Lew and Fung, 1969). The small degree of pulsatility in the flow can be ignored (Krishnakumar et al., 1976), and it can be shown mathematically that the loss of pressure head that results from deceleration of the perfusate is negligible compared to viscous losses.

Estimates of the minimal and maximal numbers of vessels in each generation (N) were obtained from the microradiographs (*see* Table 1). Flow in an "average" vessel in each generation was calculated as $q = Q/N$, where Q is the flow in the central artery. The average vessel length in each generation (L) was also determined. From simultaneous knowledge of vessel diameters (d) in each generation, it was thus possible to estimate Reynolds numbers ($\mathrm{Re} = \frac{4\rho}{\pi\mu} \frac{q}{d}$), shear stress caused by flow at the intima $\left(\frac{32\mu}{\pi} \cdot \frac{q}{d^3}\right)$, and the total hydraulic resistance ($R = \frac{1}{N} \frac{128\mu L}{\pi d^4}$) of and viscous power losses (W) in each generation (ρ = density and μ = dynamic viscosity of the buffer). Since power loss = hydraulic resistance $\times$ flow^2 (i.e., $W = RQ^2$), these parameters are equivalent if expressed relative to the central artery (i.e., with $Q = 1$).

4. Results and Discussion

4.1. Basal and Stimulated Release of EDRF. Evidence for a Complex Interaction with Vessel Tone

Acetylcholine or substance P, which dilate conduit vessels of the rabbit through an endothelium-dependent mechanism, similarly dilated small arteries and arterioles down to about 25 μm internal diameter in preparations constricted either by serotonin or histamine (Figs. 1 and 2). This dilatation was reversed by hemoglobin, an agent that inactivates EDRF specifically and does not possess a direct action on smooth muscle at the concentration used. Pharmacologically, therefore, EDRF effects in conduit arteries and in resistance vessels of the rabbit are similar. A similar conclusion has been reached in isolated small arteries (approximately 200 μm diameter) from the mesentery of the rat, in which relaxations to acetylcholine, bradykinin, and histamine, as in large vessels from the same species, also were endothelium-dependent (De Mey and Gray, 1985). In the present study, neither constriction or dilatation could be elicited in venules and veins, and their behavior was thus purely passive. Since active tone could not be induced, it was not possible to determine whether their endothelial cells release EDRF.

Spatial heterogeneity in constrictor potency was apparent at any given concentration of the constrictor agents. Thus, the degree of constriction

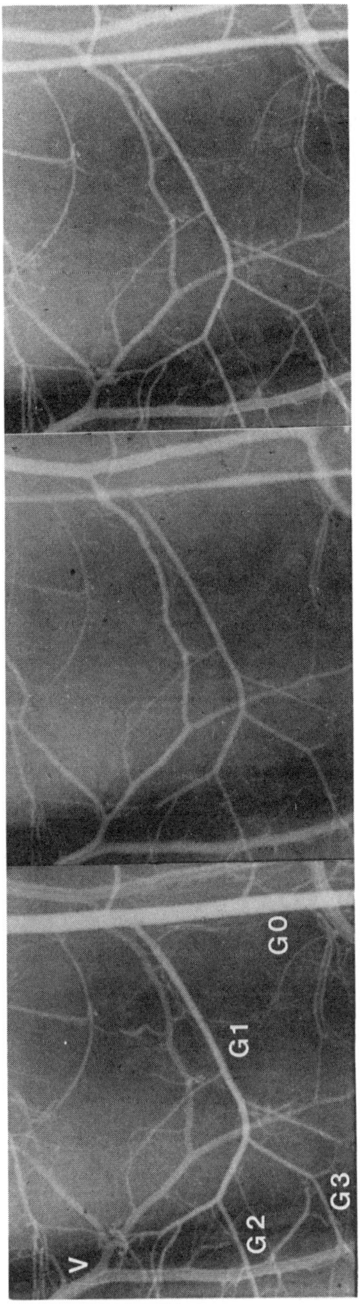

Fig. 2. Representative microradiographs illustrating diameter changes in part of the vascular bed of an isolated ear preparation perfused at a constant flow rate of 2.00 mL/min. (Left) Resting bed. (Middle) Constricted by 1 μM serotonin. (Right) Subsequently dilated by 1 μM acetylcholine. The largest arterial vessel in the field is the central artery and is labeled G0. Representative vessels of the next three major generations are also indicated (G1–G3). Constriction by serotonin and dilatation by acetylcholine (10^{-9} to $10^{-6}M$) could be demonstrated in all arteries and arterioles with an internal diameter of larger than 25 μm, which is the limit of resolution of the technique. Dilatation could also be elicited in afferent vessels by substance P (10^{-10} to $10^{-7}M$), which, like acetylcholine, is an endothelium-dependent vasodilator in large blood vessels of the rabbit and other species. The dilatation caused by these agents was inhibited by hemoglobin (*see* Fig. 1) and is therefore mediated by EDRF. There was, however, complete lack of responsiveness to both serotonin and acetylcholine in venules and small veins of any size (v).

produced by the high stimulus of 1 μM serotonin or 1 μM histamine was greatest in the central artery (G0), and was found to diminish progressively in each successive generation of branch vessels (Figs. 1–3). In contrast, 0.1 μM serotonin was most potent in the first generation of vessels (G1) (Table 1). Inhibition of basal EDRF activity accentuated these spatial differences in response in partially (Fig. 4) or highly activated preparations (Figs. 1 and 3), indicating a complex relationship between intrinsic tone,

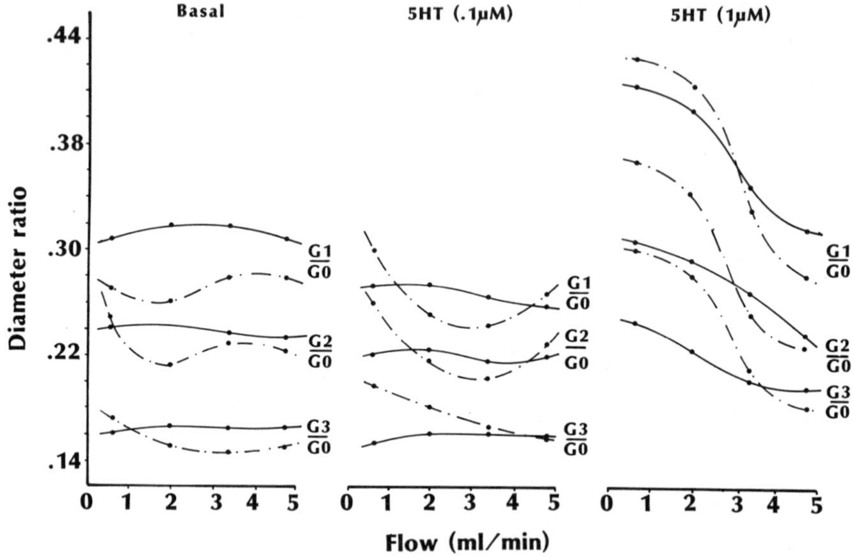

Fig. 3. This figure plots the ratio of the average vessel diameter in generations G1, G2, and G3 to that of the central artery, as a function of flow rate, in the absence of pharmacological activation (left, $n = 20$) and in the presence of either 0.1 μM (middle, $n = 8$) or 1 μM serotonin (right, $n = 8$). The data was derived from the diameter–flow curves of Fig. 4. Diameter ratios were virtually independent of flow rate in resting or partially activated preparations in the presence of basal EDRF activity (——), but not after its inhibition by hemoglobin (—·—·). The curves then vary with flow in a complex way. In contrast, in beds constricted by 1 μM serotonin, the diameter ratios exhibited marked variation with flow even in the presence of basal EDRF activity, and at low flow rates, the shape of the curves indicates greater constriction of the central artery compared to branch vessels. Such spatial heterogeneity of response was further accentuated by inhibition of basal EDRF activity. Diameter ratios in preparations constricted by 1 μM serotonin approached those of resting or partially activated preparations at high flow rates, implying that passive hydrodynamic distensive effects then tend to mask inherent differences in pharmacological responsiveness between the different generations of vessels.

developed in response to pharmacological intervention, and the vasomotor effects of basal EDRF release. This is not surprising since circumferential wall stress is an important determinant of regional differences in contractile responsiveness to pharmacological agents in microvascular networks (Gore, 1972). There is a well-documented, approximately parabolic relationship between contractile responses and circumferential wall stress in vascular smooth muscle from large vessels, with active force development increasing with distension up to a maximum and then declining (Cox, 1983); such behavior has also been demonstrated in small resistance vessels of the mesentery of the rat (Nilsson and Sjoblom, 1985). These authors also documented distension-dependent increases in sensitivity to norepinephrine that occurred even when active tension development began to fall. In conduit vessels, the sensitivity to norepinephrine similarly increases monotonically with stretch without a maximum at the length for maximum active force development (Price et al., 1983). Such physical considerations, rather than regional differences in receptor population, may underlie the regional variation in the site of maximal constriction seen with the different pharmacological stimuli in the present study, both before and after inhibition of basal EDRF activity.

There is experimental evidence that the sensitivity of resistance vessels to nitrovasodilators is less than that in large arteries (Winbury et al., 1969; Schnaar and Sparks, 1972). Such data, however, may be influenced by basal EDRF release, since this acts through the same biochemical pathway; that is, by activation of guanylate cyclase (Holzmann, 1982; Rapoport et al., 1983; Griffith et al., 1985). Thus, if basal EDRF activity were higher in resistance vessels, they might appear less responsive to nitrovasodilators. The data of Fig. 1, however, show that 1 μM acetylcholine reversed the decreases in diameter to an approximately equivalent extent (on a percent basis) in generations G0 to G3 in preparations in which a high degree of tone was induced, as did glyceryl trinitrate in constricted preparations in which basal EDRF activity first was inhibited by hemoglobin. This points to an apparently uniform underlying functional sensitivity throughout the vascular bed of the rabbit ear to the common mode of action of EDRF and glyceryl trinitrate, namely, activation of guanylate cyclase. EDRF and nitrovasodilators are, however, less effective as vasodilators at high levels of smooth muscle tone, despite equivalent elevation of cyclic GMP levels (Diamond and Chu, 1983). It would therefore seem likely that susceptibility to their action in a perfused vascular network will depend on the degree of distension of the blood vessels, and thus on both flow rate and the level of tone induced pharmacologically. In a perfused network, different generations of vessels will inevitably have different wall stresses

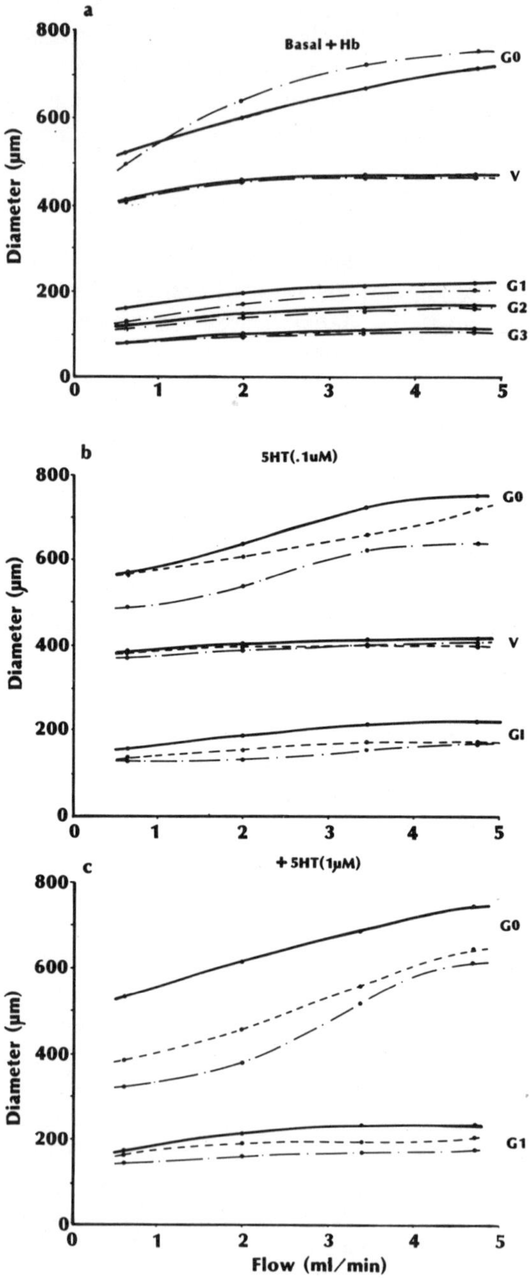
a
Basal + Hb
G0
V
G1
G2
G3
Diameter (µm)
800
600
400
200
0
0 1 2 3 4 5
b
5HT(.1uM)
G0
V
Gl
Diameter (µm)
c
+ 5HT(1µM)
G0
G1
Diameter (µm)
Flow (ml/min)

(Gore, 1972), making formal pharmacological comparison of sensitivities to dilator agents such as EDRF in vessels of different sizes difficult.

Studies of isolated resistance vessels from the rabbit ear, constricted by histamine, have suggested that acetylcholine-induced, endothelium-dependent vasodilatation increases progressively with diminishing vessel size (Owen and Bevan, 1985). This finding is clearly at variance with the data of Fig. 1, and the discrepancy almost certainly arises from differences in methodology. In perfused beds, constriction of "upstream" vessels will reduce "downstream" intraluminal pressure, which in turn is a determinant of circumferential wall stress. Similar arguments also apply in the reverse direction. Indeed, the data of Fig. 4a directly demonstrate modulation of "upstream" vascular caliber by "downstream" constriction through passive distension. Thus the responses of one part of the network will, to some extent, influence responses elsewhere in the bed, a complexity that will not arise during study of single, isolated vessels. Perhaps also relevant is the fact that conventional ring or strip preparations are necessarily studied under conditions of *extensive stress*, whereas perfused vessels, at physiological pressures, are in a state of *compressive stress*. This is a consequence of the physics of thick-walled tubes (Azuma and Oka, 1971).

Myogenic responses (i.e., spontaneous tone developing in response to stretch) have been demonstrated in isolated preparations from the first

---

← Fig. 4. Diameter measurements made under steady-state conditions at different flow rates (*see* protocol of Fig. 5) with and without inhibition of basal effects of EDRF. Experiments were performed in unactivated preparations (a; $n = 20$, ——) or with increasing degrees of tone induced either by 0.1 μM serotonin (b; $n = 8$, ---) or by 1 μM serotonin (c; $n = 8$, ----). For diagramatic clarity not all diameter data were plotted in (b) and (c), although such data are further analyzed in Fig. 3 and Table 1. Basal EDRF activity was inhibited after control experiments by addition of hemoglobin (1 μM) to the perfusate (—·—·). In (a), note that a significant ($p < 0.05$) hemoglobin-dependent constriction occurred in G1 over the complete range of flow rates. No statistically significant corresponding changes occurred in G2, G3, or venules (V). In the central artery, a small, but nevertheless significant, hemoglobin-dependent constriction occurred at the lowest flow rate, but, paradoxically, significant ($p < 0.05$) dilatation was observed at the two highest flow rates. Thus, inhibition of basal EDRF activity in G1 unmasks an intrinsic ("myogenic") constrictor response, with the result that secondary passive distension of the central artery occurs at high flow rates. This passive distension overrides the consequences of inhibition of basal EDRF activity in the central artery itself. The more complex responses that occurred in the presence of serotonin are discussed further in Fig. 3 and Table 1. In these flow experiments, venules and small veins (v) of any size again behaved passively (panels a and b).

two generations of branch vessels of the rabbit ear, but not in isolated preparations from the central artery itself (Hwa and Bevan, 1986). Interestingly, the responses of these vessels were not influenced by endothelial removal. In the present study, analagous myogenic responses occasionally resulted from increases in flow in the presence of serotonin, but were much more evident after inhibition of basal EDRF-activity (Fig. 5). Furthermore, inhibition of basal EDRF activity in resting preparations resulted in a sustained secondary constriction in G1, implying a continuous inhibition of intrinsic ("myogenic") smooth muscle tone by such basal activity (Fig. 4a). In perfused vessels there may be shear-induced amplification of basal EDRF release, and its inhibition would then be expected to result in more profound inhibiting effects on smooth muscle than in analagous experiments in which there is no flow.

4.2. Regulation of Resistance by Basal EDRF Release

Calculations based on diameter data and average vessel lengths indicate that the resistance of the rabbit ear is located mainly in the central artery and its first branch generation (Table 1). The experimental data of Fig. 6, which compare pressure drops across the intact bed and the central artery alone, support this conclusion. The geometry of the vascular network of the rabbit ear is two-dimensional, with long vessels in the proximal part of the bed. In the hind limb of the rat, however, a three-dimensional bed, most resistance is also thought to be located proximally (25% in vessels with an internal diameter of larger than 150 μm, and 75% in vessels over 50 μm in diameter) (Folkow, 1979). Basal EDRF activity was moreover shown to be maximal in those vessels that contributed most to overall resistance; namely the central artery and its first branch generation (Fig. 4, Table 1). This observation suggests that EDRF may play a central role in modulating their tone, and consequently, the overall resistance of the network. From a design point of view, this would confer obvious advantages since precise control of vessel diameter is particularly important in those vessels that regulate resistance (Speden, 1985).

4.3. Flow-Dependent EDRF-Mediated Dilatation

In resting preparations, inhibition of basal EDRF activity resulted in significant constriction of vessels in G1 at all flow rates (Fig. 4a). Flow-dependent dilatation in G1 in resting preparations, but not other vessels in the bed except G0 at low flow rates, is thus in part mediated by EDRF, although passive distensive effects also resulted from increased flow itself. Such passive distension was apparent in all vessels, including veins and venules in both resting and activated beds. Flow-dependent dilatation in

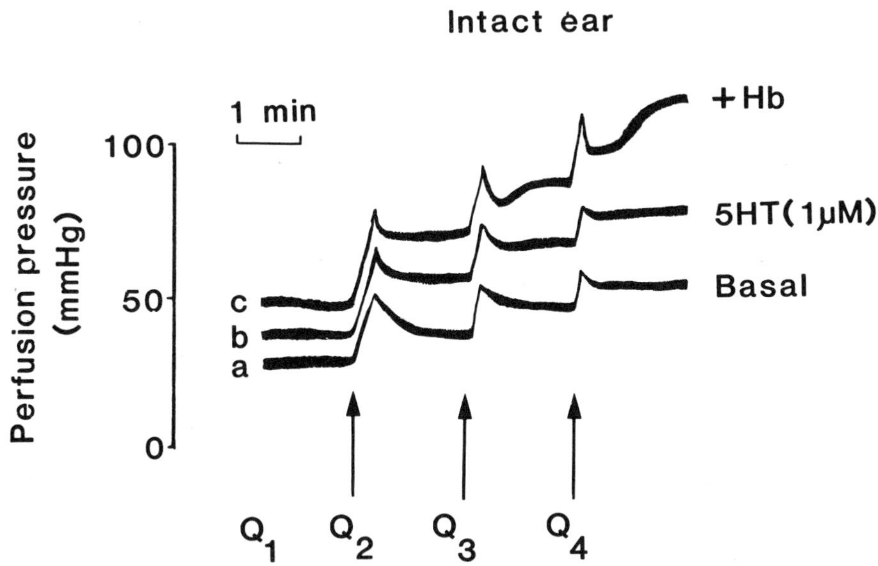

Fig. 5. Representative trace illustrating the experimental protocol used for flow studies in intact ear preparations. Four standardized flow rates were used in all experiments (Q_1 = 0.63 ± 0.08, Q_2 = 2.00 ± 0.17, Q_3 = 3.39 ± 0.18, and Q_4 = 4.70 ± 0.18 mL/min). Pressure rises resulting from positive flow increments (Q_2 to Q_4) were redrawn from the original recordings. Data was obtained (a) in the resting bed, (b) in the presence of serotonin (in this case, 1 *μM*), and also (c) after addition of 1 *μM* hemoglobin (Hb) to the buffer to inhibit basal EDRF activity. Stress–relaxation occurred in the 30 s immediately following elevation of the perfusion rates in all three traces, even when basal EDRF activity was inhibited. Note, however, that in the presence of pharmacological activation and hemoglobin, a slow, tonic rise in pressure developed after stress relaxation was complete, particularly at the higher flow rates. This was observed in about 30% of preparations, and especially in those that exhibited good contractile responses to serotonin. Such behavior represents myogenic activity of the kind described by Bayliss and was very rarely (less than 3 to 4% of all preparations) observed in the absence of hemoglobin. Basal EDRF activity can therefore inhibit expression of this fundamental property of vascular smooth muscle. Diameter measurements were made at steady-state, and it was therefore not possible to elucidate which section of the bed responded with enhanced constriction to the increase in flow. In separate experiments, however, analagous responses could be also elicited in isolated portions of the central ear artery (with side branches tied off).

canine epicardial coronary arteries in vivo depends on the presence of endothelium, and it has been speculated that the underlying mechanism is shear-dependent release of EDRF (Holtz et al., 1983a,b, 1984; Rubanyi

Table 1
Relative Hydraulic Parameters[a]

Topology				
Vessel category	G0	G1	G2	G3
Number of branches/vessel	20	3–5	3–4	2–3
Total (N)	1	20	60–100	180–400
Flow (Q)	1	0.05	0.017–0.01	0.0056–0.0025
Length (L)	1	0.15	0.05	0.02
Resting bed				
Diameter (d)	1	0.32	0.23	0.17
Shear stress	1	1.53	1.09 (1.36–0.82)	0.81 (1.1–0.51)
Reynolds number	1	0.16	0.05 (0.07–0.04)	0.02 (0.03–0.01)
Hydraulic resistance/ viscous power loss	1	0.71	0.24 (0.3–0.18)	0.09 (0.13–0.06)
Bed activated by 0.1 μM serotonin				
Diameter (d)	1	0.25	0.22	0.17
Shear stress	1	3.05	1.2 (1.5–0.9)	0.81 (1.1–0.51)
Reynolds number	1	0.2	0.06 (0.07–0.04)	0.02 (0.03–0.01)
Hydraulic resistance/ viscous power loss	1	1.9	0.27 (0.34–0.20)	0.09 (0.13–0.06)

[a]Tabulation of the important hydraulic parameters calculated on the assumption of Poiseuille flow throughout the network. These have been normalized with respect to the central artery and are therefore independent of flow rate since vessel diameters scaled geometrically as a consequence of basal EDRF activity in resting and partially activated beds (Fig. 3). Estimated maxima and minima of the flow parameters (brackets) and their mean values are given. The calculations indicate that the overall resistance of the network is mainly located in G0 and G1, with progressively smaller contributions from subsequent generations. Note that intimal shear stress is highest in G1, particularly when the bed is partially activated, and that there is a greater increase in hydraulic resistance/power loss in G1 compared to G0 after partial activation of the bed. In contrast, higher degrees of activation with 1 *μM* serotonin or 1 *μM* histamine resulted in relatively greater constriction of G0 (Figs. 1 and 3). Note also the expected rapid decline of Reynolds number in successive branch generations. Absolute Reynolds numbers for the flow in the central artery ranged from 8 to 80 and calculated absolute shear stresses in this artery corresponding to the four flow rates Q_1 to Q_4 were 1.5, 3.2, 4.0, and 4.5 N/m^2 in resting preparations. Such levels of shear stress are comparable with those that occur in the intact circulation (Dewey et al., 1981). Increased flow therefore results in sustained elevation of shear stress, in spite of flow-dependent increases in vessel caliber and the relatively more rapid inverse dependence of shear stress on diameter than direct dependence on flow (cf., $\frac{Q}{d^3}$). Similar arguments also apply in generations G1 to G3.

et al., 1986). The present data are certainly consistent with this idea, since calculated intimal shear stress was highest in G1 (Table 1), and it was in this generation of branch vessels that basal EDRF activity exerted its most marked effects.

The experimental design does not allow for distinction between shear-stimulated release of EDRF and enhanced passive distension by positive flow increments in vessels made more compliant by a constant basal release of EDRF. There is, however, evidence indicating that shear stress, particularly if pulsatile in nature, increases the basal release of prostacyclin from cultured endothelial cells (Frangos et al., 1985). A similar mechanism may also increase EDRF release (Rubanyi et al., 1986). The marked polarity of EDRF release toward the medial side of the arterial wall (Busse et al., 1985) and the inactivation of EDRF by the haptoglobin–hemoglobin complex in plasma (Edwards et al., 1985, 1986) imply that EDRF will act in vivo only on subjacent smooth muscle. Shear-induced release of EDRF would thus allow for dynamic coupling of local flow to local vessel diameter. Importantly, calculated shear stress was shown to increase with flow to a relatively greater extent than vessel diameter, implying that it can indeed function as a sustained hydrodynamic signal to endothelium (Table 1).

Changes in the cytoskeleton of endothelial cells and their geometric orientation can occur rapidly in response to shear stress (Dewey et al., 1981; Eskin et al., 1984; Franke et al., 1984). It may be speculated that not only acute changes in vascular caliber, as shown in the present study, but also the more chronic ''adaptive'' diameter changes that result from flow increases or decreases (Kamiya and Togawa, 1980; Guyton and Hartley, 1985) are somehow linked to the ambient level of shear-induced release of EDRF. Similar arguments may also be invoked to suggest a role for EDRF in the determination of vessel caliber during angiogenesis. In the present study, pressure was monitored at constant flow during various pharmacological and hydrodynamic interventions. Extrapolation of the data to the in vivo physiological situation would nevertheless imply an important contribution of EDRF to the underlying mechanisms of autoregulation, through which organs intrinsically maintain constant flow over a wide range of perfusion pressures. Reactive hyperemia would be another physiological situation in which increased distensibility of vessels, resulting from flow-stimulated release of EDRF, might be important.

4.4. Regulation of Flow Distribution by Basal EDRF Release

Basal EDRF activity was shown to preseve the geometrical similarity of internal vessel diameters over the range of flow rates studied in G0 to G3 in both resting and partially activated preparations (Fig. 3). This important property of geometrical scaling was lost, however, at high levels of pharmacological activation or if basal EDRF activity was inhibited and the ratio (branch diameter/central artery diameter) then varied in a complex way with flow. The data suggest that the topology and geometry of vessel branching are intimately linked with flow through basal EDRF release.

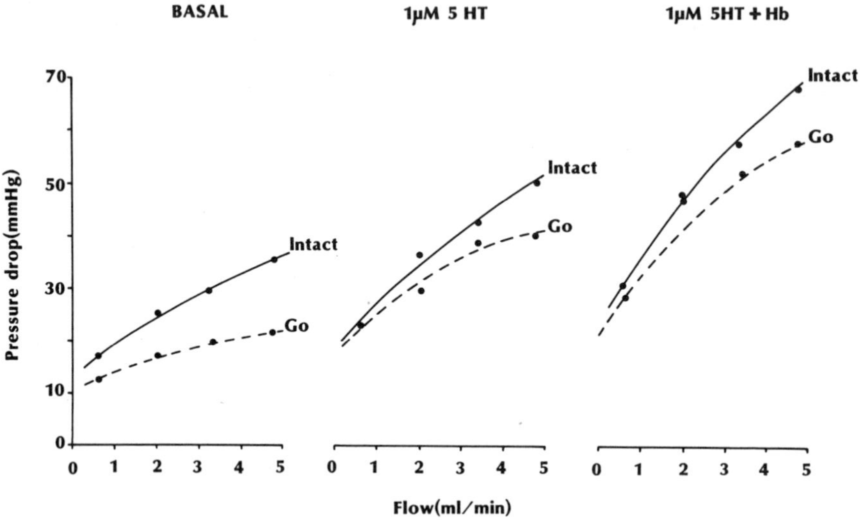

Fig. 6. Curves plotting pressure drop across the intact bed (——) and the central artery alone (G0, ---) as a function of flow rate ($n = 5$). The data were obtained in resting preparations, in the presence of 1 μM serotonin, and after addition of hemoglobin to the perfusate. The pressures were comparatively lower in this group of preparations than those illustrated in Fig. 8a. Protocol: After first obtaining data for the intact ear, the course of its central artery was identified by eye, and cuts made with a sharp scalpel blade parallel to and 2 mm from either side of its wall. No attempt was made to tie off the cut ends of the side branches. The data provide an experimental illustration of the fact that the contribution of the central artery to the overall resistance of the ear network is high. Indeed, if the cuts were made 1.5 cm to either side of the central artery, thus effectively isolating G0 and G1, pressure–flow curves imtermediate between those illustrated were obtained. This is in agreement with the conclusions of the calculations shown in Table 1; namely, that the major component of the resistance of the bed is located in proximal vessels.

From the principle of hydrodynamic similitude it follows that the flow patterns in topologically similar networks are equivalent provided that the Reynolds numbers of the flow are equal and tube diameters and lengths are scaled equally. There is, however, experimental evidence showing that the spatial distribution of flow in a series-coupled network of rigid branching tubes is virtually independent of the Reynolds number for the entry flow into the system (Isabey, 1982). By inference, if vessel diameters and lengths scaled identically with flow, the flow distribution within the network of the rabbit ear would be independent of the flow rate in the central artery. Experimentally, the internal diameters of vessels in various generations did scale geometrically in the presence of basal EDRF activity (Fig. 3). Vessel lengths did not change with flow rate however, presumably because of tethering by connective tissue. Nevertheless, since calculated entry lengths were small, Poiseuille flow would be expected to develop rapidly in each generation, and it is probable that longitudinal scaling is not a critical requirement in this system to maintain hydrodynamic similarity. It may therefore be argued that a major function of EDRF in the microcirculation is to maintain an identical spatial distribution of flow during physiological situations that result in alterations of flow rate.

4.5. Influence of Basal EDRF Release on Power Losses and Pressure–Flow Characteristics

Viscous power losses incurred at different flow rates, calculated on the basis of Poiseuille flow, were found to depend linearly on flow in both unstimulated and partially activated arteries in generations G0 to G3 as a consequence of basal EDRF activity (Fig. 7). The important theoretical implication of this finding is that the Poiseuille energy loss expended in pumping a unit volume of fluid through the network is independent of the flow rate. In highly activated beds, there was a nonlinear relationship, even in the presence of basal EDRF activity. The linear relationships of Fig. 7 can be expressed as

$$W = \frac{128\mu L}{\pi} \frac{q^2}{d^4} = aq - b$$

where a and b are positive constants. Since the pressure gradient in Poiseuille flow, $\frac{dp}{dl}$, is proportional to q/d^4, it follows that, in the central artery and first three generations of branch vessels,

$$\frac{dp}{dl} \propto \frac{q}{d^4} \propto a - b/q$$

This implies that as flow increases $\frac{dp}{dl} \rightarrow a$, and thus that the pressure gradient in vessels in G0 and G3 becomes independent of the flow rate. It also indicates that a fourth power law will then relate flow to diameter, i.e., $q \propto d^4$.

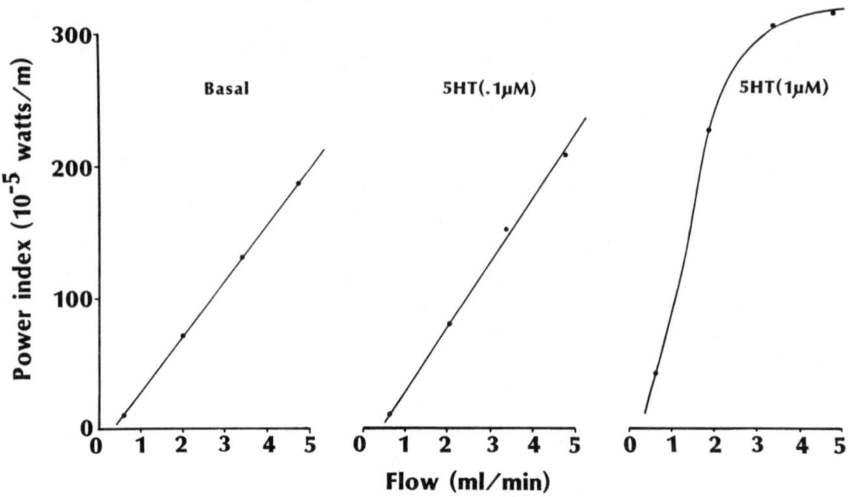

Fig. 7. Viscous power losses per unit length of the central ear artery, in the presence of basal EDRF activity, calculated as $\frac{1}{N}\frac{128\mu L}{\pi}\frac{Q^2}{d^4}$ from the data of Fig. 4, plotted against flow in the resting bed (left) and when the bed was activated by 0.1 or 1 μM serotonin (middle and right, respectively). In resting or partially activated preparations, power losses are therefore a linear function of flow, and since diameters scaled geometrically with flow (Fig. 3), this will also apply in generations G1, G2, and G3. When basal EDRF activity was inhibited, however, the curves exhibited a complex nonlinear dependency on flow. Note the very steep increases in viscous power dissipation when the bed is activated by 1 μM serotonin, although passive distensive effects come into play at high flow rates, and the curve correspondingly becomes concave toward the flow axis.

Experimentally, the rises in pressure that resulted from increased flow in intact preparations were attenuated by basal EDRF activity, and markedly so if the bed was activated (Fig. 8a). Indeed, after inhibition of basal EDRF activity by hemoglobin, pressure–flow curves obtained in the presence of serotonin became straight lines, indicating a complete loss of compliance and that the network as a whole then behaved as a rigid tube. The complex influence of anastomoses on the properties of this particular vascular bed appear to be of minor importance since they occurred in high-order generations that do not contribute greatly to the overall resistance of the network (Table 1). A contribution to the shape of the pressure–flow curves (Fig. 8a) may, however, result from the enhanced viscous losses that occur at branching points (Pedley et al., 1971), particularly those between G0 and G1, since the Reynolds number of flow in G0 ranged from 8 to 80. The influence of undeveloped flow in the entrance region of higher-order branch vessels is probably of less importance, since Reynolds numbers

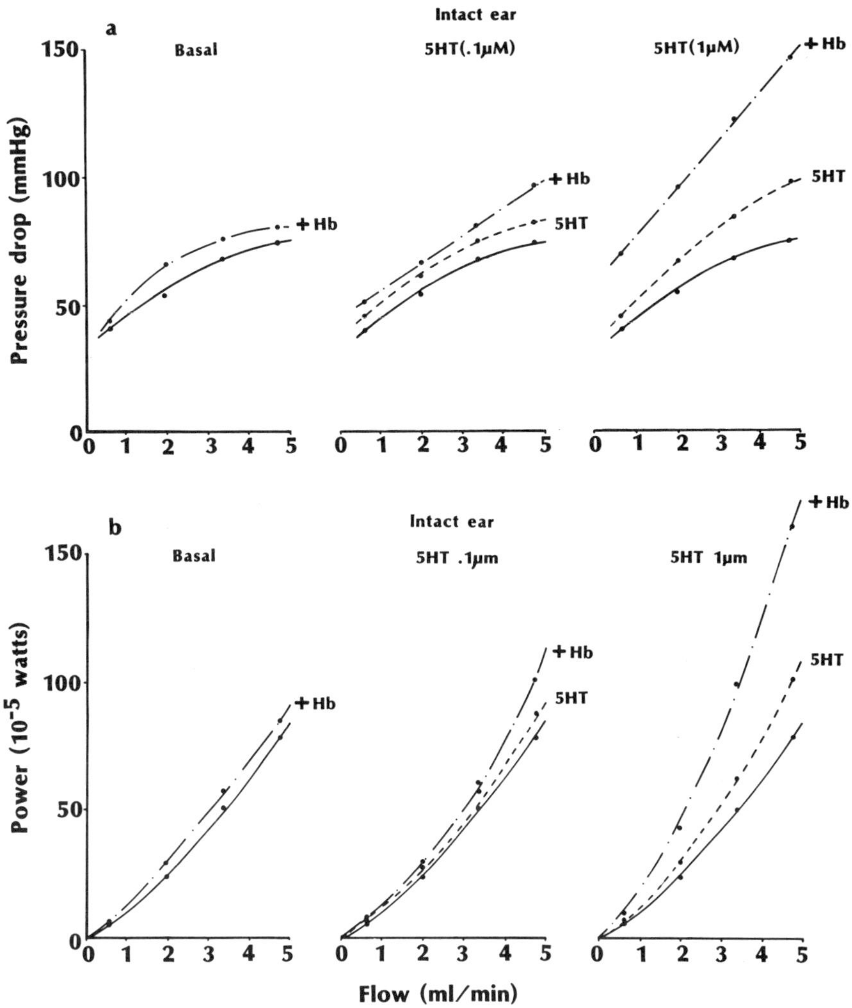

Fig. 8. (a) Perfusion pressure plotted as a function of flow in the intact ear. Data are shown for the resting bed (n = 20, left) and after activation by 0.1 or 1 μM serotonin (n = 8 in each case, middle and right, respectively). Inhibition of basal EDRF activity increased the pressure drop across the bed, and the convex upward curves then become straight lines if serotonin was simultaneously present to activate the smooth muscle. Such behavior implies a loss of compliance in the network secondary to inhibition of basal EDRF activity. Correction was made for the variation of pressure drop across the input cannula with flow rate before plotting the data. (b) Power losses (calculated as flow × pressure) derived from pressure–flow curves in intact preparations, plotted as a function of flow rate. Losses were increased by inhibition of basal EDRF activity, particularly if high tone was induced by 1 μM serotonin.

were then less than 1 (Table 1), and therefore inertial considerations are relatively unimportant (Lew and Fung, 1969).

The finding that diameter and flow are approximately related by a fourth-power law, albeit in an in vitro preparation perfused at steady flow, does not support application of the principle of optimality of minimal work to overall considerations of design in that part of the bed-regulating resistance, which has been the experimental focus of interest. Theoretically, application of this principle leads to the conclusion that intimal shear stress $\left(\frac{32\mu}{\pi} \cdot \frac{q}{d^3}\right)$ is constant throughout the vascular system and, thus, that a third-power law relates diameter to flow (Murray, 1926). Indeed, there is some experimental evidence to support this idea in vessels of a diameter smaller than 100 μm (Mayrovitz and Roy, 1983). Such small vessels, however, will presumably have design features primarily geared to the delivery of oxygen and metabolic substrates that may be quite different from those required to provide precise regulation of vascular resistance. The data of Fig. 8, nevertheless, indicate that pressure rises resulting from flow increases, as well as the associated viscous losses, were reduced in intact preparations as a result of the flow-dependent distensibility that results from basal EDRF activity. Important physiological correlates in the intact circulation would be that basal EDRF release will reduce both the pressure required to achieve adequate perfusion of the organs and the work subsequently expended. Endothelium-dependent responses elicited pharmacologically in small mesenteric vessels of the rat are impaired by hypertension (De Mey and Gray, 1985). If such pathological processes also affect basal EDRF activity in resistance vessels, and thus their flow-dependent distensibility, adverse effects on perfusion would be expected.

5. Concluding Remarks

The vasodilator action of acetylcholine and substance P was shown to be mediated by the release of EDRF in small arteries and arterioles of the rabbit ear down to 25 μm in size, analagous to findings in larger, conduit vessels of the same species. It is thus theoretically possible that such neurotransmitters influence microvascular tone secondary to an interaction with endothelial cells. Furthermore, the underlying functional sensitivity of the important resistance vessels of the rabbit ear, to both stimulated EDRF release and glyceryl trinitrate, was similar. This apparent uniformity of response is probably dependent on the spatial distribution of circumferential wall stress within the network, as well as mechanisms involving activation of guanylate cyclase, and is thus in part determined by geometry and topology.

Basal release of EDRF appears to be an important regulator of flow, since its influence was greatest in those vessels in which both shear stress and hydraulic resistance were maximal, and it is capable of inhibiting intrinsic ("myogenic") smooth muscle tone. It may also be responsible for maintaining the spatial homogeneity of flow distribution over a range of different flow rates. Experimentally, perfusion pressure and power losses were both reduced by basal EDRF release. Calculations also indicated that the Poiseuille energy loss expended in pumping unit volume of fluid through the bed was independent of flow rate. Basal EDRF release, probably amplified by shear stress, may therefore provide an important integrating link between flow and vascular topology and geometry, thus optimizing overall perfusion characteristics over a wide range of flow rates.

Caution must be exercised before extrapolation from in vitro studies that attempt to isolate a specific physiological mechanism, to the in vivo situation. Nevertheless, the principles that have emerged from the present study suggest a central role for EDRF in regulating blood flow in resistance vessels.

Acknowledgments

This work was supported by the British Heart Foundation. We are indebted to Mrs. E. Walker and Miss A. Davies for secretarial assistance, Mr. M. Stanton, Mr. N. Garrahan, and Professor E. D. Williams, Department of Pathology, University of Wales College of Medicine, for the use of image analysis facilities and computer programming, and Dr. D. A. Hutton, Department of Haematology, Cardiff Royal Infirmary, for viscosity measurements. The authors would also like to thank Mrs. M. Griffith for help with preparation of the manuscript.

References

Andersson, P. O., Bloom, S. R., and Jarhult, J.: Colonic motor and vascular response to pelvic nerve stimulation and their relation to local peptide release in the cat. *J. Physiol.* (Lond.) **334:**293–307, 1983.

Andersson, P. O., Bloom, S. R., and Mellander, S.: Haemodynamics of pelvic nerve induced penile erection in the dog: Possible mediation by vasoactive intestinal polypeptide. *J. Physiol.* (Lond.) **350:**209–224, 1984.

Angus, J. A., Campbell, G. R., Cocks, T. M., and Manderson, J. A.: Vasodilatation by acetylcholine is endothelium-dependent: A study by sonomicrometry in canine femoral artery in vivo. *J. Physiol* (Lond.) **334:**209–222, 1983.

Azuma, T. and Oka, S.: Mechanical equilibrium of blood vessel walls. *Am. J. Physiol.* **221:**1310–1318, 1971.

Bayliss, W. M.: On the local reactions of the arterial wall to changes of internal pressure. *J. Physiol.* (Lond.) **28:**220–231, 1902.

Bellman, S.: Microangiography. *Acta Radiol.* (suppl.) 102:7–104, 1953.

Brain, S. D., Williams, T. J., Tippins, J. R., Morris, H. R., and MacIntyre, I.: Calcitonin gene-related peptide is a potent vasodilator. *Nature* **313:**54–56, 1985.

Burger, W., Kellner, C., and Busse, R.: Effects of active and passive wall stress changes on the rhythmic mechanical activity of the pressurised rat tail artery. *Blood Vess.* **21:**231–245, 1984.

Busse, R., Trogisch, G., and Bassenge, E.: The role of endothelium in the control of vascular tone. *Bas. Res. Cardiol.* **80:**475–490, 1985.

Cauvin, C., Saida, K., and van Breemen, C.: Effects of Ca antagonists on Ca fluxes in resistance vessels. *J. Cardiovasc. Pharmacol.* **4:**S287–S290, 1982.

Cocks, T. M. and Angus, J. A. Endothelium-dependent relaxation of coronary arteries by noradrenaline and serotonin. *Nature* **305:**627–30, 1983.

Cocks, T. M. and Angus, J. A.: Bioassay of the Release of Endothelium-Derived Relaxing Factor (EDRF) from Isolated Endothelial Cells In Vitro, in *Vascular Neuroeffector Mechanisms* (Bevan, J. A., Godfraind, T., Maxwell, R. A., Stoclet, J. C., and Worcel, M., ed.) Elsevier, Amsterdam, New York, Oxford, 1985.

Cox, R. H.: Comparison of arterial wall mechanics using ring and cylindrical segments. *Am. J. Physiol.* **244:**H298–H303, 1983.

Davies, R. L1.: Contact microradiography in bio-medical research. Doctoral thesis. University of Wales, Cardiff, UK, 1983.

Davies, R. L1., Flores, N. A., and Evans, K. T.: Development and assessment of an image intensifier for real-time x-ray microscopy. *Br. J. Radiol.* **59:**273–276, 1986.

Davies, J. M. and Williams, K. I.: Endothelial-dependent relaxant effects of vasoactive intestinal polypeptide and archidonic acid in rat aortic strips. *Prostaglandins* **27:**195–202, 1984.

De Mey, J. G. and Gray, S. D.: Endothelium-dependent reactivity in resistance vessels. *Prog. Appl. Microcirc.* **8:**181–187, 1985.

Dewey, C. F., Bussolari, S. R., Gimbrone, M. A., and Davies, P. F.: The dynamic response of vascular endothelial cells to fluid shear stress. *J. Biomech. Eng.* **103:**177–185, 1981.

Diamond, J. and Chu, E. U.: Possible role for cyclic GMP in endothelium-dependent relaxation of rabbit aorta by acetylcholine. Comparison with nitroglycerin. *Res. Comm. Chem. Pathol. Pharmacol.* **41:**369–381, 1983.

D'Orleans-Juste, P., Dion, S., Mizrati, J., and Regoli, D.: Effects of peptides and non-peptides on isolated arterial smooth muscle: Role of endothelium. *Eur. J. Pharmacol.* **114:**9–21, 1985.

Edwards, D. H., Griffith, T. M., Henderson, A. H., Lewis, M. J., and Ryley, H. C.: Endothelium-dependent relaxation is inhibited by a high molecular weight protein fraction of whole human plasma. *Br. J. Pharm.* **85:**341P, 1985.

Edwards, D. H., Griffith, T. M., Ryley, H. C., and Henderson, A. H.: Haptoglobin-haemoglobin complex in human plasma inhibits endothelium-dependent relaxation: Evidence that endothelium derived relaxing factor acts as a local autocoid. *Cardiovasc. Res.* **20:**549–556, 1986.

Ely, R. V.: *Microfocal Radiography.* Academic, London, New York, 1980.

Eskin, S. G., Ives, C. L., McIntire, L. V., and Navarro, L. T.: Response of cultured endothelial cells to steady flow. *Microvasc. Res.* **28:**87–94, 1984.

Folkow, B.: Description of the myogenic hypothesis. *Circ. Res.* **15** (suppl. I): 279–287, 1964.

Folkow, B.: Relationships between vessel design and hemodynamics along the precapillary resistance compartment in normo and hypertension. *Blood Vess.* **16:**277–280, 1979.

Frangos, J. A., Eskin, S. G., McIntire, L. V., and Ives, C. L.: Flow effects on prostacyclin production by cultured human endothelial cells. *Science* **227:**1477–1479, 1985.

Franke, R-P., Grafe, M., Schnittler, H., Seiffge, D., and Mittermayer, C.: Induction of human vascular endothelial stress fibres by fluid shear stress. *Nature* **307:**648–649, 1984.

Furchgott, R. F.: Role of endothelium in responses of vascular smooth muscle. *Circ. Res.* **53:**557–573, 1983.

Furchgott, R. F. and Zawadzki, J. V. : The obligatory role of endothelial cells in the relaxation of arterial smooth muscle by acetylcholine. *Nature* **288:**375–376, 1980.

Gerova, M., Gero, J., Barta, E., Dolezel, S., Smiesko, V., and Levicky, V.: Neurogenic and myogenic control of conduit coronary arteries, a possible interference. *Bas. Res. Cardiol.* **76:**503–507, 1981.

Gore, R. W.: Wall stress: A determinant of regional differences in response of frog microvessels to norepinephrine. *Am. J. Physiol.* **222:**82–91, 1972.

Griffith, T. M.: Studies of endothelium-derived relaxant factor (EDRF), its nature and mode of action. *Eur. Heart J.* **6:**37–49, 1985.

Griffith, T. M., Edwards, D. H., Lewis, M. J., Newby, A., C., and Henderson, A. H.: The nature of endothelium derived vascular relaxant factor. *Nature* **308:**645–647, 1984a.

Griffith, T. M., Henderson, A. H., Hughes Edwards, D., and Lewis, M. J.: Isolated perfused rabbit coronary artery and aortic strip preparations; The role of endothelium-derived relaxant factor. *J. Physiol.* (Lond.) **351:**13–24, 1984b.

Griffith, T. M., Edwards, D. H., Lewis, M. J., and Henderson, A. H. : Evidence that cyclic guanosine monophosphate (cGMP) mediates endothelium dependent relaxation. *Eur. J. Pharm.* **112:**195–202, 1985.

Griffith, T. M., Edwards, D. H., Newby, A. C., Lewis, M. J., and Henderson, A. H.: Production of endothelium derived relaxant factor is dependent on oxidative phosphorylation and extracellular calcium. *Cardiovasc. Res.* **20:**7–12, 1986.

Guyton, J. R. and Hartley, C. J.: Flow restriction of one carotid artery in juvenile rats inhibits growth of arterial diameter. *Am. J. Physiol* **248:**H540–H546, 1985.

Habib, J. B., Wells, S. L., Williams, C. L., and Henry, P. D. Atherosclerosis impairs endothelium dependent arterial relaxation. *Circulation* **70:**(suppl.) 11:123, 1984.

Hammerson, F. and Hammerson, E.: Some structural and functional aspects of endothelial cells. *Bas. Res. Cardiol.* **80:**491–501, 1985.

Hilton, S. M.: A peripheral arterial conducting mechanism underlying dilatation of the femoral artery and concerned in functional vasodilatation in skeletal muscle. *J. Physiol. (Lond.)* **149:**93–111, 1959.

Holtz, J., Busse, R., and Giesler, M.: Flow-dependent dilation of canine epicardial coronary arteries in vivo and in vitro: Mediated by the endothelium *Naunyn Schmiedebergs Arch. Pharmacol.* **332:**R44, 1983a.

Holtz, J., Giesler, M., and Bassenge, E.: Two dilatory mechanisms of anti-anginal drugs on epicardial coronary arteries in vivo: Indirect, flow-dependent, endothelium-mediated dilation and direct smooth muscle relaxation. *Z. Cardiol.* **72** (suppl.3): 98–106, 1983b.

Holtz, J., Forstermann, U., Pohl, U., Giesler, M., and Bassenge, E.: Flow-dependent, endothelium-mediated dilation of epicardial coronary arteries in conscious dogs: Effects of cyclooxygenase inhibition. *J. Cardiovasc. Pharmacol.* **6:**1161–1169, 1984.

Holzmann, S.: Endothelium-induced relaxation by acetylcholine associated with larger rises in cyclic GMP in coronary arterial strips. *J. Cyclic. Nucl. Res.* **8:**409–419, 1982.

Hwa, J. J. and Bevan, J. A.: Stretch dependent (myogenic) tone in rabbit ear resistance arteries. *Am. J. Physiol.* **250:**H87–H95, 1986.

Isabey, D.: Steady and pulsatile flow distribution in a multiple branching network with physiological applications. *J. Biomech.* **15:**395–404, 1982.

Jarasch, E. D., Grund, C., Bruder, G., Heid, H. W., Keenan, T. W., and Franke, W. W.: Localisation of xanthine oxidase in mammary gland epithelium and capillary endothelium. *Cell* **25:**67–82, 1981.

Johnson, P. C.: The Myogenic Response, in *Handbook of Physiology* vol. 2 (Bohr, D., ed.) Waverly, Maryland, 1980.

Kalsner, S.: Cholinergic mechanisms in human coronary artery preparations: Implications of species differences. *J. Physiol.* (Lond.) **358:**509–526, 1985.

Kamiya, A. and Togawa, T.: Adaptive regulation of wall shear stress to flow change in the canine carotid artery. *Am. J. Physiol.* **239:**H14–21, 1980.

Katusic, Z. S., Shepherd, J. T., and Vanhoutte, P. M.: Vasopressin causes endothelium-dependent relaxations of the canine basilar artery. *Circ. Res.* **55:**575–579, 1984.

Konishi, M. and Su, C.: Role of endothelium in dilator responses of spontaneously hypertensive rat arteries. *Hypertension* **5:**881–886, 1983.

Krishnakumar, C. K., Rovick, A. A., and Lavan, Z.: The effect of pressure pulsatations on time mean flow rate. *Microvasc. Res.* **11:**41–49, 1976.

Lew, H. S. and Fung, Y. C.: On the low-Reynolds-number entry flow into a circular cylindrical tube. *J. Biomech.* **2:**105–119, 1969.

Martin, W., Villani, G. M., Jothianandan, D., and Furchgott, R. F.: Selective blockade of endothelium-dependent and glyceryl trinitrate induced relaxation by hemoglobin and by methylene blue in the rabbit aorta. *J. Pham. Exp. Ther.* **232:**708–716, 1985.

Mayrovitz, H. N. and Roy, J.: Microvascular blood flow: Evidence indicating a cubic dependence on arteriolar diameter. *Am. J. Physiol.* **245:**H1031–H1038, 1983.

Mulvey, R. and Newman, C. D.: Scanning electron microscopy-systems and applications. *Inst. Phys. Conf. Series* 16-21, 1973.

Murray, C. D.: The physiological principle of minimum work. *Proc. Natl. Acad. Sci. USA* **12:**207–214, 1926.

Nilsson, H. and Sjoblom, N.: Distension-dependent changes in noradrenaline sensitivity in small arteries from the rat. *Acta. Physiol. Scand.* **125:**429–435, 1985.

Owen, M. P. and Bevan, J. A.: Acetylcholine induced endothelial-dependent vasodilation increases as artery diameter decreases in the rabbit ear. *Experentia* **41:**1057–1058, 1985.

Pedley, T. J., Schroter, R. C., and Sudlow, M. F.: Flow and pressure drop in systems of repeatedly branching tubes. *J. Fluid. Mech.* **46:**365–383, 1971.

Price, J. M., Davis, D. L., and Knauss, E. B.: Length-dependent sensitivity at lengths greater than L max in vascular smooth muscle. *Am. J. Physiol.* **245:**H379–H384, 1983.

Rapoport, R. M., Draznin, M. B., and Murad, F.: Endothelium-dependent relaxation in rat aorta may be mediated through cyclic GMP dependent protein phosphorylation. *Nature* **306:**174–176, 1983.

Rubanyi, G. M., Romero, J. C., and Vanhoutte, P. M.: Flow-induced release of endothelium-derived relaxing factor. *Am. J. Physiol.* **250:**H1145–H1149, 1986.

Schnaar, R. L. and Sparks, H. V.: Response of large and small coronary arteries to nitroglycerin, $NaNO_2$ and adenosine. *Am. J. Physiol.* **223:**223–228, 1972.

Schretzenmayr, A.: Uber kreislaufregulatorische Vorgange an den grossen Arterien bei der Muskelarbeit. *Pfluegers Arch. Ges. Physiol.* **232:**743–748, 1933.

Sparks, H. V.: Effect of Local Metabolic Factors on Vascular Smooth Muscle, in *Handbook of Physiology* vol. 2. (Bohr, D., ed.) Waverly, Maryland, 1980.

Speden, R. N.: Active reactions of the rabbit ear to distension. *J. Physiol.* (Lond.) **351:**631–643, 1984.

Speden, R. N.: The use of excised, pressurized blood vessels to study the physiology of vascular smooth muscle. *Experentia* **41:**1026–1028, 1985.

Sutter, M. C., Hallback, M., Jones, J. V., and Folkow, B.: Contractile responses to noradrenaline: Varying dependance on external calcium of consecutive vascular segments of perfused rat hindquarters. *Acta. Physiol. Scand.* **99:**166–172, 1977.

Weihe, E., Reinecke, M., and Forssmann, W. G.: Distribution of vasoactive intestinal polypeptide-like immunoreactivity in the mammalian heart. *Cell. Tissue. Res.* **236:**527–540, 1984.
Winbury, M. M., Howe, B. B., and Hefner, M. A.: Effects of nitrates and other coronary dilators on large and small coronary vessels: An hypothesis for the mechanism of action of nitrates. *J. Pharm. Exp. Ther.* **168:**70–95, 1969.
Zaidi, M., Bevis, P. J. R., Girgis, S. I., Lynch, C., Stevenson, J. C., and MacIntyre, I.: Circulating CGRP coes from the perivascular nerves. *Eur. J. Pharm.* **117:**283–284, 1985.

Chapter 20

Endothelium-Dependent Vasodilatation in the Cerebral Microcirculation

Hermes A. Kontos

1. Introduction

Large and small cerebral arteries dilate in response to acetylcholine both in vitro and in vivo (Kuschinsky et al., 1974; Owman et al., 1974; Lee, 1982; Kontos et al., 1984). This response is caused by stimulation of muscarinic receptors, and it is blocked by atropine (Kuschinsky et al., 1974; Lee, 1982). With the single exception of the posterior auricular artery of the cat (Brayden and Bevan, 1985), in all other arteries tested the vasodilator response to acetylcholine is endothelium-dependent (Furchgott, 1983). In all these vessels it was shown that acetylcholine produces vasodilatation by an indirect mechanism mediated by an initial action on the endothelium, which results in the release of an unidentified vasodilator agent, endothelium-derived relaxant factor (EDRF), which then diffuses to the vascular smooth muscle and causes relaxation (Furchgott, 1983). The vasodilator responses to acetylcholine in the cerebral vessels, therefore, suggest strongly the presence of endothelium-dependent vasodilatation.

Considered below is the available evidence concerning endothelium-dependent vasodilatation in the cerebral microcirculation. It should be noted that studies of endothelium-dependent vasodilatation in the cerebral microcirculation are few and have originated from a limited number of laboratories.

2. Endothelium-Dependent Vasodilatation from Acetylcholine in Cerebral Microvessels

Removal of the endothelium is the traditional technique that has been used in large vessels to demonstrate the presence of endothelium-dependent vasodilatation. This technique has not yet been successfully applied to cerebral microvessels. Several other techniques, however, that injure the endothelium have been shown to eliminate or reverse the vasodilator response to acetylcholine in cerebral arterioles and have therefore provided evidence for the presence of endothelium-dependent vasodilatation in these vessels.

Acute severe hypertension induced for 15 min by intravenous norepinephrine in anesthetized cats converts the vasodilator effect of topical application of acetylcholine on cerebral arterioles to vasoconstriction (Wei et al., 1985). Ultraviolet (UV) light in the presence of intravascular fluorescein, a photosensitizing dye, reserved the vasodilator effect of acetylcholine on the cerebral arterioles of cats and mice (Rosenblum, 1985, 1986b; Rosenblum et al., 1985); the vasodilatation in response to arterial hypercapnia or topical application of histamine in the cat, and the vasodilatation from topical application of sodium nitroprusside or papaverine in the mouse are preserved, showing that the effect on the response to acetylcholine is not caused by a nonspecific depression of all vasodilator influences. Light from a helium-neon laser in the presence of intravascular Evans blue dye reserves the vasodilatation of cerebral arterioles to acetylcholine in mice (Rosenblum, 1986a). Topical application of arachidonate (200 μg/mL) on the brain surface of cats abolishes or reserves the vasodilatation of cerebral arterioles from topical application of acetylcholine (Wei et al., 1986).

All these techniques have the following features in common. They induce relatively minor endothelial damage, but no denudation (Kontos et al., 1980, 1981; Povlishock et al., 1983; Wei et al., 1985). After such interventions, the endothelium displays focal lesions; on transmission electron microscopy, these appear as vacuoles, probably representing swelling of the endoplasmic reticulum or localized destruction of the luminal cell membrane. On scanning electron microscopy, the vacuoles appear as dome-like protrusions into the lumen, and the lesions of the cell membrane appear as craters. These two types of lesions may be different stages of the same destructive process. At times, no demonstrable histological lesions are seen in arterioles whose response to acetylcholine is reversed.

Three of the techniques that reverse the response to acetylcholine in cerebral arterioles, namely, topical application of arachidonate (Kontos et al., 1985), acute severe hypertension (Wei et al., 1985), and UV radiation in the presence of photosensitizing dye (Rosenblum and El-Sabban, 1982), have been shown to generate oxygen radicals. The generation of superoxide anions with probable subsequent production of hydrogen peroxide and hydroxyl radical have been shown to occur after these interventions by either direct demonstration of superoxide in the extracellular space of the brain (Kontos et al., 1985; Wei et al., 1985) or by pharmacological experiments using scavengers (Kontos et al., 1984; Wei et al., 1985; Rosenblum and El-Sabban, 1982). The laser technique has not been studied in this respect, but the generation of radicals is well known to occur in photosensitizing reactions in general (Beauchamp and Fridovich, 1971). It is of interest that other agents, such as hemoglobin or methylene blue, which generate superoxide (McCord and Fridovich, 1970; Misra and Fridovich, 1971), are known to eliminate endothelium-dependent vasodilatation from acetylcholine in large vessels in vitro (Furchgott et al., 1984).

These findings support several conclusions. First, endothelium-dependent vasodilatation from acetylcholine is present in cerebral arterioles. Second, the endothelium need not be destroyed completely to eliminate endothelium-dependent vasodilatation; relatively minor injury suffices. Third, oxygen-derived free radicals may be the mediators of the endothelial injury and of the elimination of endothelium-dependent vasodilatation in several abnormal experimental situations.

3. Endothelium-Dependent Vasodilatation from Agents Other than Acetylcholine

Exposure to UV or to laser radiation, in the presence of photosensitizing dyes in the intravascular compartment, reverses the vasodilatation from bradykinin in cerebral arterioles of mice (Rosenblum, 1986a,b). Only vessels that are illuminated are affected; those outside the field of illumination retain their vasodilator responses to bradykinin.

Following acute hypertension in cats, vasodilator responses to bradykinin and arachidonate in cerebral arterioles are frequently eliminated (Kontos and Wei, unpublished observations).

These findings suggest that in cerebral arterioles of the species studied, vasodilator responses to bradykinin and arachidonate are endothelium-dependent.

4. Reversibility of Elimination of Endothelium-Dependent Vasodilatation in Cerebral Arterioles

In cats subjected to the UV light plus dye procedure, the loss of endothelium-dependent vasodilatation caused by acetylcholine is followed for 3.5 h (Rosenblum, Wei, Povlishock, and Kontos, unpublished observations). Most arterioles in which the vasodilator response to acetylcholine is eliminated shortly after exposure to the light continue to display abnormal responses for the duration of the experiment. In some arterioles, however, the vasodilator responses to acetylcholine returns spontaneously. At this time, these arterioles continue to show minor endothelial damage without progression.

In cats in which endothelium-dependent vasodilatation is eliminated after acute hypertension, topical application of superoxide dismutase and catalase reestablishes the vasodilator response to acetylcholine in the majority of arterioles (Wei et al., 1985). These enzymes scavenge superoxide and hydrogen peroxide, respectively. There is no difference in the density of endothelial lesions determined by scanning electron microscopy between those arterioles in which the response to acetylcholine remains vasoconstrictor after application of superoxide dismutase and catalase and those arterioles in which vasodilator responses to acetylcholine are reestablished. These results are interpreted to indicate that the radicals generated by acute hypertension destroy the EDRF released by acetylcholine.

It is clear from these findings that the elimination of endothelium-dependent vasodilatation can be reversible or irreversible. The reestablishment of the vasodilator response after superoxide dismutase and catalase may be explained by scavenging of oxygen-derived free radicals whose production continues long after acute hypertension subsided (Wei et al., 1985). In support of this last conclusion, it was shown that superoxide anions continue to be produced and enter the extracellular space of the brain at least 1 h after an acute hypertensive episode has subsided (Wei et al., 1985).

The radical responsible for the destruction of EDRF is likely to be the hydroxyl radical. This is based on the finding that the reversal of the vasodilator response to acetylcholine following topical application of arachidonate is prevented by simultaneous application of deferoxamine (Wei et al., 1986). This agent scavenges iron, which is essential for the production of hydroxyl radical via the iron-catalyzed Haber-Weiss reaction (McCord and Day, 1978).

In the arterioles in which superoxide dismutase and catalase were ineffective in reestablishing the vasodilator response to acetylcholine, the loss of endothelium-dependent vasodilatation can possibly be caused by the inability of the damaged endothelium to produce EDRF. This, however, could not be confirmed histologically by showing more severe damage in such vessels. Hence, other unknown factors may play a role.

5. Nature of EDRF in the Cerebral Microcirculation

In anesthetized cats, the arteriolar dilatation from topical application of arachidonate or bradykinin is blocked by cyclooxygenase inhibitors or by topical application of superoxide dismutase (Wei et al., 1980; Kontos et al., 1984). In rabbits, the cerebral arteriolar dilatation caused by bradykinin or kallikrein, an enzyme that produces bradykinin from precursors, is blocked by cyclooxygenase inhibitors (Kamitani et al., 1985) and by superoxide dismutase and catalase (Ellis, unpublished observations). These results suggest that the EDRF produced by arachidonate and bradykinin in these species is an oxygen-derived free radical generated from the metabolism of arachidonate via cyclooxygenase. Cyclooxygenase has recently been shown to produce superoxide in vitro in the presence of NADH or NADPH (Kukreja et al., 1986).

The vasodilatation in response to acetylcholine in cerebral arterioles is not affected by superoxide dismutase and catalase (Kontos et al., 1984). Hence, the EDRF induced by acetylcholine is not an oxygen-derived free radical and must therefore be different from that induced by arachidonate and bradykinin. In vitro studies in large vessels show that the EDRF from acetylcholine is an unstable substance of uncertain identity (Griffith et al., 1984). The fact that it is destroyed by a hydroxyl radical provides no clue about its identity, since this radical is a very reactive powerful oxidant, which oxidizes a large number of compounds.

6. Significance

The significance of endothelium-dependent vasodilatation in the cerebral microcirculation derives from the following considerations: It is likely that endothelium-dependent vasodilatation may be involved in physiological vasodilator responses in the cerebral circulation. The available evidence suggests that the vasodilator response to arterial hypercapnia in cerebral arterioles is not endothelium-dependent. On the other hand, the finding that the dilatation caused by reduction in arterial blood pressure is less pronounced in arterioles that displayed vasoconstriction after acute hypertension and application of superoxide dismutase and catalase than in those that showed vasodilatation under these conditions (Wei et al., 1985), suggests that endothelium-dependent vasodilatation may be involved in this response. This difference, however, may be related to the fact that arterioles that show vasoconstriction from acetylcholine are also more dilated as a result of acute hypertension, and hence closer to their maximal possible caliber.

Loss of endothelium-dependent vasodilatation may render cerebral microvessels more susceptible to vasoconstrictor influences. It was shown that cerebral arteries in vitro displayed more pronounced vasoconstrictor responses to norepinephrine after the endothelium was removed (Sercombe et al., 1985). Such circumstances may predispose to severe vasoconstriction with consequent ischemia.

Acknowledgments

Supported by grants HL21851 and NS19316 from the National Institutes of Health.

References

Beauchamp, C. and Fridovick, I.: Superoxide dismutase: Improved assays and an assay applicable to acrylamide gels. *Anal. Biochem.* **44:**276–287, 1971.

Brayden J. E. and Bevan, J. A.: Neurogenic muscarinic vasodilation in the cat. An example of endothelial cell-independent cholinergic relaxation. *Circ. Res.* **56:**205–211, 1985.

Furchgott, R. F.: Role of endothelium in responses of vascular smooth muscle. *Circ. Res.* **53:**557–573, 1983.

Furchgott, R. F., Cherry, P. D., Zawadzki, J. V., and Jothianandan, D.: Endothelial cells as mediators of vasodilation in arteries. *J. Cardiovasc. Pharmacol.* **6:**S336–S343, 1984.

Griffith, T. M., Edwards, D. H., Lewis, M. J., Newby, A. C., and Henderson, A. H.: The nature of endothelium-derived vascular relaxant factor. *Nature* **308:**645–647, 1984.

Kamitani, T., Little, M. H., and Ellis, E. F.: Evidence for a possible role of the brain kallikrein-kinin system in the modulation of the cerebral circulation. *Circ. Res.* **57:**545–552, 1985.

Kontos, H. A., Wei, E. P., Povlishock, J. T., Dietrich, W. D., Magiera, C. J., and Ellis, E. F.: Cerebral arteriolar damage by arachidonic acid and prostaglandin G_2. *Science* **209:**1242–1245, 1980.

Kontos, H. A., Wei, E. P., Dietrich, W. D., Navari, R. M., Povlishock, J. T., Ghatak, N. R., Ellis, E. F., and Patterson, J. L., Jr.: Mechanism of cerebral arteriolar abnormalities after acute hypertension. *Am. J. Physiol.* **240:** H511–H527, 1981.

Kontos, H. A., Wei, E. P., Povlishock, J. T., and Christman, C. W.: Oxygen radicals mediate the cerebral arteriolar dilation from arachidonate and bradykinin in cats. *Circ. Res.* **55:**295–303, 1984.

Kontos, H. A., Wei, E. P., Ellis, E. F., Jenkins, L. W., Povlishock, J. T., Rowe, G. T., and Hess, M. L.: Appearance of superoxide anion radical in cerebral extracellular space during increased prostaglandin synthesis in cats. *Circ. Res.* **57:**142–151, 1985.

Kukreja, R. C., Kontos, H. A., Hess, M. L., and Ellis, E. F.: Superoxide generation by prostaglandin synthetase and lipoxygenase in presence of NADH or NADPH. *Fed. Proc.* **45:**451, 1986.

Kuschinsky, W., Wahl, M., and Weiss, A.: Evidence for cholinergic dilatory receptors in pial arteries of cat: A microapplication study. *Pfluegers Arch.* **347:**199–208, 1974.

Lee, T. J. -F.: Cholinergic mechanism in the large cat cerebral artery. *Circ. Res.* **50:**870–879, 1982.

McCord, J. M. and Day, D. E., Jr.: Superoxide-dependent production of hydroxyl radical catalyzed by iron-EDTA complex. *FEBS Lett.* **86:**139–142, 1978.

McCord, J. M. and Fridovich, I.: The utility of superoxide dismutase in studying free radical reactions. *J. Biol. Chem.* **245:**1374–1377, 1970.

Misra, H. P. and Fridovich, I.: The generation of superoxide radical during the autooxidation of hemoglobin. *J. Biol. Chem.* **247:**6960–6962, 1972.

Owman, C., Edvinsoon, L., and Nielsen, K. C.: Autonomic neuroreceptor mechanisms in brain vessels. *Blood Vess.* **11:**2–31, 1974.

Povlishock J. T., Rosenblum, W. I., Sholley, M. M., and Wei, E.P.: An ultrastructural analysis of endothelial change paralleling platelet aggregation in a light/dye model of microvascular insult. *Am. J. Pathol.* **110:**148–160, 1983.

Rosenblum, W. I.: Minor endothelial damage converts relaxation of brain arterioles in vivo to constriction. *J. Neurol. Exp. Neuropath.* **44:**339, 1985.

Rosenblum, W. I.: Laser and Evan's blue induced endothelial injury causes focal elimination of EDRF from brain microvessels in vivo. *Fed. Proc.* **45:**289, 1986a.

Rosenblum, W. I.: Endothelial dependent relaxation demonstrated in vivo in cerebral arterioles. *Stroke* **17:**494, 1986b.

Rosenblum, W. I. and El-Sabban, F.: Dimethyl sulfoxide (DMSO) and glycerol, hydroxyl radical scavengers impair platelet aggregation within and eliminate the accompanying vasodilation of injured mouse pial arterioles. *Stroke* **13:**35–39, 1982.

Rosenblum, W. I., Wei, E. P., and Kontos, H. A.: Microvascular endothelial dependent relaxing factor (EDRF) in vivo: Interference by minimal endothelial injury. *Fed. Proc.* **44:**1262, 1985.

Sercombe, R., Verrecchia, C., Oudart, N., Dimitriadou, V., and Seylaz, J.: Pial artery responses to norepinephrine potentiated by endothelium removal. *J. Cereb. Blood Flow Metabol.* **5:**312–317, 1985.

Wei, E. P., Ellis, E. F., and Kontos, H. A.: Role of prostaglandins in pial arteriolar responses to CO_2 and hypoxia. *Am. J. Physiol.* **238:**H226–H230, 1980.

Wei, E. P., Kontos, H. A., Christman, C. W., De Witt, D. S., and Povlishock, J. T.: Superoxide generation and reversal of acetylcholine-induced cerebral arteriolar dilation after acute hypertension. *Circ. Res.* **57**:781–787, 1985.
Wei, E. P., Povlishock, J. T., and Kontos, H. A.: Role of hydroxyl radical in the cerebral arteriolar abnormalities from arachidonate. *Proc. Cong. Int. Union Physiol. Sci.* **16**:449, 1986.

Chapter 21

Platelets and Endothelium-Dependent Responses

Donald S. Houston and Paul M. Vanhoutte

1. Introduction

Because endothelium-dependent relaxations are rapid and potent, one can hardly doubt that they must have an important physiological role. This chapter describes the rationale and reviews the evidence for such a role: protection against platelet-induced vasospasm.

If the endothelium is to participate meaningfully in the control of vascular diameter, it should do so—responses to pressure or shear stress aside—by recognizing "messenger" substances (not necessarily either classical hormones or neurotransmitters) present in the blood. The endothelial layer would be involved more in local rather than systemic control of the circulation. Such a local role is favored by its location—endothelial cells are in intimate contact with any unstable blood-borne substances in their immediate vicinity—and by the fact that the relaxing factor they produce is so very short-lived. The list of substances that can evoke endothelium-dependent relaxations is already long and diverse; prominent on that list are some of the major factors produced during aggregation of platelets and activation of the coagulation cascade. If such substances are to affect the blood vessel wall, they must do so locally since most of these substances are cleared rapidly from the circulation, and only under the most extreme pathological circumstances does coagulation occur systemically.

This chapter will attempt to describe briefly the platelet-endothelial interaction, the evidence that platelets cause vascular responses in experimental models and human disease, and the role of the endothelium (so far as this has been determined) in these responses. Because platelet-vessel interactions are of particular clinical interest in the coronary artery, this vessel has served as the model in which much of the pertinent experimental work has been performed. Thus it will be discussed prominently, but consideration to other vascular beds will be extended when possible.

2. Platelet–Vessel Interactions

2.1. Adhesion

Platelet interaction with the endothelium is largely negative. That is to say, the endothelium inhibits platelet aggregation in a powerful way, by two different mechanisms: by prevention of activation of the platelets on the pro-aggregatory subendothelium, and by production of the potent platelet inhibitor, prostacyclin. These inhibitory effects normally prevent platelets from aggregating in the bloodstream. In contrast, almost every form of artificial vascular prosthesis suffers the drawback of constantly activating platelets, with resulting thrombosis unless the coagulation system is inhibited.

Under normal circumstances, platelets do not adhere to the endothelium (Dejana et al., 1983). When the endothelium is damaged, however, the collagen matrix of the subendothelial layer is exposed and platelets adhere because of the von Willebrand factor (synthesized by the overlying endothelial cells) it contains (Sakariassen et al., 1979). The collagen is a potent activator not only of the coagulation cascade (through Factor XII), but also of the platelets themselves. Further positive amplification occurs because platelets are a source of phospholipid, which promotes more coagulation and releases a variety of materials [adenosine diphosphate (ADP), serotonin, thromboxane A_2, and platelet-activating factor] that further stimulate platelet activation. As a result of these processes, the exposed subendothelium is rapidly carpeted with platelets (e.g., Schwartz et al., 1975; Reddick et al., 1982). The subsequent course of events depends on the anatomic location of the vessel and other factors such as flow velocity and turbulence, but may include buildup of a platelet thrombus and obstruction of the vessel or mere covering of the defect and eventual restoration of endothelial lining.

There is some evidence to suggest that thrombi may actually damage the endothelial surface (Barnhart and Chen, 1978), which could lead to a vicious circle of enhanced platelet aggregation and thrombosis. Furthermore, altered flow and shear stress resulting from partial obstruction of the coronary artery is associated with endothelial damage (Gertz et al., 1981).

Not all of the feedback cycles are positive, however. Production of prostacyclin by endothelial cells is augmented by thrombin (Weksler et al., 1978), and endothelial cells are even capable of converting endoperoxides released by the platelets into prostacyclin (Marcus et al., 1980; Schafer et al., 1984).

Not only does the endothelium inhibit the activation and aggregation of platelets, but it contains powerful mechanisms for eliminating vasoactive products released from platelets. For example, it has amine transport systems and monoamine oxidase enzymes that allow it to actively take up and destroy serotonin (Gillis, 1980), and nucleotidases on its surface that sequentially break down adenosine triphosphate (ATP) and ADP into the potent platelet inhibitor, adenosine (Pearson and Gordon, 1985). Furthermore, aside from actively destroying vasoactive materials in blood, the endothelial cells serve as a diffusion barrier to inhibit these substances from reaching the underlying smooth muscle.

2.2. Atherosclerosis

Another facet of the platelet–endothelial interaction occurs on a much longer time scale: atherosclerosis. Although the cause of atheromatous disease is far from being completely understood, many of the current theories feature platelets prominently in the evolution of the vessel damage and plaque. Endothelial damage (by hemodynamic shear stresses, exposure to injurious levels of plasma lipids, or other mechanisms), resulting in platelet adhesion, activation, and release of growth factors (such as platelet-derived growth factor, which is mitogenic and chemotactic for fibroblasts and smooth muscle cells), is thought likely to contribute to the process (Bowie et al., 1981; Ross, 1981, 1986). This too may be a vicious circle. For example, atherosclerotic vessels of rabbits have reduced capacity for prostacyclin synthesis (Gryglewski et al., 1978).

3. In Vitro Experiments: Endothelial Response to Platelets

3.1. Contractile Responses

If protective mechanisms that inhibit intravascular platelet aggregation fail and the aggregation process is initiated, the blood vessels might be expected to contract. Many studies have shown that platelets are capable of releasing potent vasoconstrictor substances. One of these, serotonin (5-hydroxytryptamine, 5-HT), was named in recognition of its powerful vasoconstrictor activity (Rapoport, 1949). It contributes to platelet-induced contraction of a wide variety of isolated vascular tissues (e.g., De Clerck and Van Nueten, 1982; Lindblad et al., 1984; McGoon et al., 1984; Moulds et al., 1984a,b).

Another potent vasoconstrictor produced by aggregating platelets is thromboxane A_2, an unstable product of the metabolism of arachidonic acid through the cyclooxygenase pathway (Ellis et al., 1976). Not only can it cause constriction in its own right, but it may amplify serotonin-induced contractions in the coronary artery (Mullane et al., 1982). Other platelet-released products can probably serve as vasoconstrictors as well, including vasopressin (Preibisz et al., 1983) and platelet-derived growth factor (Berk et al., 1986). Finally thrombin, which is produced during coagulation, constricts a variety of vascular smooth muscles (Haver and Namm, 1983; Katusic et al., 1984).

3.2. Platelet-Induced Relaxations

Cohen and coworkers (1983a,b) studied isolated canine coronary artery rings mounted for isometric tension recording in conventional tissue baths. When suspensions of canine platelets were added to the baths, aggregation promptly ensued. Although aggregating platelets induced contraction whether or not the endothelium had been removed from the rings, the contraction was markedly greater in rings without endothelium (Fig. 1). If the tone of the rings was first elevated by exposure to prostaglandin $F_{2\alpha}$, the rings relaxed on exposure to platelets if the endothelium was intact, but contracted further if the endothelium had been removed.

Finally, whole segments of canine coronary artery (with side branches ligated) were mounted in a special organ bath for perfusion with Krebs-Ringer solution, and contraction of the segments was measured by means of hooks inserted into the vessel wall and connected to a tension transducer (Cohen et al., 1984). In these experiments, if platelet aggregates were added to the perfusing solution (exposing the luminal surface to platelet-released products), a relaxation was observed, whereas if the extraluminal surface was exposed to platelets (by adding them to the solution bathing the exterior of the vessel segment), only a contraction was produced (Fig. 2). Contractions were also seen when platelets were added intraluminally, if the endothelium had been removed in the segment. This was a clear demonstration that platelets could exert opposing effects on the blood vessel wall by acting directly on the smooth muscle cells (to cause contraction) or indirectly on the endothelial cells (to cause relaxation).

To more clearly elucidate what factors released from platelets might be responsible for the endothelium-dependent responses, a pharmacological analysis of the response was undertaken (Houston et al., 1985). Platelets are commonly known to release a variety of different substances when they aggregate or are activated, including among them the adenine nucleo-

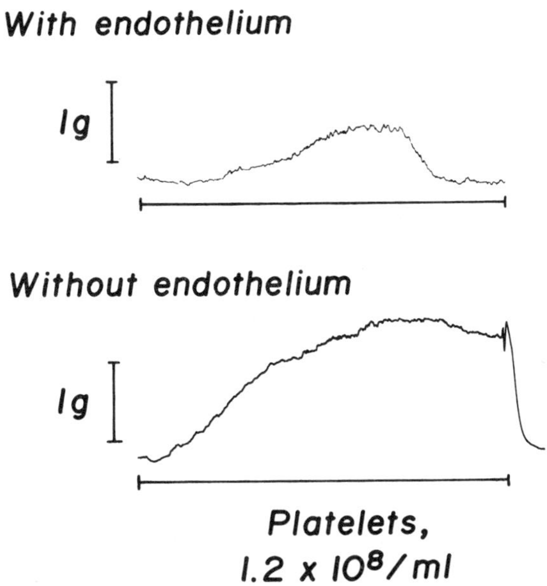

Fig. 1. Tracings of isometric tension recordings in isolated canine coronary artery rings with and without endothelium. Platelets added to the organ chamber promptly aggregated and evoked contraction. The contraction produced in rings with intact endothelium is smaller than that in rings in which the endothelium has been removed, indicating that platelets directly stimulate contractions of the smooth muscle and initiate an endothelium-dependent inhibition (from Vanhoutte and Houston, 1985, with permission of the American Heart Association).

tides, ADP and ATP; calcium and magnesium; serotonin; platelet-activating factor; thromboxane A_2 and other endoperoxides; specific proteins such as platelet Factor 4 and beta-thromboglobulin; a variety of lipids; and acid hydrolase enzyme (e.g., Meyers et al., 1982).

3.3. Serotonin

In experiments with rings of isolated coronary artery, endothelium-dependent relaxations were observed with serotonin, suggesting that this platelet-released mediator may be involved in producing the endothelium-dependent relaxations to platelets (Fig. 3; Cocks and Angus, 1983; Cohen et al., 1983a,b). Furthermore, serotonin was also capable of causing relaxations if introduced into the lumen of the perfused vessel preparation in some cases, but only contractions if added to the extraluminal bath solution (Cohen et al., 1984). If a single substance were responsible for all of the observed vasoactive effects of platelets, serotonin would be the obvious candidate.

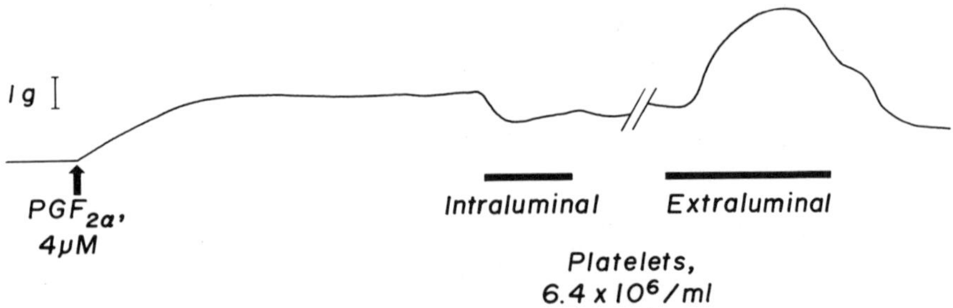

Fig. 2. Effect of intra- and extraluminal aggregating platelets in a perfused segment of canine coronary artery. Tension exerted by the smooth muscle was measured by hooks within the vessel lumen. After contraction of the smooth muscle with prostaglandin $F_{2\alpha}$, addition of aggregating platelets to the perfusate causes relaxation. Following washout of the platelets, addition of the same number of platelets to the extraluminal bath cause contraction. This experiment indicates that the route of access to the blood vessel wall is important in determining the response of the smooth muscle, and that exposure of the intima to aggregating platelets can cause relaxation (from Cohen et al., 1984, with permission).

To determine whether serotonin was the substance responsible for the relaxation, concentration–response curves to serotonin were obtained in paired rings of coronary artery with and without endothelium, in the absence and presence of serotonergic antagonists. The difference between the curves (representing the endothelial response to serotonin) was not affected by the subtype-selective $5\text{-}HT_2$-serotonergic antagonist, ketanserin, but was blocked by a nonselective ($5\text{-}HT_1$ and $5\text{-}HT_2$) antagonist, methiothepin (Fig. 4). Methiothepin, however, even at concentrations that abolished the relaxation to serotonin, did not inhibit the platelet-induced relaxation, indicating that the platelet response of the canine coronary artery must be caused by something else (Houston et al., 1985). This could already be suspected from the fact that, if large enough concentrations of platelets were added to the bath (in the physiological range), the relaxations that resulted were much larger and more rapid than was ever observed with any concentration of serotonin administered separately.

3.4. Adenine Nucleotides

The next candidates on the list of possible mediators of the endothelium-dependent relaxation to platelets were the adenine nucleotides, ADP and ATP. Both of these substances are released in large quantities from the dense granules [where they are stored in concentrations as high as 1–2 *M* (Holmsen, 1985)] when platelets aggregate, and both had been shown earlier to cause powerful endothelium-dependent relaxations in the femoral

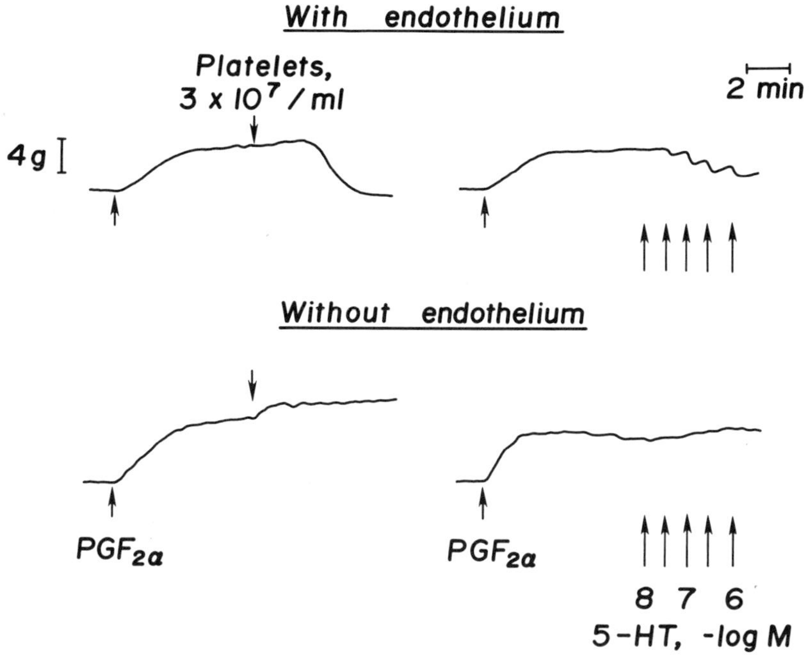

Fig. 3. Response to aggregating platelets and serotonin (5-HT) of two canine left circumflex artery rings contracted with prostaglandin $F_{2\alpha}$. In the ring with intact endothelium (upper), platelets (left) cause a transient further increase in tension, followed by relaxation, whereas only further contraction occurs in the ring denuded of endothelium (lower). The relaxation may be attributable to serotonin released by the platelets, since serotonin (right) also causes endothelium-dependent relaxation and further contraction in the denuded ring (from Cohen and Vanhoutte, 1985, with permission).

artery of the dog (De Mey and Vanhoutte, 1981a; Furchgott, 1981; De Mey et al., 1982). They can also cause such relaxations in the coronary artery (Fig. 5; Houston et al., 1985). Unfortunately, a pharmacological antagonist of responses mediated by the P_2-purinergic receptor subtype that mediates such relaxations was not available. Therefore, in order to demonstrate the role of adenine nucleotides, the enzyme apyrase was added to the organ chambers before platelets were added. It hydrolyzed the ADP and ATP as they were released from the platelets, breaking them down into adenosine monophosphate (AMP) and adenosine, and almost completely inhibited the platelet-induced relaxation (Fig. 6; Houston et al., 1985). Adenosine and AMP can relax canine large coronary arteries without endothelium, but cause relaxation only at much higher concentrations than those of ADP and ATP, indicating that ADP and ATP are the major contributors to the platelet-induced relaxation (Houston et al., 1985).

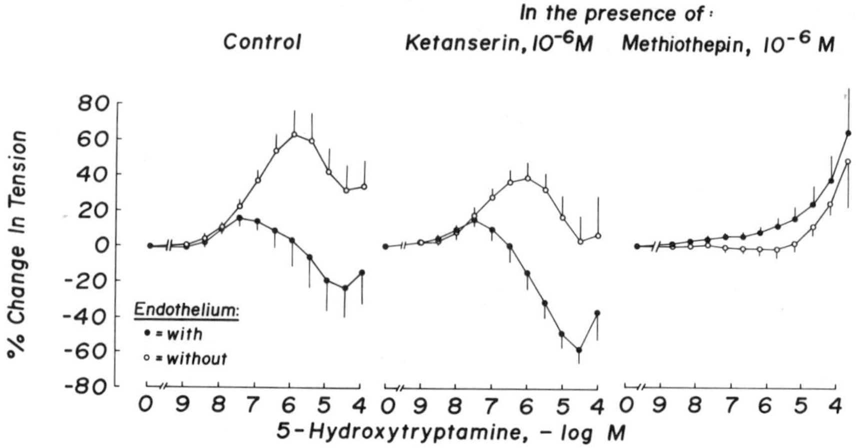

Fig. 4. Responses of rings of canine coronary artery contracted with prostaglandin $F_{2\alpha}$ ($2 \times 10^{-6}M$) to increasing cumulative concentrations of 5-hydroxytryptamine added in half-log increments. Results are means ± SEM and expressed as percent change in tension from contraction. Areas under the curves from 10^{-9} to $10^{-4}M$ 5- hydroxytryptamine were highly significantly different between intact and endothelium-denuded rings in absence of antagonists (n = 12, pooled data from experiments with antagonists). Significant differences remained in presence of ketanserin (n = 6). Difference between curves was decreased progressively by methiothepin at 10^{-8} and $10^{-7}M$ (not shown) and abolished by $10^{-6}M$ methiothepin (n = 6) (from Houston et al., 1985, with permission).

3.5. Other Possible Mediators

Although under the above experimental conditions, with the dog as the experimental model, the adenine nucleotides appear to be the chief mediators, under other circumstances other substances may possibly contribute to platelet-induced relaxations. Since vasopressin, though usually considered a vasoconstrictor, can induce endothelium-dependent relaxations in canine basilar and coronary arteries (Katusic et al., 1984), a vasopressin-receptor antagonist was tested to see whether it would affect the platelet-induced relaxation in the coronary artery; it did not (unpublished observations).

Another candidate is platelet-activating factor (PAF). This lipid mediator, released from platelets, macrophages, and other cells, is an extremely potent activator of platelets (Vargaftig and Benveniste, 1983). It has hypotensive and vasodilator effects and has furthermore been reported to cause endothelium-dependent relaxations in the isolated rat aorta (Cervoni et al., 1983; Kamitani et al., 1984). Indeed, it can induce relaxation of the canine

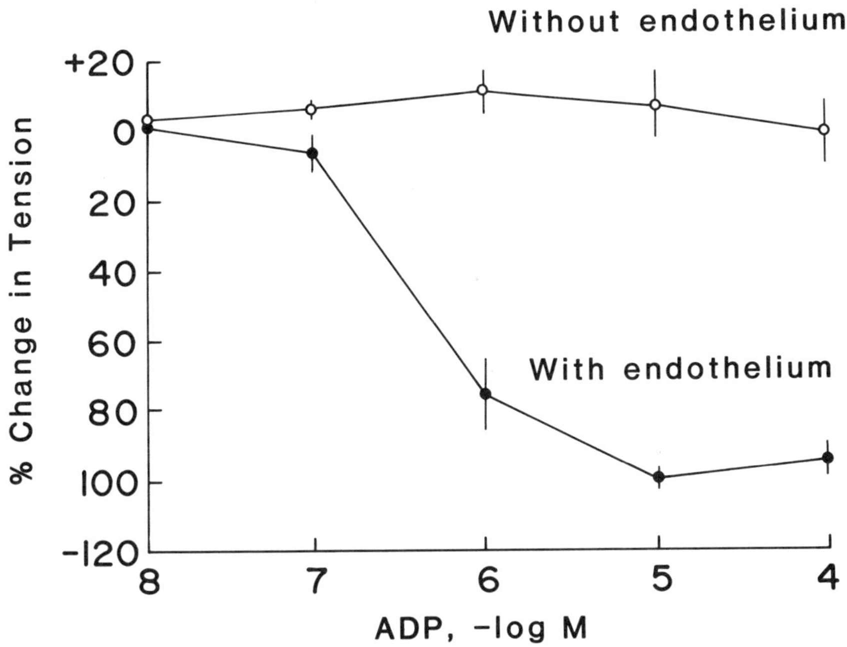

Fig. 5. Responses of rings of canine coronary artery with and without endothelium to increasing cumulative concentrations of adenosine diphosphate (ADP). Results are means ± SEM and expressed as percent change in tension from contraction with prostaglandin $F_{2\alpha}$ ($2 \times 10^{-6}M$); $n = 5$ (from Houston et al., 1985, with permission).

femoral and coronary arteries in an endothelium-dependent manner, but only at very high concentrations (above $10^{-5}M$, or 10,000–1,000,000 times the concentration required to activate platelets) (Fig. 7; Vanhoutte and Houston, 1985). Thus, it seems highly unlikely that PAF contributes to the platelet-induced relaxation.

Finally, though a product of the coagulation cascade rather than of platelets themselves, it is worth bearing in mind that thrombin potently evokes endothelium-dependent relaxations in most arteries, including the coronary (De Mey and Vanhoutte, 1981b, 1982; De Mey ct al., 1982; Ku, 1982; Katusic et al., 1984; Ku, 1986).

It is important to note that the mediators of the platelet-induced responses may vary among species. For example, in the pig as in the dog, platelets can induce a potent relaxation of the coronary arteries (Shimokawa and Vanhoutte, 1986). But in the pig, serotonin is more effective in producing endothelium-dependent relaxation and is the major element in the platelet-induced response in this species. If human platelets are used

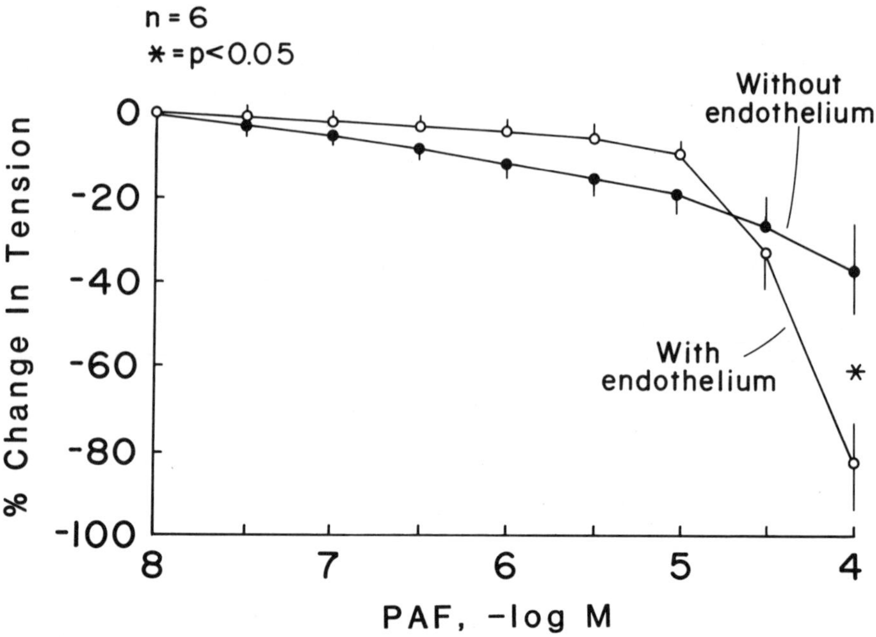

Fig. 6. Role of the endothelium in the response of rings of canine femoral artery contracted with norepinephrine ($3 \times 10^{-7}M$) to platelet-activating factor (PAF). Changes in tension are expressed as a percentage of the norepinephrine contraction; data are shown as means ± SE, $n = 6$. The difference between responses with and without endothelium is significant at $10^{-4}M$ PAF.

(assayed on canine coronary arteries), the result is much the same as with canine platelets (Houston et al., 1986), although the relaxation is, if anything, more profound, probably because human platelets release greater quantities of the adenine nucleotides (Meyers et al., 1982).

3.6. Other Vessels

Blood vessels from different anatomical locations vary greatly in their responses to platelet-released materials. In the saphenous vein, femoral artery (Houston et al., 1984), and pulmonary artery (McGoon and Vanhoutte, 1984) of the dog, aggregating platelets induce only contraction, whether or not the endothelium is intact; in the femoral vein, a brief relaxation is followed by a more powerful contraction (Houston et al., 1984).

Those studies were performed on segments of vessel suspended in an organ bath, and with that technique both the luminal and abluminal surfaces are exposed to platelet-released products simultaneously. If only

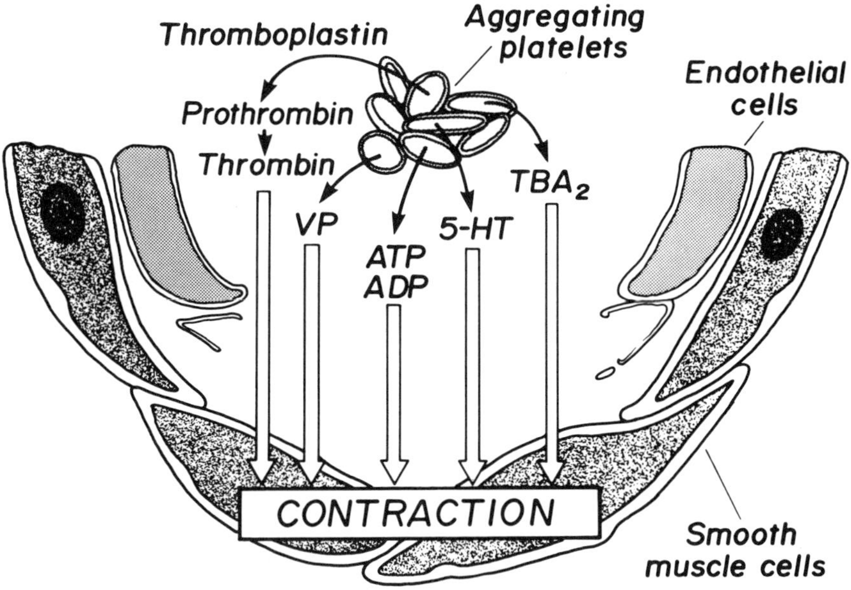

Fig. 7. Blood vessels with intact endothelium. Schematic representation of the role of endothelial cells in the prevention of vasospasm. Aggregating platelets release adenine nucleotides (ATP and ADP), serotonin (5-HT), thromboxane A_2 (TBA_2), platelet-activating factor (PAF), and vasopressin (VP), and initiate the coagulation cascade with production of thrombin. The presence of an intact endothelium prevents platelet aggregation through production of the platelet inhibitor, prostacyclin (PGI_2); PGI_2 is also a vasodilator. Serotonin is taken up by endothelial cells and degraded by monoamine oxidase (MAO). PAF (in high concentrations), 5-HT, VP, adenine nucleotides, and thrombin can all stimulate the endothelial production of a relaxing factor or factors that have an inhibitory effect on smooth muscle. Relaxation and vasodilatation would tend to flush away any developing aggregate or thrombus. Finally, an intact endothelium and basement membrane probably serve as a diffusion barrier to prevent the above-mentioned substances from reaching the smooth muscle layers of the media (from Vanhoutte and Houston, 1985, with permission of the American Heart Association).

the intimal surface had been exposed, a relaxation might have been able to outbalance the direct constrictor effect on the smooth muscle. Another approach is to antagonize the constrictor effects on the smooth muscle. This was done in the canine femoral artery, where antagonism of the constrictor effect of platelet-released serotonin with ketanserin could in some cases unmask an endothelium-dependent relaxation (unpublished observations). This is not surprising, since the endothelium of the femoral artery is sensitive to ADP and ATP (De Mey and Vanhoutte, 1981a).

A more general caveat must be made with regard to the studies described above. That is, the responses of a fragment of isolated tissue exposed to the aggregation of platelets in Krebs-Ringer solution in an organ chamber are unlikely to mimic exactly what happens in a vessel *in situ* perfused on one surface with flowing whole blood and supported on the other by surrounding tissues. The selective exposure of only one surface of the vessel to exogenous substances is not the only difference between the model and the physiological situation. For example, we have little control in such experiments over the rate of platelet aggregation and do not replicate the introduction of new platelets, the formation and lysis of thrombus, and the washout by flowing blood of the resulting products.

4. In Vivo Models

4.1. Artificial Damage

Fortunately, studies have been performed on in vivo models that support the findings described from experiments with isolated blood vessels. For example, two groups (Brum et al., 1984; Lamping et al., 1985) have found that denudation of the coronary endothelium in vivo with a balloon catheter leads to enhanced vasoconstriction by intracoronary injection of serotonin. Brum et al. (1984), but not Lamping et al. (1985), also observed a decreased resting diameter after denudation. Since endothelial denudation in vivo is accompanied by platelet adhesion and aggregation at the site of the lesion, such a vasoconstriction shortly after denudation may well be induced by platelet-released products. The enhanced responsiveness to serotonin could be caused by loss of release of endothelium-derived relaxing factor on exposure to serotonin, but it could also be caused by loss of disposition mechanisms (such as endothelial monoamine oxidase), which destroys serotonin, or to simple disruption of a diffusion barrier. In an intact isolated rabbit heart, perfused with a perfluorocarbon artificial blood solution, addition of aggregated platelets to the perfusion solution caused constriction in the epicardial coronary arteries if the endothelium had been removed, but not if it was untouched (Bing et al., 1985). In none of these models was a dilatation observed.

4.2. Atherosclerotic Damage

There is an obvious pathological correlate of endothelial damage produced mechanically by the experimenter, and that is endothelial damage produced by atherosclerosis. Experimental models of atherosclerosis have shown hypersensitivity of the smooth muscle to contractile agents. For example, after endothelial denudation of large coronary arteries and high-cholesterol diet (which induce proliferation of the regenerated endothelium, but not lipid deposition) in dogs, contractile responses to ergonovine and serotonin were enhanced, but not responses to phenylephrine (Kawachi et al., 1984). This would be in keeping with loss of endothelial response to the serotonergic agonists, whereas there is no endothelial response to alpha$_1$-adrenoceptor agonists such as phenylephrine. In a similar experiment with miniature pigs, Shimokawa et al. (1983) found augmented vasoconstrictor responses to histamine in the regions of the coronary artery with induced atherosclerosis. In perfused hindlimbs of cynomolgus monkeys, diet-induced atherosclerosis greatly enhanced the vasoconstrictor response of the large arterial tree to serotonin infusion (Heistad et al., 1984). These studies do not allow us to conclude with certainty why vascular hyperresponsiveness occurs. One possible, but by no means only, explanation is loss of production of relaxing factor from the diseased endothelial layer; other possibilities are increased sensitivity of the smooth muscle, increased access of the vasoconstrictors to it, or geometric considerations enhancing obstruction of flow when the lumen is already compromised.

Studies on smooth muscle from atherosclerotic animals in vitro have shown similar results with serotonin and ergonovine in coronary arteries and aortas of genetically hyperlipidemic rabbits (Yokoyama et al., 1983) and in aortas of cholesterol-free rabbits (Henry and Yokoyama, 1980; Heric and Tackett, 1985). These results are more thoroughly reviewed elsewhere in this volume (*see* chapter by Verbeuren and Herman).

Furthermore, endothelial function may be impaired without loss of the endothelial cells; for example, viral transformation of cultured human umbilical endothelial cells leads to a tenfold elevation in platelet adhesion to them (Gimbrone and Buchanan, 1982). In atherosclerotic rabbits, endothelial permeability is increased without loss of endothelial integrity (Stemerman, 1981). Indeed, functional impairment of endothelial cells may be as important in pathological processes as cell loss. Thus, in the pig, 1 mo after endothelial injury, and despite histologically evident full regrowth of endothelial cells, the endothelium has lost the ability to inhibit the vasoconstriction induced by aggregating platelets (Shimokawa and Vanhoutte, 1986).

4.3. Coronary Occlusion

In models of coronary obstruction (developed in hopes of mimicking the processes of coronary spasm and infarction), platelets seems to be the pivotal factor in occlusion. For example, if canine coronary arteries are occluded partially with an external constricting ring, flow varies cyclically because of buildup and then dislodgement of platelet thrombi at the site of partial occlusion; this cyclical flow reduction is inhibited by aspirin and other inhibitors of cyclooxygenase (Folts et al., 1976, 1982). In this model, active spasm does not seem to contribute to the reduction in flow. In a similar model in carotid arteries of dogs or rabbits, pronounced endothelial damage was found at the sites of the partial obstruction and a correlation drawn between the development of intimal injury (probably itself related to altered shear stresses in the blood flow) and the sites of development of platelet aggregates (Gertz et al., 1981). If the initial reduction in coronary diameter were caused by localized spasm instead of external constriction, these studies suggest a mechanism by which vasospasm could subsequently give rise to thrombosis and total coronary occlusion.

5. Clinical Studies

The involvement of coronary vasospasm in causing myocardial ischemia and infarction has been postulated most unflaggingly by Maseri and coworkers (Maseri et al., 1977, 1978a,b). They have documented that in patients with spasm, if infarction occurs it occurs in the territory subserved by the spastic vessel, and if death occurs it is associated with development of a thrombus at the site of spasm. Spasm frequently occurs in the vicinity of atherosclerotic coronary vessel disease (Chahine et al., 1975; Luchi et al., 1979), but the extent of atherosclerotic involvement at the area of spasm is highly variable. The association with atheromatous plaques, where platelets may be activated, and the finding of thrombi at the site of spasm, suggest that platelets may be involved in the disease. More direct evidence of the involvement of platelets has more recently come to light.

5.1. Evidence of Platelet Activation

The coronary sinus blood of patients with coronary artery disease contains elevated quantities of platelet products, such as thromboxane B_2 and beta-thromboglobulin, at rest or during atrial pacing (Lewy et al., 1980; Rubinstein et al., 1981; Tada et al., 1981; Cortellaro et al., 1983). Patients with coronary spasm demonstrate an acute elevation of coronary sinus thromboxane B_2 levels and increased numbers of platelet aggregates (Robertson et al., 1980; Rubinstein et al., 1981; Hirsh et al., 1981). Peripheral venous blood of patients with a positive exercise stress test contains elevated platelet Factor IV (Green et al., 1980). During evolving myocardial infarction, coronary sinus thromboxane B_2 levels are elevated (Mueller et al., 1985). These studies must be interpreted with some caution, because the elevations of these markers of platelet activation, particularly thromboxane B_2, could be caused by activation of platelets *ex vivo*. Patients dying of sudden cardiac death, in which spasm may be involved, have been found to have intraluminal thrombi in large coronary arteries in 76% of cases (Davies and Thomas, 1984); since plaque fissuring was seen in another 21%, coronary spasm alone (without thrombus formation) could account for only 5% of sudden deaths.

Platelet-endothelial interactions in other vascular beds have been studied less than in the coronary, but there is evidence of increased platelet activation in patients with peripheral vascular disease, as suggested by elevated levels of beta-thromboglobulin and platelet Factor IV (Baele et al., 1981). The subject of platelet involvement in vascular disease has been broadly reviewed by Turpie et al. (1982).

Demonstration of the release of platelet products does not tell us, however, whether platelet activation causes or is secondary to ischemia, nor if platelets cause ischemia, whether that is by release of vasoactive products causing vasospasm or by formation of a physically occlusive thrombus. Some of the available evidence links two platelet-released products, serotonin and thromboxane A_2, to coronary spasm.

5.2. Serotonin

Ergonovine, which is a serotonergic agonist in the coronary artery (Brazenor and Angus, 1981; Holtz et al., 1982), causes spasm of coronary arteries of patients with Prinzmetal's variant angina, which closely mimics the spontaneous spasm these patients experience (Schroeder et al., 1977; Curry et al., 1979).

On the other hand, the selective 5-HT_2-serotonergic antagonist, ketanserin (Van Nueten et al., 1981), has not been found to reduce the frequency of attacks in patients with variant angina (De Caterina et al., 1984). Another study reported that myocardial ischemia induced by intravenous infusion of ergonovine was not affected by ketanserin, although reanalysis of their data (presented in Table 2, Freedman et al., 1984) suggests that ketanserin did significantly increase the dose of ergonovine required to elicit ischemia (by paired Student's *t*-test between the last control ergonovine infusion and the infusion during ketanserin treatment). In any case, such discrepancies may be explained if the serotonergic receptor mediating contraction in the coronary artery is not of the 5-HT_2 subtype, which appears to be the case at least in the dog (Houston et al., 1985) though not in the pig (Cocks and Angus, 1983). The plasma concentration of ketanserin obtained in the study of Freedman and coworkers was approximately $2 \times 10^{-7}M$, and although plasma concentrations were not measured by De Caterina and coworkers, the doses they administered were somewhat lower. In contrast, we have found that ketanserin (at $10^{-6}M$) causes minimal inhibition of contraction of isolated canine coronary arteries to serotonin (Houston et al., 1985), although others report more potent antagonism (Brazenor and Angus, 1982; Frenken and Kaumann, 1985). A definitive answer on the involvement of platelet-released serotonin in coronary vasospasm in humans awaits the classification of the serotonergic receptor on the human coronary artery and the development of serotonergic blockers, devoid of agonist activity, that are selective for the appropriate subtype of receptor.

5.3. *Thromboxane A_2*

Treatment of patients with aspirin, which inhibits the platelets' production of thromboxane A_2 because it blocks cyclooxygenase, reduces mortality in patients with unstable angina (Lewis et al., 1983). Aspirin or another inhibitor of cyclooxygenase, indomethacin, however, did not reduce the number of episodes of ischemia in small groups of patients with variant (vasospastic) angina (Robertson et al., 1981; Chierchia et al., 1982). These studies are subject to the limitation that inhibitors of cyclooxygenase block the production of not only thromboxane A_2, but also of the vasodilator prostacyclin. The selective inhibitor of thromboxane, dazoxiben, but not aspirin, reduces ischemia induced by atrial pacing in patients with chronic stable angina (Thaulow et al., 1984), although another study with this drug failed to show beneficial effects in reducing anginal episodes or increasing exercise tolerance in patients with stable angina (Reuben et al., 1983). The mixed results with inhibitors of thromboxane synthetase in various other forms of cardiovascular disease have been reviewed recently (Fitzgerald et al., 1985; Harker, 1986).

Studies with inhibitors of platelet activation (which is one of the important effects of inhibition of thromboxane production) suffer the same limitation as studies reporting increased production of platelet products in coronary disease: It remains unclear whether a beneficial effect results from inhibition of vasospasm (caused by decreased release of constrictors from the platelets) or prevention of platelet thrombus formation. Clinical trials with new selective thromboxane/endoperoxide receptor antagonists (Fitzgerald et al., 1986) may help to answer this question, although these drugs too inhibit platelet aggregation.

6. Conclusions

The endothelial cells can induce a powerful relaxation of the underlying smooth muscle when stimulated by a variety of substances, including materials released from aggregating platelets (in particular, adenine nucleotides and serotonin) and during coagulation (thrombin). This response to platelets appears to be most pronounced in the coronary arteries (Fig. 8).

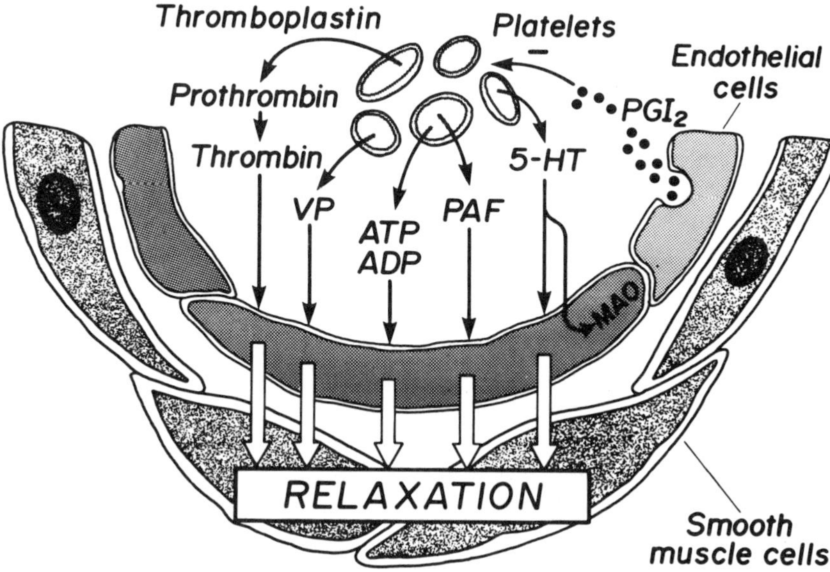

Fig. 8. Blood vessel with damaged endothelium. Platelet aggregation is enhanced by contact with collagen in the exposed vessel wall. Depending on the blood vessel, the thrombin, VP, 5-HT, adenine nucleotides, and thromboxane produced all directly activate vascular smooth muscle. These effects would be unopposed if the endothelium were damaged or dysfunctional. Vasospasm may thus result in areas of endothelial abnormality (from Vanhoutte and Houston, 1985, with permission of the American Heart Association).

If a small thrombus were to begin to develop in a normal coronary artery, the release of these mediators would cause vasodilatation through their action on the endothelial cells; at the same time, the intact endothelial layer would impair access of the various constrictor agents to the smooth muscle. The result would be a dilatation of the vessel, dispersing the developing thrombus or at least encouraging flow around it.

If, on the other hand, the endothelium were disrupted (or perhaps just dysfunctional), platelet activation would be increased, because of loss of prostacyclin production and exposure of pro-aggregatory subendothelial materials. Endothelium-dependent relaxation would be lost, and the vasoconstrictors released from the aggregating platelets (particularly serotonin and thromboxane A_2) would reach the smooth muscle unimpeded, producing spasm. Altered flow and turbulence could lead to damage of remaining endothelial cells and promote further thrombus formation. The result would be occlusion of the vessel.

If this scenario is accurate, it is notable that it is the coronary bed, among those studied, that seems to be most prone to platelet-induced ischemia. In part this is likely because it is a region of high metabolic demand with a relatively precarious circulation (no collaterals; limited flow reserve). For this reason, it makes teleological sense for a protective dilator mechanism—endothelium-dependent relaxation to platelet-released products—to have developed in the coronary bed. We might predict relaxation to predominate over contraction in other vital beds such as the cerebral and renal, though this has not been thoroughly studied. In other beds, such as the pulmonary or limb vasculature, platelet-induced relaxation has not been demonstrated. With regard to the pulmonary vessels, the lung plays an important protective role as a filter for venous emboli; in this case, dilatation by platelets would be counterproductive.

As for the peripheral beds, one should keep in mind the primary physiological function of the platelets: hemostasis. Platelet aggregation in the periphery will most often be triggered by trauma, in which case the platelets serve to prevent blood loss. Spasm of the vessel wall induced by products of aggregation would support the hemostatic process. Loss of the delicate endothelial cells during trauma would further facilitate clotting and occlusion. In contrast, hemostasis in the coronary artery is of no consequence, at least in evolutionary terms—a Cro-Magnon with a laceration of his left circumflex would have a low probability of contributing to the gene pool, with or without coronary hemostasis. Unfortunately for us, the Cro-Magnon's diet and lifestyle apparently did not prepare his descendants for our modern atherogenic behavior patterns of eating, smoking, and not exercising. We are left vulnerable to a disease that strikes at the mechanism, the endothelium, designed to protect us from the guardians of our own blood supply, the platelets.

References

Baele, G., Bogaerts, H., Clement, D. L., Pannier, R., and Barbier, F.: Platelet activation during treadmill exercise in patients with chronic peripheral arterial disease. *Thromb. Res.* **23:**215–223, 1981.

Barnhart, M. I. and Chen, S. T.: Vessel wall models for studying interaction capabilities with blood platelets. *Semin. Thromb. Hemost.* **5:**112–155, 1978.

Berk, B. C., Alexander, R. W., Brock, T. A., Gimbrone, M. A., and Webb, R. C.: Vasoconstriction: A new activity for platelet-derived growth factor. *Science* **232:87**–90,1986.

Bing, R. J., Burger, W., Chemnitius, J. M., Saeed, M., and Metz, M. Z.: Effect of intact endothelium against platelet-induced coronary artery spasm in isolated rabbit hearts. *Am. J. Cardiol.* **55:**1596–1600, 1985.

Bowie, E. J. W., Fuster, Fuster, V., Fass, D. N., and Owen, C. A.: The role of Willebrand factor in platelet-blood vessel interaction, including a discussion of resistance to atherosclerosis in pigs with von Willbrand's disease. *Philos. Trans. R. Soc. Lond.* **B294:**267–279, 1981.

Brazenor, R. M. and Angus, J. A.: Ergometrine contracts isolated canine coronary arteries by a serotonergic mechanism: No role for alpha adrenoceptors. *J. Pharmacol. Exp. Ther.* **218:**530–536, 1981.

Brazenor, R. M. and Angus, J. A.: Actions of serotonin antagonists on dog coronary artery. *Eur. J. Pharmacol.* **81:**569–576, 1982.

Brum, J. M., Sufan, Q., Lane, G., and Bove, A. A.: Increased vasoconstrictor activity of proximal coronary arteries with endothelial damage in intact dogs. *Circulation* **70:**1066–1073, 1984.

Cervoni, P., Herzlinger, H. E., Lai, F. M., and Tanikella, T. E: Aortic vascular and atrial responses to (+/−)-1-*O*-octadecyl-2-acetyl-glyceryl-3-atrial phosphorylcholine. *Br. J. Pharmacol.* **79:**667–671, 1983.

Chahine, R. A., Raizner, A. E., and Luchi, R. J.: Coronary arterial spasm in classic angina pectoris. *Cathet. Cardiovasc. Diag.* **1:**337–342, 1975.

Chierchia, S., De Caterina, R., Crea, F., Patrono, C., and Maseri, A.: Failure of thromboxane A2 blockade to prevent attacks of vasospostic angina. *Circulation* **66:**702–709, 1982.

Cocks, T. M. and Angus, J. A.: Endothelium-dependent relaxation of coronary arteries by noradrenaline and serotonin. *Nature* **305:**627–630, 1983.

Cohen, R. A. and Vanhoute, P. M.: Platelets, Serotonin, and Endothelial Cells, in *Serotonin and the Cardiovascular System* (Vanhoutte, P. M., ed.) Raven, New York, 1985.

Cohen, R. A., Shepherd, J. T., and Vanhoutte, P. M.: 5-Hydroxytryptamine can mediate endothelium-dependent relaxation of coronary arteries. *Am. J. Physiol.* **245:**H1077–H1080, 1983a.

Cohen, R. A., Shepherd, J. T., and Vanhoutte, P. M.: Inhibitory role of the endothelium in the response of isolated coronary arteries to platelets. *Science* **221:**273–274, 1983b.

Cohen, R. A., Shepherd, J. T., and Vanhoutte, P. M.: Endothelium and asymmetrical responses of the coronary arterial wall. *Am. J. Physiol.* **247:**H403–H408, 1984.

Cortellaro, M., Boschetti, C., Antoniazzi, V., Moreo, G., Repetto, S., Verna, E., Boscarini, M., Limido, A., Binaghi, G., and Polli, E. E.: Transcoronary platelet thromboxane A2 formation without platelet trapping in patients with coronary stenosis—effect of sulphinpyrazone treatment. *Thromb. Haemos.* **50:**857–859, 1983.

Curry, R. C., Pepine, C. J., Sabom, M. B., and Conti, C. R.: Similarities of ergonovine-induced and spontaneous attacks of variant angina. *Circulation* **59:**307–312, 1979.

Davies, M. J. and Thomas, A.: Thrombosis and acute coronary-artery lesions in sudden cardiac ischemic death. *N. Eng. J. Med.* **310:**1137–1140, 1984.

De Caterina, R., Carpeggiani, C., and L'Abbate, A.: A double-blind, placebo-controlled study of ketanserin in patients with Prinzmetal's angina. Evidence against a role for serotonin in the genesis of coronary vasospasm. *Circulation* **69:**889–894, 1984.

De Clerck, F. and Van Nueten, J. M.: Platelet-mediated vascular contractions: Inhibition of the serotonergic component by ketanserin. *Thromb. Res.* **27:** 713–727, 1982.

Dejana, E., Cazenave, J-P., Hatton, M. W. C., Groves, H. M., Kinlough-Rathbone, R. L., Packham, M. A., and Mustard, J. F.: The effect of thrombin on platelet accumulation on the vessel wall—influence of heparin and aspirin. *Thromb. Haemost.* **50:**567–571, 1983.

De Mey, J. G. and Vanhoutte, P. M.: Role of the intima in cholinergic and purinergic relaxation of isolated canine femoral arteries. *J. Physiol.* (Lond.) **316:**346–355, 1981a.

De Mey, J. G. and Vanhoutte, P. M.: Role of the intima in the relaxation of the canine femoral artery caused by thrombin. *Arch. Int. Pharmacodyn.* **250:**314–315, 1981b.

De Mey, J. G. and Vanhoutte, P. M.: Heterogeneous behavior of the canine arterial and venous wall. Importance of the endothelium. *Circ. Res.* **51:** 439–447, 1982.

De Mey, J. G., Claeys, M., and Vanhoutte, P. M.: Endothelium-dependent inhibitory effects of acetylcholine, adenosine triphosphate, thrombin and arachidonic acid in the canine femoral artery. *J. Pharmacol. Exp. Ther.* **222:**166–173, 1982.

Ellis, E. F., Oelz, O., Roberts, L. J., II, Payne, N. A., Sweetman, B. J., Nies, A. S., and Oates, J. A.: Coronary arterial smooth muscle contaction by a substance released from platelets: Evidence that it is thromboxane A2. *Science* **193:**1135–1137, 1976.

Fitzgerald, D. J., Doran, J., Jackson, E., and Fitzgerald, G. A.: Coronary vascular occlusion mediated via thromboxane A2-prostaglandin endoperoxide receptor activation in vivo. *J. Clin. Invest.* **77:**496–502, 1986.

Fitzgerald, G. A., Reilly, I. A. G., and Pederson, A. A. K.: The biochemical pharmacology of thromboxane synthase inhibition in man. *Circulaion* **72:**1194–1201, 1985.

Folts, J. D., Crowell, E. B., and Rowe, G. G.: Platelet aggregation in partially obstructed vessels and its elimination with aspirin. *Circulation* **54:**365–370, 1976.

Folts, J. D., Gallagher, K., and Rowe, G. G.: Blood flow reductions in stenosed canine coronary arteries: Vasospasm or platelet aggregation? *Circulation* **65:**248–255, 1982.

Freedman, S. B., Chierchia, S., Rodriguez-Plaza, L., Bugiardini, R., Smith, G., and Maseri, A.: Ergonovine-induced myocardial ischemia: No role for serotonergic receptors? *Circulation* **70:**178–183, 1984.

Frenken, M. and Kaumann, A. J.: Ketanserin causes surmountable antagonism of 5-hydroxytryptamine-induced contractions of large coronary arteries of dog. *Naunyn Schmiedebergs Arch. Pharmacol.* **328:**301–303, 1985.

Furchgott, R. F. The requirement for endothelial cells in the relaxation of arteries by acetylcholine and some other vasodilators. *Trends Pharmacol. Sci.* **2:**173–176, 1981.

Gertz, S. D., Uretsky, G., Wajnberg, R. S., Navot, N., and Gotsman, M. S.: Endothelial cell damage and thrombus formation after partial arterial constriction: Relevance to the role of coronary artery spasm in the pathogenesis of myocardial infarction. *Circulation* **63:**476–486, 1981.

Gillis, C. N.: Metabolism of Vasoactive Hormones by Pulmonary Vascular Endothelium: Possible Functional Significance, in *Vascular Neuroeffector Mechanisms* (Bevan, J. A., Godfraind, T., Maxwell, R. A., and Vanhoutte, P. M., eds) Raven Press, New York, 1980.

Gimbrone, M. A. and Buchanan, M. R.: Interactions of platelets and leukocytes with vascular endothelium: In vitro studies. *Ann. NY Acad. Sci.* **401:**171–183, 1982.

Green, L. H., Seroppian, E., and Handin, R. I.: Platelet activation during exercise-induced myocardial ischemia. *N. Engl. J. Med.* **302:**193–197, 1980.

Gryglewski, R. J., Dembinska-Kiec, A., Zmuda, A., and Gryglewska, T.: Prostacyclin and thromboxane A2 biosynthesis capacities of heart, arteries, and platelets at various stages of experimental atherosclerosis in rabbits. *Atherosclerosis* **31:**385–394, 1978.

Harker, L. A.: Clinical trails evaluating platelet-modifying drugs in patients with atherosclerotic cardiovascular disease and thrombosis. *Circulation* **73:**206–223, 1986.

Haver, V. M. and Namm, D. H.: Generation of a vasoactive substance in human plasma during coagulation. Evidence of thrombin-induced contraction of rabbit aorta and dog coronary artery. *Blood Vessels* **20:**92–98, 1983.

Heistad, D. D., Armstrong, M. L., Marcus, M. L., Piegors, D. J., and Mark, A. L.: Augmented responses to vasoconstrictor stimuli in hypercholesteolemic and atherosclerotic monkey. *Circ. Res.* **54:**711–718, 1984.

Henry, P. D. and Yokoyama, M.: Supersensitivity of atherosclerotic rabbit aorta to ergonovine. Mediation by a serotonergic mechanism. *J. Clin. Invest.* **66:**306–313, 1980.

Heric, E. and Tackett, R. L.: Altered vascular reactivity in the rabbit during hypercholesterolemia. *Pharmacology* **31:**72–81, 1985.

Hirsh, P. D., Hillis, L. D., Campbell, W. B., Firth, B. G., and Willerson, J. T: Release of prostaglandins and thromboxane into the coronary circulation in patients with ischemic heart disease. *N. Engl. J. Med.* **304:**685–691, 1981.

Holmsen, H.: Nucleotide metabolism of platelets. *Ann. Rev. Physiol.* **47:**677–690, 1985.

Holtz, J., Held, W., Sommer, O., Kuhne, G., and Bassenge, E.: Ergonovine-induced constriction of epicardial coronary arteries in conscious dogs: Alpha-adrenoceptors are not involved. *Basic Res. Cardiol.* **77:**278–291, 1982.

Houston, D. S., Shepherd, J. T., and Vanhoutte, P. M.: Platelets cause endothelium-dependent relaxations in some but not all isolated canine vessels (abstract). *Sed. Proc.* **43:**1084, 1984.

Houston, D. S., Shepherd, J. T., and Vanhoutte, P. M.: Adenine nucleotides, serotonin, and endothelium-dependent relaxations to platelets. *Am. J. Physiol.* **248:**H389–H395, 1985.

Houston, D. S., Shepherd, J. T., and Vanhoutte, P. M.: Aggregating human platelets cause direct contraction and endothelium-dependent relaxation of isolated canine coronary arteries: Role of serotonin, thromboxane A_2 and adenine nucleotides. *J. Clin. Invest.* **78:**539–544, 1986.

Kamitani, T., Katamoto, M., Tasumi, M., Katsuta, K., Ono, T., Kikuchi, H., and Kumada, S.: Mechanism(s) of the hypotensive effect of synthetic 1-*O*-octadecyl-2-*O*-acetyl-glycero-3-phosphocholine. *Eur. J. Pharmacol.* **98:**357–366, 1984.

Katusic, Z. S., Shepherd, J. T., and Vanhoutte, P. M.: Vasopressin causes endothelium-dependent relaxation of the canine basilar artery. *Circ. Res.* **55:**575–579, 1984.

Kawachi, Y., Tomoike, H., Maruoka, Y., Kikuchi, Y., Araki, H., Ishii, Y., Tanaka, K., and Nakamura, M.: Selective hypercontraction caused by ergonovine in the canine coronary artery under conditions of induced atherosclerosis. *Circulation* **69:**441–450, 1984.

Ku, D. D.: Coronary vascular reactivity after acute myocardial ischemia. *Science* **218:**576–578, 1982.

Ku, D. D.: Mechanism of thrombin-induced endothelium-dependent coronary vasodilation in dogs: Role of its proteolytic enzymatic activity. *J. Cardiovasc. Pharmacol.* **8:**29–36, 1986.

Lamping, K. G., Marcus, M. L, and Dole, W. P.: Removal of the endothelium potentiates canine large coronary artery constrictor responses to 5-hydroxytryptamine in vivo. *Circ. Res.* **57:**46–54, 1985.

Lewis, H. D., Davis, J. W., Archibald, D. G., Steinke, W. E., Smitherman, T. C., Doherty, J. E., Schnaper, W. H., LeWinte, M. M., Linares, E., Pouget, J. M., Sabharwal, S. C., Chesler, E., and DeMots, H.: Protective effects of aspirin against acute myocardial infarction and death in men with unstable angina. Results of a veterans administration cooperative study. *N. Engl. J. Med.* **309:**396–403, 1983.

Lewy, R. I., Wiener, L., Walinsky, P., Lefer, A. M., Silver, M. J., and Smith, J. B.: Thromboxane release during pacing-induced angina pectoris: Possible vasoconstrictor influence on the coronary vasculature. *Circulation* **61:**1165–1171, 1980.

Lindblad, L. E., Shepherd, J. T., and Vanhoutte, P. M.: Cooling augments platelet-induced contraction of peripheral arteries of the dog. *Proc. Soc. Exp. Biol. Med.* **176:**119–122, 1984.

Luchi, R. J.: Coronary artery spasm. *Ann. Int. Med.* **91:**441–449, 1979.

Marcus, A. J., Weksler, B. B., Jaffe, E. A., and Broekman, M. J.: Synthesis of prostacyclin from platelet derived endoperoxides by cultured human endothelial cells. *J. Clin. Invest.* **66:**979–986, 1980.

Maseri, A., Pesola, A., Marzilli, M., Severi, S., Parodi, O., L'Abbate, A., Ballestra, A. M. Maltinit, G., De Nes, D. M., and Biagini, A.: Coronary vasospasm in angina pectoris. *Lancet* **1:**713–717, 1977.

Maseri, A., L'Abbate, A., Baroldi, G., Chierchia, S., Marzilli, M., Ballestra, A. M., Serveri, S., Parodi, O., Biagini, A., Distante, A., and Pesola, A.: Coronary vasospasm as a possible cause of myocardial infarction. A conclusion derived from the study of "preinfarction" angina. *N. Engl. J. Med.* **229:**1271–1277, 1978a.

Maseri, A., Severi, S., De Nes, M., L'Abbate, A., Chierchia, S., Marzilli, M., Ballestra, A. M., Parodi, O., Biagini, A., and Distante, A.: "Variant" angina: One aspect of a continuous spectrum of vasospastic myocardial ischemia. *Am. J. Cardiol.* **42:**1019–1035, 1978b.

McGoon, M. D. and Vanhoutte, P. M.: Aggregating platelets contract isolated canine pulmonary arteries by releasing 5-hydroxytryptamine. *J. Clin. Invest.* **74:**828–833, 1984.

Meyers, K. M., Holmsen, H., and Seachord, C. L.: Comparative study of platelet dense granule constituents. *Am. J. Physiol.* **243:**R454–R461, 1982.

Moulds, R. F. W., Iwanov, V., and Medcalf, R.: The effects of platelet-derived contractile agents on human digital arteries. *Clin. Sci.* **66:**443–451, 1984a.

Moulds, R. F. W., Iwanov, V., and Yound, M. J.: Vasoactive effects of platelet aggregates (letter). *N. Engl. J. Med.* **311:**198–199, 1984b.

Mueller, H. S., Rao, P., Greenberg, M. A., Buttrick, P. M., Sussman, I. I., Levite, H. A., Grose, R. M., Perez-Davila, V., Strain, J. E., and Spaet, T. H.: Systemic and transcardiac platelet activity in acute myocardial infarction in man: Resistance to prostacyclin. *Circulation* **72:**1336–1345, 1985.

Mullane, K. M., Bradley, G., and Moncada, S.: The interaction of platelet-derived mediators on isolated canine coronary arteries. *Eur. J. Pharmacol.* **84:**115–118, 1982.

Pearson, J. D. and Gordon, J. L.: Nucleotide metabolism by the endothelium *Ann. Rev. Physiol.* **47:**617–627, 1985.

Preibisz, J. J., Sealey, J. E., Laragh, J. H., Cody, R. J., and Weksler, B. B.: Plasma and platelet vasopressin in essential hypertension and congestive heart failure. *Hypertension* **5** (supp I)**:**I129–I138, 1983.

Rapport, M. M.: Serum vasoconstrictor (sertonin) V: The presence of creatinine in the complex. A proposed structure of the vasoconstrictor principle. *J. Biol. Chem.* **180:**961–969, 1949.

Reddick, R. L., Griggs, T. R., Lamb, M. A., and Brinkhous, K. M.: Platelet adhesion to damaged coronary arteries: Comparison in normal and von Willebrand disease swine. *Proc. Natl. Acad. Sci. USA* **79:**5076–5079, 1982.

Reuben, S. R., Kuan, P., Cairns, J., and Gyde, O. H.: Effects of dazoxiben on exercise performance in chronic stable agina. *Br. J. Clin. Pharmacol.* **15:**83S–86S, 1983.

Robertson, R. M., Robertson, D., Friesinger, G. C., Timmons, S., and Hawiger, J.: Platelet aggregates in peripheral and coronary-sinus blood in patients with coronary-artery spasm. *Lancet* **2:**829–831, 1980.

Robertson, R. M., Robertson, D., Roberts, L. J., Maas, R. L., Fitzgerald, G. A., Friesinger, G. C., and Oates, J. A.: Thromboxane A2 in vasotonic angina pectoris. Evidence from direct measurements and inhibitor trials. *N. Engl. J. Med.* **304:**998-1003, 1981.

Ross, R.: Atherosclerosis: A problem of the biology of arterial wall cells and their interactions with blood components. *Arteriosclerosis* **1:**293–311, 1981.

Ross, R.: The pathogenesis of atherosclerosis—an update. *N. Engl. J. Med.* **314:**488-500, 1986.

Rubenstein, M. D., Wall, R. T., Baim, D. S., and Harisson, D. C.: Platelet activation in clinical coronary artery disease and spasm. *Am. Heart J.* **102:**363-367, 1981.

Sakariassen, K. S., Bolhuis, P. A., and Sixma, J. J.: Human blood platelet adhesion to artery subendothelium is mediated by factor VIII-von Willebrand factor bound to the subendothelium. *Nature* **279:**636–638, 1979.

Schafer, A. I., Crawford, D. D., and Gimbrone, M. A.: Unidirectional transfer of prostaglandin endoperoxides between platelets and endothelial cells. *J. Clin. Invest.* **73:**1105–1112, 1984.

Schroeder, J. S., Bolen, J. L., Quint, R. A., Clark, D. A., Hayden W. G., Higgens, C. B., and Wexler, L.: Provocation of coronary spasm with ergonovine maleate. *Am. J. Cardio.* **40:**487–491, 1977.

Schwartz, S. M., Stemerman, M. B., and Benditt, E. P.: The aortic intima II: Repair of the aortic lining after mechanical denudation. *A. J. Pathol.* **81:**15–42, 1975.

Shimokawa, H., Tomoike, H., Nabeyama, S., Yamamoto, H., Araki, H., Nakamura, M., Ishii, Y., and Tanaka, K.: Coronary artery spasm induced in atherosclerotic miniature swine. *Science* **221:**560–562, 1983.

Shimokawa, H. and Vanhoutte, P. M.: Reduced endothelium-dependent relaxations to aggregating platelets and serotonin in porcine coronary arteries after previous endothelial injury. *Circulation* **74:**II413, 1986.¾

Stemerman, M. B.: Effects of moderate hypercholesterolemia on rabbit endothelium *Arteriosclerosis* **1:**25–32, 1981.

Tada, M. Kuzuya, T., Inoue, M., Kodama, K., Mishima, M., Yamada, M., Inui, M., and Abe, H.: Elevation of thromboxane B2 levels in patients with classic and variant angina pectoris. *Circulation* **64:**1107–1115, 1981.

Thaulow, E., Dale, J., and Myhre, E.: Effects of a selective thromboxane synthetase inhibitor, dazoxiben, and of acetylsalicylic acid on myocardial ischemia in patients with coronary artery disease. *Am. J. Cardiol.* **53:**1255–1258, 1984.

Turpie, A. G. G., de Boer, A. C., and Genton, E.: Platelet consumption in cardiovascular disease. *Semin. Thromb. Hemost.* **8:**161–185, 1982.

Vanhoutte, P. M. and Houston, D. S.: Platelets, endothelium and vasospasm. *Circulation* **72:**728–734 (erratum, *Circulation* **73:**40, 1986), 1985.

Van Nueten, J. M., Janssen, P. A. J., Van Beek, J., Xhonneux, R., Verbeuren, T. J., and Vanhoutte, P. M.: Vascular effects of ketanserin (R41468), a novel antagonist of 5-HT2 serotonergic receptors. *J. Pharmacol. Exp. Ther.* **218:**217–230, 1981.

Vargaftig, B. B. and Benveniste, J.: Platelet-activating factor today. *Trends Pharmacol. Sci.* **4**:341–343, 1983.

Weksler, B. B., Ley, C. W., and Jaffe, E. A.: Stimulation of endothelial cell prostacyclin production by thrombin, trypsin, and the ionophore A23187. *J. Clin. Invest.* **62**:923–930, 1978.

Yokoyama, M., Akita, H., Mizutani, T., Fukuzaki, H., and Watanabe, Y.: Hyperreactivity of coronary arterial smooth muscles in response to ergonovine from rabbits with hereditary hyperlipidemia. *Circ. Res.* **53**:63–71, 1983.

Chapter 22

Endothelium-Dependent Responses and the Release of Endothelium-Derived Relaxing Factor in Atherosclerotic Blood Vessels

Tony J. Verbeuren and Arnold G. Herman

1. Introduction

Atherosclerosis has been defined as "a variable combination of changes in the intima of the arteries consisting of an accumulation of lipids, complex carbohydrates, blood and blood products, fibrous tissue and calcium deposits, and associated with medical changes" (*see* Kottke and Subbiah, 1978; Fuster, 1982). Several hypotheses have been introduced in order to explain the pathogenesis of atherosclerosis (*see* Ross and Glomset, 1976b; Kottke and Subbiah, 1978; Ross, 1981; Fuster, 1982; Gotto, 1985; Ross, 1986). Virchow (1856) and Rokitansky (1852) already focused attention on the possible role of the intima in the pathogenesis of the disease. Virchow believed intimal damage to be a very early step in the development of an atheroma (*see* Ross, 1981; Fuster, 1982; Gotto, 1985). More recently, these hypotheses have been translated into "the response-to-injury hypothesis"; the basis for this theory lies in the similarity between the development of atherosclerosis and the response of arteries to experimental injury (Ross and Glomset, 1976b; Ross, 1981, 1986).

It recently became evident that several vasoactive substances (mainly vasodilators, but also vasoconstrictor agents such as norepinephrine and serotonin) can evoke relaxation of blood vessels by releasing the potent vasodilator substance, endothelium-derived relaxing factor (EDRF), from endothelial cells (Furchgott and Zawadzki, 1980; De Mey and Vanhoutte, 1981a; Cocks and Angus, 1983; Cohen et al., 1983; Van de Voorde and Leusen, 1983; Griffith et al., 1984). Since atherosclerotic blood vessels are very susceptible to the development of vasospasm (Schroeder et al., 1977; Cipriano et al., 1979; Waters et al., 1983), it has been proposed that dysfunction or denudation of the endothelium in atherosclerosis may evoke impairment of EDRF release, which then in turn would lead to augmented vasoconstrictor responses (Heistad et al., 1984). In this chapter, the recent data obtained by several investigators who studied endothelium-mediated responses in hypercholesterolemic animals will be summarized.

2. Atherosclerosis and the Structure of the Blood Vessel Wall

2.1. Introduction

The arterial wall is a complex structured organ that consists of a series of three morphologically distinct tissue layers. (1) The intima: This layer is bounded on the luminal side by a monolayer of specialized polygonal cells, the endothelium, which is in direct contact with the circulating blood and on the peripheral side by the internal elastic lamina. A basement membrane separates the endothelium from the subendothelial space; the subendothelial elements (intimal smooth muscle cells and extracellular matrix components) become apparent only after the first decade of life; (2) The media: This middle layer contains the diagonally arranged smooth muscle cells, which are surrounded by a network of collagenous and elastic fibers; and (3) The adventitia: This outermost layer, which is separated from the media by the external elastic membrane, consists mainly of fibroblasts intermixed with smooth muscle cells that are arranged between collagen bundles. The adventitia contains the small blood vessels (vasa vasorum) and the nerve terminals. The intima is the cell layer principally involved in atherosclerosis; some secondary changes may be found in the media (Ross and Glosmet, 1976a).

2.2. The Atherosclerotic Lesions

Classically, three different types of atherosclerotic lesions are recognized: the fatty streak, the fibrous plaque, and the complicated lesion (*see* Ross and Glomset, 1976a; Fuster, 1982; Ross, 1986). Although cell and tissue degeneration are associated with the advanced, complicated lesion, both the fatty streak and the fibrous plaque are cellular proliferative responses (Ross, 1981). (1) The fatty streak: These intimal lesions, visible to the naked eye as yellow spots, are the earliest macroscopically detectable lesions. They are commonly found in children and adolescents and are grossly, lipid-rich lesions consisting of both lipid-laden macrophages and some smooth muscle (Ross and Glomset, 1976a; Fuster, 1982; Ross, 1986). The fatty streaks are mainly located at branches and bifurcations in areas in which the rheologic properties of the blood may disturb the endothelium; this then may lead to formation of advanced proliferative lesions (Fuster, 1982; Ross, 1986). (2) The fibrous plaque: This more permanent and severe lesion appears at branching points in the abdominal aorta in young adults and in the coronary arteries by the end of the third decade. Later it appears in other peripheral arteries including the cerebral circulation. The lesion consists of an accumulation of intimal lipid-laden smooth muscle cells surrounded by collagen elastic fibers and proteoglycans and contains variable amounts of intracellular and extracellular lipid. At the lumen of the arteries, the smooth muscle cells and the extracellular matrix components form a fibrous cap that covers a lipid core or atheroma (Ross and Glomset, 1976a; Fuster, 1982; Ross, 1986). Evidence is available that some fatty streaks may progress into fibrous plaques (Fuster, 1982; McGill, 1984; Ross, 1986). (3) The complicated lesion: This lesion develops from preexisting fibro-fatty lesions as a result of hemorrhage, calcification, cell necrosis, and mural thrombosis. This type of lesion may induce severe or even complete obstruction of the artery leading to ischemia or infarction. An important characteristic of the complicated lesion is its loss of elasticity (Fuster, 1982).

2.3. Mechanism of Atherosclerotic Plaque Formation

The development of spontaneous atherosclerosis is a slow process: It usually requires several decades before clinical symptoms become manifest. It is now believed that the pathogenesis of the disease depends upon a precise sequence of critical events based on the interaction of blood elements and lipids with the arterial wall (*see* Fuster, 1982; Ross, 1986). The concept that atherosclerosis results from a response to endothelial injury is supported by findings illustrating that intimal injury, provoked experimentally, reproduces the sequence of events noted in early atherosclerotic lesions. Endothelial injury can be diverse in nature and may be the result of mechanical forces, chemical agents (e.g., lipoproteins), toxins, immunologic injury, viruses, and so on (*see* Ross, 1981, 1986). Interaction of the endothelium with these factors may lead to stimulation of the endothelial cells, to separations between endothelial cells, to altered endothelial permeability, or even to desquamation of the endothelial cells. The response-to-injury-hypothesis has been updated taking into account the most recent experimental data available (Ross, 1986). Briefly, it is now proposed that hypercholesterolemia induces the following sequence of events leading to lesion formation in arteries: (1) Injury to the endothelium can cause growth factor secretion; (2) monocytes may then attach to the endothelium, and subendothelial migration of monocytes may lead to fatty-streak formation and release of growth factors; (3) the fatty streaks may be converted directly into fibrous plaques through release of growth factors from macrophages or endothelial cells, or the macrophages may stimulate or injure the overlying endothelium; (4) when the endothelium cover is lost, platelet attachment may occur, adding a further source (platelets) of growth factor production; (5) some of the smooth muscle cells in the proliferative lesion may also form and secrete growth factors (Ross, 1986).

Alternatively, direct stimulation of the endothelium may cause release of growth factors and may induce migration and proliferation of smooth muscle cells; in this case the endothelium may remain intact (Ross, 1986).

If the endothelial injury continues, large amounts of new connective tissue may be formed, and lipid materials may be deposited in the cells and their surrounding connective matrix (Ross, 1981; Fuster, 1982). There now is evidence that the lipid material in the lesions (cholesterol) is derived from the plasma. Endothelial cells, smooth muscle cells, and monocytes (macrophages) contain high-affinity receptors for low-density lipoproteins (LDL) on their surface (Brown and Goldstein, 1984); and LDLs are the principal class of proteins associated with atherogenesis. LDL receptors have a role in the control of the intracellular transformation of cholesterol (Bierman et al., 1974). When the number of LDL receptors decreases, this leads to a decrease in removal of LDL from the circulation and causes an increase in plasma LDL (Brown and Goldstein, 1984). When the fibrous plaque develops, extracellular cholesteryl linoleate and free cholesterol may also accumulate (Smith, 1974). Furthermore, LDLs acting with other plasma-derived elements are believed to stimulate the initial proliferation of smooth muscle cells (Ross and Glomset, 1976a), probably by providing the nutrients for the process (Ross and Harker, 1976). Removal of cholesterol may be a crucial problem, since cholesterol cannot be degraded within the arterial wall (Ross and Glomset, 1976a).

There is also increasing evidence indicating that disturbance of the balance between production of thromboxane A_2 by platelets and of prostacyclin by the vascular wall may be a factor involved in vascular disease (*see* Fuster, 1982). Thus Gryglewski et al. (1978) have illustrated that the production of prostacyclin was decreased in arteries of hypercholesterolemic rabbits, whereas the formation of thromboxane A_2 by the platelets of these rabbits was augmented. In atherosclerotic human arteries, a decreased production of prostacyclin has also been reported (Sinzinger et al., 1979). Recent information indicates that in rabbits fed a cholesterol-rich diet, the production of prostacyclin is increased at the onset of atherosclerosis, before macroscopic lesions are observed, whereas at a later stage, it decreases (Beetens et al., 1986a,b). Important evidence for the involvement of platelets in atherogenesis comes from experiments using pigs with homozygous von Willebrand disease; these pigs are resistant to spontaneous and cholesterol-induced atherosclerosis, which might be related to the impairment of platelet functions (Fuster et al., 1978).

Finally, the different processes lead to the development of the complicated lesions, which may then calcify (Ross, 1981; Fuster, 1982) and lead to acute clinical manifestations such as, for example, myocardial infarction, cerebrovascular incidents, or arterial occlusion of a limb (*see* Fuster, 1982).

2.4. Animal Models of Atherosclerosis

Many different animal models and experimental designs have been used to study diet-induced atherosclerosis. Despite several limitations, a large number of studies have been performed on cholesterol-fed rabbits; the rabbit is a convenient laboratory animal that shows a rapid formation of lesions upon cholesterol feeding, but these lesions have a different anatomic distribution from those observed in human atherosclerosis (Kottke and Subbiah, 1978). The chicken, dog, rat, and pigeon have also been used to study experimental atherosclerosis, although all these animal models also have several disadvantages (Faggiotto et al., 1984). The swine and primates are reported to develop spontaneous atherosclerosis comparable to that found in humans. Although the investigation of diet-induced atherosclerosis in these animals may provide information more applicable to humans, those animal models also have their limitations (*see* Fagiotto et al., 1984; Fagiotto and Ross, 1984; Kottke and Subbiah, 1978).

The effect of atherosclerosis on the endothelium-mediated responses thus far has only been investigated in isolated blood vessels of hypercholesterolemic animals, mostly rabbits. The question can be raised whether the limitations, expressed against the use of the rabbit as a model of atherosclerosis, also exist as regards the contractile process of isolated blood vessels. Indeed, the degree to which a particular blood vessel may be damaged by the cholesterol-feeding can vary among different species, but the effect of the lesioning on the ability of a selected blood vessel to contract or relax may not vary when blood vessels of different species are compared. Since endothelium-dependent relaxations have been reported in a wide variety of arteries obtained from several species, including humans (Furchgott, 1983; Vanhoutte and Rimele, 1983), the selection of the rabbit as a model to study the effects of hypercholesterolemia on these relaxations seems justified.

In male rabbits fed a cholesterol-rich diet containing only 0.3% cholesterol, for 8 or 16 wk, the development of fatty streaks in different arterial tissues was investigated (Verbeuren et al., 1986a). When examined by visual inspection, the aortic arch of the 16-wk hypercholesterolemic rabbits was the most severely lesioned tissue; the brachiocephalic artery and the thoracic aorta were somewhat less affected and the abdominal aorta and the pulmonary artery showed much less lesioning. Planimetry of stained lesions in the thoracic aorta (Beetens et al., 1986a) and microscopic examination of the lesions in proximal and distal portions of the thoracic aorta (Verbeuren et al., 1986a) confirmed the results of these visual inspections and validated the method of visual scoring (Verbeuren et al., 1986a). The microscopic evaluation also revealed less fatty streaks in the 8- than in the 16-wk hypercholesterolemic rabbits (Verbeuren et al., 1986a); these results were in agreement with those obtained in thoracic aortas of rabbits fed also a 0.3% cholesterol-diet, in which a correlation was described between the fatty streak formation and the time (2–10 wk) the rabbits were fed the atherogenic diet (Beetens et al., 1986a). In a morphologic investigation in nonhuman primates, the development of atherosclerotic lesions was studied in chronic, diet-induced hypercholesterolemia (0.5% of cholesterol); these investigations illustrate the time-dependent formation of fatty streaks (Faggiotto et al., 1984) and of fibrous plaques (Faggiotto and Ross, 1984).

2.5. *Effects of Drugs on the Pathogenesis and Regression of Atherosclerosis*

The precise cause of atherosclerosis is not known, which complicates the design of methods for its treatment and prevention. A potential way of protecting the arteries may be through reducing the overload of intracellular calcium using Ca^{2+} antagonists (*see* Gotto, 1985). Certain Ca^{2+} antagonists such as nifedipine, diltiazem, and verapamil can interfere with atherogenesis in experimental animals; these agents did not influence serum cholesterol levels, but they normalized serum Ca^{2+} levels and decreased Ca^{2+} deposition in aortic tissues (*see* Henry and Bentley, 1981; DeFeudis, 1983). Whether the Ca^{2+} antagonists exert these effects in part by acting at the level of the platelets remains to be investigated; platelet aggregation is a Ca^{2+}-dependent phenomenon, and certain Ca^{2+} antagonists can inhibit platelet aggregation both in vitro and in vivo (*see* DeFeudis, 1983). Some of the findings with the Ca^{2+} entry blockers still are controversial, since in a recent study no significant effect of nifedipine on atherogenesis in cholesterol-fed rabbits could be detected (Stender et al., 1984).

Since platelets are believed to play a crucial role in the genesis of atherosclerosis (Mustard et al., 1964; Ross and Glomset, 1976b; Ross, 1981; Hess et al., 1985), it has been suggested that drugs that can inhibit platelet functions may have a beneficial effect on atherosclerotic disease (Harker et al., 1976; *see also* Hess et al., 1985). Dipyridamole, an inhibitor of phosphodiesterase, is known for its ability to stabilize platelets (Moncada et al., 1976; Moncada and Korbut, 1978). It was recently shown that dipyridamole can delay the progression of peripheral occlusive arterial disease in humans, particularly when administered together with aspirin, a compound that by blocking cyclooxygenase inhibits the formation of thromboxane in platelets, thereby stabilizing them (Hess et al., 1985). The combination of aspirin and dipyridamole was also effective in preventing occlusion after coronary bypass operations (Chesebro et al., 1982, 1984). Furthermore the compound was able to slow down the formation of the lesions when administered orally in rabbits that were fed a 2% cholesterol-rich diet (Hallermayer, 1982). The suppression of the formation of atherosclerotic plaques caused by dipyridamole was even more pronounced when aspirin was also added to the diet (Hallermayer, 1982). In a recent study the effect of dipyridamole treatment in rabbits fed a cholesterol-rich diet containing 0.3% cholesterol was investigated; the addition of dipyridamole to the atherogenic diet markedly slowed down the increase in plasma cholesterol levels and the formation of fatty streaks in different arteries of these rabbits (Verbeuren et al., 1985b, 1986b).

3. Vascular Responses in Atherosclerotic Blood Vessels

Until recently, only a limited number of studies were performed to investigate the effects of atherosclerosis on vascular responsiveness. Some evidence was available to suggest that atherosclerotic blood vessels are susceptible to the development of vasospasm (Schroeder et al., 1977; Cipriano et al., 1979; Waters et al., 1983; *see also* Heistad et al., 1984). Thus ergonovine, which causes constrictions of the coronary artery that closely resemble those occurring during spontaneous spasms, has been used routinely as a provocative agent for the angiographic diagnosis of coronary vasospasm (Curry et al., 1979; Heupler et al., 1978).

3.1. *In Vivo Observations*

Heistad et al. (1984) recently suggested that four lines of evidence exist to illustrate that vasoconstrictor responses in vivo may be augmented by hypercholesterolemia or atherosclerosis. (1) In hypercholesterolemic dogs, coronary vasoconstrictor responses to norepinephrine are moderately increased (Rosendorff et al., 1981); (2) reflex coronary vasoconstriction is greater in patients with coronary artherosclerosis than in those with normal coronary arteries (Mudge et al., 1976); (3) in the atherosclerotic swine, histamine produces coronary artery spasm (Shimokawa et al., 1983); and (4) in patients with coronary atherosclerotic disease, spasm is common (Schroeder et al., 1977); it also occurs in limb vessels of patients with peripheral vascular disease (Juergens et al., 1980).

In atherosclerotic monkeys, a pronounced potentiation of hindlimb vasoconstrictor responses to serotonin was observed at a stage when the responses to norepinephrine were unaltered, whereas in hypercholesterolemic monkeys, prior to the development of atherosclerosis, constrictor responses to norepinephrine were potentiated and those to serotonin were unaltered (Heistad et al., 1984). From these studies on the primate model, the authors concluded that during early stages of hypercholesterolemia, vasoconstrictor responses of small vessels to norepinephrine are potentiated, whereas at a later stage, when atherosclerosis develops, vasoconstrictor responses of large arteries to serotonin are greatly augmented.

3.2. In Vitro Observations

Because patients with coronary vasospastic disease are more susceptible to the coronary constrictor effects of ergonovine, Henry and Yokoyama (1980) compared the contractile responses to ergonovine, serotonin, phenylephrine, and K^+ in isolated aortas obtained from control or cholesterol-fed (2%) rabbits. These studies illustrated that the aortas from rabbits fed the cholesterol-rich diet were supersensitive to the contractile effects of ergonovine and suggested that this supersensitivity was mainly mediated by a serotonergic mechanism (Henry and Yokoyama, 1980). Support for that conclusion came from the findings that the contractions to low concentrations of serotonin were augmented by hypercholesterolemia, whereas the responses to phenylephrine and K^+ were not or only minimally affected (Henry and Yokoyama, 1980). Similar results were obtained when the responsiveness of aortas and coronary arteries obtained from control rabbits and rabbits with hereditary hyperlipidemia were compared (Yokoyama et al., 1983).

In a recent investigation, the vascular responsiveness to a variety of constrictor substances in control and hypercholesterolemic rabbits has been compared (Verbeuren et al., 1986a). In the aortic arch, the abdominal aorta, and the pulmonary artery of rabbits fed a 0.3% cholesterol-diet for 8 wk, the contractile responses to serotonin were not significantly altered when compared to those noted in tissues obtained from control rabbits (Verbeuren et al., 1986a). In the aortic arch, the abdominal aorta, the brachiocephalic artery, and the pulmonary artery of rabbits fed the 0.3% cholesterol diet for 16 wk, the contractions to prostaglandin $F_{2\alpha}$ and to high concentrations of acetylcholine were comparable to those noted in control tissues; the contractile responses to norepinephrine and clonidine were reduced, and those to low concentrations of serotonin were augmented in the arteries (aortic arch, brachiocephalic artery), which were the most severely affected by the 16-wk hypercholesterolemia (Verbeuren et al., 1986a). Those observations illustrate that a cholesterol-rich diet, which already evokes structural changes to the intima of the arteries, does not influence the responsiveness of the vascular smooth muscle cells to all stimuli (acting via different mechanisms) in the same manner. Certain responses, i.e., the contractions to prostaglandin $F_{2\alpha}$ and those to acetylcholine, are unaltered. Similarly the contractions to K^+ were reported not to be significantly affected in aortas (Henry and Yokoyama, 1980) and ear arteries (Maggi et al., 1985) of hypercholesterolemic rabbits and in aortas and coronary arteries of rabbits with hereditary hyperlipidemia (Yokoyama et al., 1983).

As the degree of intimal lesioning augments, an increased responsiveness to serotonin becomes apparent; indeed the contractions to low concentrations of serotonin were augmented only in the severely affected arteries, after 16 wk of cholesterol feeding (Verbeuren et al., 1986a). These observations are in agreement with those described earlier for aortas and coronary arteries of hypercholesterolemic and hyperlipidemic rabbits (Henry and Yokoyama, 1980; Yokoyama et al., 1983). In basilar arteries of rabbits fed a 2% cholesterol diet for 4 or 8 wk, a decreased responsiveness to serotonin was reported, whereas in the ear arteries of the same rabbits, the contractions to serotonin were not significantly altered (McCalden and Nath, 1984).

Somewhat more conflicting data have thus far been presented for the alpha-adrenergic responses. In vivo, augmented vasoconstrictor responses for norepinephrine have been reported (Rosendorff et al., 1981; Heistad et al., 1984). In aortas of rabbits fed a cholesterol-rich diet, an increased number of alpha-adrenergic receptors has been detected (Nanda and Henry, 1982). The contractions to the more specific alpha_1-adrenergic agonist phenylephrine noted in aortas and coronary arteries of hypercholesterolemic and hyperlipidemic rabbits were not altered when compared to those noted in arteries obtained from control rabbits (Henry and Yokoyama, 1980; Yokoyama et al., 1983). The contractions to norepinephrine and those to the more specific alpha_2-adrenoceptor agonist clonidine were decreased in the aortic arch and the brachiocephalic artery of hypercholesterolemic rabbits (Verbeuren et al., 1986a). These results are in agreement with those reported by Maggi et al. (1985), who illustrated a decreased responsiveness to norepinephrine in ear arteries of hypercholesterolemic rabbits, and partly with those presented by McCalden and Nath (1984), who found contractions to norepinephrine to be increased after 4 but decreased after 8 wk of hypercholesterolemia in basilar arteries of the rabbit; the responses to norepinephrine in ear arteries of the same species were not affected by the cholesterol feeding (McCalden and Nath, 1984).

The findings that in arteries of hypercholesterolemic animals the contractions caused by the more specific alpha$_1$-adrenergic agonist phenylephrine are not altered, whereas those to the more specific alpha$_2$-adrenergic agonist clonidine are decreased and those to the nonspecific agonist norepinephrine tend to be decreased, suggest that the two subpopulations of postjunctional alpha-adrenoceptors (*see* De Mey and Vanhoutte, 1981b) may be differently affected by the atherosclerotic process.

Isolated coronary arteries obtained from cardiac patients were hyperreactive to histamine, serotonin, and norepinephrine (Kalsner and Richards, 1984) and atherosclerotic human coronary arteries showed a greater degree of spontaneous activity (Godfraind and Miller, 1983).

4. Endothelium-Dependent Relaxations in Atherosclerotic Blood Vessels

4.1. Introduction

Since endothelial injury in arteries may be a trigger for the development of atherosclerosis (*see* Ross 1981, 1986; Fuster, 1982) and endothelium removal in arteries leads to the inhibition of a potent vasodilator response mediated by EDRF (Furchgott and Zawadzki, 1980), it has been speculated that the endothelium-mediated relaxations might be impaired during atherosclerosis (Heistad et al., 1984). The authors' laboratory performed a series of investigations aimed at studying endothelium-dependent relaxations and the release of EDRF in rabbits fed a cholesterol-rich diet (Verbeuren et al., 1986a,b). During the course of our studies, several abstracts and a brief communication describing similar experiments in rabbits and monkeys were presented (Habib et al., 1984; Chappell et al., 1985; Coene et al., 1985; Freiman et al., 1985; Herman et al., 1985; Jayakody et al., 1985; Verbeuren et al., 1985a,b).

4.2. Endothelium-Dependent Relaxations

4.2.1. Acetylcholine

During contractions caused by norepinephrine or prostaglandin $F_{2\alpha}$, increasing concentrations of acetylcholine (10^{-9} to $3 \times 10^{-6}M$) evoked relaxations in arterial segments of aortic arches, brachiocephalic arteries, abdominal aortas, and pulmonary arteries of rabbits fed a control diet (Fig. 1); these relaxations were absent when the endothelium of the arteries had been mechanically removed (Verbeuren et al., 1986a) and confirm the original observations describing the endothelium dependency of the acetylcholine-induced relaxations in arteries of the rabbit and the dog (Furchgott and Zawadzki, 1980; De Mey and Vanhoutte, 1981a).

In segments of the arteries obtained from rabbits fed a 0.3% cholesterol diet for 8 or 16 wk, the relaxations to acetylcholine were significantly attenuated compared to those noted in tissues of control animals (Fig. 1).

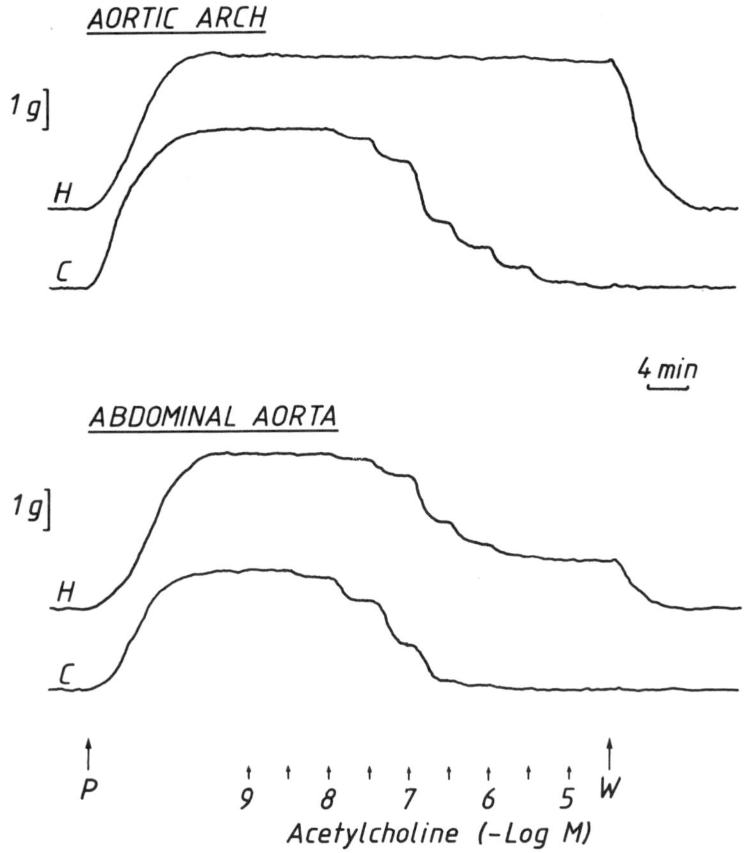

Fig. 1. Inhibition of the relaxations to acetylcholine in aortic tissues obtained from cholesterol-fed rabbits. In the aortic segments contractions were evoked with $2 \times 10^{-6} M$ of prostaglandin $F_{2\alpha}$ (P); during these contractions increasing concentrations of acetylcholine were administered. C, aorta obtained from a control rabbit; H, aorta obtained from a hypercholesterolemic rabbit. Note the complete inhibition of the relaxations in the more severely lesioned aortic arch.

The shifts of the concentration–response curves to acetylcholine noted in the 16-wk hypercholesterolemic arteries were more pronounced than those noted in the 8-wk hypercholesterolemic tissues (Verbeuren et al., 1986a). When the magnitude of the maximal relaxations obtained with acetylcholine in the arteries used for the 16-wk experiments was plotted against the degree of fatty streak formation noted in the preparations, a significant correlation was obtained indicating that less relaxation could be evoked in the more severely affected arteries (Verbeuren et al., 1986a; *see* Fig. 2). In aortas of rabbits fed 2% cholesterol for 4, 8, 10, and 11 wk, the relaxations to acetylcholine were also reduced (Habib et al., 1984; Chappell et

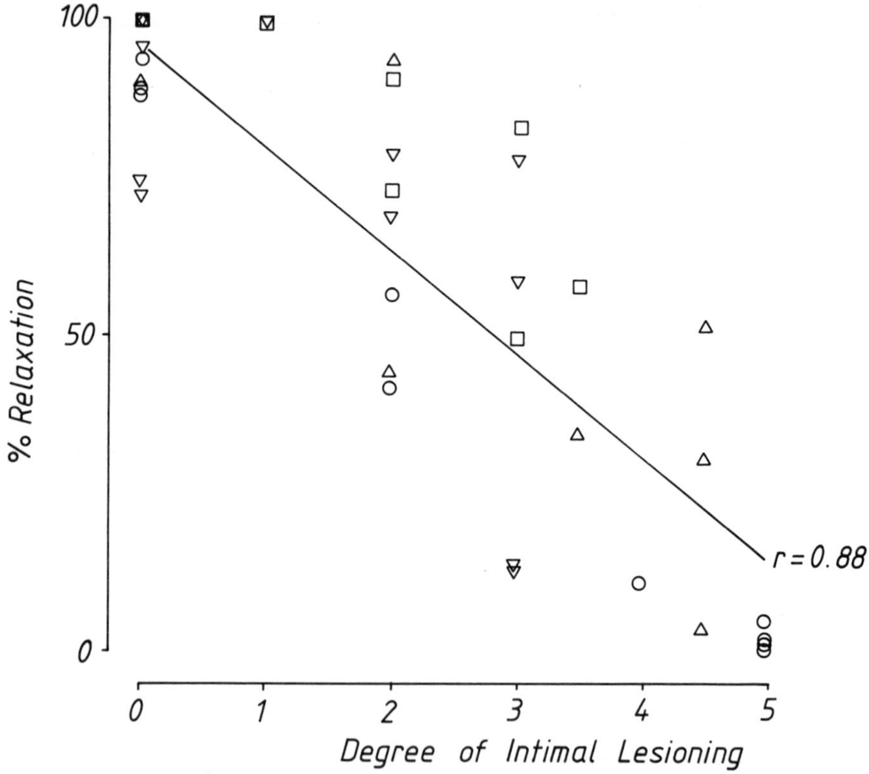

Fig. 2. Correlation between the maximal relaxation evoked by acetylcholine and the degree of fatty streak formation in control arteries and arteries obtained from 16-wk hypercholesterolemic rabbits. The response to acetylcholine is expressed as percent of the initial contraction caused by prostaglandin $F_{2\alpha}$ ($2 \times 10^{-6} M$). The correlation obtained was significant. Note that 17 of the unlesioned arteries relaxed completely to acetylcholine. ○, aortic arch; □, abdominal aorta; △, brachiocephalic artery; ▽, pulmonary artery (from Verbeuren et al. 1986a, by permission of the American Heart Association).

al., 1985; Jayakody et al., 1985), and the decreased ability to relax to acetylcholine correlated positively with the morphometrically quantitated severity of the atherosclerosis (Habib et al., 1984). In iliac arteries of cynomolgus monkeys with atherosclerosis, induced by an 18-mo atherogenic diet, the relaxations to acetylcholine were markedly inhibited as compared to those noted in control arteries (Freiman et al., 1985).

4.2.2. Relaxations to Adenosine Triphosphate

In the aortic arch, the abdominal aorta, the brachiocephalic artery, and the pulmonary artery of rabbits fed a control diet, adenosine triphosphate (ATP) (10^{-8} to $10^{-4}M$) evokes concentration-dependent relaxations (Fig. 3) that are significantly attenuated (but not abolished) when the endothelium is removed (Verbeuren et al., 1986a). These experiments with ATP confirm the results in canine arteries (De Mey and Vanhoutte, 1981a) and illustrate that ATP causes its relaxation partly by acting on the endothelium and partly by directly stimulating the vascular smooth muscle cells. In the arteries of both the 8- and 16-wk hypercholesterolemic rabbits, the relaxations to ATP were significantly attenuated compared to those noted in the control preparations (Verbeuren et al., 1986a, *see* Fig. 3). In the most severely affected arteries of the 16-wk hypercholesterolemic rabbits, the aortic arch, the relaxations to ATP were inhibited more than after removal of the endothelium in the control animals (Verbeuren et al., 1986a).

4.2.3. Relaxations to Thrombin

In canine arteries, thrombin can cause endothelium-dependent relaxations (De Mey et al., 1982). In iliac arteries of cynomolgus monkeys, thrombin evoked relaxations that were significantly attenuated in the arteries obtained from atherosclerotic animals (Freiman et al., 1985).

4.3. Endothelium-Independent Relaxations to Nitroglycerin

In aortic arches, abdominal aortas, and brachiocephalic and pulmonary arteries of rabbits fed control diet, increasing concentrations of nitroglycerin caused relaxations (Fig. 3) that were not affected by endothelium removal (Verbeuren et al., 1986a). In abdominal and pulmonary arteries of rabbits fed the 0.3% cholesterol diet for 16 wk, the relaxations to nitroglycerin were not significantly altered. In the least affected arteries of the 16-wk hypercholesterolemic rabbits, the relaxations to acetylcholine and ATP were significantly attenuated, whereas those to nitroglycerin were not altered. In the more severely affected tissues, the aortic arch and the brachiocephalic artery, the relaxations to nitroglycerin were significantly attenuated; complete relaxation could still be obtained, however, with the highest concentration of the drug (Verbeuren et al., 1986a; Fig. 3). In

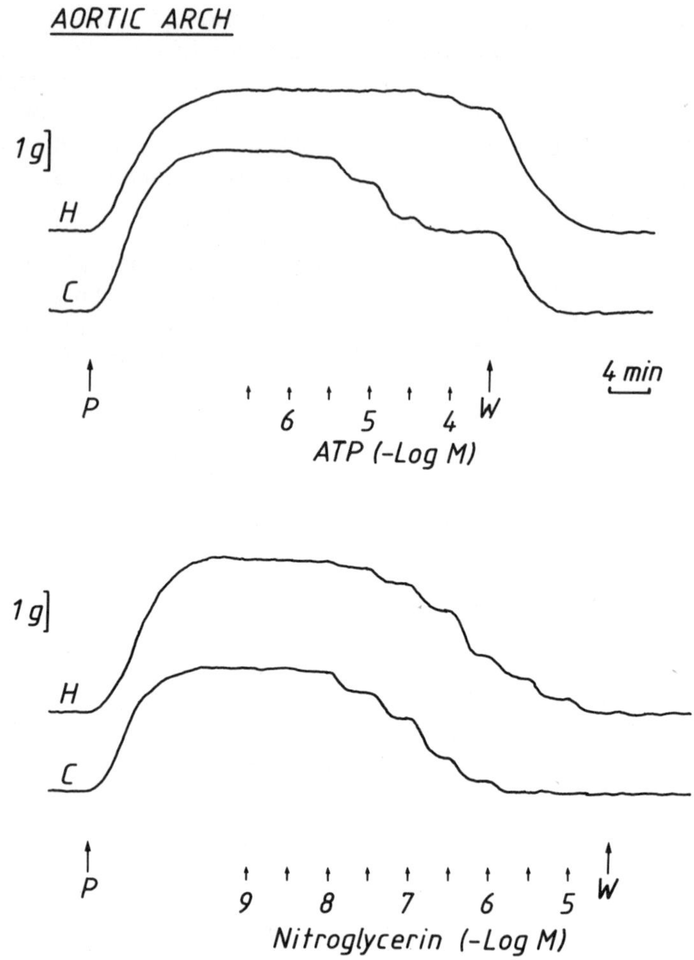

Fig. 3. Inhibition of the relaxations to adenosine triphosphate (ATP) and nitroglycerin in aortic arch segments obtained from cholesterol-fed rabbits. In the segments, contractions were evoked with $2 \times 10^{-6} M$ prostaglandin $F_{2\alpha}$ (P); during these contractions increasing concentrations of ATP (upper) or nitroglycerin (lower) were administered. C, aortic arch obtained from control rabbit; H, aortic arch obtained from hypercholesterolemic rabbit. Note that the hypercholesterolemic artery still relaxed maximally to nitroglycerin.

aortas of rabbits fed a 2% cholesterol diet, the ED_{50} and ED_{15} to nitroglycerin were not significantly altered by the hypercholesteremia (Habib et al., 1984), and in iliac arteries of atherosclerotic monkeys, no change in the nitroglycerin dose–response relationship was detected (Freiman et al., 1985). These findings show that the cholesterol-rich diet not only alters the endothelium-dependent relaxations, but that in the more severely lesioned arteries, the endothelium-independent relaxations to nitroglycerin may be altered also; thus the cholesterol-rich diet may directly affect the ability of the smooth muscle cells to relax. Such an effect can explain why the relaxations to ATP, which are in part caused by a direct action on the smooth muscle cells, are inhibited more in the 16-wk hypercholesterolemic aortic arches than those noted in deendothelialized aortic arches obtained from control rabbits (Verbeuren et al., 1986a).

4.4. Effect of Dipyridamole Treatment on Endothelium-Dependent Relaxations in Atherosclerotic Arteries

Addition of 0.60 g of dipyridamole daily to the control diets of the rabbits did not alter contractile responses or relaxations in the different arteries investigated (Verbeuren et al., 1986b). When the dipyridamole was added to the atherogenic diet (0.3% cholesterol), it reduced plasma cholesterol levels and the degree of fatty streak formation; it also restored the endothelium-independent relaxations to nitroglycerin and partially restored the endothelium-dependent relaxations to acetylcholine and ATP (Verbeuren et al., 1986b). The latter effects of dipyridamole most likely are the result of the decreased fatty streak formation; indeed, in aortic arches of control and hypercholesterolemic rabbits, whether treated or not with dipyridamole, a positive correlation was found between the degree of fatty streak formation and the maximal relaxation obtained with acetylcholine (Verbeuren et al., 1986b).

4.5. Release of EDRF

Histological examination of arterial segments obtained from rabbits fed a 0.3% cholesterol diet for 16 wk revealed that, at this stage of the disease, the arteries still contained endothelial cells (Verbeuren et al., 1986a). One of the possible explanations for the reduced endothelium-mediated relaxations noted in the atherosclerotic arteries is a decreased release of EDRF.

In segments of the rabbit abdominal aorta, the effect of 16-wk cholesterol feeding on the production of EDRF was investigated using a bioassay technique; in these segments acetylcholine evoked release of EDRF, and no difference was detected when aortas of control and the hypercholesterolemic rabbits were compared (Verbeuren et al., 1986a; Fig. 4). Although the abdominal aorta is not the artery that is the most severely affected by the hypercholesterolemia, it should be noted that in segments of the aortas

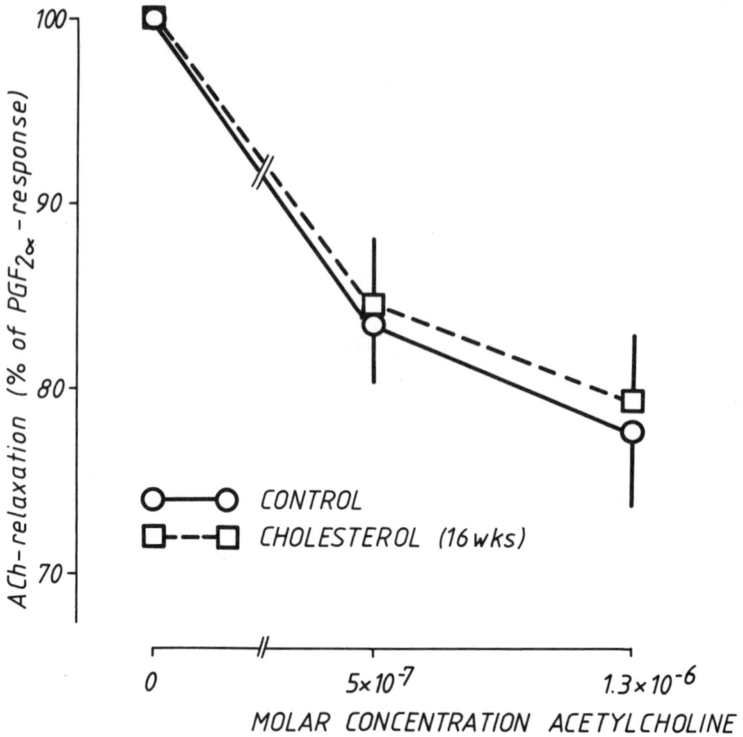

Fig. 4. Release of EDRF in abdominal aortas of control and 16-wk hypercholesterolemic rabbits. The release of the relaxing factor is evaluated by determining the relaxation of an aortic arch segment without endothelium (bioassay tissue) that was contracted with prostaglandin $F_{2\alpha}$ ($2 \times 10^{-6}M$) in the presence of indomethacin ($3 \times 10^{-6}M$), phentolamine ($3 \times 10^{-6}M$), and propranolol ($10^{-6}M$). The response of the bioassay tissue is shown as percent of the initial contraction. Increasing concentrations of EDRF were evoked in the abdominal aorta with two concentrations of acetylcholine. The results are shown as mean $\pm$ SEM ($n = 4$). Acetylcholine caused a similar release of EDRF in abdominal aortas of control and hypercholesterolemic rabbits (from Verbeuren et al., 1986a, by permission of the American Heart Association).

adjacent to the ones used for the bioassay experiments, a marked inhibition of the acetylcholine-induced relaxations was observed (*see* Fig. 1).

Preliminary data from our laboratory illustrate that after 16 wk of hypercholesterolemia, the release of EDRF induced by acetylcholine in the rabbit thoracic aorta is decreased (Herman et al., 1987); similar results were obtained in Dr. Kappagoda's laboratory (personal communication).

Taken together, these observations indicate that more than one step in the endothelium-smooth muscle coupling is affected by the atherosclerotic disease. In the most affected arteries (like the thoracic aorta), both the relaxations caused by the EDRF and the release of the factor are inhibited (Herman et al., 1987); in less affected blood vessels (like the abdominal aorta), the relaxations to the EDRF are already attenuated despite the fact that a normal release of the substance is detected (Verbeuren et al., 1986a).

5. Conclusions

Only a few studies are presently available describing the vascular reactivity of isolated atherosclerotic blood vessels; almost all of these investigations were performed on vessels obtained from animals (rabbits or monkeys) fed a cholesterol-rich diet. The results obtained illustrate that during early stages of the atherosclerotic disease (fatty streak formation), the contractile process of the vascular smooth muscle cells is not significantly altered; indeed contractions to K^+, prostaglandin $F_{2\alpha}$, and acetylcholine are unaltered. The evidence available indicates that supersensitivity to the contractile action of serotonin occurs as the effects of the cholesterol-rich diet become obvious, whereas the alpha-adrenergic responses are either unchanged (alpha$_1$-adrenoceptors?) or decreased (alpha$_2$-adrenoceptors?). The processes in vascular smooth muscle leading to relaxation may become impaired as the hypercholesterolemia progresses, but this stage of decreased ability of the smooth muscle to relax is preceded by the appearance of decreased endothelium-dependent relaxations. Since in the abdominal aorta of the atherosclerotic rabbit the release of EDRF was not altered at a time that the endothelium-mediated relaxations were decreased, a step subsequent to the release of EDRF must be altered, thus preventing the relaxing factor from evoking relaxation of smooth muscle. A likely possibility is that the thickening of the intimal layer of the arterial wall, caused by the infiltration and subendothelial accumulation of monocytes, macrophages, and smooth muscle cells (forming the fatty streaks), prevents EDRF, a substance with a short half-life (± 6 s), from reaching the vascular smooth muscle cells; the lesions will indeed increase the diffusion distances and the sequestration characteristics of the vascular segment. Obviously the atherosclerotic disease process might also alter steps subsequent to the diffusion of the EDRF. When, however, the atherosclerotic process becomes more important (as in the rabbit thoracic aorta), the release of the EDRF is decreased, which then results in a more pronounced inhibition of the endothelium-mediated relaxations (Verbeuren et al., 1986a; Herman et al., 1987); ultimately the EDRF-mediated relaxations are completely blocked (Verbeuren et al., 1986a). In patients with vascular diseases such as coronary atherosclerosis and peripheral vascular disease, the incidence of vasospasm is augmented. The fact that in atherosclerotic blood vessels the vasoconstrictor responses to serotonin are enhanced, while at the same time a process leading to relaxation is impaired (dysfunction of the endothelium), may play an important role in the pathogenesis of vasospasm.

Acknowledgments

The authors wish to thank Dr. M.-C. Coene, F. Jordaens, M. Van Diest, C. Van Hove, and L. Zonnekeyn for their valuable collaboration with our studies reported in this manuscript. They are also indebted to Mrs. L. Van den Eynde for typing the manuscript.

The investigations from our laboratory were supported by grants of the FGWO and by a BEKALES award.

References

Beetens, J. R., Coene, M.-C., Verheyen, A., Zonnekeyn, L., and Herman, A. G.: Biphasic response of endothelial prostacyclin production during the development of experimental atherosclerosis. *Prostaglandins* **32:**319–334, 1986a.

Beetens, J. R., Coene, M.-C., Verheyen, A., Zonnekeyn, L., and Herman, A. G.: Vitamin C increases the prostacyclin production and decreases the vascular lesions in experimental atherosclerosis in rabbits. *Prostaglandins* **32:**335–352, 1986b.

Bierman, E. L., Stein, O., and Stein, Y.: Lipoprotein uptake and metabolism by rat aortic smooth muscle cells in tissue culture. *Circ. Res.* **5:**136–150, 1974.

Brown, M. S. and Goldstein, J. L.: How LDL receptors influence cholesterol and atherosclerosis. *Sci. Am.* **251:**58–66, 1984.

Chappell, S. P., Griffith, T. M., Henderson, T. M., and Lewis, A. H.: Influence of cholesterol feeding on endothelium-dependent vasomotor response in rabbit aortic strips. *Br. J. Pharmacol.* **85:**266P, 1985.

Chesebro, J. H., Clements, I. P., Fuster, V., Elveback, L. R., Smith, H. C., Bardsley, W. T., Frye, R. L., Holmes, D. R., Jr., Vlietstra, R. E., Pluth, J. R., Wallace, R. B., Puga, F. J., Orszulak, T. A., Piehler, J. M., Schaff, H. V., and Danielson, G. K.: A platelet inhibitor drug trial in coronary artery bypass operations. Benefit of perioperative dipyridamole and aspirin therapy on early postoperative vein-graft patency. *N. Engl. J. Med.* **307:**73–78, 1982.

Chesebro, J. H., Fuster, V., Elveback, L. R., Clements, I. P., Smith, H. C., Holmes, D. R., Bardsley, W. T., Pluth, J. R., Wallace, R. B., Puga, F. J., Orszulak, T. A., Piehler, J. M., Danielson, G. K., Schaff, H. V., and Frye,

R. L.: Effect of dipyridamole and aspirin on late vein-graft patency after coronary bypass operations. *N. Engl. J. Med.* **310:**209–214, 1984.
Cipriano, P. R., Guthaner, D. F., Orlick, A. E., Ricci, D. R., Wexler, L., and Silverman, J. F.: The effects of ergonovine maleate on coronary arterial size. *Circulation* **59:**82–89, 1979.
Cocks, T. M. and Angus, J. A.: Endothelium-dependent relaxation of coronary arteries by noradrenaline and serotonin. *Nature* **305:**627–630, 1983.
Coene, M.,-C., Herman, A. G., Jordaens, F., Van Hove, C., Verbeuren, T. J., and Zonnekeyn, L.: Endothelium-dependent relaxations in isolated arteries of control and hypercholesterolemic rabbits. *Br. J. Pharmacol.* **85:**267P, 1985.
Cohen, R. A., Shepherd, J. T., and Vanhoutte, P. M.: Inhibitory role of the endothelium in the response of isolated coronary arteries to platelets. *Science* **221:**273–274, 1983.
Curry, R. C., Jr., Pepine, C. J., Shaban, M. B., and Conti, C. R.: Similarities of ergonovine-induced and spontaneous attacks of variant angina. *Circulation* **59:**307–312, 1979.
DeFeudis, F. V.: Calcium-antagonists and atherosclerosis—Basic studies and therapeutic implications. *Life Sci.* **32:**557–563, 1983.
De Mey, J. G., Claeys, M., and Vanhoutte, P. M.: Endothelium-dependent inhibitory effects of acetylcholine, adenosine triphosphate, thrombin and arachidonic acid in the canine femoral artery. *J. Pharmacol. Exp. Ther.* **222:**166–173, 1982.
De Mey, J. G. and Vanhoutte, P. M.: Role of the intima in cholinergic and purinergic relaxation of isolated canine femoral arteries. *J. Physiol.* (Lond.) **316:**347–355, 1981a.
De Mey, J. G. and Vanhoutte, P. M.: Uneven distribution of postjunctional $alpha_1$- and $alpha_2$-like adrenoceptors in canine arterial and venous smooth muscle. *Circ. Res.* **48:**875–884, 1981b.
Faggiotto, A. and Ross, R.: Studies of hypercholesterolemia in the nonhuman primate. II. Fatty streak conversion to fibrous plaque. *Arteriosclerosis* **4:**341–356, 1984.
Faggiotto, A., Ross, R., and Harker, L.: Studies of hypercholesterolemia in the nonhuman primate. I. Changes that lead to fatty streak formation. *Arteriosclerosis* **4:**323–340, 1984.
Freiman, P. C., Orgren, K. I., Armstrong, M. L., Heistad, D. D., and Harrison, D. G.: Atherosclerosis alters endothelium-dependent vasodilatation to acetylcholine and thrombin. *Fed. Proc.* **44:**523, 1985.
Furchgott, R. F.: Role of endothelium in responses of vascular smooth muscle. *Circ. Res.* **53:**557–573, 1983.
Furchgott, R. F. and Zawadzki, J. V.: The obligatory role of endothelial cells in the relaxation of arterial smooth muscle by acetylcholine. *Nature* **288:**373–376, 1980.
Fuster, V: Pathogenesis of Atherosclerosis, in *Cardiovascular Pharmacology of the Prostaglandins* (Herman, A. G., Vanhoutte, P. M., Denolin, H., and Goossens, A., eds.) Raven, New York, 1982.

Fuster, V., Bonie, E. J. W., Lewis, J. C., Fass, D. N., Owen, C. A., Jr., and Brown, A. L.: Resistance to arteriosclerosis in pigs with von Willebrand's disease. Spontaneous and high cholesterol diet-induced arteriosclerosis. *J. Clin. Invest.* **61:**722–730, 1978.

Godfraind, T. and Miller, R. C.: Specificity of action of Ca^{++} entry blockers. A comparison of their actions in rat arteries and in human coronary arteries. *Circ. Res.* **52** (suppl. I), 81–91, 1983.

Gotto, A. M., Jr.: Some reflections on arteriosclerosis: Past, present, and future. *Circulation* **72:**8–17, 1985.

Griffith, T. M., Edwards, D. H., Lewis, M. J., Newby, A. C., and Henderson, A. H.: The nature of endothelium-derived vascular relaxant factor. *Nature* **308:**645–647, 1984.

Gryglewski, R. J., Dembinska-Kiec, A., Zmuda, A., and Gryglewska, T.: Prostacyclin and thromoxane A_2 biosynthesis capacities of heart, arteries and platelets at various stages of experimental atherosclerosis in rabbits. *Atherosclerosis* **31:**385–394, 1978.

Habib, J. B., Wells, S.L., Williams, C. L., and Henry, P. D.: Atherosclerosis impairs endothelium-dependent arterial relaxation. *Circulation* **70** (suppl. II)**:**123, 1984.

Hallermayer, G.: Animal models of atherosclerosis. *Diab. Croat.* **11** (suppl. III)**:**7–15, 1982.

Harker, L. A., Ross, R., Slichter, S. J., and Scott, C. R.: Homocystine-induced arteriosclerosis: The role of endothelial cell injury and platelet response in its genesis. *J. Clin. Invest.* **58:**731–741, 1976.

Heistad, D. D., Armstrong, M. L., Marcus, M. L., Piegors, D., and Mark, A. L.: Augmented responses to vasoconstrictor stimuli in hypercholesterolemic and atherosclerotic monkeys. *Circ. Res.* **54:**711–718, 1984.

Henry, P. D. and Bentley, K.: Suppression of atherogenesis in cholesterol-fed rabbits treated with nifedipine. *J. Clin. Invest.* **68:**1366–1369, 1981.

Henry, P. D. and Yokoyama, M.: Supersensitivity of atherosclerotic rabbit aorta to ergonovine. Mediation by a serotonergic mechanism. *J. Clin. Invest.* **66:**306–313, 1980.

Herman, A. G., Coene, M.-C., Jordaens, F., Van Hove, C., Zonnekeyn, L., and Verbeuren, T. J.: Endothelium-dependent Relaxations are Modified in Isolated Arteries from Hypercholesterolemic Rabbits, in *Abstracts of the 50th Jubilee Congress of the Hungarian Physiological Society* Budapest, Hungary, 1985.

Herman, A. G., Jordaens, F. H., and Verbeuren, T. J.: Adrenergic, serotonergic and EDRF-mediated responses in the atherosclerotic thoracic aorta of the rabbit. *Br. J. Pharmacol*, in press, 1987.

Hess, H., Mietaschk, A., and Deichsel, G.: Drug-induced inhibition of platelet function delays progression of peripheral occlusive arterial disease. A prospective double-blind arteriographically controlled trial. *Lancet* **1:**415–419, 1985.

Heupler, J. F. A., Proudfit, W. L., Razais, M., Shirry, E. K., Greenstreet, R., and Sheldon, W. C.: Ergonovine maleate provocative test for coronary arterial spasm. *Am. J. Cardiol.* **41:**631–640, 1978.

Jayakody, R. L., Senaratne, M. P. J., Thomson, A. B. R., and Kappagoda, C. T.: Cholesterol feeding impairs endothelium-dependent relaxation of rabbit aorta. *Can. J. Physiol. Pharmacol.* **63:**1206–1209, 1985.

Juergens, J. L., Spittle, J. A., and Fairbarin, J. F., II: *Peripheral Vascular Disease* Saunders, Philadelphia, Pennsylvania, 1980.

Kalsner, S. and Richards, R.: Coronary arteries of cardiac patients are hyperreactive and contain stores of amines: A mechanism for coronary spasm. *Science* **223:**1435–1437, 1984.

Kottke, B. A. and Subbiah, M. T. R.: Pathogenesis of atherosclerosis. Concepts based on animal models. *Mayo Clin. Proc.* **53:**35–48, 1978.

Maggi, C. A., Manzini, S., Grimaldi, G., Evangelista, S., and Meli, A.: High-cholesterol diet induces altered responsiveness of rabbit arterial smooth muscle to noradrenaline. *Pharmacology* **30:**273–280, 1985.

McCalden, T. A. and Nath, R. G.: Hypercholesterolemia and the response of the rabbit ear and basilar arteries to biogenic amines. *Proc. West. Pharmacol. Soc.* **27:**411–413, 1984.

Moncada, S. and Korbut, R.: Dipyridamole and other phosphodiesterase inhibitors act as antithrombotic agents by potentiating endogenous prostacyclin. *Lancet* **1:**1286–1289, 1978.

Moncada, S., Gryglewski, R., Bunting, S., and Vane, J. R.: An enzyme isolated from arteries transforms prostaglandin endoperoxides to an unstable substance that inhibits aggregation. *Nature* **263:**663–665, 1976.

Mudge, G. H., Jr., Grossman, W., Mills, R. M., Jr., Lesch, M., and Braunwald, E.: Reflex increase in coronary vascular resistance in patients with ischemic heart disease. *N. Eng. J. Med.* **295:**1333–1337, 1976.

Mustard, J. F., Murphy, E. A., Rowsell, H. C., and Downie, H. G.: Platelets and atherosclerosis. *J. Atheroscler. Res.* **4:**1–28, 1964.

Nanda, V. and Henry, P. D.: Increased serotonergic and alpha-adrenergic receptors in aortas from rabbits fed a high cholesterol died. *Clin. Res.* **30:**209A, 1982.

Rokitansky, C.: A manual of pathological anatomy. Translated by G. E. Day, vol. 4, p. 261, Sydenham Society, London, UK, 1852.

Rosendorff, C., Hoffman, J. I. E., Verrier, E. D., Rouleau, J., and Boerboom, L. E.: Cholesterol potentiates the coronary artery response to norepinephrine in anesthetized and conscious dogs. *Circ. Res.* **48:**320–329, 1981.

Ross, R.: Atherosclerosis—a problem of the biology of arterial wall cells and their interaction with blood components. *Arteriosclerosis* **1:**293–311, 1981.

Ross, R.: The pathogenesis of atherosclerosis—an update. *N. Engl. J. Med.* **8:**488–500, 1986.

Ross, R. and Glomset, J. A.: 1. The pathogenesis of atherosclerosis. *N. Engl. J. Med.* **295:**369–377, 1976a.

Ross, R. and Glomset, J. A.: 2. The pathogenesis of atherosclerosis. *N. Engl. J. Med.* **295:**420–425, 1976b.

Ross, R. and Harker, L.: Hyperlipidemia and atherosclerosis. *Science* **193:**1094–1100, 1976.

Schroeder, J. S., Bolen, J. L., Quint, R. A., Clark, D. A., Hayden, W. G., Higgins, C. B., and Wexler, L.: Provocation of coronary spasm with ergonovine maleate. *Am. J. Cardiol.* **40:**487–491, 1977.

Shimokawa, H., Tomoike, H., Nabeyama, S., Yamamoto, H., Araki, H., and Nakamura, M.: Coronary artery spasm induced in atherosclerotic miniature swine. *Science* **221:**560–562, 1983.

Sinzinger, H., Winter, M., and Feigl, W.: Prostacyclin (PGI_2)-generation by different types of human atherosclerotic lesions. *Lancet* **2:**469, 1979.

Smith, E. B.: The relationship between plasma and tissue lipids in human atherosclerosis. *Adv. Lipid Res.* **12:**1–49, 1974.

Stender, S., Stender, I., Nordestgaard, B., and Kjeldsen, K.: No effect of nifedipine on atherogenesis in cholesterol-fed rabbits. *Arteriosclerosis* **4:**389–394, 1984.

Van de Voorde, J. and Leusen, I.: Role of the endothelium in the vasodilator response of rat thoracic aorta to histamine. *Eur. J. Pharmacol.* **87:**113–120, 1983.

Vanhoutte, P. M. and Rimele, T. J.: Role of the endothelium in the control of vascular smooth muscle function. *J. Physiol.* (Paris):681–686, 1983.

Verbeuren, T. J., Jordaens, F. H., Zonnekeyn, L. L., Van Hove, C. E., Coene, M.-C., and Herman, A. G.: Effect of hypercholesterolemia on vascular reactivity in the rabbit. I. Endothelium-dependent and endothelium-independent contractions and relaxations in isolated arteries of control and hypercholesterolemic rabbits. *Circ. Res.* **58:**552–564, 1986a.

Verbeuren, T. J., Coene, M.-C., Jordaens, F. H., Van Hove, C. E., Zonnekeyn, L. L., and Herman, A. G.: Effect of hypercholesterolemia on vascular reactivity in the rabbit. II. Influence of treatment with dipyridamole on endothelium-dependent and endothelium-independent responses in isolated aortas of control and hypercholesterolemic rabbits. *Circ. Res.* **59:**496–504, 1986b.

Verbeuren, T. J., Coene, M.-C., Jordaens, F. H., Van Hove, C., Zonnekeyn, L., and Herman, A. G.: Effect of hypercholesterolemia on endothelium-dependent relaxations in isolated arteries of the rabbit. *Arch. Int. Pharmacodyn. Ther.* **277:**333–334, 1985a.

Verbeuren, T. J., Coene, M.-C., Jordaens, F., Van Hove, C., Zonnekeyn, L., and Herman, A. G.: Influence of dipyridamole on the changes in vascular reactivity induced by hypercholesterolemia in rabbit arteries. *Naunyn Schmiedebergs Arch. Pharmacol.* **330** (suppl.)**:**R46, 1985b.

Virchow, R.: Phlogose und Thrombose im Gefasssystem, Gesammelte Abhandlungen zur Wissenschaftlichen Medicin, Meidinger Sohn, Frankfurt-am-Main, 1856.

Waters, D. D., Szlachcic, J., Bonan, R., Miller, D. D., Dauwe, F., and Theroux, P.: Comparative sensitivity of exercise, cold pressor and ergonovine testing in provoking attacks of variant angina in patients with active disease. *Circulation* **67:**310–315, 1983.

Yokoyama, M., Akita, H., Mizutani, T., Fukuzaki, H., and Watanabe, Y.: Hyperreactivity of coronary arterial smooth muscles in response to ergonovine from rabbits with hereditary hyperlipidemia. *Circ. Res.* **53:**63–71, 1983.

Chapter 23

Endothelium-Dependent Relaxations in Hypertensive Blood Vessels

Raymond J. Winquist

1. Introduction

A diminished ability to relax has been advanced as a characteristic attribute of vascular preparations from animal models of hypertension (Cohen and Berkowitz, 1976; Shibata and Cheng, 1978; Cohen et al., 1982; Winquist et al., 1982). The discovery that the endothelium is of paramount importance for the action of many vasodilators (Furchgott and Zawadzki, 1980; Furchgott, 1983) has prompted several studies to assess whether endothelium and/or abnormalities of smooth muscle underlie the depressed relaxation of hypertensive blood vessels. Indeed, the expression of an endothelium-derived relaxing factor (EDRF) in resistance arteries (De Mey and Gray, 1985; Owen and Bevan, 1985) implies that endothelium-dependent relaxation responses are important modulators of peripheral vascular tone and hence resistance in vivo. Abnormalities in endothelium-dependent relaxations might be expected from the structural alterations in endothelial cells that have been documented in various models of hypertension. Therefore, this chapter will examine the data that suggest that endothelium-dependent relaxations are compromised in hypertensive blood vessels. Analysis of these studies requires, as a preamble, consideration of factors that influence interpretation of these data.

2. Factors Influencing Vasodilator Responsiveness

2.1. Contractile State of Vascular Preparation

An important consideration when studying vasodilator responsiveness is the amount of contractile force or preload of the vascular preparation before challenge with the vasodilator agent. The relaxation of vascular segments to nitrovasodilators (Furchgott and Bhadrakom, 1953; Cohen and Berkowitz, 1974) as well as to papaverine (Bevan et al., 1978) shows an inverse relationship to the contractile preload (i.e., effectiveness $\propto \frac{1}{\text{preload}}$). Interestingly, not all vasodilators show this inverse relationship (e.g., isoproterenol, Bevan et al., 1978). The force generated to various spasmogens in vascular smooth muscle is typically altered in hypertension (Winquist et al., 1982), which necessitates care in comparing vasodilator responsiveness in hypertensive versus normotensive preparations. Similar levels of preload should be used in order to ascertain whether an altered relaxation is a function of the disease entity and not a result of a different level of contractile state.

2.2. Age

Both contractile and relaxant properties of vascular preparations can vary substantially with age (Cohen and Berkowitz, 1974; Fleisch, 1980). In general, vasodilator responsiveness appears to decline with senescence. Arteries may be more susceptible, however, than veins to age-related decreases in vasodilator responsiveness (e.g., Duckles and Hurlbert, 1986), and, as discussed for the effects of preload, not all vasodilators exhibit similar declines in efficacy with advancing age (e.g., Cohen and Berkowitz, 1974). Although most studies examine age-matched hypertensive and normotensive rats, this clearly becomes a complicating issue when comparing results from different laboratories.

2.3. Origin of Vasodilator Abnormality

Comparison of endothelium-dependent vs endothelium-independent relaxations is a necessity in order to discern if abnormal endothelial function exists in hypertension. This requires the inclusion of appropriate compounds that act either via release of EDRF or directly on the vascular smooth muscle. Interpretation of data obtained with acetylcholine, the prototype compound used for the former criterion, may be complicated by any alteration in receptor sensitivity in hypertension (Winquist et al., 1982). This complexity should be circumvented by the incorporation of experiments using the calcium ionophore A23187. Nitrovasodilators have been typically used as agents that act directly on vascular smooth muscle (e.g., *see* Winquist et al., 1984; Lockette et al., 1986), since these compounds (1) relax endothelium-denuded vascular preparations and (2) appear to elicit relaxation of the smooth muscle via the same cellular mechanism (i.e., increased production of cyclic GMP) as EDRF (Rapoport et al., 1983). A recent study by Shirasaki and Su (1985), however, showed that both sodium nitrite and sodium nitroprusside elicited a greater relaxation in endothelium-denuded aortic rings compared to intact preparations. This augmentation was more pronounced in aortic rings from spontaneously hypertensive rats (SHR) than in the Wistar-Kyoto normotensive control rats (WKY). Although these data question the use of nitrovasodilators as endothelium-independent relaxing agents, the findings imply that their use as reference compounds would underestimate any abnormality in endothelium-dependent relaxation. It remains to be determined if this endothelium-dependency of the nitrovasodilators exists in other tissues and if the augmentation (upon rubbing) is more pronounced in other forms of hypertension.

Incorporation of suitable reference compounds is further complicated by the possibility of both species and strain differences in the importance of the endothelium for vasodilator responsiveness. For example, adenosine was initially described as an endothelium-independent relaxing agent in both rabbit (Furchgott, 1981) and canine (De Mey and Vanhoutte, 1981) vascular preparations, but a subsequent study showed that adenosine exhibits an endothelium-dependent component in relaxing aortic rings from SHR (Konishi and Su, 1983). The studies on nitrovasodilators (Shirasaki and Su, 1985) and adenosine (Konishi and Su, 1983) in SHR and WKY imply strain differences in the dependence on the endothelium. These results must be kept in mind when interpreting relaxant responses of hypertensive versus normotensive vascular preparations.

3. Relaxations in Hypertensive Blood Vessels

3.1. Prior to Discovery of EDRF

Most of the studies that examined vasodilator responsiveness in preparations from hypertensive animals were conducted prior to the discovery of endothelium-dependent (i.e., via an EDRF) and endothelium-independent mechanisms (e.g., Cohen and Berkowitz, 1976; for reviews, *see* Cohen et al., 1982; Winquist et al., 1982). Although these studies were not designed for evaluation of altered endothelium-dependent responses, some of the data are pertinent for discussion, since acetylcholine was evaluated by the investigators.

Shibata and Cheng (1978) examined the relaxant effects of acetylcholine (Fig. 1) and isoproterenol on helically cut strips of aortas from SHR and normotensive Wistar rats (NWR). Both drugs were less effective in relaxing thoracic aortic strips from SHR compared to NWR, which suggests a generalized defect in the ability of the vascular smooth muscle from SHR to relax. The authors state, however, that nitroglycerin and papaverine, relaxants that act primarily on vascular smooth muscle, elicited similar responses in the vascular preparations from SHR and NWR. These latter findings raise the question of whether the results could be explained by changes in receptor sensitivity in both the endothelium (acetylcholine) and vascular smooth muscle (isoproterenol), as described above (section 2.3). The defect occurred in thoracic aortic strips from young, prehypertensive (3–5 wk) as well as in older (12–16 wk, 22–25 wk) hypertensive SHR, which led the authors to conclude that the depressed response was not a function of the elevated mean arterial pressure. The interpretation of these data are complicated because the preparations were contracted by an equivalent concentration of serotonin ($10^{-5}M$), which is a supramaximal concentration (e.g., *see* Webb and Bohr, 1984). Moreover, the preparation of helically cut vascular strips is tacitly assumed to damage the vascular endothelium, which may be more pronounced when preparing strips of vessels from hypertensive animals. Shibata and Cheng found, however, that vascular strips prepared from abdominal aortas from SHR and NWR gave comparable relaxations to acetylcholine (Fig. 1B) or isoproterenol. These data imply that regional differences exist in the comparison of relaxations between vessels from hypertensive and normotensive animals and should serve as a cautionary note when interpreting observations (such as in thoracic aortic preparations) as a consequence solely of elevated arterial pressure. It is possible, for example, that the regional difference noted by Shibata and Cheng could be a result of a greater shear force (and hence endothelium damage) in the thoracic compared to the abdominal aorta.

Cauvin and Pegram (1983) found a depressed relaxation to acetylcholine (Fig. 2) in small (175 μm internal diameter) mesenteric arteries from two-kidney, one-clip (2K-1C) renal hypertensive rats compared to the response in arteries from sham-operated controls. These authors also reported that adenosine was less effective in relaxing the small vessels from the 2K-1C versus control rats and concluded that a functional defect exists in the smooth muscle from the hypertensive animals. It is not clear whether adenosine has an endothelium-dependent component to its relaxation response in these vessels; if so, a functional (or structural) defect in the endothelium may explain these observations, since the relaxation

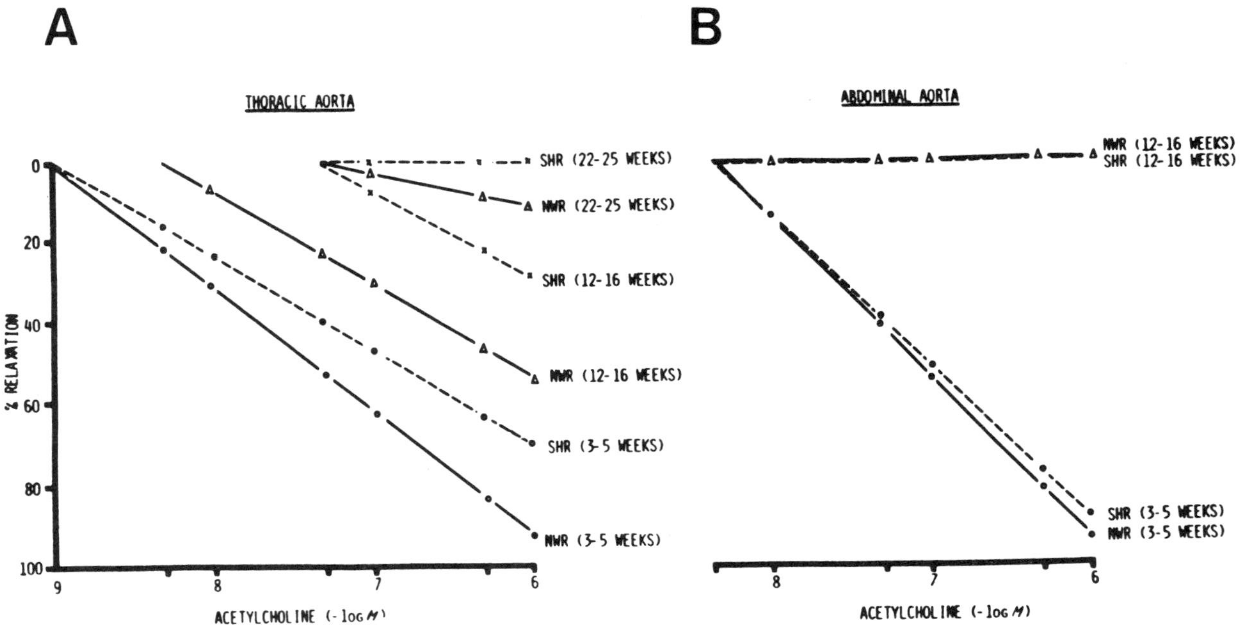

Fig. 1. The relaxing effects of acetylcholine on thoracic (A) or abdominal (B) aortic strips from spontaneously hypertensive (SHR) or normotensive Wistar (NWR) rats at different ages. Aortic strips were contracted with serotonin ($10^{-5}M$) with the responses calculated as percent relaxations of the serotonin-induced contraction. The acetylcholine-induced relaxations of thoracic aortic strips from SHR were significantly ($p < 0.05$) less than the relaxations in thoracic strips from NWR. No difference was observed in abdominal aortic strips [reproduced from Shibata and Cheng (1978) by courtesy of S. Karger AG, Basel].

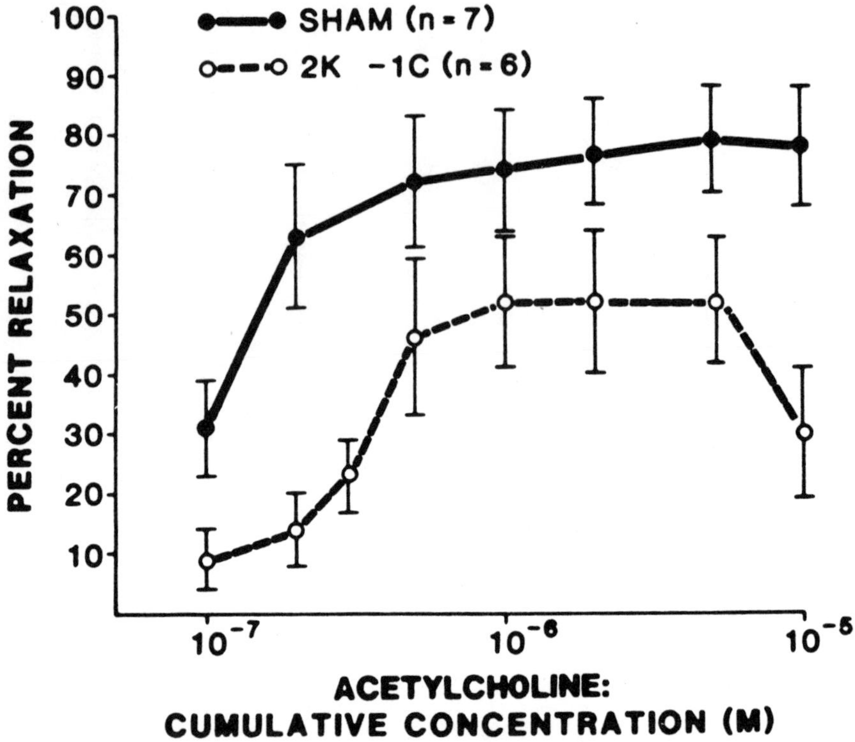

Fig. 2. The relaxing effects of acetylcholine on small (150 μm) mesenteric arteries from two-kidney, one-clip (2K-1C) renal hypertensive rats and sham-operated controls. Arteries were contracted with norepinephrine ($5 \times 10^{-6}M$) with the ordinate expressed as percent relaxation of the norepinephrine-induced contraction. Relaxations were significantly less ($p < 0.05$) in vessels from the 2K-1C rats [reproduced from Cauvin and Pegram (1983, Fig. 4, p. 393) by courtesy of Marcel Dekker].

to acetylcholine, ostensibly acting completely via the release of EDRF, appeared to be depressed to a greater extent than the relaxation to adenosine (having both endothelium-dependent and -independent components). Clearly, additional studies with other vasodilator drugs would be necessary to substantiate such a conclusion. Nonetheless, the study by Cauvin and Pegram is of interest in providing evidence for a depressed relaxation to acetylcholine in resistance vessels from an animal model of hypertension.

3.2. Studies Analyzing Endothelium-Dependent Vs Independent Relaxations in Hypertensive Blood Vessels

There are now several laboratories that have provided a more focused analysis of endothelium-dependent versus -independent relaxations in vascular preparations from animal models of hypertension. These studies have typically documented the endothelium-dependence of the relaxation to acetylcholine and/or A23187 in the particular model and have contrasted these responses with putative endothelium-independent relaxations such as those obtained with nitrovasodilators, isoproterenol, potassium ions, or adenosine. In retrospect, some of the conclusions reached by these authors must be viewed with the perspective that very few compounds are truly endothelium-independent.

3.2.1. Genetic Hypertension

Konishi and Su (1983) reported that acetylcholine (Fig. 3A), but neither isoproterenol nor adenosine, exhibited a depressed maximal relaxation in thoracic aortic rings from SHR versus WKY (15–16 wk old). These results suggest the presence of an endothelial abnormality in the aortas from SHR. The findings could also be a result, however, of changes in (endothelial) receptor sensitivity to acetylcholine. The discrepancy in the responsiveness to isoproterenol between this study (no difference between SHR and WKY) and the study by Shibata and Cheng (diminished relaxation in SHR versus NWR) may be a function of (1) different strains of normotensive rats (Wistar Kyoto versus Wistar), (2) different conditions of active preload (submaximal norepinephrine versus supramaximal serotonin), and/or (3) ring versus strip preparations (conditions in Konishi and Su versus Shibata and Cheng, respectively).

Konishi and Su also found evidence for regional vascular differences, since the maximal relaxation to acetylcholine was actually more pronounced in femoral arteries from SHR versus WKY (Fig. 3B). The reason for this qualitative difference is not clear, but the exact conditions of preload (e.g., level of contractile force of SHR and WKY) were not specified. Nonetheless, the data would favor the interpretation that the greater shear force in the thoracic versus abdominal aorta (as in Shibata and Cheng) or femoral artery (as in Konishi and Su) is instrumental in modulating endothelium-dependent relaxations in hypertension.

As mentioned above, Konishi and Su reported that adenosine contained an endothelium-dependent component in its relaxation of aortas from SHR, but not WKY. Therefore, this study provides a rather paradoxical finding: the endothelium in the SHR appears (structurally or functionally) damaged, since the relaxation to acetylcholine is diminished, but, unlike the situation in WKY, it develops a component of the overall response to adenosine. The fascinating implication is that under pathophysiological conditions the vascular endothelium may play an important role in the vasodilator response to compounds that normally act directly on the vascular smooth muscle.

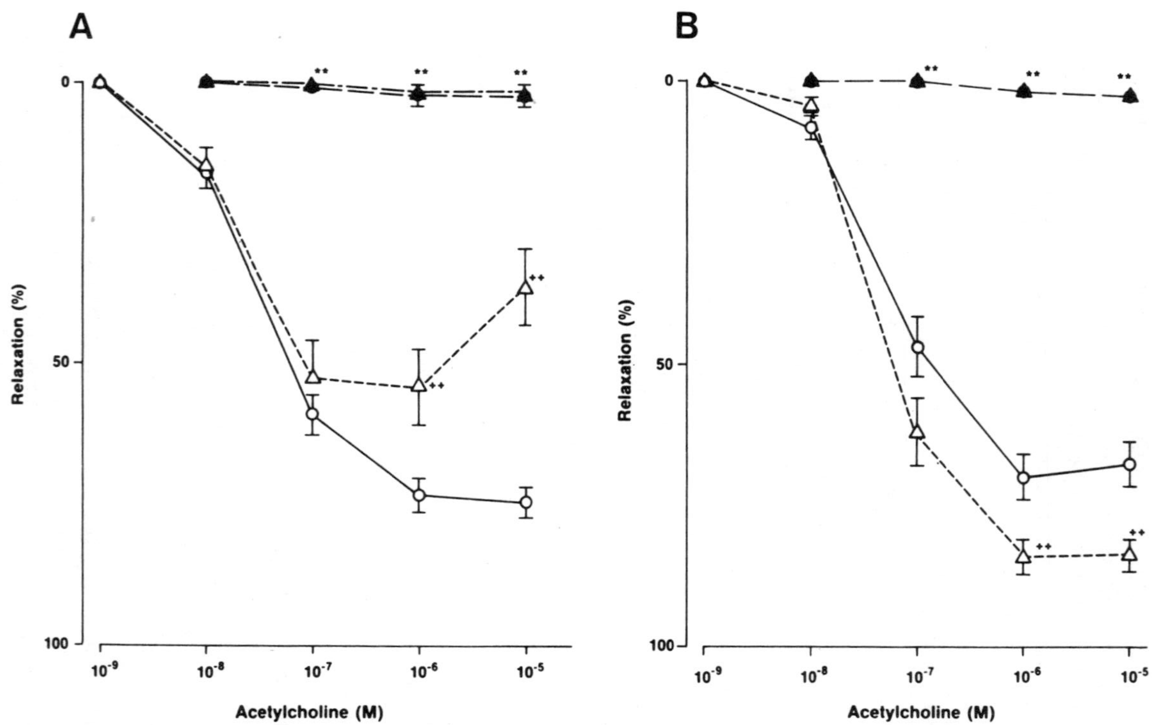

Fig. 3. The relaxing effects of acetylcholine in either intact (open symbols) or endothelium-denuded (closed symbols) thoracic aortic (A) and femoral arterial (B) rings from spontaneously hypertensive (SHR, △, ▲) or normotensive Wistar-Kyoto (WKY, ○, ●) rats. The asterisks or crosses mark a significant difference ($p < 0.01$) between SHR and WKY (**) or intact and denuded (+ +) rings. Rings were contracted by the EC_{50} of norepinephrine (10^{-8} to $10^{-7}M$) [reproduced from Konishi and Su (1983), by permission of the American Heart Association, Inc.].

The author's laboratory studied endothelium-dependent (acetylcholine, A23187) and endothelium-independent (sodium nitroprusside) relaxations in thoracic aortic rings from age-matched (130 d) New Zealand genetic hypertensive rats (GH) and their normotensive controls (N) (Winquist et al., 1984). The apparent sensitivity to all of the vasodilators was decreased to a similar extent, but only the endothelium-dependent vasodilators were associated with a diminished maximal relaxation (Fig. 4). The depressed maximal relaxation obtained to acetylcholine (Fig. 4A) appears similar to the findings by Konishi and Su (1983). The identical finding with A23187 rules unlikely the possibility that the observations could be caused by changes in receptor sensitivity and points out that depressed endothelium-dependent relaxations are not specific for cholinergic agonists.

As reported by Konishi and Su (1983), Winquist et al. showed that the endothelium-dependent vasodilator drugs failed to cause contractile responses in endothelium-denuded aortic rings initially contracted with an alpha-adrenoceptor agonist. Recently, Luscher and Vanhoutte (1986) have shown that acetylcholine, in the absence of any active preload, can elicit endothelium-dependent contractions in thoracic aorta from SHR, which may account, in part, for the depressed maximal relaxation to acetylcholine in the referenced studies. Such experiments have not been published for A23187. We have observed, however, depressed relaxations to acetylcholine in renal arteries from SHR versus WKY in the absence of any appreciable contractile responses (to explain the depressed relaxation) to this agent in noncontracted preparations (Fig. 5). Therefore, although endothelium-dependent contractile responses represent a complicating, but exciting, development in the interpretation of vasodilator responsiveness, it is not at present likely that such phenomena can account fully for the depressed endothelium-dependent relaxations in hypertension.

It is of interest to note that although regional differences were reported for large conduit arteries, a depressed endothelium-dependent relaxation has been documented in resistance arteries. De Mey and Gray (1985) reported on the vasodilator responsiveness of small mesenteric arteries (200 μM) from SHR and WKY. These authors used acetylcholine, bradykinin, and histamine as endothelium-dependent and sodium nitroprusside and potassium ions as relaxants acting directly on smooth muscles. The maximal relaxation to the endothelium-dependent vasodilators was depressed in SHR versus WKY, which is similar to the results found for the larger conduit vessels discussed above. Sodium nitroprusside and potassium ion gave comparable relaxations in the small arterial preparations. The depressed endothelium-dependent relaxation was not observed in very young SHR (2 wk of age), which these authors maintain is an age at which blood pressure is already elevated compared to WKY. Therefore,

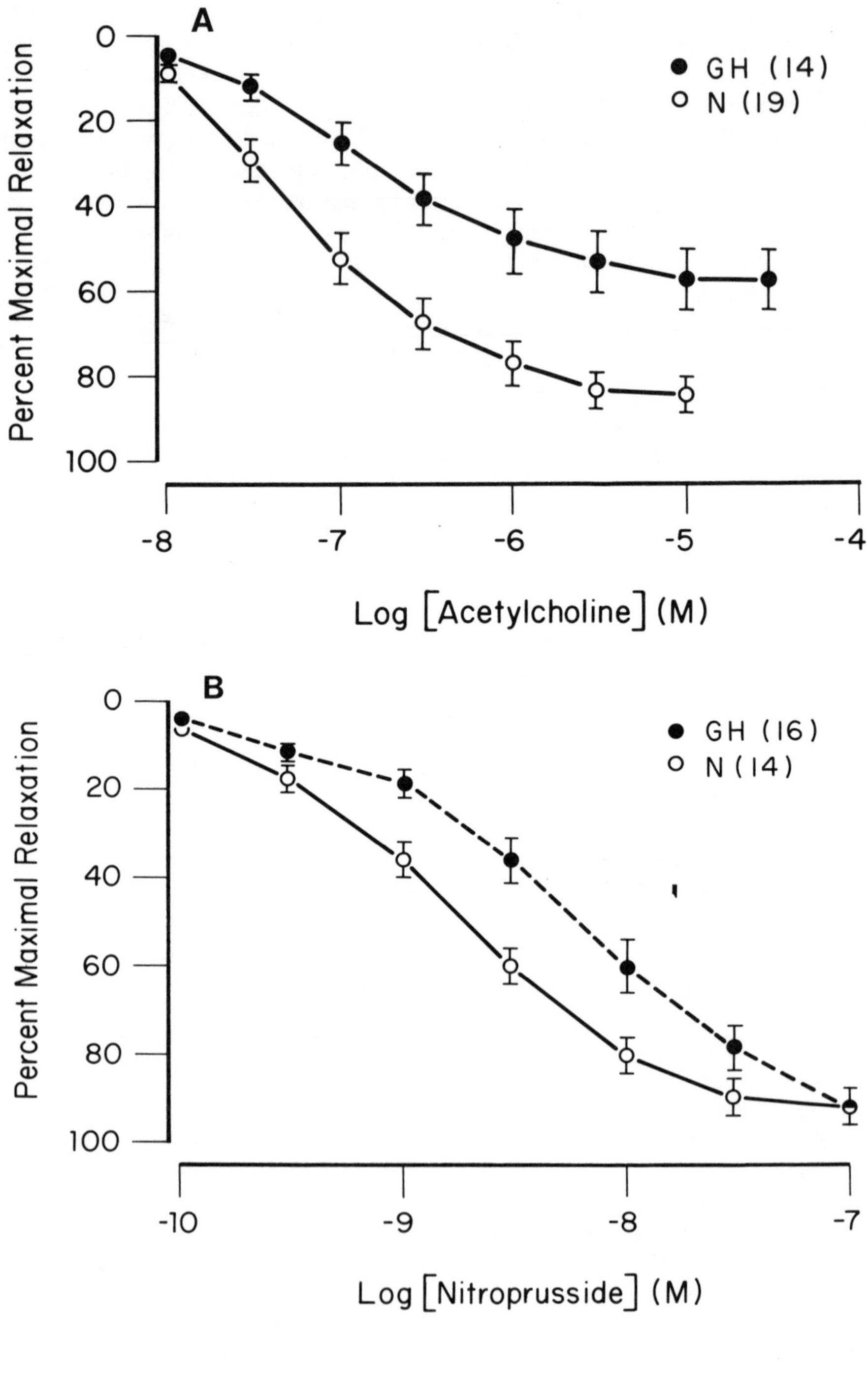
A
Percent Maximal Relaxation
0
20
40
60
80
100
GH (14)
N (19)
-8
-7
-6
-5
-4
Log [Acetylcholine] (M)
B
Percent Maximal Relaxation
0
20
40
60
80
100
GH (16)
N (14)
-10
-9
-8
-7
Log [Nitroprusside] (M)

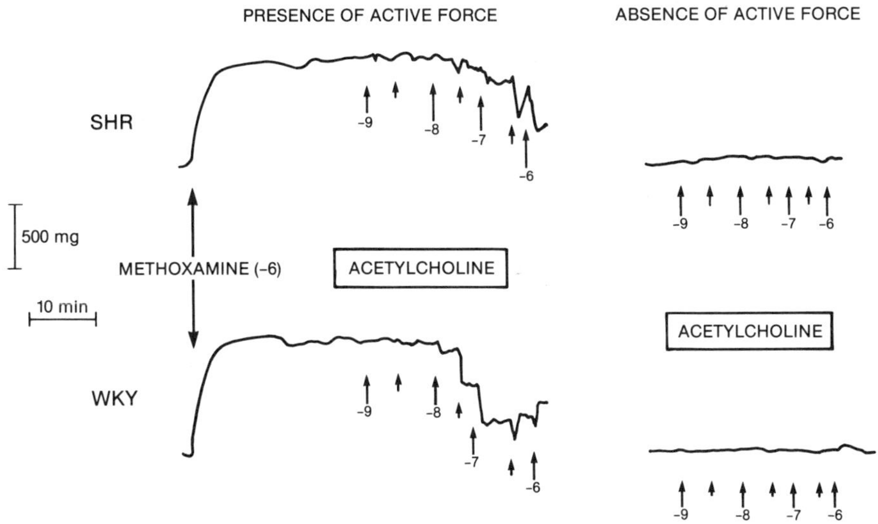

Fig. 5. Recordings of isometric force in renal arterial rings from spontaneously hypertensive (SHR, top) and normotensive Wistar-Kyoto (WKY, bottom) rats showing the effects of exogenously added acetylcholine in the presence (left panel) and absence (right panel) of active force (methoxamine, $10^{-6}M$). Acetylcholine is less effective in relaxing rings from SHR versus WKY rats (left panel). The depressed relaxation does not appear to be a function of acetylcholine-induced contractile responses (right panel). Concentrations are expressed as logarithms of cumulative molar concentrations. Truncated arrows represent half-log concentrations (Winquist and Baskin, unpublished observations).

De Mey and Gray concluded that the altered endothelium-dependent relaxation was a consequence and not part of the etiology of genetic hypertension.

3.2.2. Mineralocorticoid, Renal, and Coarctation Hypertension

Van de Voorde and Leusen (Van de Voorde et al., 1984; Leusen and Van de Voorde, 1985) examined endothelium-dependent relaxations in thoracic aortic rings from both renal (one-kidney, wrapped; Fig. 6) and deoxycorticosterone (DOCA-salt) hypertensive rats. The relaxation to his-

---

← Fig. 4. The concentration–relaxation curves to acetylcholine (A) and sodium nitroprusside (B) in thoracic aortic rings from New Zealand genetic hypertensive (GH) and age-matched normotensive (N) rats. Rings were contracted to similar levels of force (500 mg) by methoxamine. The maximal relaxation to acetylcholine, but not that to sodium nitroprusside, was significantly less ($p < 0.05$) in aortic rings from GH compared to N rats [reproduced from Winquist et al. (1984) by permission].

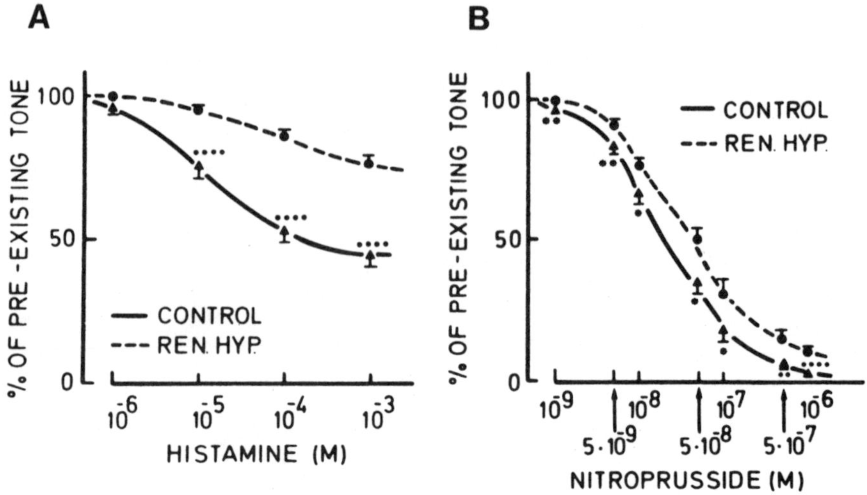

Fig. 6. The relaxant dose–response curve to histamine (A) and sodium nitroprusside (B) in thoracic aortic rings from renal hypertensive (one-kidney, wrap) and control rats. Rings were precontracted with norepinephrine ($10^{-7}M$) with the relaxation expressed as percent decreases of the norepinephrine-induced contraction. The maximal relaxation to histamine, but not that to sodium nitroprusside, was depressed in the aortic rings from the hypertensive rats. There was a small but significant shift to the right in the concentration–relaxation curve to sodium nitroprusside [reproduced from Leusen and Van de Voorde (1985), by permission].

tamine, which in the rat aorta requires the presence of an intact endothelium (Van de Voorde and Leusen, 1983), was depressed in aortas from both groups of hypertensive rats compared to the normotensive controls (Fig. 6A). The maximal relaxation obtained with sodium nitroprusside was similar in aortas from hypertensive and normotensive rats, although there was a shift to the right in the apparent potency in the hypertensive group (Fig. 6B). Thus, these data are very similar to those obtained in genetic hypertension by Winquist et al. (1984; Fig. 4). Leusen and Van de Voorde observed that the vascular endothelium in the hypertensive rat aortas exhibited ultrastructural abnormalities that may underlie the depressed relaxation to histamine. Ultrastructural abnormalities in the vascular endothelium with development of hypertension will be discussed below.

Lockette et al. (1986) investigated endothelium-dependent versus -independent relaxations in rat thoracic aortic strips from DOCA-salt, renal (one-kidney, one-clip; 1K-1C) and aortic-coarctation (between the renal arteries) models of hypertension. These authors utilized both receptor (acetylcholine)- and nonreceptor (A23187)-mediated endothelium-dependent relaxations and compared these to the relaxations obtained with sodium nitroprusside. The relaxations to acetylcholine (Fig. 7A) and A23187 were almost nonexistent in the aortic strip from each model of hypertension. The relaxation to sodium nitroprusside was, as found in the previous citations, similar in maximum, but displaced to the right (Fig. 7B). Lockette et al. also observed that upon reversal of the mineralocorticoid or renal hypertension, the relaxation to acetylcholine in the aortic strips approached that obtained in tissues from the control rats (Fig. 8). These data imply that the diminished endothelium-dependent relaxation in hypertension is a function of an elevated blood pressure.

Preliminary data from Mullane and coworkers (Pinto et al., 1985), who examined responses in aortic segments from rabbits made hypertensive via aortic coarctation, also show a depressed relaxation to acetylcholine in rings that were removed from the aortic region that was subjected to elevated blood pressure. In addition, these investigators found that aortic segments isolated distally to the coarctation (i.e., segments not subjected to elevated pressure) exhibited similar relaxations to acetylcholine compared to tissues from sham-operated controls. These data support the tenet that the defect in endothelium-dependent relaxant responses is a result of elevated blood pressure.

The study by Lockette et al. should help dispel criticism of the use of helical strips of the aorta (from normotensive rats) in examining endothelium-dependent relaxant responses. Aortic strips from control rats typically achieved close to an 80% maximal relaxation to acetylcholine or A23187 that was abolished in rubbed preparations. The possibility exists, however, that the aortic endothelium from rats with elevated blood pressure is more fragile and susceptible to damage upon cutting helical strips compared to tissues removed from normotensive rats. This may explain, in part, the almost qualitative difference (i.e., absence of relaxation) obtained by Lockette et al. The likelihood of this possibility also depends on whether such susceptibility to damage can be reversed upon normalization of blood pressure. In light of this, it would be of interest to examine the same relaxing agents in ring preparations from these animals. Other variables (i.e., age, level of preload) were well controlled in the study by Lockette et al.

Therefore, there are a number of studies in the literature that report on a diminished endothelium-dependent relaxation in vascular preparations from hypertensive versus normotensive animals. The most common observation is a depressed maximal relaxation, whereas endothelium-independent concentration–relaxation curves are typically shifted in a parallel fashion to the right.

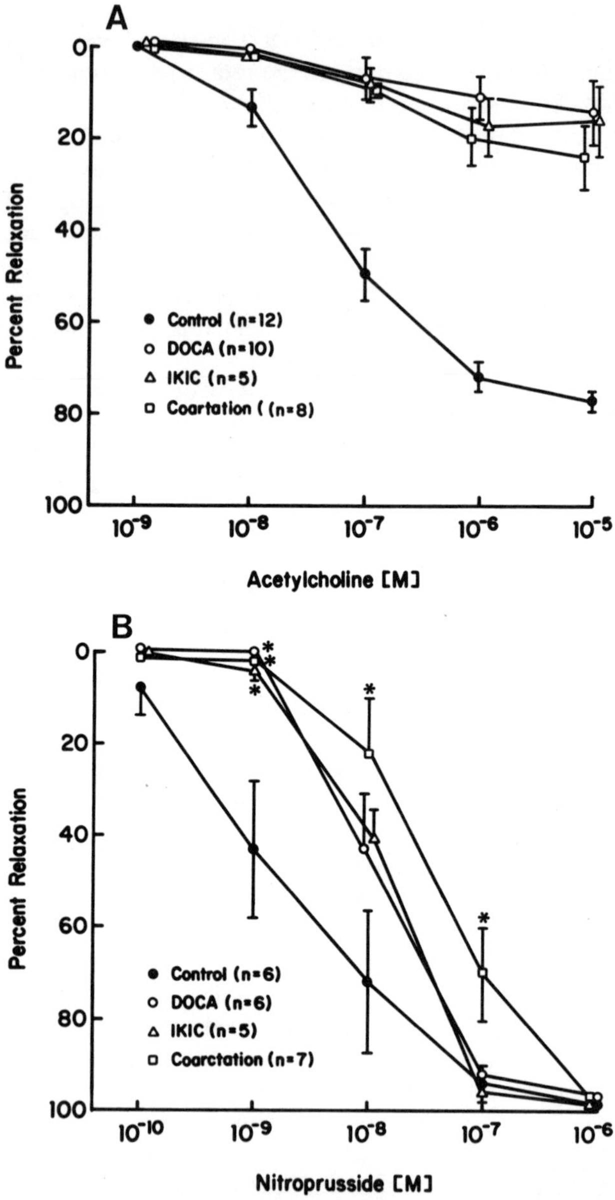

Fig. 7. The concentration–relaxation curve to acetylcholine (A) and sodium nitroprusside (B) in thoracic aortic strips from hypertensive [DOCA-salt, renal (one-kidney, one-clip) and coarctation hypertension] and normotensive control rats. Strips were contracted by the EC_{50} of phenylephrine. The relaxation to acetylcholine was significantly depressed ($p < 0.05$) in all of the aortic strips

4. Structural and/or Biochemical Endothelial Alterations in Hypertension

This survey of endothelium-dependent relaxations in hypertensive blood vessels will include a brief review of known structural and biochemical alterations of the endothelium that occur upon acute or long-term elevation of blood pressure. These data may provide helpful insight in the explanation of the pharmacological results, as well as aid in formulating the therapeutic implications of this research focus.

4.1. Structural Alterations

4.1.1. Genetic Hypertension

The aortic wall of genetic hypertensive rats undergoes a pronounced subendothelial thickening caused by an accumulation of both blood-borne and medial material (Haudenschild et al., 1980; Limas et al., 1980a,b; Hadjiisky and Peyri, 1982). The intimal changes were more apparent in the aorta versus small (20–100 μm) renal arteries (Limas et al., 1980a), but Lee et al. (1983) have observed substantial subendothelial thickening in small mesenteric arteries. Haudenschild et al. noted an increased height of aortic endothelial cells, which is believed to be an index of cell contraction that could lead to an increased permeability. The expansion of the subendothelial space could serve as a permeability barrier, and the accumulated material could act to trap or bind to vasoactive substances. Morphological abnormalities were noticeable at 10 wk of age and progressed with age (Limas et al., 1980a), with the endothelium of older SHR (i.e., 60–70 wk) displaying a hyperreactive profile (i.e., increased phagocytic and contractile activity) compared to controls (Hadjiisky and Peyri, 1982).

It is important to note, however, that there is no evidence for endothelial denudation in chronic hypertension. An increased subendothelial thickening, if acting as either a permeability barrier or a sponge to EDRF, would be expected to shift the lower end (i.e., threshold level of relaxation) of the concentration–response curve to acetylcholine. Such has not been the case, since several studies (Konishi and Su, 1983; Winquist et al., 1984) have

← Fig. 7. (*Continued*)

from the hypertensive versus the normotensive rats. The concentration–relaxation curve to sodium nitroprusside was significantly shifted to the right, but reached the same maximum in the hypertensive aortic strips [reproduced from Lockette et al. (1986), by permission of the American Heart Association, Inc.].

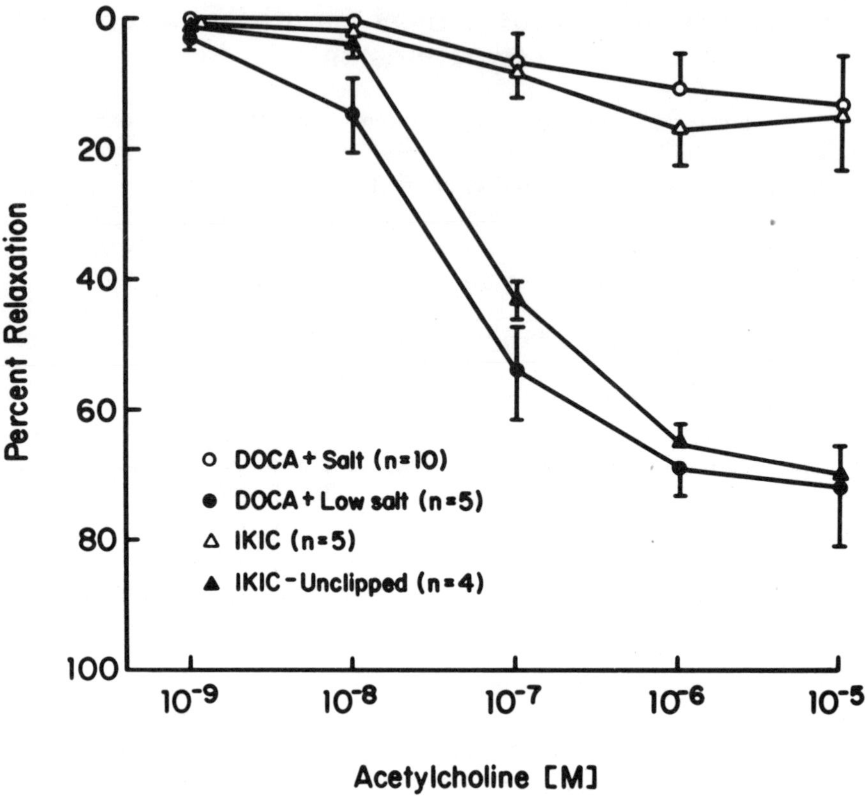

Fig. 8. The relaxing effects of acetylcholine in thoracic aortic strips from hypertensive rats (DOCA-salt, one-kidney, one-clip renal hypertension, open symbols) and from rats in which the hypertension was reversed (closed symbols) by either unclipping (renal) or a low salt diet (DOCA-salt). Strips were contracted by the EC_{50} of phenylephrine. Reversal of the hypertension restored the relaxation to acetylcholine [reproduced from Lockette et al. (1986), by permission of the American Heart Association, Inc.].

found a depressed maximum relaxation with little or no effect on the threshold relaxation to acetylcholine in aortas from genetic hypertensive versus normotensive rats. Moreover, it is unlikely that such a profile would result from medial smooth muscle hypertrophy (Winquist et al., 1984). The increased cell height or projections may be a sign of cell injury (Bjorkerud and Bondjers, 1972), which could underlie the compromised maximal expression of endothelium-dependent relaxation.

4.1.2. Mineralocorticoid, Renal, and Coarctation Hypertension

Subendothelial thickening has been observed in aortas from mineralocorticoid (Haudenschild et al., 1980), renal (Grunwald et al., 1982), and coarctation (Gabbiani et al., 1979) models of hypertension. In addition, an increased proportion of endothelial cytoplasmic projections compared to control levels were noted in these models (Still and Dennison, 1974; Svendson and Tindall, 1981; Gabbiani et al., 1979). In a chronic study on DOCA-salt hypertension, Haudenschild et al. (1980) withdrew mineralocorticoid treatment (after 7 wk of treatment), which normalized blood pressure, but failed to regress the intimal changes. These latter data would imply that the observed ultrastructural abnormalities are not causal to the depressed endothelium-dependent relaxations in hypertension, since Lockette et al. (1986; section 3.2.2.) found a return of the relaxation to acetylcholine upon normalization of blood pressure. Haudenschild et al., however, treated rats with the DOCA-salt regimen for a longer period of time (7 wk versus 4 wk in the study by Lockette et al.), which could induce irreversible alterations in endothelial structure as well as function.

Interestingly, Svendson and Tindall (1981) found that the strongest correlation between endothelial cell injury (as indicated by protruding endothelial cells) and systolic blood pressure was in the aortic arch. This observation, which ostensibly is a function of mechanical forces of the bloodstream, may provide an explanation for the regional differences in endothelium-dependent relaxations reported by Konishi and Su (1983).

4.2. Biochemical Alterations

Biochemical abnormalities were observed in aortic endothelium from SHR (stroke-prone strain) versus WKY by Yamada et al. (1983). These authors found an increased acid phosphatase and *N*-acetyl-3-glucosaminidase activity in SHR endothelium at 18 wk (but not at 12 wk of age) compared to age-matched tissues from WKY. These biochemical markers of lysosomal activity imply an increased degradation in the aortic endothelium from hypertensive rats, which may translate into a decreased production of EDRF. Unfortunately, there are no functional data available concerning endothelium-dependent versus -independent relaxation using this particular strain of genetic hypertensive rat.

Schwartz and Standaert (1981) found an increased rate of aortic endothelial replication in acute, but not chronic, renal hypertensive rats. The increased rate of endothelial replication may be an index of endothelial injury. The replication rate was not elevated in SHR, which suggests that an increased endothelial replication could be an adaptive phenomenon with acute elevations in blood pressure. It is unknown if chronic renal hypertension is associated with depressed endothelium-dependent relaxation, since the aforementioned studies under section 3.2.2 utilized rats with acute exposure to elevated blood pressure.

It is important to note that regional vascular differences exist in the biochemical properties of endothelial cells (Gerlach et al., 1985). These differences include the relative ability to degrade purines because of the presence or absence of xanthine dehydrogenase. It is therefore possible that in acute or chronic hypertension, xanthine dehydrogenase could be converted to xanthine oxidase, as occurs under ischemic conditions (Roy and McCord, 1983), which causes the liberation of oxygen-derived free radicals while catalyzing the formation of uric acid from hypoxanthine and xanthine. Formation of oxygen-derived free radicals has been associated with endothelium damage and loss of endothelium-dependent relaxation following acute hypertensive episodes (Kontos, 1985). Moreover, superoxide dismutase has been shown to prolong the half-life of EDRF (Rubanyi and Vanhoutte, 1986; Gryglewski et al., 1986). It remains to be determined if scavengers of free radicals normalize the depressed relaxation observed in hypertensive blood vessels.

5. Conclusions

Both functional and structural abnormalities exist in the endothelium from animal models of hypertension. In general, a depressed maximal relaxation to endothelium-dependent vasodilators in the hypertensive versus normotensive vascular preparation appears to be the most common observation. This functional abnormality occurs after both acute (e.g., DOCA-salt, renal hypertension) and chronic (e.g., genetic hypertension) exposure to elevated blood pressure and has been reversed following normalization of the blood pressure in the acute models. In addition, there are data that show that elevations in arterial pressure precede development of depressed endothelium-dependent relaxation (De Mey and Gray, 1985). These findings argue for a causative relationship between elevated blood pressure and diminished endothelium-dependent relaxation. If true, this causative relationship does not appear to be uniform throughout the vascular tree, which may be a function of differences in (1) mechanical forces such as shear stress, (2) structural abnormalities such as endothelial injury, or (3) biochemical abnormalities related to the presence or absence of metabolic enzymes. Alternatively, regional differences in the production of endothelium-derived constricting factors may explain some of the data reported in the literature.

An altered coupling between EDRF and its (vascular smooth muscle) receptor must also be considered as a possible explanation for the depressed relaxation in hypertensive blood vessels. The lack of a stable analog of EDRF has precluded such an analysis. Indirect evidence does not appear consistent with this hypothesis, however, since typically high, but not low-to-moderate, concentrations of endothelium-dependent vasodilators lack the efficacy (as compared to endothelium-independent vasodilators) in hypertensive versus normotensive preparations (e.g., Konishi and Su, 1983; Winquist et al., 1984). An altered EDRF-receptor coupling would be expected to produce a more parallel shift to the right in the concentration–relaxation curve. Nonetheless, recent data suggest that an impaired coupling between EDRF and the smooth muscle cells, and not impaired release of EDRF, may be responsible for the depressed relaxation in hypertensive blood vessels (Van de Voorde and Leusen, 1986).

Finally, the depressed endothelium-dependent relaxation in hypertension may be a corollary to the documented structural and/or biochemical abnormalities in the endothelium. It is clear that we need a systematic evaluation of duration of hypertensive exposure (using the acute models) versus reversibility of both structural and functional abnormalities. In addition, elaboration of any biochemical indices related to abnormal endothelium-dependent relaxation may aid ultimately in the elucidation and stabilization of EDRF. It is anticipated that the knowledge gained from examining the mechanisms subserving the diminished relaxation, both endothelium-dependent and -independent, in pathophysiological models will certainly be of value in formulating a more rationale approach to drug design for these disease states.

References

Bevan, J. A., Pegram, B. L., Prehn, J. L., and Winquist, R. J.: β-Adrenergic Receptor Mediated Vasodilatation, in *Mechanisms of Vasodilatation* (Vanhoutte, P. M. and Leusen, I., eds.) Karger, Basel, 1978.

Bjorkerud, S. and Bondjers, G.: Endothelial integrity and viability in the aorta of normal rabbit and rat as evaluated with dye exclusion tests and interference contrast microscopy. *Atherosclerosis* **15**:285–300, 1972.

Cauvin, C. and Pegram, B.: Decreased relaxation of isolated mesenteric resistance vessels from two-kidney, one-clip Goldblatt hypertensive rats. *Clin. Exp. Hypertens.* **A5**:383–400, 1983.

Cohen, M. L. and Berkowitz, B. A.: Age-related changes in vascular responsiveness to cyclic nucleotides and contractile agonists. *J. Pharmacol. Exp. Ther.* **191**:147–155, 1974.

Cohen, M. L. and Berkowitz, B. A.: Decreased vascular relaxation in hypertension. *J. Pharmacol. Exp. Ther.* **196**:396–406, 1976.

Cohen, D. M., Webb, R. C., and Bohr, D. F.: Nitroprusside-induced vascular relaxation in DOCA hypertensive rats. *Hypertension* **4**:13–19, 1982.

De Mey, J. G. and Gray, S. D.: Endothelium-dependent reactivity in resistance vessels. *Prog. Appl. Microcirc.* **8**:181–187, 1985.

De Mey, J. G. and Vanhoutte, P. M.: Role of the intima in cholinergic and purinergic relaxation of isolated canine femoral arteries. *J. Physiol.* (Lond.) **316:**347-355, 1981.

Duckles, S. P. and Hurlbert, J. S.: Effect of age on beta adrenergic relaxation of the rat jugular vein. *J. Pharmacol. Exp. Ther.* **236:**71–74, 1986.

Fleisch, J. H.: Age-related changes in the sensitivity of blood vessels to drugs. *Pharmacol. Ther.* **8:**477–487, 1980.

Furchgott, R. F.: The requirement for endothelial cells in the relaxation of arteries by acetylcholine and some other vasodilators. *Trends Pharmacol. Sci.* **2:**173–176, 1981.

Furchgott, R. F.: Role of endothelium in responses of vascular smooth muscle. *Circ. Res.* **53:**557–573, 1983.

Furchgott, R. F. and Bhadrakom, S.: Reactions of strips of rabbit aorta to epinephrine, isopropylarterenol, sodium nitrite and other drugs. *J. Pharmacol. Exp. Ther.* **108:**129–143, 1953.

Furchgott, R. F. and Zawadzki J. V.: The obligatory role of endothelial cells in the relaxation of arterial smooth muscle by acetylcholine. *Nature* **288:**273–276, 1980.

Gabbiani, G., Elemer, G., Guelpa, C., Vallotton, M. B., Badonnel, M.-C., and Huttner, I.: Morphologic and functional changes of the aortic intima during experimental hypertension. *Am. J. Pathol.* **96:**399–422, 1979.

Gerlach, E., Nees, S., and Becker, B. F.: The vascular endothelium: A survey of some newly evolving biochemical and physiological features. *Basic Res. Cardiol.* **80:**459–474, 1985.

Grunwald, J., Robenek, H., Mey, J., and Hauss, W. H.: *In vivo* and *in vitro* cellular changes in experimental hypertension: Electromicroscopic and morphometric studies of aortic smooth muscle cells. *Exp. Mol. Pathol.* **36:**164–176, 1982.

Gryglewski, R. J., Palmer, R. M. J., and Moncada, S.: Superoxide anion is involved in the breakdown of endothelium-derived vascular relaxing factor. *Nature* **320:**454–456, 1986.

Hadjiisky, P. and Peyri, N.: Hypertensive arterial disease and atherogenesis. *Atherosclerosis* **44:**181–199, 1982.

Haudenschild, C. C., Prescott, M. F., and Chobanian, A. V.: Effects of hypertension and its reversal on aortic intima lesions of the rat. *Hypertension* **2:**33–44, 1980.

Konishi, M. and Su, C.: Role of endothelium in dilator responses of spontaneously hypertensive rat arteries. *Hypertension* **5:**881–886, 1983.

Kontos, H. A.: Oxygen radicals in cerebral vascular injury. *Circ. Res.* **57:**508–516, 1985.

Lee, R. M. K. W., Forrest, J. B., Garfield, R. E., and Daniel, E. E.: Ultrastructural changes in mesenteric arteries from spontaneously hypertensive rats. A morphometric study. *Blood Vess.* **20:**72–91, 1983.

Leusen, I. and Van de Voorde, J.: Role of Endothelium in the Vascular Reactivity to Histamine, in *Vascular Neuroeffector Mechanisms* (Bevan, J. A., Godfraind, T., Maxwell, R. A., Stoclet, J. C., and Worcel, M., eds.) Elsevier, Amsterdam, 1985.

Limas, C., Westrum, B., and Limas, C. J.: The evolution of vascular changes in the spontaneously hypertensive rat. *Am. J. Pathol.* **98:**357–384, 1980a.

Limas, C., Westrum, B., Limas, C. J., and Cohn, J. N.: Effect of salt on the vascular lesions of spontaneously hypertensive rats. *Hypertension* **2:**477–489, 1980.

Lockette, W., Otsuka, Y., and Carretero, O.: The loss of endothelium-dependent vascular relaxation in hypertension. *Hypertension* **8:**(suppl. II)**:**II61–II66, 1986.

Luscher, T. F. and Vanhoutte, P. M.: Endothelium-dependent contractions in acetylcholine in the aorta of spontaneously hypertensive rats. *Hypertension* **8:**344–348, 1986.

Owen, M. P. and Bevan, J. A.: Acetylcholine-induced endothelium-dependent vasodilation increases as artery diameter decreases in the rabbit ear. *Experientia* **41:**1057–1058, 1985.

Pinto, A., Miller, M., McGiff, J. C., and Mullane, K.: Endothelial-dependent contractions and relaxations of rabbit aorta following coarctation-induced hypertension. *Circulation* **72:**III–264, 1985.

Rapoport, R. M., Draznin, M. B., and Murad, F.: Endothelium-dependent vasodilator- and nitrovasodilator-induced relaxation may be mediated through cyclic GMP formation and cyclic GMP-dependent protein phosphorylation. *Trans. Assoc. Am. Physiol.* **96:**19–30, 1983.

Roy, R. S. and McCord, J. M.: Superoxide and Ischemia: Conversion of Xanthine Dehydrogenase to Xanthine Oxidase, in *Oxygen Radicals and Their Scavenger Systems* vol. 2 *Cellular and Molecular Aspects* (Greenwald, R. and Cohen, G., eds.) Elsevier, New York, 1983.

Rubanyi, G. M. and Vanhoutte, P. M.: Superoxide anions and hyperoxia inactivate endothelium-derived relaxing factor(s). *Eur. J. Pharmacol.* **141:**93-96, 1985.

Schwartz, S. M. and Standaert, D. M.: Endothelial Injury in Hypertension, in *Frontiers in Hypertension Research* (Laragh, J. H., Buhler, F. R., and Seldin, D. W., eds.) Springer-Verlag, New York, 1981.

Shibata, S. and Cheng, J. B.: Vascular Relaxation in Hypertensive Rats, in *Vasodilatation* (Vanhoutte, P. M., ed.) Karger, Basel, 1978.

Shirasaki, Y. and Su, C.: Endothelium removal augments vasodilation by sodium nitroprusside and sodium nitrite. *Eur. J. Pharmacol.* **141:**93-96, 1985.

Still, W. J. S. and Dennison, S.: The arterial endothelium of the hypertensive rat. *Arch. Pathol.* **97:**337–342, 1974.

Svendson, E. and Tindall, A. R.: Raised blood pressure and endothelial cell injury in rabbit aorta. *Acta Pathol. Microbiol. Scand.* **89:**325–334, 1981.

Van de Voorde, J. and Leusen, I.: Endothelium-dependent and independent relaxation of aortic rings from hypertensive rats. *Am. J. Physiol.* **250:**H711–H717, 1986.

Van de Voorde, J. and Leusen, I.: Role of endothelium in the vasodilator response of rat thoracic aorta to histamine. *Eur. J. Pharmacol.* **87:**113–120, 1983.

Van de Voorde, J., Cuvelier, C., and Leusen, I.: Endothelium-dependent relaxation effects in aorta from hypertensive rats. *Arch. Int. Physiol. Biochim.* **92:**P10–P11, 1984.

Webb, R. C. and Bohr, D. F.: The Membrane of the Vascular Smooth Muscle Cell in Experimental Hypertension and Its Response to Serotonin, in *Smooth Muscle Contractions* (Stephens, N., ed.) Marcel Dekker, New York, 1984.

Winquist, R. J., Bunting, P. B., Baskin, E. P., and Wallace, A. A.: Decreased endothelium-dependent relaxation in New Zealand genetic hypertensive rats. *J. Hypertens.* **2:**541–545, 1984.

Winquist, R. J., Webb, R. C., and Bohr, D. F.: Vascular smooth muscle in hypertension. *Fed. Proc.* **41:**2387–2393, 1982.

Yamada, E., Hazama, F., Amano, S., Sashara, M., Kataoka, H., and Kayembe, K.: Semi-quantitative histochemical investigation of lysosomal enzyme activities in the aortic endothelial cells of spontaneously hypertensive rats. *Atherosclerosis* **47:**19–26, 1983.

Chapter 24

Mechanisms of Altered Endothelium-Dependent Responses in Hypertensive Blood Vessels

Thomas F. Luscher and Paul M. Vanhoutte

1. Introduction

The endothelium is a monolayer of squamous cells covering the inner surface of the blood vessel wall of arteries, capillaries, and veins. Although it represents only a thin layer of flat cells, its overall volume in the entire cardiovascular system is comparable to that of large organs such as the liver (Huttner and Gabbiani, 1983). The functional importance of the endothelium has been grossly underestimated for a long time in spite of its strategic anatomical position between the circulating blood and vascular smooth muscle cells. In recent years, however, it has been recognized that the vascular endothelium may play an important role in the local control of vascular smooth muscle tone, in hemostasis, and as a primary target organ in cardiovascular diseases such as atherosclerosis and hypertension (Moncada and Vane, 1978; Schwartz and Standaert, 1981; Chobanian et al., 1986; Ross, 1986; Vanhoutte et al., 1986).

This chapter focuses on alterations in the function of the endothelium in hypertension, with special reference to the production of endothelium-derived contracting factor(s).

2. Morphological Changes

In experimental hypertension, although the integrity of the endothelium as a cell layer is usually well preserved, alterations in permeability, shape, replication rate, and density of endothelial cells occur (Fig. 1; Huttner and Gabbiani, 1983). The endothelial cells appear more voluminous and bulge into the lumen. This may indicate contraction of the endothelial cells. An increased number of actin microfilaments has been observed in the endothelial cells of hypertensive rats (Gabbiani et al., 1979). The subendothelial layer of the blood vessel wall is thickened, particularly in latter stages of the disease. This might at least in part be caused by an increased permeability resulting in the accumulation of substances from the blood into the arterial wall. An increased accumulation of horseradish peroxidase is characteristic for the hypertensive endothelium (Gabbiani et al., 1979). The cell replication rate and number of endothelial cells in the aorta of hypertensive rats is increased (Huttner and Gabbiani, 1983; Chobanian et al., 1986). This may represent a response to injury or an adaptational process to the increase in size of large conduit arteries. Hypertension is also associated with an increased adherence of circulating blood cells to the endothelial cell layer. The adhering cells appear to be granulocytes, monocytes, and lymphocytes, but not platelets (Chobanian et al., 1986).

Similar morphological changes also occur during atherosclerosis induced by cholesterol feeding. Intimal lipid accumulation does not typically occur, however, in experimental hypertension in the absence of hyperlipoproteinemia (Chobanian et al., 1986). The absence of an increased predisposition to the development of atherosclerosic lesions in most experimental models of hypertension is a major difference from human hypertensive vascular disease.

Electronmicroscopic studies suggest that the number of cytoplasmatic organelles such as rough endoplasmatic reticulum, polyribosomes, microtubules, mitochondria, and Golgi complexes is increased in experimental hypertension (Huttner and Gabbiani, 1983). Also, spontaneously hypertensive rats exhibit an increased length and complexity of tight junctions at all ages (McGuire and Twietmeyer, 1985). These morphological changes are not related to the high blood pressure alone. In the various models of experimental hypertension studied, the morphological changes are more pronounced in the presence of high than low levels of mineralocorticoid hormones (Huttner and Gabbiani, 1983). In the spontaneously hypertensive rat, high sodium intake *per se* aggravates morphological changes of the intima without major changes in blood pressure (Limas et al., 1980). Thus, the increase in blood pressure and various hormonal and dietary factors may interact in the development of morphological changes of the endothelium and the vascular complications of hypertension.

3. Endothelium-Dependent Responses in Hypertensive Blood Vessels

3.1. Endothelium-Dependent Relaxations

There is a considerable heterogeneity of endothelium-dependent responses among different vascular beds from normotensive animals and among different species (De Mey and Vanhoutte, 1982; Katusic et al., 1984; Luscher et al., 1986). Blood vessels of hypertensive animals exhibit further heterogeneity in endothelium-dependent responses. Konishi and Su (1983) and Winquist et al. (1984) described decreased endothelium-dependent relaxations to acetylcholine and calcium ionophore A23187 in the thoracic aorta of genetically hypertensive rats (Fig. 2). Similar results were obtained in small mesenteric arteries of the spontaneously hypertensive rat (De Mey and Gray, 1985). Endothelium-dependent relaxations to acetylcholine are also depressed in renal hypertension, salt-induced hypertension, coarctation, and DOCA-salt hypertension of the rat (Van de Voorde and Leusen, 1984; Lockette et al., 1986; Luscher et al., 1987a). Decreased endothelium-dependent relaxations to acetylcholine are more prominent in forms of hypertension associated with a rapid and severe increase in blood pressure than in those with a more progressive moderate hypertension. In the cerebral circulation of the cat, acute severe increases in blood pressure abolish endothelium-dependent relaxations to acetylcholine; under these conditions the muscarinic agonist causes contractions (Kontos, 1985). Coarctation of the aorta causes a major reduction of endothelium-dependent relaxations (Lockette et al., 1986). On the other hand, in mice, which develop mild hypertension in response to psychosocial stress, endothelium-dependent relaxations to acetylcholine are enhanced (Webb et al., 1986, personal communication). Similar results were obtained in the femoral artery of the spontaneously hypertensive rat (Konishi and Su, 1983). Thus the endothelium may exhibit either depressed or increased responsiveness to acetylcholine depending on the species, the form and degree of hypertension, and the vascular bed studied.

Endothelium-dependent responses to substances released from aggregating platelets or formed during activation of the coagulation cascade can also be altered in hypertension. Thus, endothelium-dependent relaxations to adenosine diphosphate are reduced in the aorta of the spontaneously hypertensive and the Dahl hypertensive rat (Luscher and Vanhoutte, 1986b; Luscher et al., 1987a). Endothelium-dependent relaxations to thrombin are reduced in Dahl hypertensive, but not in spontaneously hypertensive rats. Similarly, endothelium-dependent relaxations to histamine are reduced in renal hypertensive rats, but slightly enhanced in spontaneously hypertensive rats (Van de Voorde and Leusen, 1983, 1984; Luscher and Vanhoutte,

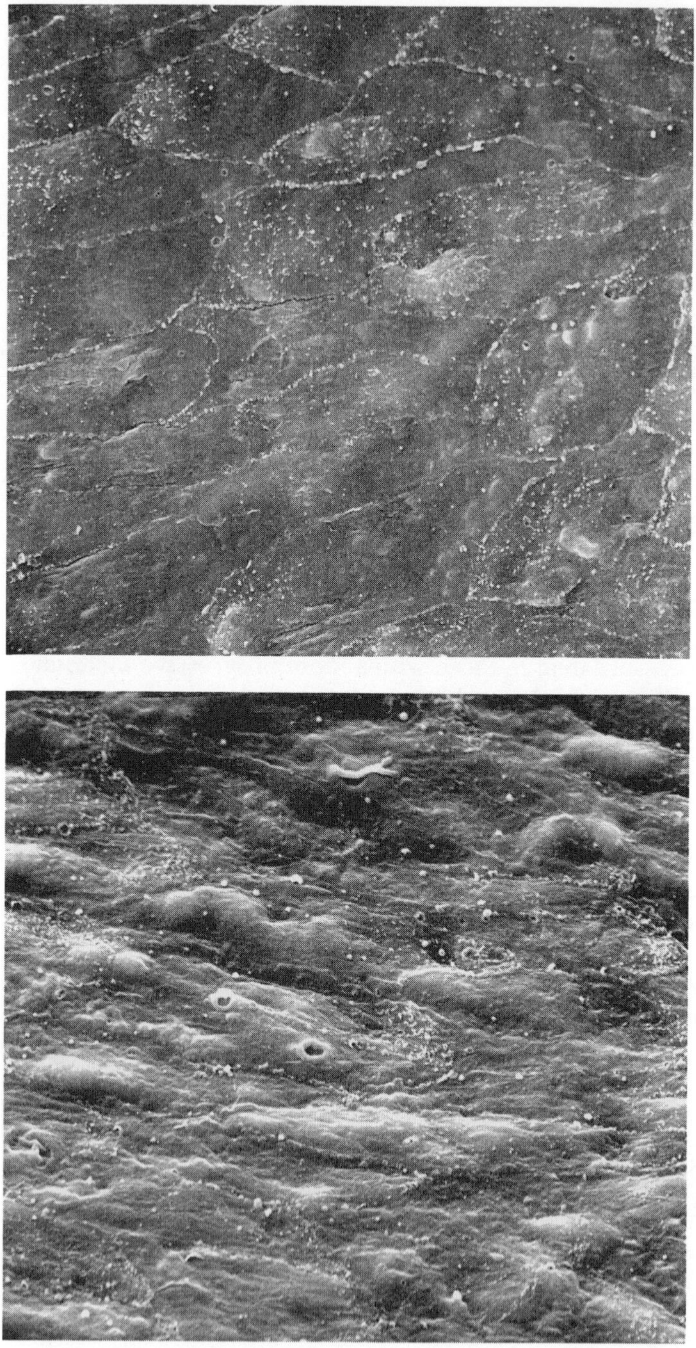

unpublished observation). These data further illustrate the heterogeneity of endothelium-dependent relaxations in different models of hypertension.

3.2. Endothelium-Dependent Contractions

Acetylcholine causes endothelium-dependent contractions in the aorta of the adult spontaneously hypertensive rat, but not in that of normotensive Wistar-Kyoto rats (Fig. 3; Luscher and Vanhoutte, 1986a). The contractions occur with higher concentrations of acetylcholine than those required to trigger the release of endothelium-derived relaxing factor(s). As for endothelium-dependent relaxations, the receptor involved is muscarinic in nature. Typically, these endothelium-dependent contractions are associated with rhythmic oscillations in tension. These oscillations in tension are blocked by ouabain or potassium-free solution, suggesting an involvement of the Na^+/K^+ ATPase (Luscher and Vanhoutte, unpublished observation). Inhibitors of H_1- and H_2-histaminergic receptors and of the converting enzyme do not affect the contractions, indicating that histamine or angiotensin II produced in the vascular wall (Dzau, 1984; De Forrest and Hollis, 1978) are not involved. The endothelium-dependent contractions to acetylcholine are abolished by inhibitors of phospholipase A_2 and cyclooxygenase, suggesting that the metabolism of arachidonic acid is responsible (Fig. 4). Since inhibitors of prostacyclin synthetase and thromboxane synthetase do not reduce the contractions induced by acetylcholine, the metabolite of arachidonic acid is not likely to be prostacyclin or thromboxane A_2. A similar conclusion has been reached regarding the endothelium-dependent contractions in canine veins (Miller and Vanhoutte, 1985). Most prostaglandins cause contraction of the rat aorta; prostacyclin causes contractions only at higher concentrations (Luscher and Vanhoutte, 1986a).

Bioassay experiments demonstrate that the major prostaglandin released from the aorta of normotensive Wistar-Kyoto rats and spontaneously hypertensive rats is prostacyclin (Luscher et al., 1987c). The aorta of the spontaneously hypertensive rat, but not that of the Wistar-Kyoto rat, may also release small amounts of prostaglandin $F_{2\alpha}$, prostaglandin E_2, and thromboxane B_2. To judge from concentration–response curves to these prostanoids given exogenously, the amounts released, however, do not appear to be large enough to fully explain endothelium-dependent contractions.

← Fig. 1. Scanning electronmicroscopy of the endothelium of the aorta of a normotensive Wistar-Kyoto rat (upper) and a spontaneously hypertensive rat (lower). Note the more prominent endothelial cells in the hypertensive animal.

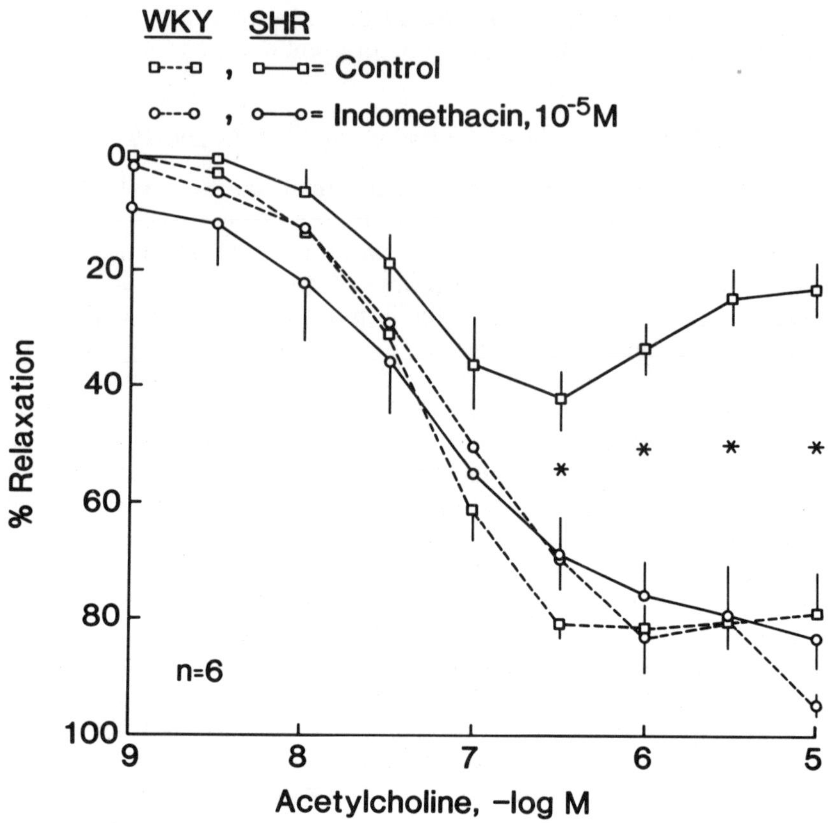

Fig. 2. Endothelium-dependent relaxations to acetylcholine in aortic rings with endothelium from normotensive Wistar-Kyoto rats (------) and spontaneously hypertensive rats (———). The rings were contracted with norepinephrine. The effect of indomethacin in either animal is represented by the dotted lines. Results are expressed as percent relaxation of the contraction to norepinephrine. Asterisk indicates significant difference from control (from Luscher and Vanhoutte, 1986a, with permission of the American Heart Association).

3.3. Mechanisms of Altered Endothelium-Dependent Responses in Hypertensive Blood Vessels

Theoretically, endothelium-dependent responses of hypertensive blood vessels could be altered because of: (1) a decreased release of endothelium-derived relaxing factor(s); (2) a decreased release of other endothelium-derived vasodilator substances such as prostacyclin; (3) the release of endothelium-derived constrictor factor(s); (4) an impaired diffusion of these factors from the endothelium to the vascular smooth muscle cells; and (5) a decreased responsiveness of the vascular smooth muscle cells to vasodilator substances.

In the aorta of the spontaneously hypertensive rat, endothelium-dependent relaxations to acetylcholine can be normalized by inhibitors of cyclooxygenase such as indomethacin (Fig. 2; Luscher and Vanhoutte, 1986a). This strongly suggests that the decreased endothelium-dependent

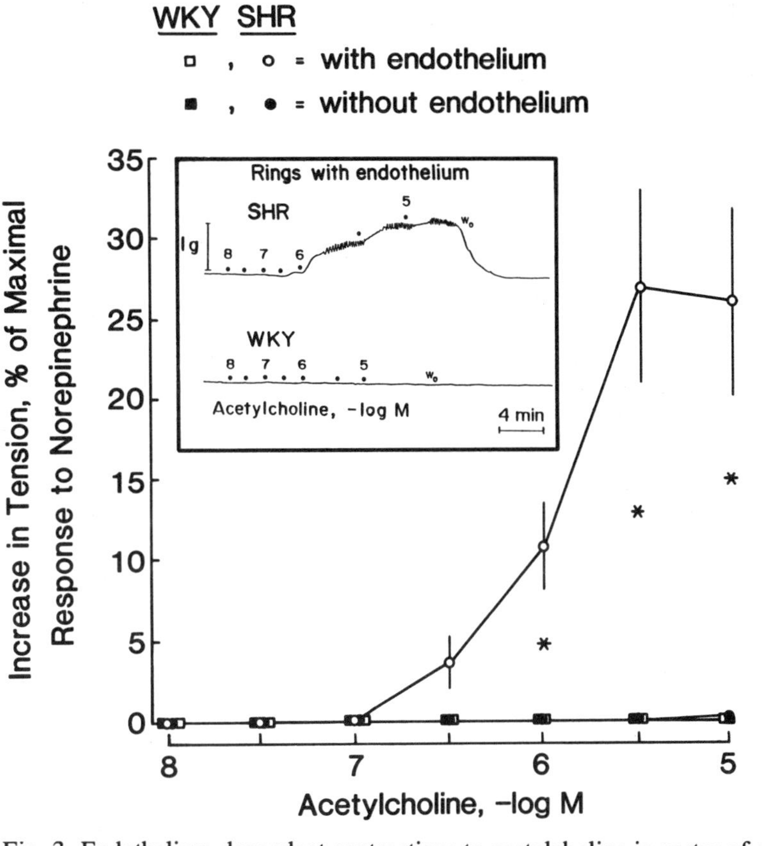

Fig. 3. Endothelium-dependent contractions to acetylcholine in aortas of spontaneously hypertensive rats (SHR). Quiescent rings with and without endothelium were exposed to increasing concentrations of the muscarinic agonist. Results are expressed as percent of the maximal response to norepinephrine. Asterisk indicates significant difference between rings from SHR with and without endothelium. WKY, normotensive Wistar-Kyoto rat. Inset: Original recordings of isometric tension in rings with endothelium of aortas of SHR and WKY (from Luscher and Vanhoutte, 1986a, with permission of the American Heart Association).

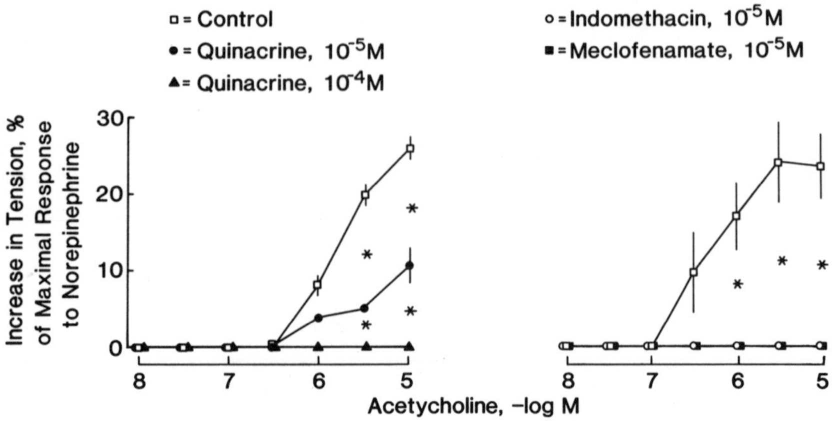

Fig. 4. Effects of quinacrine, meclofenamate, and indomethacin on endothelium-dependent contractions evoked by acetylcholine in aortas of SHR. Results are expressed as percent of the maximal response to norepinephrine. Asterisk indicates significant difference from control (from Luscher and Vanhoutte, 1986a, with permission of the American Heart Association).

relaxations in this model of hypertension are caused by the simultaneous release of endothelium-derived relaxing and contracting factors (Fig. 5). Bioassay experiments demonstrate that the release of endothelium-derived relaxing factor(s) in response to acetylcholine is normal in the aorta of the spontaneously hypertensive rat (Luscher et al., 1987c). The possibility that acetylcholine might also release endothelium-derived constrictor factor(s) in normotensive Wistar-Kyoto rats has not been excluded. Indeed, even if the release of the constrictor factor(s) would be similar in both strains, the vascular smooth muscle cells of hypertensive animals could be more sensitive to the factor(s), as they are to many neurohumoral mediators, including vasoconstrictor prostaglandins (*see* Luscher and Vanhoutte, 1986a).

Subendothelial thickening occurring during the hypertensive process could impair the diffusion of substances released from the endothelium to vascular smooth muscle cells (Gabbiani et al., 1979). In the aorta of the adult spontaneously hypertensive rat, where extensive subintimal thickening occurs, endothelium-dependent relaxations to acetylcholine are normal in the presence of indomethacin (Fig. 2; Luscher and Vanhoutte, 1986a). This then would suggest that these morphological changes are of minor importance or that the release of endothelium-derived relaxing factor(s) is actually increased. Bioassay experiments demonstate that the intraluminal stimulated release of endothelium-derived relaxing factor(s) is normal in this model of hypertension, making the latter explanation unlikely (Luscher

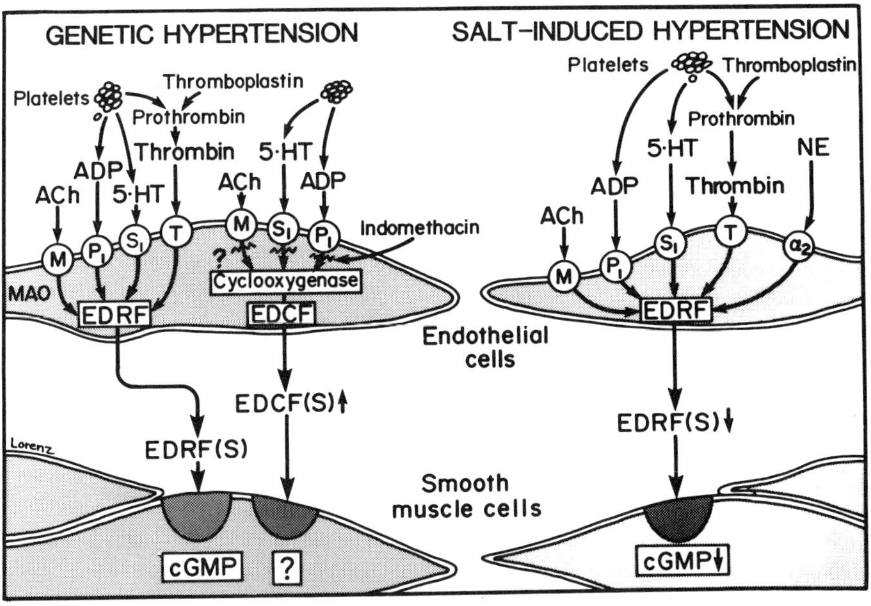

Fig. 5. Mechanisms of altered endothelium-dependent relaxations in genetic and salt-induced hypertension of the rat. In spontaneously hypertensive rats, endothelium-dependent relaxations are depressed because of a concomitant release of endothelial relaxing and contracting factors. In salt-induced hypertension, a decreased release of endothelium-derived relaxing factor(s) appears responsible for the decreased relaxations. Abbreviations: ACh, acetylcholine; ADP, adenosine diphosphate; EDCF(s), endothelium-derived constrictor factor(s); EDRF(s), endothelium-derived relaxing factor(s); 5-HT, 5-hydroxytryptamine (serotonin). Receptors are represented by open circles: M, muscarinic; S_1, S_1-serotonergic; P_1, P_1-purinergic; T, thrombin.

et al., 1987c). In the aorta of the hypertensive Dahl rat, however, endothelium-dependent relaxations are also decreased in the presence of indomethacin (to block cyclooxygenase), suggesting a decreased release of endothelium-derived relaxing factor(s) and/or a decreased responsiveness of vascular smooth muscle cells to the factor(s) (Luscher et al., 1987a).

The relaxations induced by endothelium-derived relaxing factor(s) are mediated through an increase of cyclic guanosine 3′,5′-monophosphate (cyclic GMP) levels in vascular smooth muscle cells (Rapoport and Murad, 1983). Cyclic GMP levels may be lower in hypertensive blood vessels under both basal and stimulated conditions. The relaxations to sodium nitroprusside, a substance that also acts through the activation of cyclic GMP, are reduced in hypertensive blood vessels (Winquist et al., 1984; Luscher et al., 1987a). In the rat with salt-induced hypertension, however, the relaxations to sodium nitroprusside are less impaired than those to acetylcholine, suggesting that although the release of endothelium-derived relaxing factor(s) may be decreased, structural changes and a decreased responsiveness of vascular smooth muscle cells to the endothelium-derived relaxing factor(s) may contribute. Similar conclusions have been reached in mineralocorticoid hypertension, coarctation of the aorta in the rat, and acute hypertension in the cat (Lockette et al., 1986; Kontos, 1985).

Thus, in the rat, the mechanisms of altered endothelium-dependent responses in hypertensive blood vessels differs in salt-induced and genetic hypertension. Although in the former a decreased release of endothelium-derived relaxing factor(s) seems to be responsible, a simultaneous release of endothelium-derived relaxing and constrictor factors appears to be involved in the latter (Fig. 5).

4. Possible Importance of Altered Endothelial Function in Hypertension

Endothelial dysfunction may result in an enhanced responsiveness of vascular smooth muscle cells to circulating vasoconstrictor hormones and substances released from aggregating platelets or formed during activation of the coagulation cascade. Hence an altered endothelial function in hypertension could play a role in the pathophysiology of the increase in peripheral vascular resistance, in the development of vascular complications of hypertension, or in the action and effectiveness of antihypertensive drugs.

4.1. Peripheral Vascular Resistance

It is likely that the alterations of endothelial function in secondary hypertension (e.g., salt-induced hypertension) are a consequence rather than a cause of high blood pressure. Therefore, endothelial dysfunction probably plays no role in the initiation of the hypertensive process. At later stages of the disease, however, endothelial dysfunction could contribute to the increase in peripheral vascular resistance. In mesenteric resistance vessels of the genetically hypertensive rat, decreased endothelium-dependent responses to bradykinin, acetylcholine, and histamine have been demonstrated as early as 4 wk after birth (De Mey and Gray, 1985). The loss of endothelial cells enhances the contractions caused by various agonists in both large conduit vessels and in resistance vessels (Luscher et al., 1987a;

Criscione et al., 1984). Endothelial cell dysfunction may have a comparable effect (Luscher and Vanhoutte, 1986b; Luscher et al., 1986; Shimokawa and Vanhoutte, 1986). In Dahl salt-induced hypertension, contractile responses to serotonin and norepinephrine are enhanced in rings with, but not in those without, endothelium (Luscher et al., 1987a). The release of endothelium-derived constrictor factor(s) may also contribute to an increased peripheral vascular resistance in spontaneously hypertensive rats. Contractions to serotonin are enhanced in rings with endothelium in these animals, presumably because the monoamine triggers the release of endothelium-derived contracting factor(s) (Luscher and Vanhoutte, 1986b). Such a mechanism would also explain the exaggerated vasoconstrictor responses caused by serotonin in perfused kidneys and in the isolated heart of the spontaneously hypertensive rat (Collis and Vanhoutte, 1977; Luscher et al., 1987b); in the latter, the abnormal vasoconstrictor response is prevented by inhibitors of cyclooxygenase, which supports the involvement of endothelium-derived contracting factor (Luscher and Vanhoutte, 1986a; Luscher et al., 1987b).

4.2. Vascular Complications

Vascular complications in hypertension are probably related to structural vascular lesions, functional and vascular alterations, and interactions between the arterial wall and platelets, platelet-derived substances, and substances involved in hemostasis (Ross, 1986).

In the aorta of normotensive Wistar-Kyoto rats, aggregating platelets cause marked contractions in rings without, but only minor changes in tension in rings with endothelium, illustrating the protective role of the vascular endothelium when platelet aggregation occurs (Vanhoutte and Cohen, 1983; Vanhoutte and Houston, 1985; *see* chapter by Houston and Vanhoutte, in this volume). In contrast, in the spontaneously hypertensive rat, rings with, but not those without, endothelium exhibit significantly enhanced contractions as compared to normotensive controls (Fig. 6; Luscher and Vanhoutte, 1986b). This effect seems to be mainly related to the endothelium-dependent responses to serotonin, one of the major substances released from aggregating platelets, although depressed relaxations to adenosine diphosphate might contribute. Similarly, in Dahl hypertensive rats, decreased relaxations to adenosine diphosphate and thrombin, as well as increased contractions to serotonin, may reflect a depressed protective function of the endothelium in hypertension. Possibly, this could be of importance for the development of complications of the hypertensive process such as myocardial infarction, vasospasm, peripheral vascular disease, and stroke.

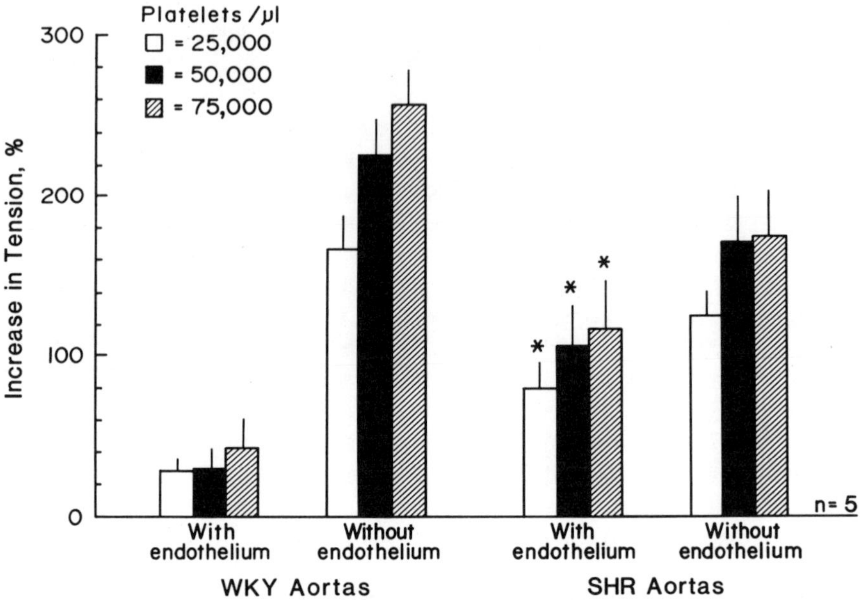

Fig. 6. Endothelium-dependent responses to platelets in spontaneously hypertensive rats (SHR). Increases in tension evoked by aggregating platelets in aortic rings (with and without endothelium) contracted with prostaglandin $F_{2\alpha}$ from SHR and normotensive Wistar-Kyoto rats (WKY) are given. The data are expressed as percent of that contraction. In both SHR and WKY, rings without endothelium were significantly more responsive than rings with endothelium. Asterisk denotes significant difference between rings with endothelium of SHR and WKY (from Luscher and Vanhoutte, 1986b; with permission of the American Heart Association).

4.3. Antihypertensive Therapy

The vasodilator action of hydralazine, but not that of nitroglycerine, is facilitated by the presence of the endothelium (Spokas et al., 1983). If in patients with severe hypertension endothelium-dependent responses were altered, this may in part explain treatment resistance to hydralazine. Conversely, an improvement of endothelium-dependent responses with antihypertensive treatment might be of primary importance for the ability of these drugs to prevent or reduce cardiovascular events. In Dahl salt-induced hypertension of the rat, antihypertensive therapy normalizes endothelium-dependent relaxations to acetylcholine, adenosine diphosphate, and thrombin (Luscher et al., 1987d).

References

Chobanian, A. V., Brecher, P. I., and Haudenschild, C. C.: Effects of hypertension and of antihypertensive therapy on atherosclerosis. *Hypertension* **8** (suppl. I):15–21, 1986.

Collis, M. G. and Vanhoutte, P. M.: Vascular reactivity of isolated perfused kidneys from male and female spontaneously hypertensive rats. *Circ. Res.* **41**:759–767, 1977.

Criscione, L., Muller, K., and Prescott, M. F.: Endothelial cell loss enhances the pressor response in resistance vessels. *J. Hypertension* **2** (suppl. 3): 441–444, 1984.

De Forrest, J. M. and Hollis, T. M.: Shear stress and aortic histamine synthesis. *Am. J. Physiol.* **234**:H701–H705, 1978.

De Mey, J. G. and Gray, S. D.: Endothelium-dependent reactivity in resistance vessels. *Prog. Appl. Microcirc.* **8**:181–187, 1985.

De Mey, J. G. and Vanhoutte, P. M.: Heterogeneous behavior of the canine arterial and venous wall; importance of the endothelium. *Circ. Res.* **51**: 439–447, 1982.

Dzau, V. J.: Vascular wall renin angiotensin pathway in control of the circulation. *Am. J. Med.* **77**:31–36, 1984.

Gabbiani, G., Glemes, G., Guelpa, C., Valloton, M. B., Badonnel, M. C., and Huttner, I.: Morphologic and functional changes of the aortic intima during experimental hypertension. *Am. J. Pathol.* **96**:399–422, 1979.

Huttner, I. and Gabbiani, G.: Vascular Endothelium and Hypertension, in *Hypertension* (Genest, J., Kuchel, O., Hamet, P., and Cantin, M., eds.) McGraw-Hill, New York, 1983.

Katusic, Z. S., Shepherd, J. T., and Vanhoutte, P. M.: Vasopressin causes endothelium-dependent relaxations of the canine basilar artery. *Circ. Res.* **55**:575–579, 1984.

Konishi, M. and Su, C.: Role of endothelium in dilator responses of spontaneously hypertensive rat arteries. *Hypertension* **5**:881–886, 1983.

Kontos, H. A.: Oxygen radicals in cerebral vascular injury. *Circ. Res.* **57**:508–516, 1985.

Limas, C., Westrum, B., Limas, C. J., and Cohn, J. N.: Effect of salt on the vascular lesions of spontaneously hypertensive rats. *Hypertension* **2**:477–489, 1980.

Lockette, W., Otsuha, Y., and Carretero, O.: The loss of endothelium-dependent relaxation in hypertension. *Hypertension* **8** (suppl II):61–66, 1986.

Luscher, T. F., Cooke, J. P., Houston, D. S., Neves, R., and Vanhoutte, P. M.: Endothelium-dependent relaxations in human peripheral and renal arteries. *Blood Vessels* **23**:104 (Abstract), 1986.

Luscher, T. F. and Vanhoutte, P. M.: Endothelium-dependent contractions to acetylcholine in the aorta of the spontaneously hypertensive rat. *Hypertension* **8**:344–348, 1986a.

Luscher, T. F. and Vanhoutte, P. M.: Endothelium-dependent responses to platelets and serotonin in spontaneously hypertensive rats. *Hypertension* **8** (suppl. II):II55–II60, 1986b.

Luscher, T. F., Raij, L., and Vanhoutte, P. M.: Endothelium-dependent vascular responses in normotensive and hypertensive Dahl rats. *Hypertension*, in press, 1987a.

Luscher, T. F., Rubanyi, G. M., Aarhus, L. L., and Vanhoutte, P. M.: Serotonin reduces coronary flow in isolated hearts of the spontaneously hypertensive rat. *J. Hypertension*, in press, 1987b.

Luscher, T. F., Romero, J. C., and Vanhoutte, P. M.: Bioassay of endothelium-derived vasoactive substances in the aorta of normotensive and spontaneously hypertensive rats. *J. Hypertension*, in press, 1987c.

Luscher, T. F., Vanhoutte, P. M., and Raij, L.: Antihypertensive treatment normalizes endothelium-dependent relaxations in salt-induced hypertension of the rat. *Hypertension*, in press, 1987d.

McGuire, P. G. and Twietmeyer, T. A.: Aortic endothelial junctions in developing hypertension. *Hypertension* **7**:483–490, 1985.

Miller, V. M. and Vanhoutte, P. M.: Endothelium-dependent contractions to arachidonic acid are mediated by products of cyclooxygenase in canine veins. *Am. J. Physiol.* **248**:H432–H437, 1985.

Moncada, S. and Vane, J. R.: Prostacyclin (PGI_2), The Vascular Wall and Vasodilation, in *Mechanisms of Vasodilatation* (Vanhoutte, P. M. and Leusen, I., eds.) S. Karger, Basel, Switzerland, 1978.

Rapoport, R. and Murad, F.: Agonist induced endothelium-dependent relaxation in rat thoracic aorta may be mediated through cGMP. *Circ. Res.* **52**:352–367, 1983.

Ross, R.: The pathogenesis of atherosclerosis—an update. *N. Engl. J. Med.* **314:** 488–500, 1986.

Schwartz, S. M. and Standaert, D. M.: Endothelial Injury in Hypertension, in *Frontiers in Hypertension Research* (Laragh, J. H., Buhler, F. R., and Seldin, D., eds.) Springer, New York, 1981.

Shimokawa, H. and Vanhoutte, P. M.: Reduced endothelium-dependent relaxations to aggregating platelets and serotonin in porcine coronary arteries after previous endothelial injury. *Circulation* **74**:II413, 1986.

Spokas, E. G., Folco, G., Quilley, J., Chander, P., and McGiff, J. C.: Endothelial mechanism in the vascular action of hydralazine. *Hypertension* **5** (suppl. I):107–111, 1983.

Van de Voorde, J. and Leusen, I.: Role of the endothelium in the vasodilator response of rat thoracic aorta to histamine. *Eur. J. Pharmacol* **87**:118–120, 1983.

Van de Voorde, J. and Leusen, I.: Endothelium-dependent and independent relaxation effects on aorta preparations of renal hypertensive rats. *Arch. Int. Physiol. Biochem.* **92**:P35–P36, 1984.

Vanhoutte, P. M. and Cohen, R. A.: The elusory role of serotonin in vascular function and disease. *Biochem. Pharmacol.* **32**:3671–3674, 1983.

Vanhoutte, P. M. and Houston, D. S.: Platelets, endothelium and vasospasm. *Circulation* **72**:729–734, 1985.

Vanhoutte, P. M.. Rubanyi, G. M., Miller, V. M., and Houston, D. S.: Modulation of vascular smooth muscle contraction by the endothelium. *Ann. Rev. Physiol.* **48**:307–320, 1986.

Webb, R. C., Vander, A. J., and Henry, P. J.: Increased vasodilator responses to acetylcholine in psychosocial hypertensive mice. *Hypertension*, in press, 1986.

Winquist, R. J., Bunting, P. B., Baskin, E. P., and Wallace, A. A.: Decreased endothelium-dependant relaxation in New Zealand hypertensive rats. *J. Hypertension* **2**:541–536, 1984.

Chapter 25

Endothelium-Dependent Responses in Human Arteries

S. Thom, A. Hughes, and P. S. Sever

1. Introduction

This series of experiments details the presence of endothelium-dependent relaxation responses in the human vasculature, defines the similarities between human EDRF and the factor first discovered in the rabbit aorta by Furchgott and Zawadzki (1980), and highlights several interspecies differences.

2. Source Materials and Methods

Segments of human blood vessel were obtained at the time of surgical resection from 89 patients (52 male; median age 56 yr; range, 3–89 yr). Arteries were taken from renal, splenic, gastric, mesenteric, pulmonary, transverse cervical, lumbar, brachial, and cerebral sites. Human coronary arteries were obtained from hearts removed at transplantation. Segments of saphenous vein tributaries were obtained when this blood vessel was ligated for control of varices or from lengths surplus to requirements for coronary bypass grafting. The patients suffered from a wide variety of diseases and had received a number of drugs, including anesthetic agents and preoperative medication. Many, however, could reasonably be expected to have a ''normal'' cardiovascular system. The vessels were collected directly into ice-cold Kreb's solution and used for studies within 3 h of removal. They were cleaned of excess fat and connective tissue and mounted as rings (2.5 mm in length, 1–6 mm internal diameter) in 6 mL organ baths containing Kreb's buffer [composition (m*M*): NaCl, 118; KCl, 4.7; $MgSO_4$, 1.2; $CaCl_2$, 2.5; NaH_2, PO_4, $7H_2O$, 1.0; glucose, 11.1; Na HCO_3, 25.0; Na_2 EDTA, 0.03) under resting tension between 1 and 4 g. The Kreb's solution was maintained at 37 °C and aerated with 95% O_2 and 5% CO_2. After equilibrium for at least 1 h, vessels were constricted with an approximate EC_{50} of either norepinephrine or prostaglandin $F_{2\alpha}$ ($PGF_{2\alpha}$). Compounds under investigation were added cumulatively once a stable contraction ensued. Antagonists, when used, were added 20 min prior to the constriction. All experiments were performed in the presence of indomethacin ($10^{-5}M$). A response was determined to be endothelium-dependent if it occurred in a vessel relaxing either to acetylcholine or the calcium ionophore A23187 and was subsequently abolished by rubbing the lumen with a wooden stick to remove endothelium. The vessels were examined histologically subsequent to completion of the in vitro studies using hematoxylin plus eosin stains and elastic Van Gieson stains to assess the morphology of the arterial wall, and a factor VIII immunocytochemical stain (rabbit antibody) to assess the integrity of the endothelial layer. In some cases scanning electron microscopy was used to view the endothelial cells and the consequence of luminal rubbing.

3. Outline

The following sections illustrate the vascular responses. It was possible to demonstrate endothelium-dependent relaxation to either acetylcholine or A23187 in vessels from only approximately half the patients studied. Relaxation was not seen in vessels with gross intimal hypertrophy and atheromatous disease, even though in some of these specimens a continuous endothelial layer could be identified. A detailed assessment of whether any relationship exists between the extent of disease or intimal thickness and endothelium-dependent relaxation has not yet been made. No attempt was made to divide the donor patients into normotensive or hypertensive groups on the basis of available surgical pre-operative clerking. Surprisingly, several seemingly normal arteries from young subjects have also failed to show endothelium-dependent responses. There are clearly numerous possible contributory factors to this variability—tissue handling, ischemic time prior to collection, possible exposure to hemolyzed blood, time to study in organ bath, age, and drugs. Recently, halothane (Muldoon et al., 1985) and inhibitors of the Na^+K^+ pump (ouabain, digoxin) (Rapoport et al., 1985) have been demonstrated to inhibit endothelium-dependent relaxation, although such effects were not evident in this series, and it has not been possible to ascribe the failure rate to any particular factor. Further work is necessary to determine what factor or factors are important in this respect. Unfortunately, such a study is complicated by the nature and irregularity of the supply of tissue.

It should be emphasized that all these observations apply to large-caliber arteries, namely first or second generation branches from the sites described. Endothelium-dependent relaxation to acetylcholine does occur in small human vessels of 70–100 μm internal diameter (Aalkjaer et al., 1987), and this feature of resistance vessels is supported by the vasodilatation produced in the forearm by cholinergic agonists infused via the brachial artery (Duff et al., 1953). Owen and Bevan (1985) have shown that in the ear arteries of the rabbit endothelium-dependent relaxation is more prominent with distal progression along the arterial tree. It is highly likely that release mechanisms, both receptor-mediated and mechanical, will be found to vary with site and size. Such changes may be an integral part of interspecies variation.

4. Response to Acetylcholine

Acetylcholine is capable of producing 100% endothelium-dependent relaxation in all the arteries studied from noncoronary sites (Fig. 1), and after luminal rubbing the response is abolished or converted to a constrictor response as illustrated for the pulmonary artery. This acetylcholine-induced relaxation is mediated by a muscarinic receptor (Fig. 1). It was more difficult to demonstrate endothelium-dependent relaxation by acetylcholine or A23187 in human veins. De May and Vanhoutte (1982) have reported similar results using canine saphenous vein. The response is present in veins (Fig. 2), but to a considerably lesser extent than in arteries. The possibility that other substances may be more effective at releasing endothelium-derived relaxing factor (EDRF) in the venous circulation has not been pursued.

In the human coronary arteries the response to acetylcholine is considerably more complex than two contradictory reports would suggest—one describing endothelium-dependent relaxation (Ginsburg and Zera, 1984) and the other, consistent constriction (Kalsner, 1985). Spontaneous phasic activity is a prominent feature of the coronary artery, and acetylcholine seems particularly effective in triggering such activity (Fig. 3). The response is biphasic in some cases (Fig. 4); possibly this results from the opposing actions of acetylcholine causing relaxation by an action on the endothelium and contraction by a direct effect on coronary vascular smooth muscle. This might be particularly likely if the inner muscle is relatively insensi-

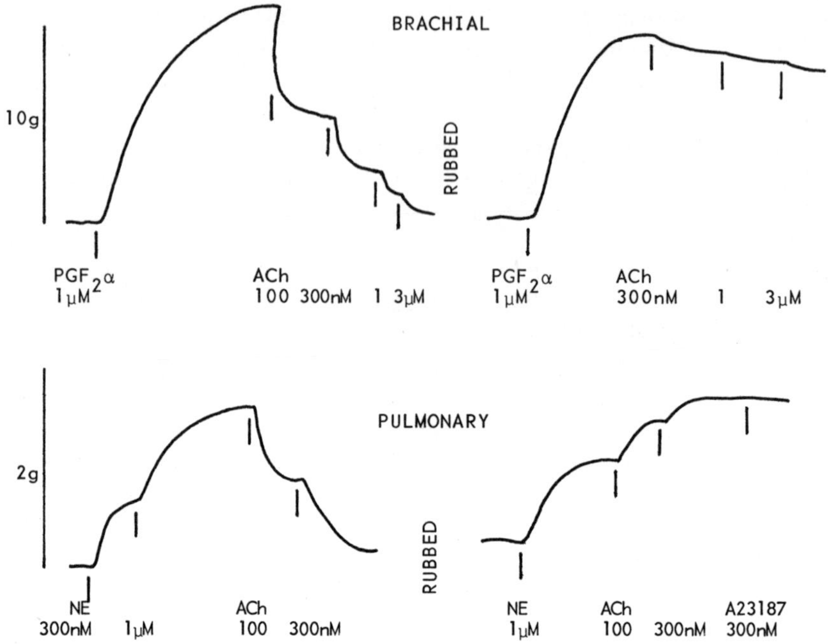
BRACHIAL
10g
RUBBED
PGF$_{2}$α
1µM
ACh
100 300nM 1 3µM
PGF$_{2}$α
1µM
ACh
300nM 1 3µM
PULMONARY
2g
RUBBED
NE
300nM 1µM
ACh
100 300nM
NE
1µM
ACh
100 300nM
A23187
300nM

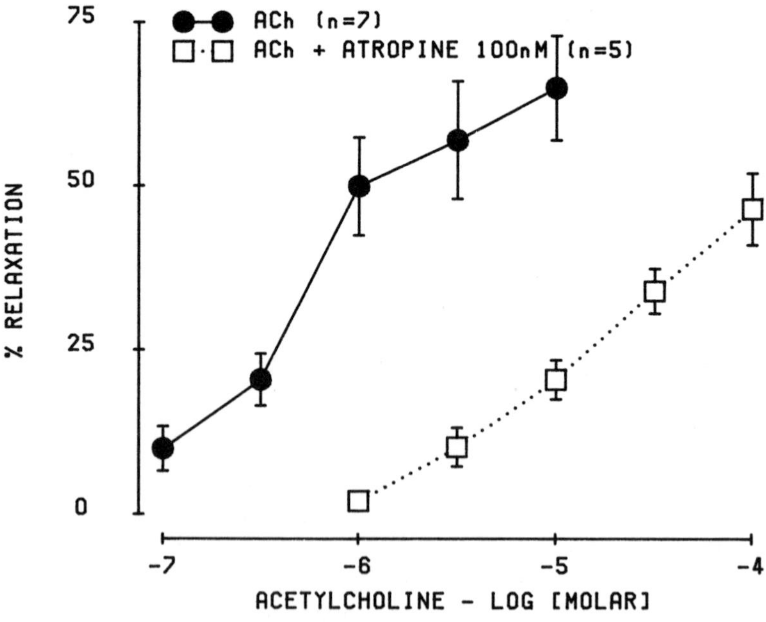
ACh (n=7)
ACh + ATROPINE 100nM (n=5)
75
50
25
0
% RELAXATION
-7
-6
-5
-4
ACETYLCHOLINE - LOG [MOLAR]

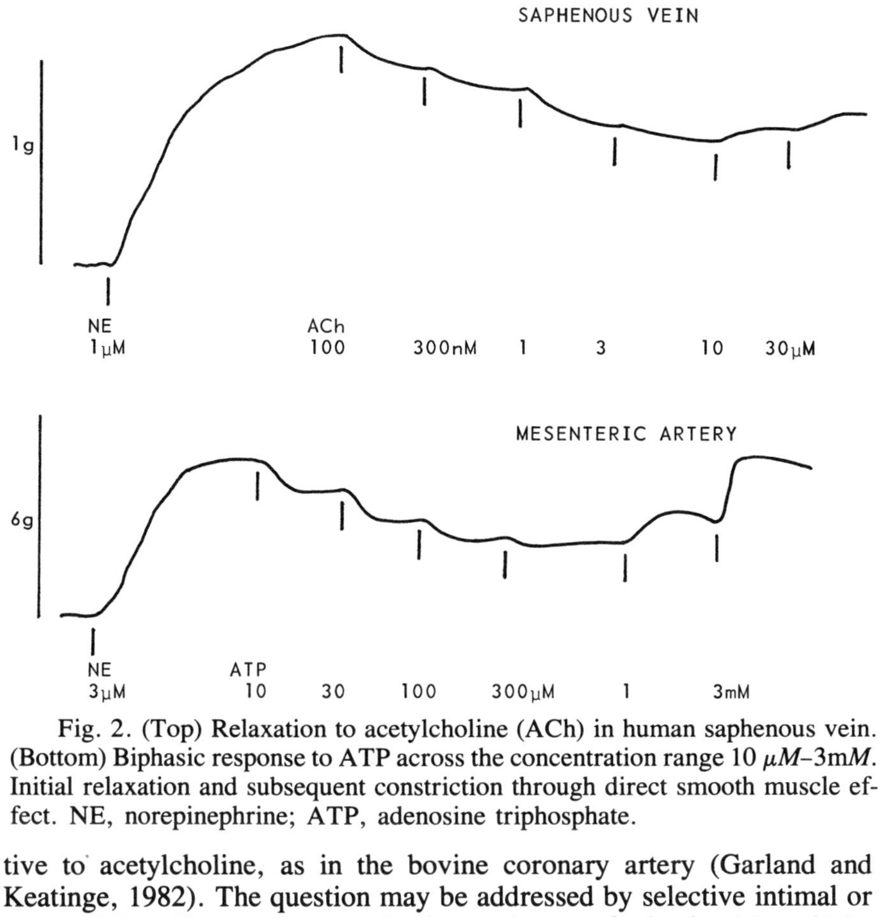

Fig. 2. (Top) Relaxation to acetylcholine (ACh) in human saphenous vein. (Bottom) Biphasic response to ATP across the concentration range 10 μM–3mM. Initial relaxation and subsequent constriction through direct smooth muscle effect. NE, norepinephrine; ATP, adenosine triphosphate.

tive to acetylcholine, as in the bovine coronary artery (Garland and Keatinge, 1982). The question may be addressed by selective intimal or adventitial application of acetylcholine or the use of selective M_1 and M_2-antagonists, since it is probable that the endothelial muscarinic receptor for EDRF release is of the M_2 subtype (Eglen and Whiting, 1985), whereas the receptor on the vascular smooth muscle may be of the M_1 subtype.

Seventy-one coronary artery segments have been studied from 15 patients, and endothelium-dependent relaxations are definitely present, as manifest by the responses to A23187 (Figs. 3 and 5), calcitonin-gene related

← Fig. 1. (Top) Typical responses of arterial segments from noncoronary sites to acetylcholine (ACh) before and after removal of the endothelium by luminal rubbing. NE, norepinephrine; $PGF_{2\alpha}$, prostaglandin $F_{2\alpha}$. (Bottom) The effect of incubation with atropine (100 nM) on the concentration–response curve to ACh in gastric and splenic arteries.

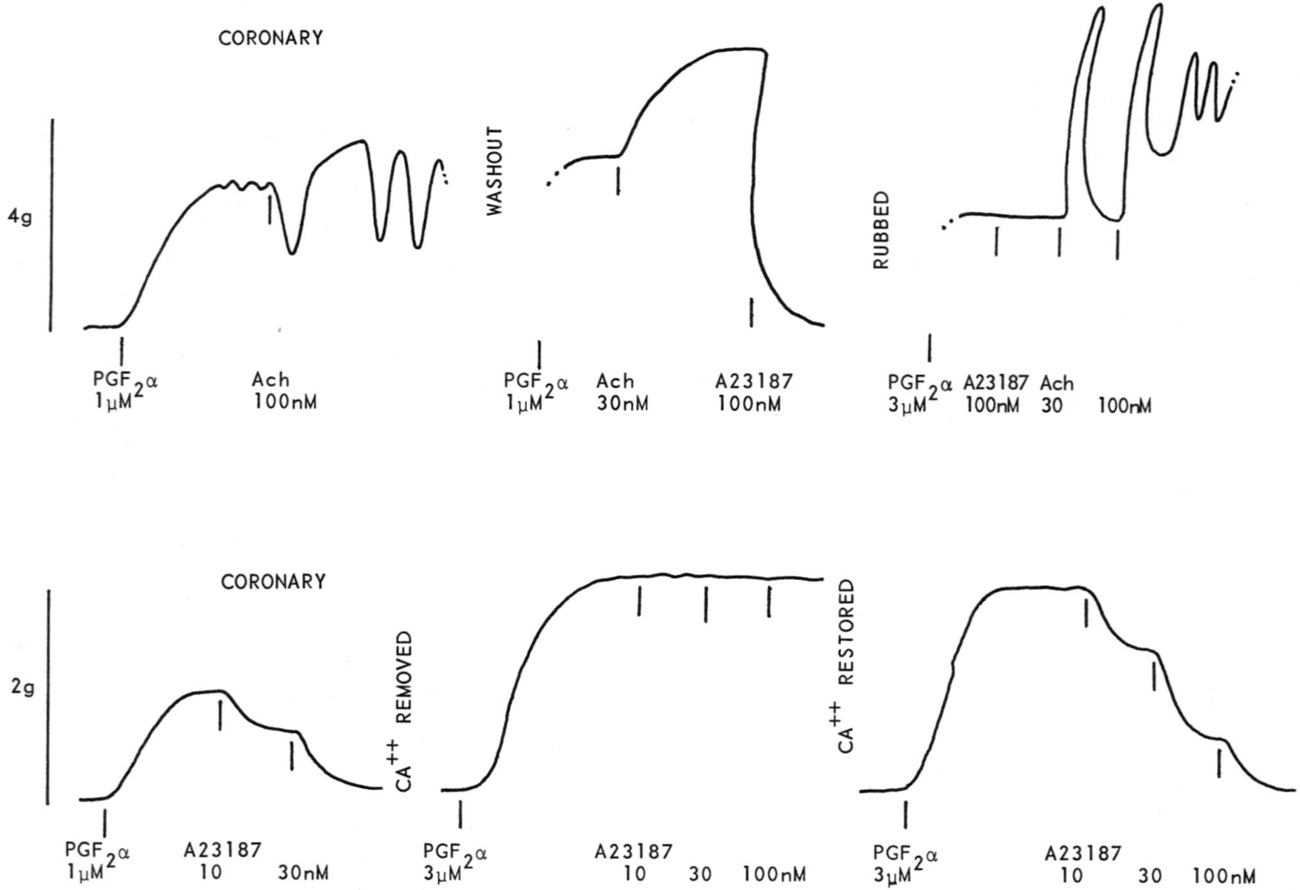

Fig. 3. (Top) The effect of acetylcholine in a normal coronary artery—predominantly a constrictor response. 100% relaxation by 100 n*M* A23187, subsequently abolished after endothelial removal. (Bottom) The effect of Ca^{2+} removal on A23187-induced endothelium-dependent relaxation in the same normal coronary artery. $PGF_{2\alpha}$, prostaglandin $F_{2\alpha}$.

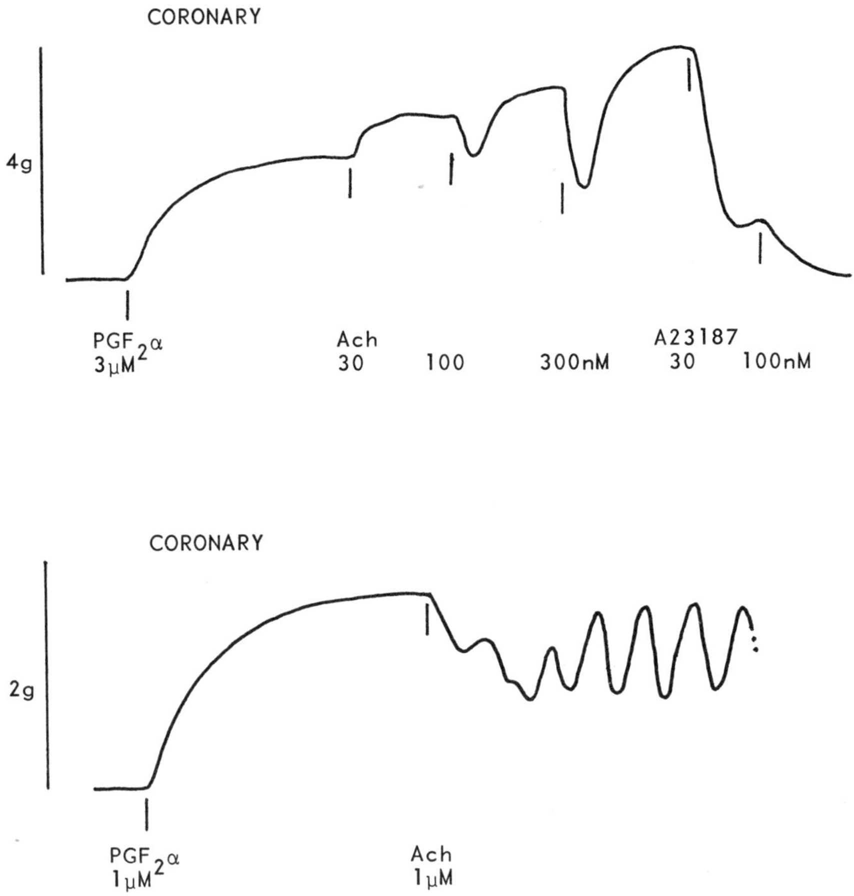

Fig. 4. (Top) Biphasic response to acetylcholine (ACh) or just spontaneous phasic activity in a normal coronary artery? (Bottom) Spontaneous phasic activity triggered by ACh in an atheromatous coronary artery. $PGF_{2\alpha}$, prostaglandin $F_{2\alpha}$.

peptide (CGRP), and insulin (Figs. 6 and 7). The relaxation to A23187 has only been seen in 23 of these 71 segments (32%), and this partly correlates with the finding of extensive atheromatous disease on subsequent histological examination of these blood vessels.

5. Other EDRF Releasers

A23187 proved to be the most consistent of all agents at inducing endothelium-dependent relaxation in the arteries studied and was used as

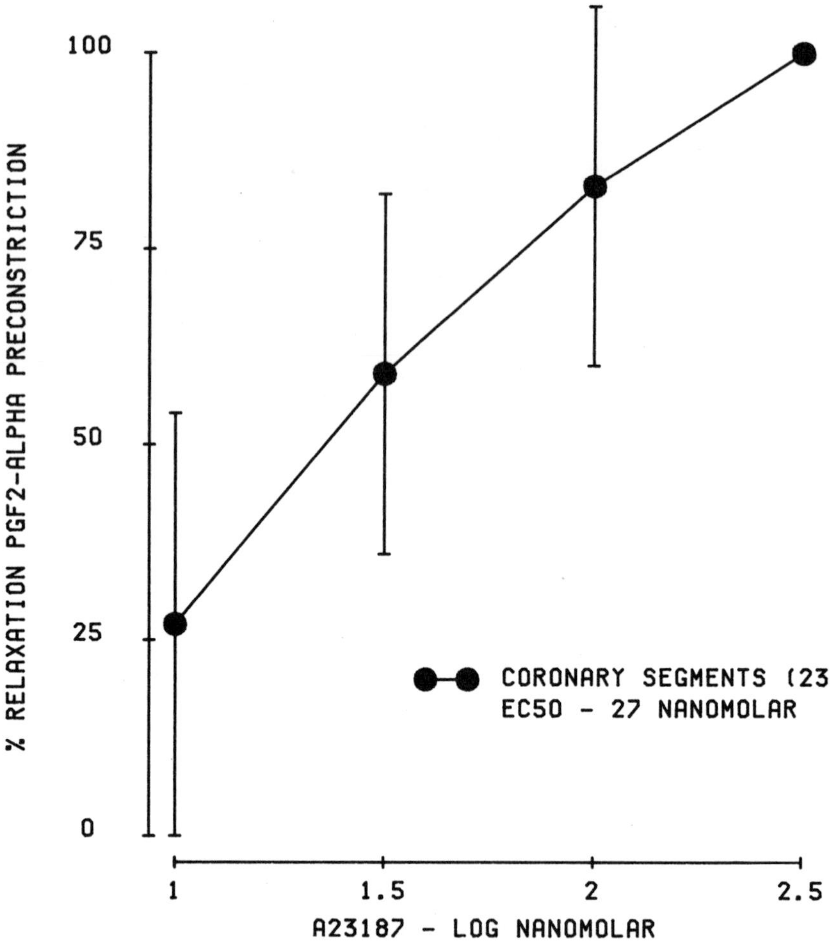

Fig. 5. Concentration–response curve for the endothelium-dependent relaxation caused by A23187 in coronary arteries. EC_{50}, concentration causing 50% inhibition.

the reference compound for the functional demonstration of endothelial integrity. In the concentration range used (10–300 n*M*), relaxation to A23187 was always endothelially dependent, and constrictor responses were never observed. CGRP was shown to be a potent stimulant of EDRF release in arteries from all sites including the cerebral and coronary vasculatures (Fig. 6). No evidence was found in these studies for endothelium-independent relaxation by CGRP. In the isolated rat aorta, CGRP-induced relaxations are also endothelium-dependent (Brain et al., 1985); however, in smaller pial arteries of internal diameter 250–600 μm taken from rab-

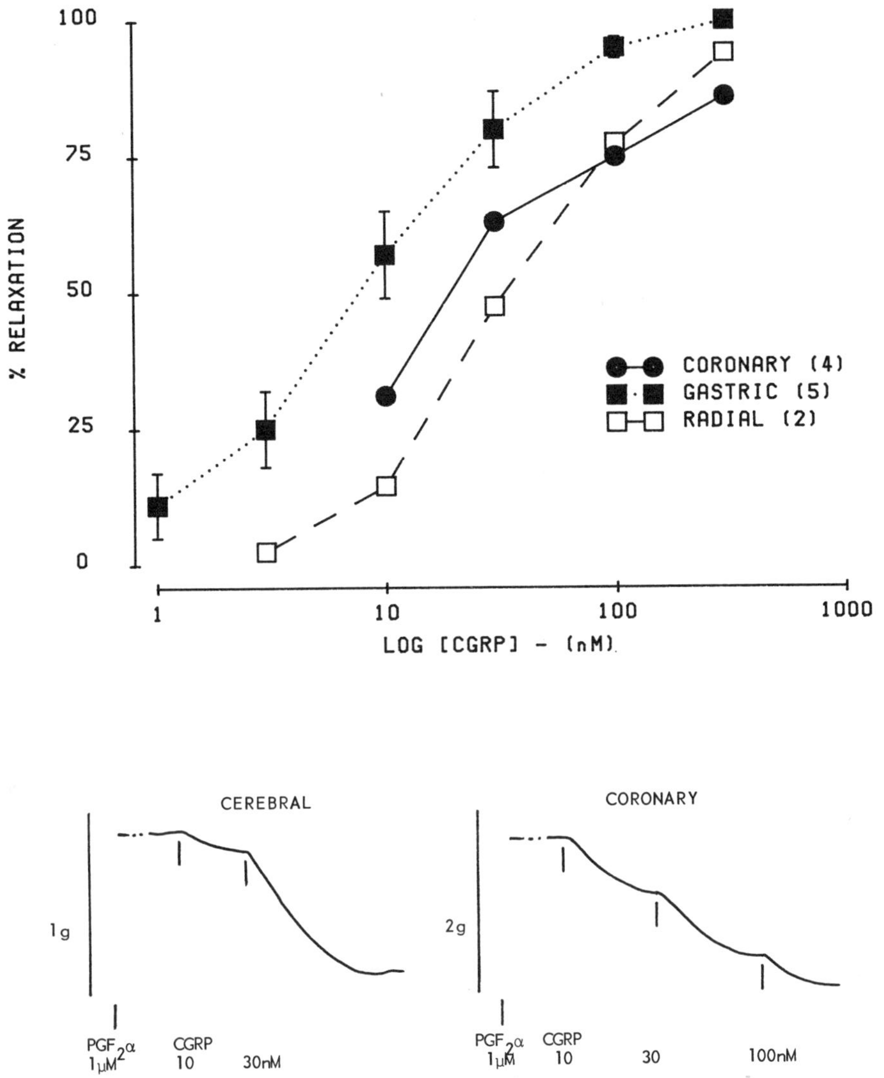

Fig. 6. (Top) Endothelium-dependent relaxation to calcitonin gene-related peptide (CGRP) in coronary, gastric, and radial arteries. (Bottom) Trace examples of CGRP-induced relaxation in human cerebral and coronary arteries. $PGF_{2\alpha}$, prostaglandin $F_{2\alpha}$.

bit, cat, and humans, although acetylcholine and adenosine triphosphate (ATP) remain endothelium-dependent dilators, the relaxation in response to CGRP is preserved after endothelial removal (Hardebo et al., 1985).

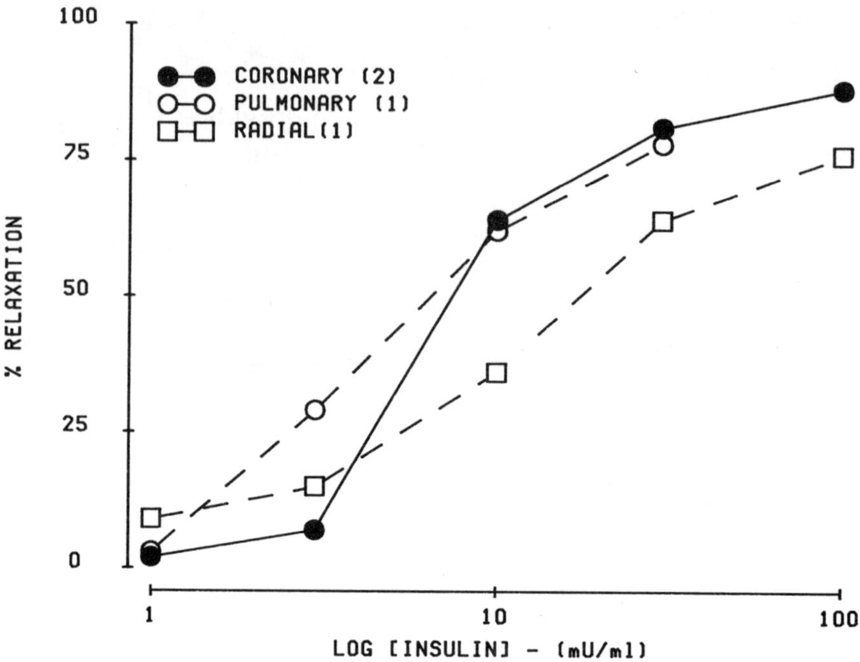

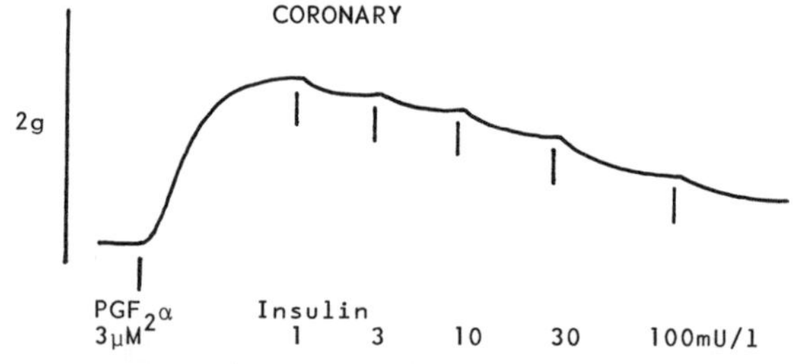

Fig. 7. (Top) Endothelium-dependent relaxation of coronary, pulmonary, and radial arteries to insulin. (Bottom) Trace example of insulin-induced relaxation in human coronary artery. $PGF_{2\alpha}$, prostaglandin $F_{2\alpha}$.

Vasoactive intestinal polypeptide (VIP) has been shown to relax human gastric and splenic arteries by an endothelium-dependent mechanism, but in the pulmonary artery this peptide acts independently of the endothelium (Hughes et al., 1986). VIP is an endothelium-dependent dilator in the aorta of the rat (Davies and Williams, 1983), but not in the canine carotid artery (D'Orleans-Juste et al., 1985); again, in the pial arteries of rabbit, cat, and humans it is endothelium-independent (Hardebo et al., 1985). In parallel studies in vivo, measuring forearm blood flow, both CGRP and VIP infused via the brachial artery produced striking vasodilatation (Thom et al., 1987; Hughes et al., 1986) (Fig. 8). The role of the endothelium in mediating this response has not been established in vivo, however.

In the context of postprandial hypotension in patients with autonomic neuropathy (Da Costa et al., 1985) and insulin-induced euglycemic syn-

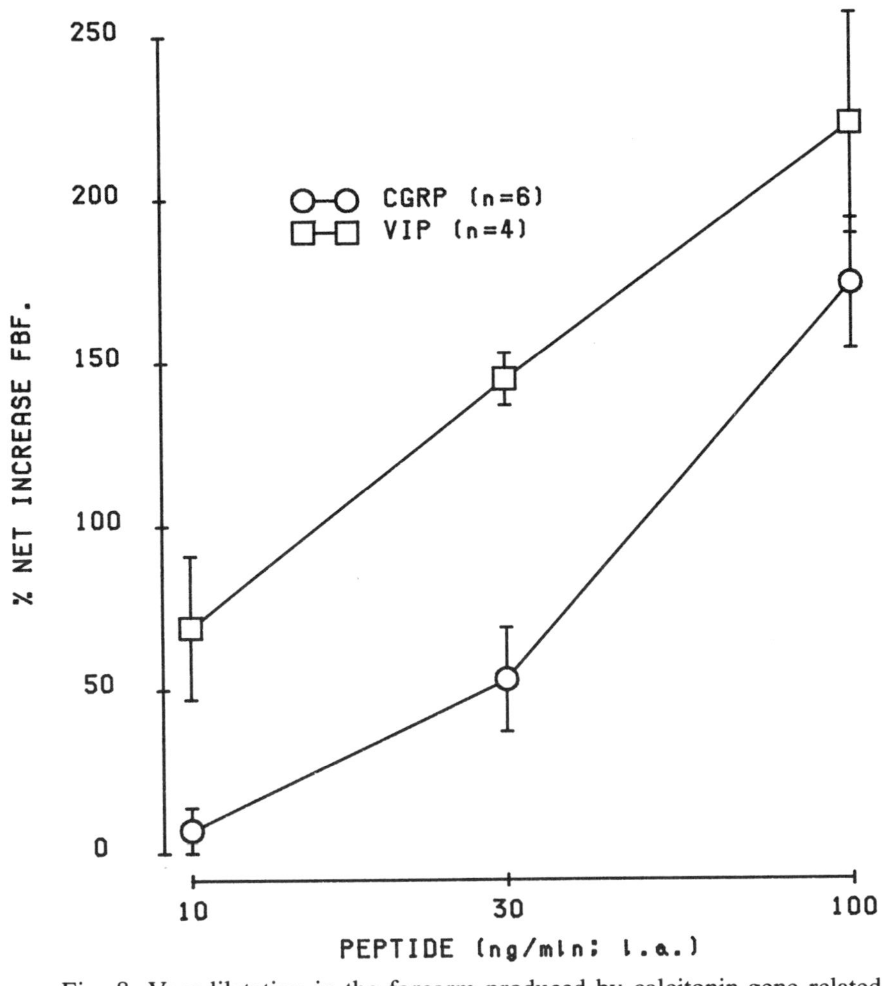

Fig. 8. Vasodilatation in the forearm produced by calcitonin-gene related peptide (CGRP) and vasoactive intestinal peptide (VIP). Forearm blood flow (FBF) measurements were made by venous occlusion plethysmography. The effects of CGRP (10–100 ng/min) and VIP (10–100 ng/min) infused via the brachial artery on FBF are shown in six and four subjects, respectively (points as means ±SEM).

cope in diabetics (Page and Watkins, 1976), attention has been focused on the vasoactive properties of insulin. The vasodilator action of insulin in the forearm is partially mediated by a beta-adrenergic mechanism (Creager et al., 1985), and an immunoreactively insulin-like peptide may be released from peripheral nerves of the cat in response to electrical stimulation (Posloncec et al., 1985). From the in vitro studies reported here, a direct endothelium-dependent relaxation to insulin is evident (Fig. 7). On the other hand, in the canine carotid artery, this direct effect is endothelium-independent (D'Orleans-Juste et al., 1985).

In humans, histamine relaxes isolated pulmonary and splenic arteries by an endothelium-dependent mechanism through activation of H_1-histaminergic receptor; this effect is antagonized by mepyramine ($10^{-5}M$). In some species the response is variable, constriction or dilatation being observed at different levels of the vasculature. Histamine has been reported to relax the aorta of the rat, but not that of the rabbit, by an endothelium-dependent mechanism (Van de Voorde and Leusen, 1983); and in the canine carotid artery, it has a dilator action that is independent of the endothelium (D'Orleans-Juste et al., 1985).

ATP has received considerable attention since Burnstock and Kennedy's demonstration (1986) of its role as a cotransmitter and awareness of the complexity of purinergic influence on vascular tone. Observations in the current series of experiments have been limited, but a biphasic response is clearly demonstrable (Fig. 2), the initial relaxation being endothelium-dependent.

Responses to bradykinin in human arteries have been inconsistent, but it does appear to relax these arteries, at least in part through the release of EDRF, as others have reported (Cherry et al., 1982). This response has also been documented in smaller human vessels (Hardebo et al., 1985).

6. Inhibitors of EDRF

Human EDRF shares several of the biochemical properties described in animal species (Martin et al., 1985; Griffith et al., 1984). It is not inhibited by indomethacin, and it is calcium-dependent (Fig. 3). Nordihydroguaretic acid (NDGA), an antioxidant and lipoxygenase inhibitor, hemoglobin, an inhibitor of EDRF, and methylene blue, an inhibitor of guanylate cyclase, reverse the endothelium-dependent relaxations of human blood vessels (Fig. 9). It is of interest that all three compounds have no effect when added alone on resting even on contracted arterial segments after constriction. Yet, in reversing the relaxation produced by EDRF releasers, the tension generated far exceeds the preceeding constriction. These in-

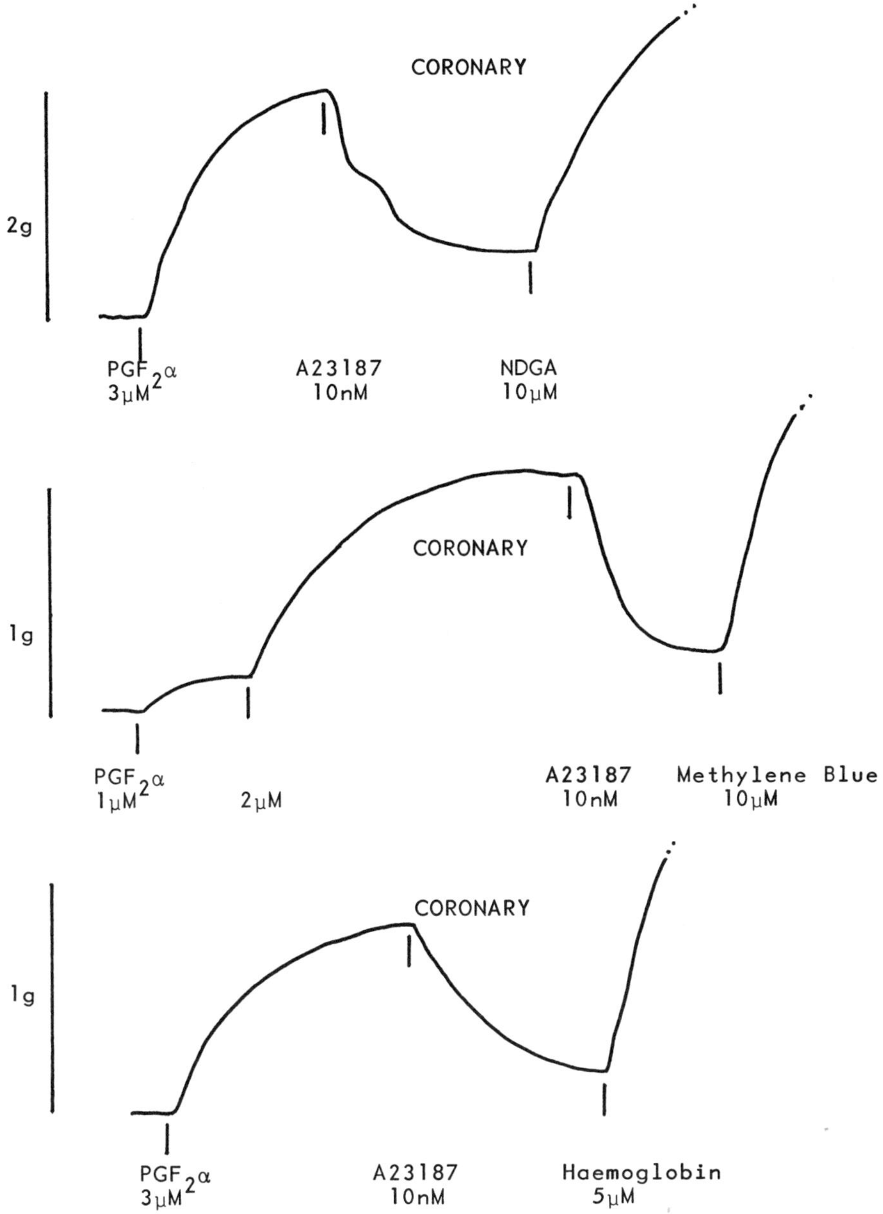

Fig. 9. Reversal of A23187-induced relaxation in adjacent segments of normal human coronary artery by NDGA (10 *μM*), methylene blue (10 *μM*), and hemoglobin (5 *μM*). $PGF_{2\alpha}$, prostaglandin $F_{2\alpha}$; NDGA, nordihydroguaretic acid.

hibitory characteristics do not show any variation between differing substances releasing EDRF—there is no inherent evidence here for more than one factor.

7. "Interesting Negatives"

Attempts to identify an alpha$_2$-adrenergic or an S_1-serotonergic receptor linked to release of EDRF in isolated human arteries have been unsuccessful (Fig. 10). These experiments have been performed in a variety of arteries, including coronaries, with the same methodology as described

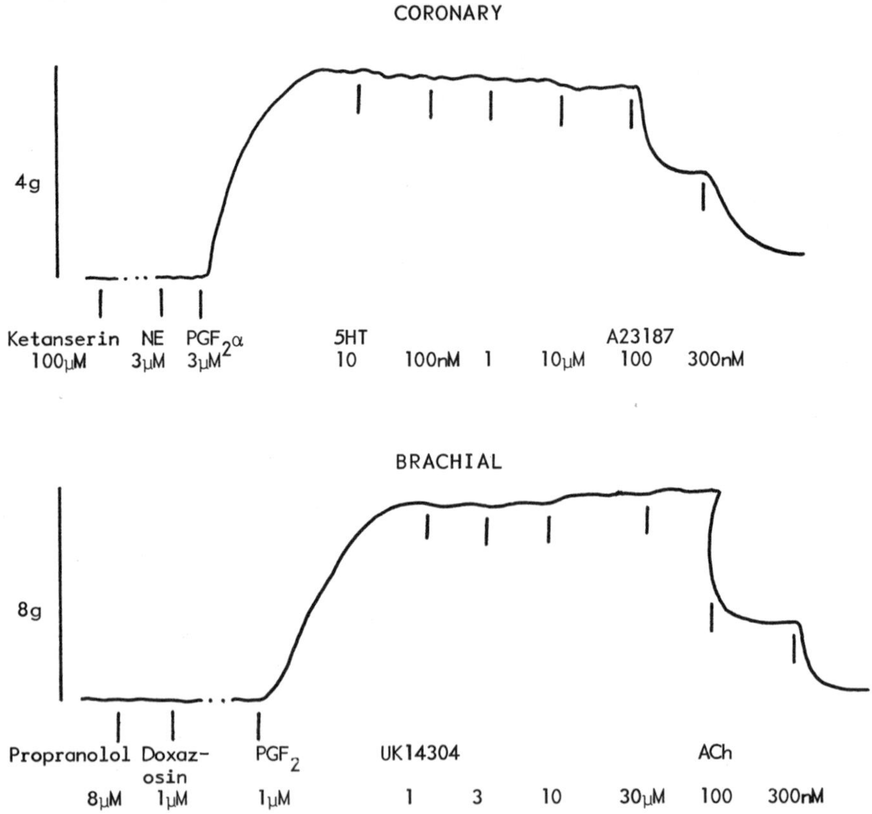

Fig. 10. (Top) Absence of endothelium-dependent relaxation to serotonin (5-HT)-induced release of EDRF in a coronary artery [contracted with norepinephrine (NE)] in the presence of the 5-HT$_2$-serotonergic blocker ketanserin. (Bottom) UK14304, a selective alpha$_2$-adrenergic agonist, fails to induce relaxation in brachial artery-contracted with prostaglandin $F_{2\alpha}$ ($PGF_{2\alpha}$) in the presence of the alpha$_1$-adrenergic blocker doxazosin.

by Cocks and Angus, who made the original observations in coronary arteries of the pig and the dog (Cocks and Angus, 1983). It will be important to look for these possible mechanisms in smaller vessels.

8. Concluding Comments

The importance of endothelium-dependent vasodilatation in clinical terms is hard to assess, but the realization that VIP and CGRP, the most potent endogenous vasodilator substances yet discovered, probably produce their effect by releasing EDRF puts it well into perspective. The diversity of the many substances releasing EDRF has been confusing. Physiologically, how and when are they likely to encounter the endothelium? Several, including ATP, acetylcholine, VIP, serotonin, and CGRP have been identified as cotransmitters—they may conceivably reach the endothelium after release from perivascular nerves in small blood vessels, but in the case of ATP, acetylcholine, and serotonin, the opposing constrictor linked receptors on the smooth muscle makes this unlikely. The localization of choline acetyltransferase in the endothelial cells of rat cerebral arteries (Parnavelas et al., 1985) with the implication that acetylcholine may be generated locally offers a more logical explanation if this holds true in humans. Porcine endothelial cells have a significant capacity for storage and metabolism of adenine nucleotides (Pearson et al., 1980).

In vivo studies in the dog (Angus et al., 1983) show that the EDRF is not simply a tissue bath phenomenon. In vivo human studies are more different to devise, but are increasingly important particularly in the light of the growing vogue of balloon catheter angioplasty.

If basal release of EDRF (Martin et al., 1986) is also an in vivo feature, the loss of this protective mechanism must be of importance in the progression of atherosclerosis as the endothelial layer is gradually distanced from the smooth muscle and in the acceleration of malignant hypertension in which the endothelium is physically disrupted.

Several groups have reported diminished EDRF responses in varieties of hypertensive rats (Winquist et al., 1984), although the depression of the response does not appear to be caused by a diminished release of EDRF (Van de Voorde and Leusen, 1986). The question posed in humans is whether an abnormality in endothelium-dependent relaxation has a primary or secondary role in the pathogenesis of the hypertensive disease process. Mechanical changes in the vessel wall make it difficult to unravel the morphological hypertensive changes occurring in both smooth muscle (Folkow, 1982) and endothelium (McGuire and Twietmeyer, 1985) from the possibility of a primary or even acquired defect in any local vasodilator mechanism. It may be rewarding to study the biochemical aspects of endothelium-dependent relaxation in smooth muscle and endothelial cell cultures derived from normotensive and hypertensive subjects.

Acknowledgments

We gratefully acknowledge the assistance of the surgeons of St. Mary's Hospital, London, W.2., and Dr. Peter Stovin, Papworth Hospital, Cambridgeshire, for providing tissue specimens, and Mr. Gerald Haffenden, Department of Histology, St. Mary's Hospital, London, W.2., for histological preparations. This study was supported by a grant from the British Heart Foundation.

References

Aalkjaer, C., Heagarty, A. M., Swales, J. D., and Thurston, M.: Endothelial-dependent relaxation in human subcutaneous resistance vessels. *Blood Vess.*, **24:**85–88, 1987.

Angus, J. A., Campbell, G. R., Cocks, T. M., and Manderson, J. A.: Vasodilatation by acetylcholine is endothelium dependent: A study by sonomiorometry in canine femoral artery in vivo. *J. Physiol.* **344:**209–222, 1983.

Brain, S. D., Williams, T. J., Tippins, J. R., Morris, H. R., and MacIntyre, I.: Calcitonin gene-related peptide is a potent vasodilator. *Nature* **313:**54–56, 1985.

Burnstock, G. and Kennedy, C.: A dual functional for adenosine 5′-triphosphate in the regulation of vascular tone. *Circ. Res.* **58:**319–330, 1986.

Cherry, P. D., Furchgott, R. F., Zawadzki, J. V., and Jothianandan, D.: Role of endothelial cells in relaxation of isolated arteries by bradykinin. *Proc. Natl. Acad. Sci. USA* **79:**2106–2110, 1982.

Cocks, J. M. and Angus, J. A.: Endothelium dependent relaxation of coronary arteries by noradrenaline and serotonin. *Nature* **305:**627–630, 1983.

Creager, M. A., Liang, C. S., and Coffman, J. D.: Beta-adrenergic mediated vasodilator response to insulin in the human forearm. *J. Pharmacol. Exp. Ther.* **235:**709–714, 1985.

Da Costa, D. F., McIntosh, C., Bannister, R., Christensen, N. J., and Mathias, C. J.: Unmasking of the cardiovascular effects of carbohydrate in subjects with sympathetic denervation. *J. Hyperten.* **3:**S447–S448, 1985.

Davies, J. M. and Williams, K. I.: Relaxation of the rat aorta by vasoactive intestinal polypeptide is endothelial cell dependent. *J. Physiol.* **343:**65P, 1983.

De Mey, J. G. and Vanhoutte, P. M.: Heterogeneous behaviour of the canine arterial and venous wall. Importance of the endothelium. *Circ. Res.* **51:** 439–447, 1982.

D'Orleans-Juste, P., Dion, S., Mizrahi, J., and Regoli, D.: Effects of peptides and non-peptides on isolated arterial smooth muscles. *Eur. J. Pharmacol.* **114:**9–21, 1985.

Duff, F., Greenfield, A. D. M., Shepherd, J. T., and Thompson, I. D.: A quantitative study of the response to acetylcholine and histamine of the blood vessels of the human hand and forearm. *J. Physiol.* **120:**160–170, 1953.

Eglen, R. M. and Whiting, R. L.: Determination of the muscarinic receptor subtype mediating vasodilatation. *Br. J. Pharmacol.* **84:**3–5, 1985.

Folkow, B.: Physiological aspects of primary hypertension. *Physiol. Rev.* **62:**347–428, 1982.

Furchgott, R. F. and Zawadzki, J.: The obligatory role of endothelial cells in the relaxation of arterial smooth muscle by acetylcholine. *Nature* **288:** 373–376, 1980.

Garland, C. J. and Keatinge, W. R.: Constrictor actions of acetylcholine, 5-hydroxytryptamine and histamine on bovine coronary artery inner and outer muscle. *J. Physiol.* **327:**363–376, 1982.

Ginsburg, R. and Zera, P.H.: Endothelial relaxant factor in the human epicardial coronary artery. *Circulation* **70**(suppl. II):122, 1984.

Griffith, T. M., Edwards, D. H., Lewis, M. J., Newby, A. C., and Henderson, A. H.: The nature of endothelium derived vascular relaxant factor. *Nature* **308:**645–647, 1984.

Hardebo, J. E., Hanko, J., Kahrstrom, J., and Owman, C.: Endothelium-dependent relaxation in cerebral arteries. *J. Cereb. Blood Flow* Metab. **5:**(suppl. 1):S533–S534, 1985.

Kalsner, S.: Cholinergic mechanisms in human coronary artery preparations: Implications of species differences. *J. Physiol.* **358:**509–526, 1985.

Martin, W., Furchgott, R. F., Villani, G. M., and Jothianandan, D.: Depression of contractile responses in rat aorta by spontaneously released endothelium derived relaxing factor. *J. Pharmacol. Exp. Ther.* **237:**529–538, 1986.

Martin, W., Villani, G. M., Jothianandan, D., and Furchgott, R. F.: Selective blockade of endothelium dependent and glyceryl trinitrate induced relaxation by haemoglobin and by methylene blue in the rabbit aorta. *J. Pharmacol. Exp. Ther.* **232:**708–716, 1985.

McGuire, P. G. and Twietmeyer, A. T.: Aortic endothelial functions in developing hypertension. *Hypertension* **7:**483–490, 1985.

Muldoon, S. M., Hart, J. L., Bowen, K., and Freas, W.: Effects of halothane on canine arteries: Role of the endothelium. *Fed. Proc.* **44:**1835, 1985.

Owen, M. P. and Bevan, J. A.: Acetylcholine endothelial dependent vasodilatation increases as artery diameter decreases in the rabbit ear. *Experientia* **41:**1057–1058, 1985.

Page, M. M. and Watkins, P. J.: Provocation of postural hypotension by insulin in diabetic autonomic neuropathy. *Diabetes* **25:**90–95, 1976.

Parnavelas, J. G., Kelly, W., and Burnstock, G.: Ultrastructural localisation of choline acetyltransferase in vascular endothelial cells in rat brain. *Nature* **316:**724–725, 1985.

Pearson, J. D., Carleton, J. S., and Gordon, J. C.: Metabolism of adenine nucleotides by ecto-enzymes of vascular endothelial and smooth muscle cells in culture. *Biochem. J.* **190:**421–429, 1980.

Posloncec, B., Uvnas-Moberg, K., Hagerman, M., Castennsson, S., and Uvnas, B.: Release of an insulin-like peptide from perfused extirpated cat legs in response to electrical stimulation of the sciatic and brachial nerves and to administration of ACh, bombesin, oxytocin and glibenclamide. *Acta Physiol. Scand.* **125:**195–203, 1985.

Rapoport, R. M., Schwartz, K., and Murad, F.: Effects of Na^+, K^+-pump inhibitors and membrane depolarizing agents on acetylcholine induced endothelium dependent relaxation and cyclic GMP accumulation in rat aorta. *Eur. J. Pharmacol.* **110:**203–209, 1985.

Thom, S., Hughes, A. D., Goldberg, P., Martin, G., Schachter, M., and Sever, P. S. The actions of calcitonin gene related peptide and vasoactive intestinal peptide as vasodilators in man in vivo and in vitro. *Br. J. Pharmacol.* **24:**139–144, 1987.

Van de Voorde, J. and Leusen, I.: Role of the endothelium in vasodilator response of rat thoracic aorta to histamine. *Eur. J. Pharmacol.* **87:**113–120, 1983.

Van de Voorde, J. and Leusen, I.: Endothelium dependent and independent relaxation of aortic rings from hypertensive rats. *A. J. Physiol. 250:*H711–H717, 1986.

Winquist, R. J., Bunting, P. B., Baskin, E. P., and Wallace, A. A.: Decreased endothelium dependent relaxation in New Zealand genetic hypertensive rats. *J. Hyperten.* **2:**541–545, 1984

Index